Techniques and Concepts of
High-Energy Physics IX

NATO ASI Series

Advanced Science Institutes Series

A series presenting the results of activities sponsored by the NATO Science Committee, which aims at the dissemination of advanced scientific and technological knowledge, with a view to strengthening links between scientific communities.

The series is published by an international board of publishers in conjunction with the NATO Scientific Affairs Division

A	Life Sciences	Plenum Publishing Corporation
B	Physics	New York and London
C	Mathematical and Physical Sciences	Kluwer Academic Publishers Dordrecht, Boston, and London
D	Behavioral and Social Sciences	
E	Applied Sciences	
F	Computer and Systems Sciences	Springer-Verlag
G	Ecological Sciences	Berlin, Heidelberg, New York, London,
H	Cell Biology	Paris, Tokyo, Hong Kong, and Barcelona
I	Global Environmental Change	

PARTNERSHIP SUB-SERIES

1. Disarmament Technologies	Kluwer Academic Publishers
2. Environment	Springer-Verlag
3. High Technology	Kluwer Academic Publishers
4. Science and Technology Policy	Kluwer Academic Publishers
5. Computer Networking	Kluwer Academic Publishers

The Partnership Sub-Series incorporates activities undertaken in collaboration with NATO's Cooperation Partners, the countries of the CIS and Central and Eastern Europe, in Priority Areas of concern to those countries.

Recent Volumes in this Series:

Volume 363 — Masses of Fundamental Particles: *Cargèse 1996*
edited by Maurice Lévy, Jean Iliopoulos, Raymond Gastmans, and Jean-Marc Gérard

Volume 364 — Quantum Fields and Quantum Space Time
edited by Gerard 't Hooft, Arthur Jaffe, Gerhard Mack, Pronob K. Mitter, and Raymond Stora

Volume 365 — Techniques and Concepts of High-Energy Physics IX
edited by Thomas Ferbel

Series B: Physics

Techniques and Concepts of High-Energy Physics IX

Edited by
Thomas Ferbel
University of Rochester
Rochester, New York

Plenum Press
New York and London
Published in cooperation with NATO Scientific Affairs Division

Proceedings of a NATO Advanced Study Institute on
Techniques and Concepts of High-Energy Physics,
held July 11–22, 1996,
in St. Croix, U.S. Virgin Islands

NATO-PCO-DATA BASE

The electronic index to the NATO ASI Series provides full bibliographical references (with keywords and/or abstracts) to about 50,000 contributions from international scientists published in all sections of the NATO ASI Series. Access to the NATO-PCO-DATA BASE is possible in two ways:

—via online FILE 128 (NATO-PCO-DATA BASE) hosted by ESRIN, Via Galileo Galilei, I-00044 Frascati, Italy

—via CD-ROM "NATO Science and Technology Disk" with user-friendly retrieval software in English, French, and German (©WTV GmbH and DATAWARE Technologies, Inc. 1989). The CD-ROM contains the AGARD Aerospace Database.

The CD-ROM can be ordered through any member of the Board of Publishers or through NATO-PCO, Overijse, Belgium.

ISBN 0-306-45709-1

© 1997 Plenum Press, New York
A Division of Plenum Publishing Corporation
233 Spring Street, New York, N.Y. 10013

http://www.plenum.com

10 9 8 7 6 5 4 3 2 1

All rights reserved

No part of this book may be reproduced, stored in a retrieval system, or transmitted in any form or by any means, electronic, mechanical, photocopying, microfilming, recording, or otherwise, without written permission from the Publisher

Printed in the United States of America

We were all saddened this past year to learn of the tragic death of one of our young colleagues at the Institute. George Michail was killed in an automobile collision with a drunk driver. George was a bright and warm individual, on his way to a highly successful career in physics. It was a terrible loss, and we dedicate this volume to his memory.

PREFACE

The ninth Advanced Study Institute (ASI) on Techniques and Concepts of High Energy Physics was almost canceled before if began! A certain visitor to the area (Hurricane Bertha) arrived unexpectedly early in 1996. It was the first hurricane in memory to menace the Caribbean in early July! Fortunately, it passed St. Croix several days before our meeting, and left very little damage. (The Altarellis survived the eye of the storm in the in the British West Islands!)

The meeting was held once again at the hotel on the Cay, on that spec of land in the harbor of Chrirtiansted, St. Croix, U. S. Virgin Islands. After the first two days of, at times, outrageous downpour, the 71 participants from 26 countries began to relax and enjoy the lectures and the lovely surroundings of the Institute. The primary support for the meeting was provided by the Scientific Affairs Division of the North Atlantic Treaty Organization (NATO). The ASI was cosponsored by the U. S. department of Energy, by the Fermi National Accelerator Laboratory (Fermi-lab), by the U. S. National Science Foundation, and by the University of Rochester. In addition, the International Science Foundation contributed to the support of a participant from Russia.

As in the case of the previous ASIs, the scientific program was designed for advanced graduate students and recent Ph.D. recipients in experimental particle physics. The present volume of lectures should complement the material published in the first eight ASIs and prove to be of value to a wider audience of physicists.

It is a pleasure to acknowledge the encouragement and support that I have continued to receive from colleagues and friends in organizing this meeting. I am indebted to Chris Quigg and to the other members of my Advisory Committee for their infinite patience and excellent advice. I am grateful to the distinguished lecturers for their enthusiastic participation in the ASI, and, of course, for their hard work in preparing the lectures and providing the manuscripts for the Proceedings. I thank Ivan Korolko for organizing the student presentations, and Zandy-Marie Hillis of the National Park Service for another fascinating description of the geology, marine life and the gigantic nesting turtles of St. Croix.

I thank P. K. Williams for support from the Department of Energy, and Willy Chinowsky for assistance from the National Science Foundation. I am grateful to John Peoples for providing the talents of Angela Gonzales for designing the poster for the school. At Rochester, I am indebted to Ovide Corriveau for help with budgeting issues and to Connie Jones for her exceptional organizational assistance.

I owe thanks to Ann Downs and earl Powell for their and their staff's efficiency and hospitality at the Hotel on the Cay, and to Herchel Greenaway and his staff at the Harbormaster for keeping us well fed and entertained. I wish to acknowledge the support from Bert Yost and his colleagues at the LeCroy Research Systems Corporation. Finally, I thank Pat Vann from Plenum and Luis da Cunha of NATO for their cooperation and confidence.

T. Ferbel
Rochester, New York

CONTENTS

1. Status of Precision Tests of the Standard Model 1
   G. Altarelli

2. SUSY and Such 33
   S. Dawson

3. Challenges of the LHC 81
   N. Ellis

4. Statistical Issues in Data Analysis 131
   H. B. Prosper

5. Muon-Muon and Other High Energy Colliders 183
   R. B. Palmer and J. C. Gallardo

6. Electroweak and Top Physics at Haldron Colliders 273
   M. Strovink

7. Advancements in Tracking Chambers 331
   R. Bellazzini and M. A. Spezziga

8. Electroweak Studies at LEP and SLD 381
   A. Blondel

9. The Physics of Massive Neutrinos 429
   F. Vannucci

10. Prospects for B-Physics in the Next Decade 465
    S. Stone

First $e^+ e^- \psi W^+ W^-$ 533
   (Original Art By C. Parkes)

Participants 535

Index 537

STATUS OF PRECISION TESTS OF THE STANDARD MODEL

Guido ALTARELLI

Theoretical Physics Division, CERN
1211 Geneva 23, Switzerland and
Terza Università di Roma, Rome, Italy

INTRODUCTION

The running of LEP1 was terminated in 1995 and close-to-final results of the data analysis are now available and were presented at the Warsaw Conference in July 1996[1],[2]. LEP and SLC started in 1989 and the first results from the collider run at the Tevatron were also first presented at about that time. I went back to my rapporteur talk at the Stanford Conference in August 1989[3] and I found the following best values quoted there for some of the key quantities of interest for the Standard Model (SM) phenomenology: $m_Z = 91120(160)$ MeV; $m_t = 130\ (50)$ GeV; $\sin^2\theta_{eff} = 0.23300(230)$ and $\alpha_s(m_Z) = 0.110(10)$. Now, after seven years of experimental and theoretical work (in particular with 16 million Z events analysed altogether by the four LEP experiments) the corresponding numbers, as quoted at the Warsaw Conference, are: $m_Z = 91186.3(2.0)$ MeV; $m_t = 175(6)$ GeV; $\sin^2\theta_{eff} = 0.23165(24)$ and $\alpha_s(m_Z) = 0.118(3)$. The progress is quite evident. The top quark has been at last found and the errors on m_Z and $\sin^2\theta_{eff}$ went down by two and one orders of magnitude respectively. At the start the goals of LEP, SLC and the Tevatron were to: a) perform precision tests of the SM at the level of a few per mille accuracy; b) count neutrinos ($N_\nu = 2.989(12)$); c) search for the top quark ($m_t = 175(6)$ GeV); d) search for the Higgs ($m_H > 65$ GeV); e) search for new particles (none found). While for most of the issues the results can be summarized in very few bits, as just shown, it is by far more complex for the first one. The validity of the SM has been confirmed to a level that I can say was unexpected at the beginning. This is even more true after Warsaw. Contrary to the situation presented at the winter '96 Conferences we are now left with no significant evidence for departures from the SM. The discrepancy on R_c has completely disappeared, that on R_b has been much reduced, and so on, and no convincing hint of new physics is left in the data (also including the first results from LEP2). The impressive success of the SM poses strong limitations on the possible forms of new physics. Favoured are models of the Higgs sector and of new physics that preserve the SM structure and only very delicately improve it, as is the case for fundamental Higgs(es) and Supersymmetry. Disfavoured are models with a nearby strong non-perturbative regime that almost inevitably would affect the radiative corrections, as for composite Higgs(es) or for technicolor and its variants.

STATUS OF THE DATA

The relevant new electroweak data together with their SM values are presented in table 1. The SM values correspond to a fit in terms of m_t, m_H and $\alpha_s(m_Z)$, described later in sect. 3, eq. (14), of all the available data including the CDF/D0 value of m_t. A number of comments on the novel aspects of the data are now in order.

Table 1

Quantity	Data (Warsaw '96)	Standard Model	Pull
m_Z (GeV)	91.1863(20)	91.1861	0.1
Γ_Z (GeV)	2.4946(27)	2.4960	-0.5
σ_h (nb)	41.508(56)	41.465	0.8
R_h	20.788(29)	20.757	0.7
R_b	0.2178(11)	0.2158	1.8
R_c	0.1715(56)	0.1723	-0.1
A_{FB}^l	0.0174(10)	0.0159	1.4
A_τ	0.1401(67)	0.1458	-0.9
A_e	0.1382(76)	0.1458	-1.0
A_{FB}^b	0.0979(23)	0.1022	-1.8
A_{FB}^c	0.0733(48)	0.0730	0.1
A_b	SLD direct 0.863(49) LEP indir. 0.895(23) Average 0.889(21)	0.935	-2.2
A_c	SLD direct 0.625(84) LEP indir. 0.670(44) Average 0.660(39)	0.667	-0.2
$\sin^2\theta_{eff}$(LEP-combined)	0.23200(27)	0.23167	1.2
$A_{LR} \to \sin^2\theta_{eff}$	0.23061(47)	0.23167	-2.2
m_W (GeV)	80.356(125)	80.353	0.3
m_t (GeV)	175(6)	172	0.5

What happened to R_c? The tagging method for charm is based on the reconstruction of exclusive final channels. This is rather complicated and depends on branching ratios and on the probability that a charm quark fragments into given hadrons. A shift in the measured value of the branching ratio for $D^0 \to K^-\pi^+$ and the measurement at LEP of $P(c \to D^*)$, acting on R_c in the same direction, have been sufficient to restore a perfect agreement with the SM.

What happened to R_b? The old result at the winter '96 Conferences was (assuming the SM value for R_c) $R_b = 0.2202(16)$. The present official average, shown in table 1, is much lower and only 1.8σ away from the SM value. The essential difference is the result of a new-from-scratch, much improved analysis from ALEPH, which is given by [1],[2]

$$R_b = 0.2161 \pm 0.0014 \qquad (ALEPH). \qquad (1)$$

In fact if one combines the average of the "old" measurements, given above, with the "new" ALEPH result one practically finds the official average given by the electroweak LEP working group and reported in table 1. This happens to be true in spite of the fact that in the correct procedure one has to take away the ALEPH contribution, now superseded, from the "old" average and add to it some newly presented refinements to some of the "old" analyses. In view of this, it is clear that the change is mainly due to the new ALEPH result. There are objective improvements in this new analysis. Five mutually exclusive tags are simultaneously used in order to decrease the sensitivity to individual sources of systematic error. Separate vertices are reconstructed in the two hemispheres of each event to minimize correlations between the hemispheres. The implementation of a mass tag on the tracks from each vertex reduces the charm background that dominates the systematics. As a consequence it appears to me that the weight of the new analysis in the combined value should be larger than what is obtained from the stated errors. In view of the ALEPH result the necessity of new physics in R_b has disappeared, while the possibility of some small deviation (more realistic than before) of course is still there. In view of the importance of this issue the other collaborations will go back to their data and freshly reconsider their analyses with the new improvements taken into account.

It is often stated that there is a 3σ deviation on the measured value of A_b with respect to the SM expectation[1],[2]. But in fact that depends on how the data are combined. In my opinion one should rather talk of a 2σ effect. Let us discuss this point in detail. A_b can be measured directly at SLC, taking advantage of the beam longitudinal polarization. SLD finds

$$A_b = 0.863 \pm 0.049 \qquad \text{(SLD direct} : -1.5\sigma) \,, \tag{2}$$

where the discrepancy with respect to the SM value, $A_b^{SM} = 0.935$, has also been indicated. At LEP one measures $A_b^{FB} = 3/4 \, A_e A_b$. As seen in table 1, the value found is somewhat below the SM prediction. One can then derive A_b by using the value of A_e obtained, using lepton universality, from the measurements of A_l^{FB}, A_τ, A_e: $A_e = 0.1466(33)$:

$$A_b = 0.890 \pm 0.029 \qquad \text{(LEP, } A_e \text{ from LEP} : -1.6\sigma) \,. \tag{3}$$

By combining the two above values one obtains

$$A_b = 0.883 \pm 0.025 \qquad \text{(LEP + SLD,} A_e \text{ from LEP} : -2.1\sigma) \,. \tag{4}$$

The LEP electroweak working group combines the SLD result with the LEP value for A_b modified by adopting for A_e the SLD+LEP average value, which also includes A_{LR} from SLD, $A_e = 0.1500(25)$:

$$A_b = 0.867 \pm 0.020 \qquad \text{(LEP + SLD,} A_e \text{ from LEP + SLD} : -3.1\sigma) \,. \tag{5}$$

There is nothing wrong with that but, in this case, the well-known $\sim 2\sigma$ discrepancy of A_{LR} with respect to A_e measured at LEP and also to the SM, which is not related to the b couplings, further contributes to inflate the number of σ's. Since the b couplings are more suspect than the lepton couplings it is perhaps wiser to obtain A_b from LEP by using the SM value for A_e: $A_e^{SM} = 0.1458(16)$, which gives

$$A_b = 0.895 \pm 0.023 \qquad \text{(LEP, } A_e = A_e^{SM} : -1.7\sigma) \,. \tag{6}$$

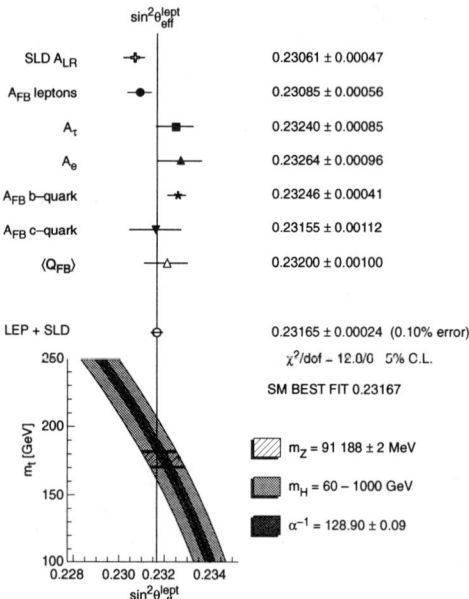

Figure 1

Finally, combining the last value with SLD we have

$$A_b = 0.889 \pm 0.021 \quad (\text{LEP} + \text{SLD}, \ A_e = A_e^{\text{SM}} : -2.2\sigma) \, . \tag{7}$$

Note that these are the values reported in table 1.

Finally if one looks at the values of $\sin^2 \theta_{eff}$ obtained from different observables, shown in fig. 1, one notices that the value obtained from A_l^{FB} is somewhat low (indeed quite in agreement with its determination by SLD from A_{LR}). Looking closer, this is due to the FB asymmetry of the τ lepton that, systematically in all four LEP experiments, has a central value above that of e and μ [1],[2]. The combined value for the τ channel is $A_\tau^{FB} = 0.0201(18)$ while the combined average of e and μ is $A_{e/\mu}^{FB} = 0.0162(11)$. On the other hand A_τ and Γ_τ appear normal. In principle these two facts are not incompatible, because the FB lepton asymmetries are very small. The extraction of A_τ^{FB} from the data on the angular distribution of τ's could be biased if the imaginary part of the continuum was altered by some non-universal new physics effect[4]. But a more trivial experimental problem is at the moment more plausible.

The distribution of measured values of $\sin^2 \theta_{eff}$, as it is summarized in fig. 1, is somewhat wide ($\chi^2/\text{d.o.f.} = 2.13$) with A_l^{FB} and A_{LR} far on one side and A_b^{FB} on the other side. In view of this it would perhaps be appropriate to enlarge the error on the average from ± 0.00024 up to $\pm\sqrt{2.13} \, 0.00024 = \pm 0.00034$, according to the recipe adopted by the Particle Data Group. Thus from time to time in the following we will use the average

$$\sin^2 \theta_{eff} = 0.23165 \pm 0.00034 \tag{8}$$

PRECISION ELECTROWEAK DATA AND THE STANDARD MODEL

For the analysis of electroweak data in the SM one starts from the input parameters: some of them, α, G_F and m_Z, are very well measured, some other, $m_{f_{light}}$, m_t and $\alpha_s(m_Z)$ are only approximately determined, while m_H is largely unknown. With respect to m_t the situation has much improved since the CDF/D0 direct measurement of the top quark mass[5]. From the input parameters one computes the radiative corrections[6],[7] to a sufficient precision to match the experimental capabilities. Then one compares the theoretical predictions and the data for the numerous observables that have been measured, checks the consistency of the theory and derives constraints on m_t, $\alpha_s(m_Z)$, and hopefully also on m_H.

Some comments on the least known of the input parameters are now in order. The only practically relevant terms where precise values of the light quark masses, $m_{f_{light}}$, are needed are those related to the hadronic contribution to the photon vacuum polarization diagrams that determine $\alpha(m_Z)$. This correction is of order 6%, much larger than the accuracy of a few per mille of the precision tests. Fortunately, one can use the actual data to in principle solve the related ambiguity. But we shall see that the left-over uncertainty is still one of the main sources of theoretical error. As is well known[8]–[18], the QED running coupling is given by:

$$\alpha(s) = \frac{\alpha}{1 - \Delta\alpha(s)}$$
$$\Delta\alpha(s) = \Pi(s) = \Pi_\gamma(0) - \text{Re}\Pi_\gamma(s) , \qquad (9)$$

where $\Pi(s)$ is proportional to the sum of all 1-particle irreducible vacuum polarization diagrams. In perturbation theory $\Delta\alpha(s)$ is given by

$$\Delta\alpha(s) = \frac{\alpha}{3\pi} \sum_f Q_f^2 N_{Cf} \left(\log \frac{2}{m_f^2} - \frac{5}{3} \right) , \qquad (10)$$

where $N_{Cf} = 3$ for quarks and 1 for leptons. However, the perturbative formula is only reliable for leptons, not for quarks (because of the unknown values of the effective quark masses). Separating the leptonic, the light quark and the top quark contributions to $\Delta\alpha(s)$ we have:

$$\Delta\alpha(s) = \Delta\alpha(s)_l + \Delta\alpha(s)_h + \Delta\alpha(s)_t \qquad (11)$$

with[18]:

$$\Delta\alpha(s)_l = 0.0331421 \; ; \; \Delta\alpha(s)_t = \frac{\alpha}{3\pi} \frac{4}{15} \frac{m_Z^2}{m_t^2} = -0.000061 . \qquad (12)$$

Note that in QED there is decoupling so that the top quark contribution approaches zero in the large m_t limit. For $\Delta\alpha(s)_h$ one can use (9) and the Cauchy theorem to obtain the representation:

$$\Delta\alpha(m_Z^2)_h = -\frac{\alpha m_Z^2}{3\pi} \text{Re} \int_{4m_\pi^2}^{\infty} \frac{ds}{s} \frac{R(s)}{s - m_Z^2 - i\epsilon} \qquad (13)$$

where $R(s)$ is the familiar ratio of the hadronic to the point-like $\ell^+\ell^-$ cross-section from photon exchange in e^+e^- annihilation. At s large, one can use the perturbative expansion for $R(s)$ while at small s one can use the actual data.

Recently there has been a lot of activity on this subject and a number of independent new estimates of $\alpha(m_Z)$ have appeared in the literature[8]. In table 2 we report

5

the results of these new computations together with the most significant earlier determinations (previously the generally accepted value was that of Jegerlehner in 1991[12]).

Table 2

Author	Year and Ref.	$\Delta\alpha(m_Z^2)_h$	$\alpha(m_Z^2)^{-1}$
Jegerlehner	1986 [9]	0.0285 ± 0.0007	128.83 ± 0.09
Lynn et al.	1987 [10]	0.0283 ± 0.0012	128.86±0.16
Burkhardt et al.	1989 [11]	0.0287 ± 0.0009	128.80 ± 0.12
Jegerlehner	1991 [12]	0.0282 ± 0.0009	128.87±0.12
Swartz	1994 [13]	0.02666± 0.00075	129.08±0.10
Swartz (rev.)	1995 [14]	0.0276 ± 0.0004	128.96 ± 0.06
Martin et al.	1994 [15]	0.02732± 0.00042	128.99 ± 0.06
Nevzorov et al.	1994 [16]	0.0280 ± 0.0004	128.90 ± 0.06
Burkhardt et al.	1995 [17]	0.0280 ± 0.0007	128.89 ± 0.09
Eidelman et al.	1995 [18]	0.0280 ± 0.0007	128.90 ± 0.09

The differences among the recent determinations are due to the procedures adopted for fitting the data and treating the errors, for performing the numerical integration, etc. The differences are also due to the threshold chosen to start the application of perturbative QCD at large s and to the value adopted for $\alpha_s(m_Z)$. For example, in its first version Swartz[13] used parametric forms to fit the data, while most of the other determinations use a trapezoidal rule to integrate across the data points. It was observed that the parametric fitting introduces a definite bias[14]. In fact Swartz gets systematically lower results for all ranges of s. In its revised version[14] Swartz improves his numerical procedure. Martin et al.[15] use perturbative QCD down to $\sqrt{s} = 3$ GeV (except in the upsilon region) with $\alpha_s(m_Z) = 0.118\pm0.007$. Eidelman et al.[18] only use perturbative QCD for $\sqrt{s} > 40$ GeV and with $\alpha_s(m_Z) = 0.126\pm 0.005$, i.e. the value found at LEP. They use the trapezoidal rule. Nevzorov et al.[16] make a rather crude model with one resonance per channel plus perturbative QCD with $\alpha_s(m_Z) = 0.125 \pm 0.005$. Burkhardt et al.[17] use perturbative QCD for $\sqrt{s} > 12$ GeV, but with a very conservative error on $\alpha_s(m_Z) = 0.124 \pm 0.021$. This value was determined [19] from e^+e^- data below LEP energies. The excitement produced by the original claim by Swartz[13] of a relatively large discrepancy with respect to the value obtained by Jegerlehner[12] resulted in a useful debate. As a conclusion of this re-evaluation of the problem the method of Jegerlehner has proved its solidity. As a consequence I think that the recent update by Eidelman and Jegerlehner[18] gives a quite reliable result (which is the one used by the LEP groups and in the following). Also, I do not think that a smaller error than quoted by these authors can be justified.

As for the strong coupling $\alpha_s(m_Z)$ we will discuss in detail the interesting recent developments in sect. 4. The world average central value is quite stable around 0.118, before and after the most recent results. The error is going down because the dispersion among the different measurements is much smaller in the most recent set of data. The error is taken to be between ±0.003 and ±0.005, depending on how conservative one wants to be. Thus in the following our reference value will be $\alpha_s(m_Z) = 0.118 \pm 0.005$.

Finally a few words on the current status of the direct measurement of m_t. The error is rapidly going down. It was ±9 GeV before the Warsaw Conference, it is now ±6 GeV [5]. I think one is soon approaching a level where a more careful investigation

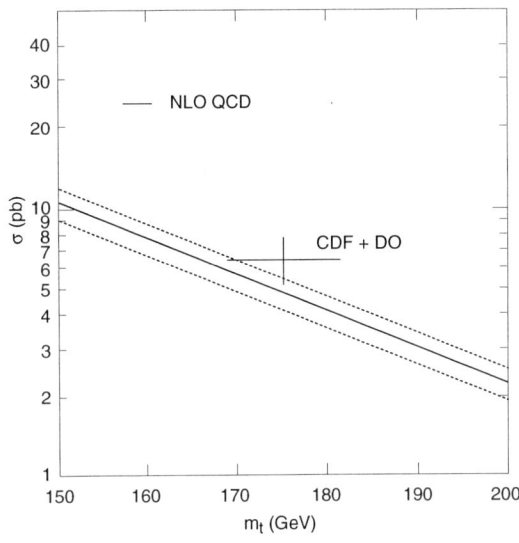

Figure 2

of the effects of colour rearrangement on the determination of m_t is needed. One wants to determine the top quark mass, defined as the invariant mass of its decay products (i.e. $b + W +$ gluons $+ \gamma$'s). However, due to the need of colour rearrangement, the top quark and its decay products cannot be really isolated from the rest of the event. Some smearing of the mass distribution is induced by this colour crosstalk, which involves the decay products of the top, those of the antitop and also the fragments of the incoming (anti)protons. A reliable quantitative computation of the smearing effect on the m_t determination is difficult because of the importance of non-perturbative effects. An induced error of the order of a few GeV on m_t is reasonably expected. Thus further progress on the m_t determination demands tackling this problem in more depth.

The measured top production cross section is in fair agreement with the QCD prediction, but the central value is a bit large (see fig. 2)[20]. The world average for the cross section times branching ratio is $\sigma B = 6.4 \pm 1.3$ pb and the QCD prediction for σ is $\sigma_{QCD} = 4.75 \pm 0.65$ pb [21]. Thus the branching ratio $B = B(t \to bW)$ cannot be far from 100% unless there is also some additional production mechanism from new physics.

In order to appreciate the relative importance of the different sources of theoretical errors for precision tests of the SM, I report in table 3 a comparison for the most relevant observables, evaluated using ref. [22].

What it is important to stress is that the ambiguity from m_t, once by far the largest one, is by now smaller than the error from m_H. We also see from table 3 that the error from $\Delta\alpha(m_Z)$ is especially important for $\sin^2\theta_{eff}$ and, to a lesser extent, is also sizeable for Γ_Z and ϵ_3.

We now discuss fitting the data in the SM. As the mass of the top quark is now rather precisely known from CDF and D0 one must distinguish between two different types of fits. In one type one wants to answer the question: Is m_t from radiative corrections in agreement with the direct measurement at the Tevatron? For answering

Table 3: Errors from different sources: Δ^{exp}_{now} is the present experimental error; $\Delta\alpha^{-1}$ is the impact of $\Delta\alpha^{-1} = \pm 0.09$; Δ_{th} is the estimated theoretical error from higher orders; Δm_t is from $\Delta m_t = \pm 6\mathrm{GeV}$; Δm_H is from $\Delta m_H = 60{-}1000$ GeV; $\Delta\alpha_s$ corresponds to $\Delta\alpha_s = \pm 0.005$. The epsilon parameters are defined in ref. [23].

Parameter	Δ^{exp}_{now}	$\Delta\alpha^{-1}$	Δ_{th}	Δm_t	Δm_H	$\Delta\alpha_s$
Γ_Z (MeV)	±2.7	±0.7	±0.8	±1.4	±4.6	±2.7
σ_h (pb)	56	1	4.3	3.3	4	2.7
$R_h \cdot 10^3$	29	4.3	5	2	13.5	34
Γ_l (keV)	110	11	15	55	120	6
$A^l_{FB} \cdot 10^4$	10	4.2	1.3	3.3	13	0.3
$\sin^2\theta \cdot 10^4$	~3	2.3	0.8	1.9	7.5	0.15
m_W (MeV)	125	12	9	37	100	4
$R_b \cdot 10^4$	11	0.1	1	2.1	0.25	0
$\epsilon_1 \cdot 10^3$	1.3		~0.1			0.4
$\epsilon_3 \cdot 10^3$	1.4	0.6	~0.1			0.25
$\epsilon_b \cdot 10^3$	3.2		~0.1			2

this interesting but somewhat limited question, clearly one must exclude the CDF/D0 measurement of m_t from the input set of data. Fitting the data in terms of m_t, m_H and $\alpha_s(m_Z)$ one finds the results shown in table 4[2].

Table 4

Parameter	LEP	LEP + SLD	All $\neq m_t$
$\alpha_s(m_Z)$	0.1211(32)	0.1200(32)	0.1202(33)
m_t (GeV)	155(14)	156(11)	157(10)
m_H (GeV)	86(+202 − 14)	48(+83 − 26)	149(+148 − 82)
$(m_H)_{MAX}$ at 1.64σ	417	184	392
χ^2/dof	5/8	18/11	18/13

The extracted value of m_t is typically a bit too low. For example, from LEP data alone one finds $m_t = 155(14)$ GeV. But this is simply due to R_b being taken from the official average: $R_b = 0.2178(11)$. If m_H is not fixed the fit prefers lower values of m_t to adjust R_b. In fact by removing R_b from the input data one increases the central value of m_t from 155 to 171 GeV. In this context it is important to remark that fixing m_H at 300 GeV, as is often done, is by now completely obsolete, because it introduces a strong bias on the fitted value of m_t. The change induced on the fitted value of m_t when moving m_H from 300 to 65 or 1000 GeV is in fact larger than the error on the direct measurement of m_t.

In a more general type of fit, e.g. for determining the overall consistency of the SM or the best present estimate for some quantity, say m_W, one should of course not ignore the existing direct determination of m_t. Then, from all the available data, including $m_t = 175(6)$ GeV, by fitting m_t, m_H and $\alpha_s(m_Z)$ one finds (with $\chi^2/\mathrm{d.o.f.} = 19/14$) [2] (see also[24]):

$$m_t = 172 \pm 6 \text{ GeV},$$

$$m_H = 149 + 148 - 82 \text{ (or } m_H < 392 \text{ GeV at } 1.64\sigma)$$
$$\alpha_s(m_Z) = 0.1202 \pm 0.0033 \ . \tag{14}$$

This is the fit reported in table 1. The corresponding fitted values of $\sin^2\theta_{eff}$ and m_W are:

$$\sin^2\theta_{eff} = 0.23167 \pm 0.0002$$
$$m_W = 80.352 \pm 0.034 \text{ GeV} \ . \tag{15}$$

The error of 34 MeV on m_W clearly sets up a goal for the direct measurement of m_W at LEP2 and the Tevatron.

STATUS OF $\alpha_s(m_Z)$

There are important developments in the experimental determination of $\alpha_s(m_Z)$[25]. There is now a much better agreement between the different methods of measuring $\alpha_s(m_Z)$. In fact the value of $\alpha_s(m_Z)$ from the Z line shape went down and the values from scaling violations in deep inelastic scattering and from lattice QCD went up. We will discuss these developments in detail in the following.

The value of $\alpha_s(m_Z)$ from the Z line shape (assuming that the SM is valid for Γ_h, which is not completely evident in view of R_b) went down for two reasons[1],[2]. First the value extracted from R_h only, which was $\alpha_s(m_Z) = 0.126(5)$, is now down to $\alpha_s(m_Z) = 0.124(5)$. Second the value from all the Z data changed from $\alpha_s(m_Z) = 0.124(5)$ down to $\alpha_s(m_Z) = 0.120(4)$, which corresponds to the fit in eq. (14). The main reason for this decrease is the new value of σ_h (with a sizeably smaller error than in the past) that prefers a smaller $\alpha_s(m_Z)$. However this determination depends on the assumption that Γ_b is given by the SM. We recall that R_b itself with good approximation is independent of α_s, but its deviation from the SM would indicate an anomaly in Γ_b hence in Γ_h. Taking a possible anomaly in R_b into account the Z line shape determination of $\alpha_s(m_Z)$ becomes approximately:

$$\alpha_s(m_Z) = (0.120 \pm 0.004) - 4\delta R_b \ . \tag{16}$$

If the ALEPH value for R_b (see eq .(1)) is adopted, the central value of $\alpha_s(m_Z)$ is not much changed, but of course the error on δR_b is transferred on $\alpha_s(m_Z)$, which becomes

$$\alpha_s(m_Z) = 0.119 \pm 0.007 \ . \tag{17}$$

If, instead, one takes R_b from table 1 one obtains a much smaller central value:

$$\alpha_s(m_Z) = 0.112 \pm 0.006 \tag{18}$$

Summarizing: the Z line shape result for $\alpha_s(m_Z)$, obtained with the assumption that Γ_h is given by the SM, went down a bit. The central value could be shifted further down if R_b is in excess with respect to the SM.

While $\alpha_s(m_Z)$ from LEP goes down, $\alpha_s(m_Z)$ from the scaling violations in deep inelastic scattering goes up. To me the most surprising result from Warsaw was the announcement by the CCFR collaboration that their well-known published analysis of $\alpha_s(m_Z)$ from xF_3 and F_2 in neutrino scattering off Fe target is now superseded by a re-analysis of the data based on better energy calibration[26]. We recall that their previous result, $\alpha_s(m_Z) = 0.111(3\ \mathrm{exp})$, being in perfect agreement with the value obtained from e/μ beam data by BCDMS and SLAC combined[27], $\alpha_s(m_Z) = 0.113(3\ \mathrm{exp})$, convinced most of us that the average value of $\alpha_s(m_Z)$ from deep inelastic scattering was close to 0.112. Now the new result presented in Warsaw is [26], [25]

$$\alpha_s(m_Z) = 0.119 \pm 0.0015(\mathrm{stat}) \pm 0.0035(\mathrm{syst}) \pm 0.004(th) \quad (\mathrm{CCFR-revised})\ , \quad (19)$$

where the error also includes the collaboration estimate of the theoretical error from scale and renormalization scheme ambiguities. As a consequence the new combined value of $\alpha_s(m_Z)$ from scaling violations in deep inelastic scattering is given by

$$\alpha_s(m_Z) = 0.115 \pm 0.006\ , \qquad (20)$$

with my more conservative estimate, of the common theoretical error (Schmelling, the rapporteur in Warsaw quotes ± 0.005 [25]). If we compare eq. (20) with LEP eq. (14), we see that, whatever our choice of theoretical errors is, there is no need for any new physics in R_b to fill the gap between the two determinations of $\alpha_s(m_Z)$.

Finally $\alpha_s(m_Z)$ from lattice QCD is also going up [28]. The main new development is a theoretical study of the error associated with the extrapolation from unphysical values of the light quark masses, which is used in the lattice extraction of $\alpha_s(m_Z)$ from quarkonium splittings. According to ref. [29] this effect amounts to a shift upward of $+0.003$ in the value of $\alpha_s(m_Z)$. From the latest unquenched determinations of $\alpha_s(m_Z)$, Flynn, the rapporteur in Warsaw[28], gives an average of 0.117(3). But the lattice measurements of $\alpha_s(m_Z)$ moved very fast over the last few years. At the Dallas conference in 1992, the quoted value (from quenched computations) was $\alpha_s(m_Z) = 0.105(4)$ [30], while at Beijing in 1995 the claimed value was $\alpha_s(m_Z) = 0.113(2)$ but the error was estimated to be ± 0.007 by the rapporteur Michael[31]. So, with the present central value, I will keep this more conservative error in the following:

$$\alpha_s(m_Z) = 0.117 \pm 0.007\ . \qquad (21)$$

To my knowledge, there are no other important new results on the determination of $\alpha_s(m_Z)$. Adding a few more well-established measurements of $\alpha_s(m_Z)$ we have table 5, where the errors denote my personal view of the weights the different methods should have in the average (in brackets Th and Exp are labels that indicate whether the dominant error is theoretical or experimental).

The average value given as

$$\alpha_s(m_Z) = 0.118 \pm 0.003 \qquad (22)$$

is very stable. The same value was quoted by Schmelling, a rapporteur at the Warsaw Conference[25], with a different treatment of errors. Had we used $\alpha_s(m_Z)$ from the Z line shape assuming the SM value for R_b, i.e. $\alpha_s(m_Z) = 0.120 \pm 0.004$, the average value would have been 0.119. To be safe one could increase the error to ± 0.005.

A MORE MODEL-INDEPENDENT APPROACH

We now discuss an update of the epsilon analysis[23]. The epsilon method is more complete and less model-dependent than the similar approach based on the variables

Table 5

Measurements	$\alpha_s(m_Z)$	
R_τ	0.122 ± 0.007	(Th)
Deep Inelastic Scattering	0.115 ± 0.006	(Th)
Y_{decay}	0.112 ± 0.010	(Th)
Lattice QCD	0.117 ± 0.007	(Th)
$Re^+e^-(\sqrt{s} < 62 \text{ GeV})$	0.124 ± 0.021	(Exp)
Fragmentation functions in e^+e^-	0.124 ± 0.010	(Th)
Jets in e^+e^- at and below the Z	0.121 ± 0.008	(Th)
Z line shape (taking R_b from ALEPH)	0.119 ± 0.007	(Exp)

S, T and U [32]-[35] which, from the start, necessarily assumes dominance of vacuum polarization diagrams from new physics and truncation of the q^2 expansion of the corresponding amplitudes. In a completely model-independent way we define[23] four variables, called ϵ_1, ϵ_2, ϵ_3 and ϵ_b, that are precisely measured and can be compared with the predictions of different theories. The quantities ϵ_1, ϵ_2, ϵ_3 and ϵ_b are defined in ref. [23] in one-to-one correspondence with the set of observables m_W/m_Z, Γ_l, A_l^{FB} and R_b. The four epsilons are defined without need of specifying m_t and m_H. In the SM, for all observables at the Z pole, the whole dependence on m_t and m_H arising from one-loop diagrams only enters through the epsilons. The same is true for any extension of the SM such that all possible deviations only occur through vacuum polarization diagrams and/or the $Z \to b\bar{b}$ vertex.

The epsilons represent an efficient parametrization of the small deviations from what is solidly established, in a way that is unaffected by our relative ignorance of m_t and m_H. The variables S, T, U depend on m_t and m_H because they are defined as deviations from the complete SM prediction for specified m_t and m_H. Instead the epsilons are defined with respect to a reference approximation, which does not depend on m_t and m_H. In fact the epsilons are defined in such a way that they are exactly zero in the SM in the limit of neglecting all pure weak loop-corrections (i.e. when only the predictions from the tree level SM plus pure QED and pure QCD corrections are taken into account). This very simple version of improved Born approximation is a good first approximation according to the data. Values of the epsilons in the SM are given in table 6 [22],[23].

By combining the value of m_W/m_Z with the LEP results on the charged lepton partial width and the forward-backward asymmetry, all given in table 1, and following the definitions of ref. [23], one obtains:

$$\begin{aligned} \epsilon_1 &= \Delta\rho = (4.3 \pm 1.4) \times 10^{-3} \\ \epsilon_2 &= (-6.9 \pm 3.4) \times 10^{-3} \\ \epsilon_3 &= (3.0 \pm 1.8) \times 10^{-3} \, . \end{aligned} \quad (23)$$

Finally, by adding the value of R_b listed in table 1 and using the definition of ϵ_b given in ref. [23] one finds (note that ϵ_b is defined through R_b and the expression of R_b as a function of ϵ_b is practically independent of α_s):

$$\epsilon_b = (-1.1 \pm 2.8) \times 10^{-3} \quad (R_b \text{ from table 1}) \, . \quad (24)$$

Table 6: Values of the epsilons in the SM as functions of m_t and m_H as obtained from recent versions[22] of ZFITTER and TOPAZ0. These values (in 10^{-3} units) are obtained for $\alpha_s(m_Z) = 0.118$, $\alpha(m_Z) = 1/128.87$, but the theoretical predictions are essentially independent of $\alpha_s(m_Z)$ and $\alpha(m_Z)$ [23].

m_t (GeV)	ϵ_1 m_H (GeV) =			ϵ_2 m_H (GeV) =			ϵ_3 m_H (GeV) =			ϵ_b All m_H
	65	300	1000	65	300	1000	65	300	1000	
150	3.47	2.76	1.61	−6.99	−6.61	−6.4	4.67	5.99	6.66	−4.45
160	4.34	3.59	2.38	−7.29	−6.9	−6.69	4.6	5.91	6.55	−5.28
170	5.25	4.46	3.21	−7.6	−7.2	−6.97	4.52	5.82	6.43	−6.13
180	6.2	5.37	4.1	−7.93	−7.51	−7.24	4.42	5.72	6.34	−7.02
190	7.2	6.33	5.07	−8.29	−7.81	−7.49	4.31	5.6	6.26	−7.95
200	8.26	7.34	6.1	−8.65	−8.12	−7.75	4.19	5.49	6.19	−8.92

This is the value that corresponds to the official average reported in table 1 which I have criticized. Here in this epsilon analysis we prefer to use the ALEPH value for R_b, ($R_b = 0.2161(14)$), which leads to

$$\epsilon_b = (-5.7 \pm 3.4) \times 10^{-3} \qquad (R_b \text{ from ALEPH}) \qquad (25)$$

To proceed further and include other measured observables in the analysis, we need to make some dynamical assumptions. The minimum amount of model dependence is introduced by including other purely leptonic quantities at the Z pole such as $A_{\tau_{pol}}$, A_e (measured from the angular dependence of the τ polarization) and A_{LR} (measured by SLD). For this step, one is simply relying on lepton universality. Note that the choice of A_l^{FB} as one of the defining variables appears at present not particularly lucky, because the corresponding determination of $\sin^2\theta_{eff}$ markedly underfluctuates with respect to the average value (see fig. 1). We then use the combined value of $\sin^2\theta_{eff}$ obtained from the whole set of asymmetries measured at LEP and SLC, with the error increased according to eq. (8) and the related discussion. At this stage the best values of ϵ_1, ϵ_2, ϵ_3 and ϵ_b are modified according to

$$\begin{aligned} \epsilon_1 &= \Delta\rho = (4.7 \pm 1.3) \times 10^{-3} \\ \epsilon_2 &= (-7.8 \pm 3.3) \times 10^{-3} \\ \epsilon_3 &= (4.8 \pm 1.4) \times 10^{-3} \\ \epsilon_b &= (-5.7 \pm 3.4) \times 10^{-3} \end{aligned} \qquad (26)$$

In fig. 3 we report the 1σ ellipse in the ϵ_1–ϵ_3 plane that correspond to this set of input data.

All observables measured on the Z peak at LEP can be included in the analysis, provided that we assume that all deviations from the SM are only contained in vacuum polarization diagrams (without demanding a truncation of the q^2 dependence of the corresponding functions) and/or the $Z \to b\bar{b}$ vertex. From a global fit of the data on m_W/m_Z, Γ_T, R_h, σ_h, R_b and $\sin^2\theta_{eff}$ (for LEP data, we have taken the correlation matrix for Γ_T, R_h and σ_h given by the LEP experiments[2], while we have considered the additional information on R_b and $\sin^2\theta_{eff}$ as independent), we obtain:

$$\epsilon_1 = \Delta\rho = (4.7 \pm 1.3) \times 10^{-3}$$

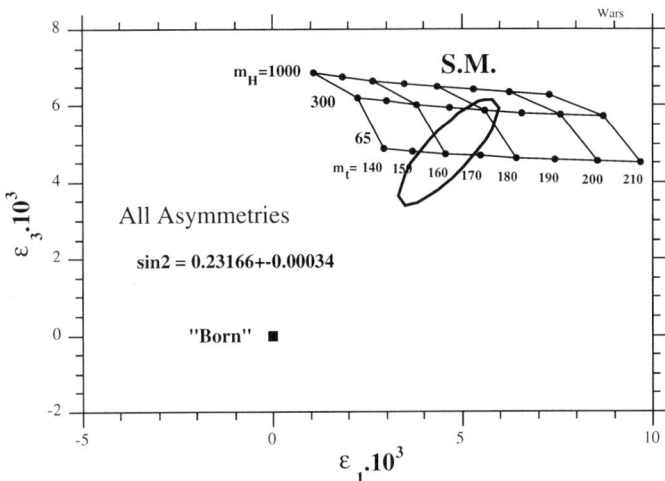

Figure 3

$$\epsilon_2 = (-7.8 \pm 3.3) \times 10^{-3}$$
$$\epsilon_3 = (4.7 \pm 1.4) \times 10^{-3}$$
$$\epsilon_b = (-4.8 \pm 3.2) \times 10^{-3} \qquad (27)$$

The comparison of theory and experiment in the planes ϵ_1–ϵ_3 and ϵ_b–ϵ_3 is shown in figs. 4 and 5, respectively. Note that adding the hadronic quantities hardly makes a difference in the ϵ_1–ϵ_3 plot in comparison with fig. 3 which only included the leptonic variables. In other words the inclusive hadronic quantities do not show any peculiarity. A number of interesting features are clearly visible from this plot. First, the good agreement with the SM and the evidence for weak corrections, measured by the distance of the data from the improved Born approximation point (based on tree level SM plus pure QED or QCD corrections). Second, we see the preference for light Higgs or, equivalently, the tendency for ϵ_3 to be rather on the low side (both features are now somewhat less pronounced than they used to be). Finally, if the Higgs is light the preferred value of m_t is somewhat lower than the Tevatron result (which in this analysis is not included in the input data). The data ellipse in the ϵ_b–ϵ_3 plane is consistent with the SM and the CDF/D0 value of m_t. This is because we have taken the ALEPH value for R_b. For comparison, we also show in figs. 6 and 7 the same plots as in figs. 4 and 5, but for the official average values of R_b and $\sin^2 \theta_{eff}$ as reported in table 1. The main difference is the obvious displacement of ϵ_b and the smaller errors in the ϵ_1–ϵ_3 plot. Finally, the status of ϵ_2 is presented in fig. 8. The agreement is very good. ϵ_2 is sensitive to m_W and a more precise test will only be possible when the measurement of m_W will be much improved at LEP2 and the Tevatron.

To include in our analysis lower-energy observables as well, a stronger hypothesis needs to be made: vacuum polarization diagrams are allowed to vary from the SM only in their constant and first derivative terms in a q^2 expansion[33]–[35], a likely picture, e.g. in technicolor theories[36]–[38]. In such a case, one can, for example, add to the analysis the ratio R_ν of neutral to charged current processes in deep inelastic neutrino scattering on nuclei[39], the "weak charge" Q_W measured in atomic parity violation experiments on

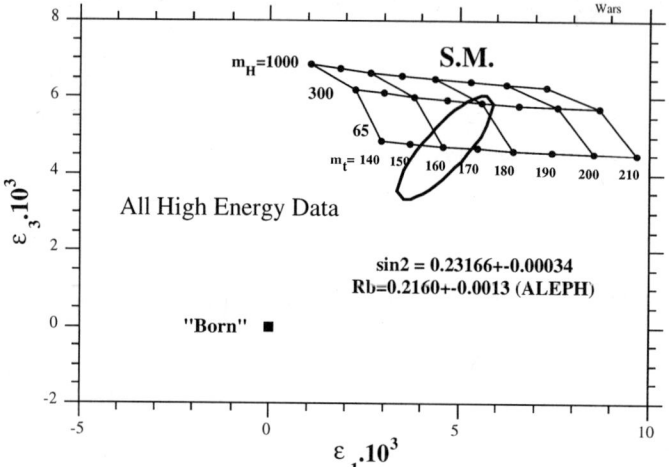

Figure 4

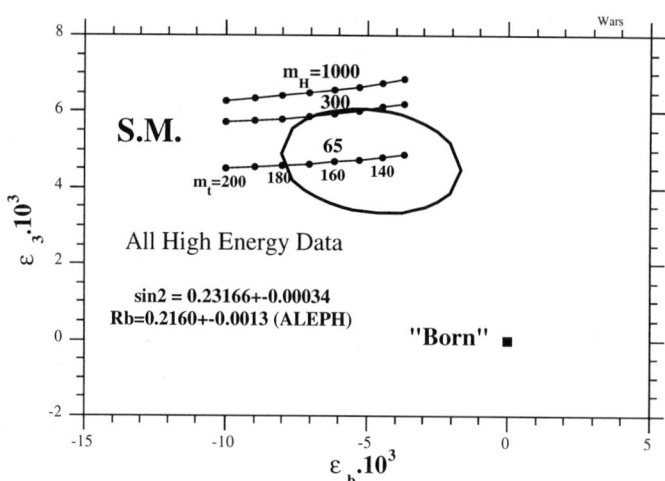

Figure 5

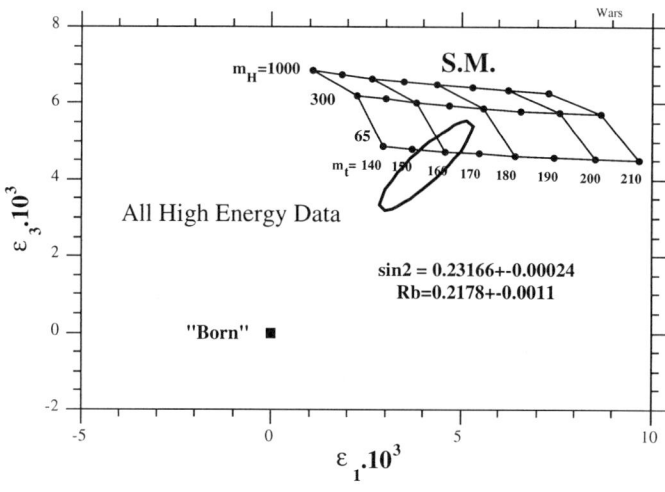

Figure 6

Cs [40], and the measurement of g_V/g_A from $\nu_\mu e$ scattering[41]. In this way one obtains the global fit (R_b from ALEPH, $\sin^2\theta_{eff}$ with enlarged error as in eq. (8)):

$$\begin{align}\epsilon_1 &= \Delta\rho = (4.3 \pm 1.2) \times 10^{-3}\\ \epsilon_2 &= (-8.0 \pm 3.3) \times 10^{-3}\\ \epsilon_3 &= (4.4 \pm 1.3) \times 10^{-3}\\ \epsilon_b &= (-4.6 \pm 3.2) \times 10^{-3}\ .\end{align} \qquad (28)$$

With the progress of LEP, the low-energy data, while important as a check that no deviations from the expected q^2 dependence arise, play a lesser role in the global fit. Note that the present ambiguity on the value of $\delta\alpha^{-1}(m_Z) = \pm 0.09$ [18] corresponds to an uncertainty on ϵ_3 (the other epsilons are not much affected) given by $\Delta\epsilon_3\,10^3 = \pm 0.6$ [23]. Thus the theoretical error is still comfortably less than the experimental error.

To conclude this section I would like to add some comments. As is clearly indicated in figs. 3 to 8 there is by now solid evidence for departures from the "improved Born approximation" where all the epsilons vanish. In other words a strong evidence for the pure weak radiative corrections has been obtained, and LEP/SLC are now measuring the various components of these radiative corrections. For example, some authors [42] have studied the sensitivity of the data to a particularly interesting subset of the weak radiative corrections, i.e. the purely bosonic part. These terms arise from the virtual exchange of gauge bosons and Higgses. The result is that indeed the measurements are sufficiently precise to require the presence of these contributions in order to fit the data.

CONCEPTUAL PROBLEMS WITH THE STANDARD MODEL

Given the striking success of the SM, why are we not satisfied with that theory? Why not just find the Higgs particle, for completeness, and declare that particle physics

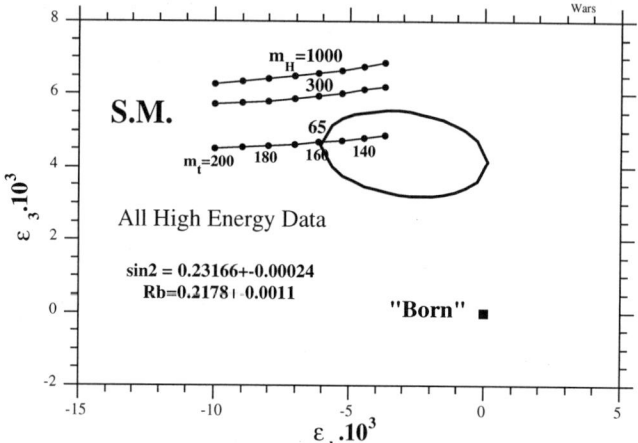

Figure 7

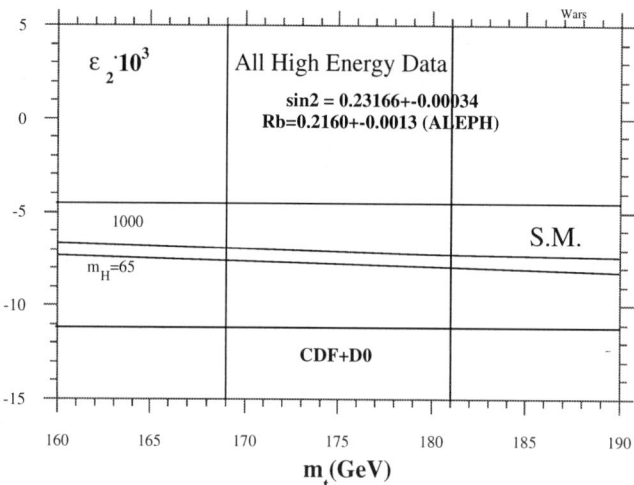

Figure 8

is closed? The main reason is that there are strong conceptual indications for physics beyond the SM.

It is considered highly implausible that the origin of the electroweak symmetry breaking can be explained by the standard Higgs mechanism, without accompanying new phenomena. New physics should be manifest at energies in the TeV domain. This conclusion follows fron an extrapolation of the SM at very high energies. The computed behaviour of the $SU(3) \otimes SU(2) \otimes U(1)$ couplings with energy clearly points towards the unification of the electroweak and strong forces (Grand Unified Theories: GUTs) at scales of energy $M_{GUT} \sim 10^{14}$–10^{16} GeV, which are close to the scale of quantum gravity, $M_{Pl} \sim 10^{19}$ GeV [43]. One can also imagine a unified theory of all interactions also including gravity (at present superstrings [44] provide the best attempt at such a theory). Thus GUTs and the realm of quantum gravity set a very distant energy horizon that modern particle theory can no longer ignore. Can the SM without new physics be valid up to such large energies? This appears unlikely because the structure of the SM could not naturally explain the relative smallness of the weak scale of mass, set by the Higgs mechanism at $m \sim 1/\sqrt{G_F} \sim 250$ GeV, G_F being the Fermi coupling constant. The weak scale m is $\sim 10^{17}$ times smaller than M_{Pl}. Even if the weak scale is set near 250 GeV at the classical level, quantum fluctuations would naturally shift it up to where new physics starts to apply, in particular up to M_{Pl} if there was no new physics up to gravity. This so-called hierarchy problem[45] is related to the presence of fundamental scalar fields in the theory with quadratic mass divergences and no protective extra symmetry at $m = 0$. For fermions, first, the divergences are logaritmic and, second, at $m = 0$ an additional symmetry, i.e. the chiral symmetry, is restored. Here, when talking of divergences we are not worried of actual infinities. The theory is renormalizable and finite once the dependence on the cut off is absorbed in a redefinition of masses and couplings. Rather the hierarchy problem is one of naturalness. If we consider the cut off as a manifestation of new physics that will modify the theory at large energy scales, then it is relevant to look at the dependence of physical quantities on the cut off and to demand that no unexplained enormously accurate cancellation arise.

According to the above argument the observed value of $m \sim 250$ GeV is indicative of the existence of new physics nearby. There are two main possibilities. Either there exist fundamental scalar Higgses, but the theory is stabilized by supersymmetry, the boson–fermion symmetry that would downgrade the degree of divergence from quadratic to logarithmic. For approximate supersymmetry the cut off is replaced by the splitting between the normal particles and their supersymmetric partners. Then naturalness demands that this splitting (times the size of the weak gauge coupling) is of the order of the weak scale of mass, i.e. the separation within supermultiplets should be of the order of no more than a few TeV. In this case the masses of most supersymmetric partners of the known particles, a very large menagerie of states, would fall, at least in part, in the discovery reach of the LHC. There are consistent, fully formulated field theories constructed on the basis of this idea, the simplest one being the MSSM[46]. Note that all normal observed states are those whose masses are forbidden in the limit of exact $SU(2) \otimes U(1)$. Instead, for all SUSY partners the masses are allowed in that limit. Thus when supersymmetry is broken in the TeV range, but $SU(2) \otimes U(1)$ is intact only spartners take mass while all normal particles remain massless. Only at the lower weak scale the masses of ordinary particles are generated. Thus a simple criterion exists to understand the difference between particles and sparticles.

The other main avenue is compositeness of some sort. The Higgs boson is not elementary but either a bound state of fermions or a condensate, due to a new strong force, much stronger than the usual strong interactions, responsible for the attraction. A plethora of new "hadrons", bound by the new strong force, would exist in the LHC range. A serious problem for this idea is that nobody so far has been able to build up a realistic model along these lines, which could eventually be explained by a lack of ingenuity on the theorists side. The most appealing examples are technicolor theories[36],[37]. These models where inspired by the breaking of chiral symmetry in massless QCD induced by quark condensates. In the case of the electroweak breaking new heavy techniquarks must be introduced and the scale analogous to Λ_{QCD} must be about three orders of magnitude larger. The presence of such a large force relatively nearby has a strong tendency to clash with the results of the electroweak precision tests[38]. Another interesting idea is to replace the Higgs by a $t\bar{t}$ condensate[47]. The Yukawa coupling of the Higgs to the $t\bar{t}$ pair becomes a four-fermion $\bar{t}t\bar{t}t$ coupling with the corresponding strength. The strong force is in this case provided by the large top mass. At first sight this idea looks great: no fundamental scalars, no new states. But, looking closely, the advantages are largely illusory. First, in the SM the required value of m_t is too large: $m_t \geq 220$ GeV or so. Also a tremendous fine-tuning is required, because m_t would naturally be of the order of M_{GUT} or M_{Pl} if no new physics is present (the hierarchy problem in a different form!). Supersymmetry could come to the rescue in this case also. In a minimal SUSY version the required value of the top mass is lowered[48], $m_t \sim 195 \sin \beta$ GeV. But the resulting theory is physically indistinguishable from the MSSM with small $\tan \beta$, at least at low energies[49]. This is because a strongly coupled Higgs looks the same as a $t\bar{t}$ pair.

The hierarchy problem is certainly not the only conceptual problem of the SM. There are many more: the proliferation of parameters, the mysterious pattern of fermion masses and so on. But while most of these problems can be postponed to the final theory that will take over at very large energies, of order M_{GUT} or M_{Pl}, the hierarchy problem arises from the instability of the low-energy theory and requires a solution at relatively low energies. A supersymmetric extension of the SM provides a way out that is well defined, computable and that preserves all virtues of the SM. The necessary SUSY breaking can be introduced through soft terms that do not spoil the stability of scalar masses. Precisely those terms arise from supergravity when it is spontaneoulsly broken in a hidden sector[50]. But alternative mechanisms of SUSY breaking are also being considered[94]. As we shall now discuss, there are also experimental and phenomenological hints that point in this direction.

At present the most important phenomenological evidence in favour of supersymmetry is obtained from the unification of couplings in GUTs. Precise LEP data on $\alpha_s(m_Z)$ and $\sin^2 \theta_W$ confirm what was already known with less accuracy: standard one-scale GUTs fail in predicting $\sin^2 \theta_W$ given $\alpha_s(m_Z)$ (and $\alpha(m_Z)$), while SUSY GUTs[51] are in agreement with the present, very precise, experimental results. According to the recent analysis of ref. [52], if one starts from the known values of $\sin^2 \theta_W$ and $\alpha(m_Z)$, one finds for $\alpha_s(m_Z)$ the results:

$$\alpha_s(m_Z) = 0.073 \pm 0.002 \quad \text{(Standard GUTS)}$$
$$\alpha_s(m_Z) = 0.129(+0.010, -0.008) \quad \text{(SUSY GUTS)} \quad (29)$$

to be compared with the world average experimental value $\alpha_s(m_Z) = 0.118(5)$.

A very elegant feature of the GUT-extended supersymmetric version of the SM is that the occurrence of the $SU(2) \otimes U(1)$ electroweak symmetry breaking is naturally and automatically generated by the large mass of the top quark[53]. Assuming that all scalar masses are the same at the GUT scale, the effect of the large Yukawa coupling of the top quark in the renormalization group evolution down to the weak energy scale, drives one of the Higgs squared masses negative (that Higgs which is coupled to the up-type quarks). The masses of sleptons and of the Higgs coupled to the down-type quark are much less modified, while the squark masses are increased due to the strongly interacting gluino exchange diagrams. The negative value of the squared mass corresponds to the onsetting of the electroweak symmetry breaking. That the correct mass for the weak bosons is obtained as a result of the breaking implies constraints on the model, more stringent if no fine-tuning is allowed to a given level of accuracy. Various fine-tuning criteria have been analysed in the literature[54],[55]. Typically no more than a factor 10 fine tuning is allowed. With this assumption and realistic values of m_t one obtains the bounds shown in fig. 9[56]. These upper bounds give a quantitative specification of the constraints implied by a natural solution of the hierarchy problem in the context of the GUT-extended MSSM. They look very promising for LEP2 (but the bounds scale with the inverse square root of the fine-tuning factor...).

Many of the simpler GUTs predict the unification at M_{GUT} of the b and τ Yukawa couplings, or, equivalently, that for the running masses $m_b(M_{GUT}) = m_\tau(M_{GUT})$ [57]. The observed difference of the b and τ masses arises from the evolution due to the different interactions of quarks and leptons. Many authors studied the combined constraints from coupling unification and b and τ Yukawa unification[58]. The result is that there are a small $\tan\beta$ solution (typically in the range $\tan\beta$ =0.5-3) and a large $\tan\beta$ solution (with $\tan\beta$ =40-60). However the large $\tan\beta$ solution is somewhat disfavoured by a natural implementation of the electroweak symmetry breaking, according to the mechanism discussed above. In fact at large values of $\tan\beta \geq m_t/m_b$, the dominance of the top over the bottom Yukawa coupling, which is an important ingredient for that mechanism, is erased or even inverted. A closer look at the small $\tan\beta$ solution shows that the top mass is close to its fixed-point solution $m_t \sim 195 \sin\beta$ GeV so that $m_t \sim 175$ GeV corresponds to $\tan\beta \sim 2$. Correspondingly the mass of the lightest Higgs is relatively small[59], as discussed in sect. 9, which is good for LEP2.

In the MSSM the lightest neutralino is stable and provides a very good cold dark matter candidate. It is interesting that if the constraint $\Omega = 1$, which corresponds to the critical density for closure of the Universe, is added to the previous ones, consistency can still be achieved in a sizeable domain of the parameter space[58]-[60].

In conclusion, gauge coupling unification, natural $SU(2) \otimes U(1)$ electroweak symmetry breaking, b and τ Yukawa unification and a plausible amount of dark matter all fit together for realistic values of m_t, $\alpha_s(m_Z)$ and m_b. SUSY GUTs, with a single step of symmetry breaking from m_{GUT} down to m_W, appear to work well.

PRECISION ELECTROWEAK TESTS AND THE SEARCH FOR NEW PHYSICS

We now concentrate on some well-known extensions of the SM, which not only are particularly important per se but also are interesting in that they clearly demonstrate the constraining power of the present level of precision tests.

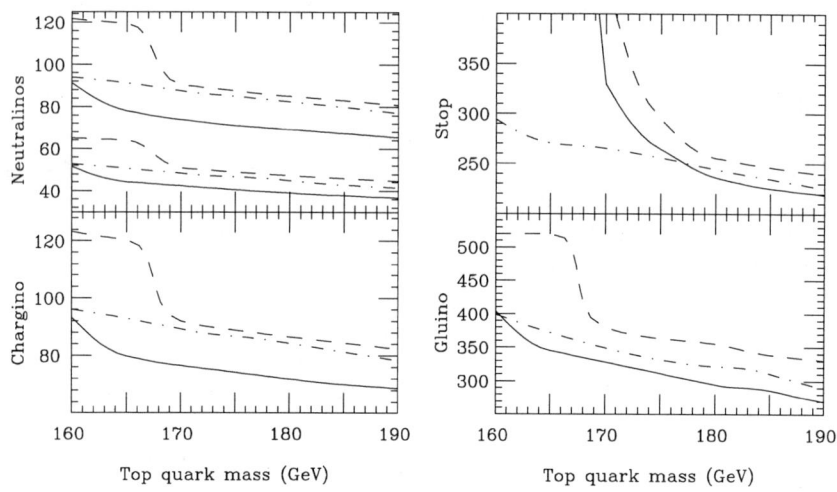

Figure 9

Upper bounds on gluino, lightest and next-to-lightest neutralino, and lightest chargino and stop masses based on the requirement of no fine tuning larger than 10%. The solid (dashed) lines refer to the minimal supersymmetric standard model with universal boundary conditions at M_{GUT} for the soft supersymmetry-breaking terms, without (with) the inclusion of the one-loop effective potential. The dot-dashed lines show the mass upper limits, for non-universal boundary conditions at M_{GUT}, without the includion of the one-loop effective potential.

Minimal Supersymmetric Standard Model

The MSSM[46] is a completely specified, consistent and computable theory. There are too many parameters to attempt a direct fit of the data to the most general framework. So one can consider two significant limiting cases: the "heavy" and the "light" MSSM.

The "heavy" limit corresponds to all sparticles being sufficiently massive, still within the limits of a natural explanation of the weak scale of mass. In this limit a very important result holds[61]: for what concerns the precision electroweak tests, the MSSM predictions tend to reproduce the results of the SM with a light Higgs, say $m_H \lesssim 100$ GeV.

In the "light" MSSM option, some of the superpartners have a relatively small mass, close to their experimental lower bounds. In this case the pattern of radiative corrections may sizeably deviate from that of the SM. The most interesting effects occur in vacuum polarization amplitudes and/or the $Z \to b\bar{b}$ vertex and therefore are particularly suitable for a description in terms of the epsilons (because in such a case, as explained in ref. [23], the predictions can be compared with the experimental determination of the epsilons from the whole set of LEP data). They are:

i) a threshold effect in the Z wave function renormalization[61], mostly due to the vector coupling of charginos and (off-diagonal) neutralinos to the Z itself. Defining the vacuum polarization functions by $\Pi_{\mu\nu}(q^2) = -ig_{\mu\nu}[A(0) + q^2 F(q^2)] + q_\mu q_\nu$ terms, this is a positive contribution to $\epsilon_5 = m_Z^2 F'_{ZZ}(m_Z^2)$, the prime denoting a derivative with respect to q^2 (i.e. a contribution to a higher derivative term not included in the usual S, T, U formalism). The ϵ_5 correction shifts ϵ_1, ϵ_2 and ϵ_3 by $-\epsilon_5$, $-c^2\epsilon_5$ and $-c^2\epsilon_5$ respectively, where $c^2 = \cos^2\theta_W$, so that all of them are reduced by a comparable amount. Correspondingly all the Z widths are reduced without affecting the asymmetries. This effect falls down particularly fast when the lightest chargino mass increases from a value close to $m_Z/2$. Now that we know, from the LEP1.5 and LEP2 runs, that the chargino mass is not so light, its possible impact is drastically reduced.

ii) A positive contribution to ϵ_1 from the virtual exchange of the scalar top and bottom superpartners[62], analogous to the contribution of the top–bottom left-handed quark doublet. The needed isospin splitting requires one of the two scalars (in the MSSM the stop) to be light. From the value of m_t, not much space is left for this possibility. If the stop is light then it must be mainly a right-handed stop.

iii) A negative contribution to ϵ_b, due to the virtual exchange of a charged Higgs[63]. If one defines, as customary, $\tan\beta = v_2/v_1$ (v_1 and v_2 being the vacuum expectation values of the Higgs doublets giving masses to the down and up quarks, respectively), then, for negligible bottom Yukawa coupling or $\tan\beta \geq m_t/m_b$, this contribution is proportional to $m_t^2/\tan^2\beta$.

iv) A positive contribution to ϵ_b due to virtual chargino–stop exchange[64], which in this case is proportional to $m_t^2/\sin^2\beta$ and prefers small $\tan\beta$. This effect again requires the chargino and the stop to be light in order to be sizeable.

range are then favoured. The second interpretation is based on the newly revived alternative approach in which SUSY breaking is mediated by ordinary gauge rather than gravitational interactions[46],[94]. In the most familiar approach of the MSSM, SUSY is broken in a hidden sector and the scale of SUSY breaking is very large, of order $\Lambda \sim \sqrt{G_F^{-1/2} M_P}$, where M_P is the Planck mass. But since the hidden sector only communicates with the visible sector through gravitational interactions the splitting of the SUSY multiplets is much smaller, in the TeV energy domain, and the goldstino is practically decoupled. In the alternative scenario the (not so much) hidden sector is connected to the visible one by ordinary gauge interactions. As these are much stronger than the gravitational interactions, Λ can be much smaller, as low as 10–100 TeV. It follows that the goldstino is very light in these models (with mass of order or below 1 eV typically) and is the lightest, stable SUSY particle, but its couplings are observably large. Then, in the CDF event, N' is a neutralino and N is the goldstino. The signature of photons comes out more naturally in this SUSY breaking pattern than in the MSSM. If the event is really due to selectron production it would be a manifestation of nearby SUSY that could be confirmed at LEP2. This is what we all wish. We shall see!

THE LEP2 PROGRAMME

The LEP2 programme started at the end of June'96. At first the energy was fixed at 161 GeV, which is the most favourable energy for the measurement of m_W from the cross-section for $e^+e^- \to W^+W^-$ at threshold. Then gradually the energy will be increased up to a maximum of about 193 GeV to be reached in mid '98. An average integrated luminosity of about 150 pb^{-1} per year is foreseen. LEP2 will run until the end of 1999 at least, before the shutdown for the installation of the LHC. The main goals of LEP2 are the search for the Higgs and for new particles, the measurement of m_W and the investigation of the triple gauge vertices WWZ and $WW\gamma$. A complete updated survey of the LEP2 physics is collected in two volumes[95].

An important competitor of LEP2 is the Tevatron collider. By and around the year 2000 the Tevatron will have collected about 1 fb^{-1} of integrated luminosity at 1.8–2 TeV. The competition is especially on the search of new particles, but also on m_W and the triple gauge vertices. For example, for supersymmetry while the Tevatron is superior for gluinos and squarks, LEP2 is strong on Higgses, charginos, neutralinos and sleptons.

Concerning the Higgs it is interesting to recall that the large value of m_t has important implications on m_H both in the minimal SM [96]-[98] and in its minimal supersymmetric extension[99],[100]. I will now discuss the restrictions on m_H that follow from the CDF value of m_t.

It is well known[96]-[98] that in the SM with only one Higgs doublet a lower limit on m_H can be derived from the requirement of vacuum stability. The limit is a function of m_t and of the energy scale Λ where the model breaks down and new physics appears. Similarly an upper bound on m_H (with mild dependence on m_t) is obtained[101] from the requirement that up to the scale Λ no Landau pole appears. The lower limit on m_H is particularly important in view of the search for the Higgs at LEP2. Indeed the issue is whether one can reach the conclusion that if a Higgs is found at LEP2, i.e. with $m_H \leq m_Z$, then the SM must break down at some scale $\Lambda > 1$ TeV.

The possible instability of the Higgs potential $V[\phi]$ is generated by the quantum loop corrections to the classical expression of $V[\phi]$. At large ϕ the derivative $V'[\phi]$ could become negative and the potential would become unbound from below. The one-loop corrections to $V[\phi]$ in the SM are well known and change the dominant term at large ϕ according to $\lambda \phi^4 \to (\lambda + \gamma \log \phi^2/\Lambda^2)\phi^4$. The one-loop approximation is not enough for our purposes, because it fails at large enough ϕ, when $\gamma \log \phi^2/\Lambda^2$ becomes of order 1. The renormalization group improved version of the corrected potential leads to the replacement $\lambda \phi^4 \to \lambda(\Lambda)\phi'^4(\Lambda)$, where $\lambda(\Lambda)$ is the running coupling and $\phi'(\mu) = \exp \int^t \gamma(t')dt' \phi$, with $\gamma(t)$ being an anomalous dimension function and $t = \log \Lambda/v$ (v is the vacuum expectation value $v = (2\sqrt{2}G_F)^{-1/2}$). As a result, the positivity condition for the potential amounts to the requirement that the running coupling $\lambda(\Lambda)$ never becomes negative. A more precise calculation, which also takes into account the quadratic term in the potential, confirms that the requirements of positive $\lambda(\Lambda)$ leads to the correct bound down to scales Λ as low as ~ 1 TeV. The running of $\lambda(\Lambda)$ at one loop is given by:

$$\frac{d\lambda}{dt} = \frac{3}{4\pi^2}[\lambda^2 + 3\lambda h_t^2 - 9h_t^4 + \text{gauge terms}] , \quad (31)$$

with the normalization such that at $t = 0, \lambda = \lambda_0 = m_H^2/2v^2$ and the top Yukawa coupling $h_t^0 = m_t/v$. We see that, for m_H small and m_t large, λ decreases with t and can become negative. If one requires that λ remains positive up to $\Lambda = 10^{15}$–10^{19} GeV, then the resulting bound on m_H in the SM with only one Higgs doublet is given by[97]:

$$m_H > 135 + 2.1\,[m_t - 174] - 4.5\,\frac{\alpha_s(m_Z) - 0.118}{0.006} . \quad (32)$$

Summarizing, we see from Eq. (32) that indeed for $m_t > 150$ GeV the discovery of a Higgs particle at LEP2 would imply that the SM breaks down at a scale Λ below M_{GUT} or M_{Pl}, smaller for lighter Higgs. Actually, for $m_t \sim 174$ GeV, only a small range of values for m_H is allowed, $130 < m_H < \sim 200$ GeV, if the SM holds up to $\Lambda \sim M_{GUT}$ or M_{Pl} (where the upper limit is from avoiding the Landau pole[101]). As is well known[96] the lower limit is not much relaxed, even if strict vacuum stability is replaced by some sufficiently long metastability. Of course, the limit is only valid in the SM with one doublet of Higgses. It is enough to add a second doublet to avoid the lower limit. A particularly important example of theory where the bound is violated is the MSSM, which we now discuss.

As is well known[46], in the MSSM there are two Higgs doublets, which implies three neutral physical Higgs particles and a pair of charged Higgses. The lightest neutral Higgs, called h, should be lighter than m_Z at tree-level approximation. However, radiative corrections[102] increase the h mass by a term proportional to m_t^4 and logarithmically dependent on the stop mass. Once the radiative corrections are taken into account the h mass still remains rather small: for $m_t = 174$ GeV one finds the limit (for all values of $\tan \beta$) $m_h < 130$ GeV [100]. Actually there are reasons to expect that m_h is well below the bound. In fact, if h_t is large at the GUT scale, which is suggested by the large observed value ot m_t and by a natural onsetting of the electroweak symmetry breaking induced by m_t, then at low energy a fixed point is reached in the evolution of m_t. The fixed point corresponds to $m_t \sim 195 \sin \beta$ GeV (a good approximate relation for $\tan \beta = v_{up}/v_{down} < 10$). If the fixed-point situation is realized, then m_h is considerably below the bound, as shown in Ref. 56.

In conclusion, for $m_t \sim 174$ GeV, we have seen that, on the one hand, if a Higgs is found at LEP the SM cannot be valid up to M_{Pl}. On the other hand, if a Higgs is found at LEP, then the MSSM has good chances, because this model would be excluded for $m_h > 130$ GeV.

For the SM Higgs, which plays the role of a benchmark, also important for a more general context, the LEP2 reach has been studied in detail. Accurate simulations have shown[95] that at LEP2 with 500 pb^{-1} per experiment, or with 150 pb^{-1} if the four experiments are combined, one can reach the 5σ discovery range given by $m_H \leq 82, 95$ GeV for $\sqrt{s} = 175, 192$ GeV respectively, and the 95% exclusion range $m_H \leq 83, 98$ GeV. On the basis of these ranges we understand why a few GeV make a lot of difference. With $\sqrt{s} = 175$ GeV there would be practically no overlap with the LHC, which, even in the most optimistic projections, cannot see the Higgs below $m_H = 80$ GeV or so. With $\sqrt{s} = 185$ GeV, there starts to be some overlap, but only limited to $m_H \leq 85$–90 GeV, which is still a very difficult, time-consuming and debatable range for the LHC. With $\sqrt{s} = 195$ GeV there is already a quite reasonable overlap, up to $m_H \leq 95$–100 GeV. The issue is not only that of avoiding a gap between LEP2 and the LHC, but also of providing an essential independent and complementary channel to study the new particle in a range of mass that is certainly rather marginal for the LHC.

In the MSSM a more complicated discussion is needed because there are several Higgses and the parameter space is multidimensional. Also, through the radiative corrections, the Higgs masses at fixed values of all MSSM parameters sensitively depend on the top quark mass. For decreasing top quark masses the upper bound on the light Higgs mass decreases. We note that the discovery range for LEP2 can be specified in terms of the light Higgs mass with little model dependence. On the contrary the same analysis for the LHC depends very much on the detailed quantitative pattern of the decay branching ratios. The usual plots that are seen in the experimental discussions are based on some typical choice of parameters, which is to some extent indicative.

In Ref. [95], the analysis for the MSSM is presented in great detail, as this case is rather complicated and was not deeply studied previously. With the typical choice of parameters, in the sense specified above, the domains of the $\tan\beta, m_A$ plane which are most difficult for the LHC are a "hole" at moderate values of $\tan\beta$ and m_A (say $\tan\beta < 10, m_A = 100$–$200$ GeV) and a "strip" at small $\tan\beta$ and large m_A (typically $\tan\beta = 1$–3 and $m_A > 300$ GeV). If m_t is not too small, these difficult regions can probably be covered at the LHC, but only with very large integrated luminosities $L = 3 \times 10^5$ pb^{-1}. LEP2 potentially can reduce the "hole" and completely cover the "strip", especially for m_t rather small. But while for $\sqrt{s} = 175$ GeV this is only true for rather extreme values of m_t and the squark mixing, at $\sqrt{s} = 192$ GeV only the central values are required (always with 150 pb^{-1} of integrated luminosity and the four experiments combined). Thus, as in the case of the SM, $\sqrt{s} = 192$ GeV is needed for a reasonable overlap, while less than that appears risky.

We now consider the search for supersymmetry. For charginos the discovery range at LEP2 is only limited by the beam energy for practically all values of the parameters. Thus every increase of the beam energy is directly translated into the upper limit in chargino mass for discovery or exclusion. For the Tevatron the discovery range is much more dependent on the position in parameter space. For some limited regions of this space, with 1 fb^{-1} of integrated luminosity, the discovery range for charginos at the Tevatron goes well beyond $m_\chi = 90$–100 GeV, i.e. the boundary of LEP2, while in most of the parameter space one would not be able to go that far and only LEP2, with sufficient energy, would find the chargino.

The stop is probably the lightest squark. For a light stop the most likely decay modes are $\tilde{t} \to b\chi^+$ if kinematically allowed, otherwise $\tilde{t} \to c\chi$. A comparative study of these modes at LEP2 and at the Tevatron is presented in Ref. [95]. The result is that in either case at LEP2 the discovery range is up to about $(E_{beam} - 10)$ GeV. At the Tevatron there is some difference between the two possible decay modes and some dependence on the position in the $\tilde{t}$–χ or the $\tilde{t}$–χ^+ planes, but it is true that very soon, at the end of the present run, with 100 pb^{-1}, a large region of the potential LEP2 discovery range will be excluded (in particular for the $\tilde{t} \to c\chi$ mode). Some limited regions will require more luminosity at the Tevatron and could be accessible to LEP2.

While on the stop the chances are better at the Tevatron than at LEP2 the converse is true for the sleptons. Here the Tevatron can only compete for a particularly favourable pattern of branching ratios. Finally, for neutralinos there is only a small region of the parameter space where these particles would be the first spartners to be discovered. The discovery ranges are very much parameter-dependent both at the Tevatron and at LEP2. For these reasons no detailed quantitative comparison still exists, although the channel $e^+e^- \to \chi\chi'$ has been extensively studied at the LEP2 workshop[95].

The measurement of m_W will be done at LEP2 from the cross-section at threshold and from direct reconstruction of the jet–jet final state in W decay. At present m_W is known with an error of ± 150 MeV from the direct measurement (see table 1). From the fit to all electroweak data one finds $m_W = 80352 \pm 34$ MeV (see eq. (10)), in agreement with the direct measurement. As a consequence the goal for LEP2 is to measure m_W with an accuracy $\delta m_W \leq \pm(30-40)$ MeV, in order to provide an additional significant check of the theory.

For the threshold method[95] the minimum of the statistical error is obtained for $\sqrt{s} = 2m_W + 0.5$ GeV $= 161$ GeV, which in fact was the initial operating energy of LEP2. The total error of this method is dominated by the statistics. If each of the four experiments will eventually collect 50 pb^{-1} of integrated luminosity (10 already collected and the rest in a possible future comeback at low energy) and the results are combined, then the statistical error will be $\delta m_W = \pm 95$ MeV and the total error $\delta m_W = \pm 108$ MeV. After ~ 10 pb^{-1} the present combined result is $m_W = (80.4 \pm 0.2 \pm 0.1)$ GeV [103]. Thus with realistic luminosity this method is not sufficient by itself.

In principle the direct reconstruction method can use the totally hadronic or the semileptonic final states $e^+e^- \to W^+W^- \to jjjj$ or $jjl\nu$. The total branching ratio of the hadronic modes is 49%, while that of the $\ell = e, \mu$ semileptonic channels is 28%. The hadronic channel has more statistics but could be severely affected by non-perturbative strong interaction effects: colour recombination among the jets from different W's and Bose correlations among mesons in the final state from WW overlap. Colour recombination is perturbatively small. But gluons with $E < \Gamma_W$ are important and non-perturbative effects could be relatively large, of the order of 10–100 MeV. Similarly for Bose correlations. One is not in a position to really quantify the associated uncertainties. Fortunately the direct reconstruction from the semileptonic channels can, by itself, lead to a total error $\delta m_W = \pm 44$ MeV, for the combined four experiments, each with 500 pb^{-1} of luminosity collected at $\sqrt{s} \geq 175$ GeV. Thus the goal of measuring m_W with an accuracy below $\delta m_W = \pm 50$ MeV can be fulfilled, and it is possible to do better if one learns from the data how to limit the error from colour recombination and Bose correlations.

The study of triple gauge vertices is another major task of LEP2. The capabilities of LEP2 in this domain are comparable to those of the LHC and go well below the level of deviations from the tree-level couplings that in the SM are expected from one-loop radiative corrections. LEP2 can push down the existing direct limits considerably. For given anomalous couplings the departures from the SM are expected to increase with energy. For the energy and the luminosity available at LEP2, given the accuracy of the SM established at LEP1, it is however not very likely, to find signals of new physics in the triple gauge vertices.

It is a pleasure for me to thank Tom Ferbel for his kind invitation and warm hospitality in St.Croix. It is with great sadness that I recall the memory of George Michail, a student at this School, a fine young man who died shortly afterwards in a tragic accident.

REFERENCES

1. A. Blondel, Proceedings of ICHEP '96, Warsaw, 1996;
   M. Pepe-Altarelli, Proceedings of the Symposium on Radiative Corrections, Cracow, 1996.
2. The LEP Electroweak Working Group, LEPEWWG/96-02.
3. G. Altarelli, Proceedings of the Int. Symposium on Lepton and Photon Interactions, Stanford, 1989.
4. F. Caravaglios, hep-ph/9610416.
5. P. Tipton, Proceedings of ICHEP '96, Warsaw, 1996.
6. G. Altarelli, R. Kleiss and C. Verzegnassi (eds.), Z Physics at LEP1 (CERN 89-08, Geneva, 1989), Vols. 1–3.
7. Precision Calculations for the Z Resonance, eds. D. Bardin, W. Hollik and G. Passarino, Report CERN 95-03 (1995).
8. B. Pietrzyk, Proceedings of the Symposium on Radiative Corrections, Cracow, 1996.
9. F. Jegerlehner, *Z. Phys.* **C32** (1986) 195.
10. B.W. Lynn, G. Penso and C. Verzegnassi, *Phys. Rev.* **D35** (1987) 42.
11. H. Burkhardt et al., *Z. Phys.* **C43** (1989) 497.
12. F. Jegerlehner, *Progr. Part. Nucl. Phys.* **27** (1991) 32.
13. M.L. Swartz, Preprint SLAC-PUB-6710, 1994.
14. M.L. Swartz, *Phys. Rev.* **D53** (1996) 5268.
15. A.D. Martin and D. Zeppenfeld, *Phys. Lett.* **B345**(1995) 558.
16. R.D. Nevzorov and A.V. Novikov, Preprint FTUV/94-27, 1994, Preprint MAD/PH/855, 1994.
17. H. Burkhardt and B. Pietrzyk, Preprint LAPP-EXP-95.05, 1995.
18. S. Eidelman and F. Jegerlehner, *Z. Phys.* **C67** (1995) 585.
19. D. Haidt in "Precision tests of the Standard Electroweak Model", ed. P. Langacker, World Scientific, Singapore, 1993.
20. S. Willenbrock,Proceedings of ICHEP '96, Warsaw, 1996.
21. G. Altarelli, M. Diemoz, G. Martinelli and P. Nason, *Nucl. Phys.* **B308** (1988) 724;
    R.K. Ellis, *Phys. Lett.* **259B** (1991) 492;
    E. Laenen, J. Smith and W.L. van Neerven, *Nucl. Phys.* **B369** (1992) 543, *Phys. Lett.* **321B** (1994) 254;
    E. Berger and H. Contopanagos, *Phys. Lett.* **361B** (1995) 115; hep-ph 9512212;
    S. Catani, M. Mangano, P. Nason and L. Trentadue, hep-ph 9602208.
22. ZFITTER: D. Bardin et al., CERN-TH. 6443/92 and refs. therein;
    TOPAZ0: G. Montagna et al., *Nucl. Phys.* **B401** (1993) 3; *Comp. Phys. Comm.* **76** (1993) 328. BHM: G. Burgers et al., LEPTOP: V.A. Novikov et al., WOH, W. Hollik : see ref. [7].
23. G. Altarelli, R. Barbieri and S. Jadach, *Nucl. Phys.* **B369** (1992)3;
    G. Altarelli, R. Barbieri and F. Caravaglios, *Nucl. Phys.* **B405** (1993) 3; *Phys. Lett.* **B349** (1995) 145.
24. A. Gurtu, Preprint TIFR-EHEP/96/01.
25. M. Schmelling, Proceedings of ICHEP '96, Warsaw, 1996.
26. D. Harris (CCFR Collaboration),Proceedings of ICHEP '96, Warsaw, 1996.

27. M. Virchaux and A. Milsztajn, *Phys. Lett.* **B274** (1992) 221.
28. J. Flynn, Proceedings of ICHEP '96, Warsaw, 1996.
29. B. Grinstein and I.Z. Rothstein, hep-ph/9605260.
30. A.X. El-Khadra, Proceedings of ICHEP '92, Dallas, 1992.
31. C. Michael, Proceedings of the Int. Symposium on Lepton and Photon interactions, Beijing, 1995.
32. M.E. Peskin and T. Takeuchi, *Phys. Rev. Lett.* **65** (1990) 964 and *Phys. Rev.* **D46** (1991) 381.
33. G. Altarelli and R. Barbieri, *Phys. Lett.* **B253** (1990) 161;
    B.W. Lynn, M.E. Peskin and R.G. Stuart, SLAC-PUB-3725 (1985); in Physics at LEP, Report CERN 86-02, Vol. I, p. 90.
34. B. Holdom and J. Terning, *Phys. Lett.* **B247** (1990) 88;
    D.C. Kennedy and P. Langacker, *Phys. Rev. Lett.* **65** (1990) 2967 and preprint UPR-0467T;
    B. Holdom, Fermilab 90/263-T (1990);
    A. Ali and G. Degrassi, DESY preprint DESY 91-035 (1991);
    E. Gates and J. Terning, *Phys. Rev. Lett.* **67** (1991) 1840;
    E. Ma and P. Roy, *Phys. Rev. Lett.* **68** (1992) 2879;
    G. Bhattacharyya, S. Banerjee and P. Roy, *Phys. Rev.* **D45** (1992) 729.
35. M. Golden and L. Randall, *Nucl. Phys.* **B361** (1991) 3;
    M. Dugan and L. Randall, *Phys. Lett.* **B264** (1991) 154;
    A. Dobado et al., *Phys. Lett.* **B255** (1991) 405;
    J. Layssac, F.M. Renard and C. Verzegnassi, Preprint UCLA/93/TEP/16 (1993).
36. S. Weinberg, *Phys. Rev.* **D13** (1976) 974 and *Phys. Rev.* **D19** (1979) 1277;
    L. Susskind, *Phys. Rev.* **D20** (1979) 2619;
    E. Farhi and L. Susskind, *Phys. Rep.* **74** (1981) 277.
37. R. Casalbuoni et al., *Phys. Lett.* **B258** (1991) 161;
    R.N. Cahn and M. Suzuki, LBL-30351 (1991);
    C. Roisnel and Tran N. Truong, *Phys. Lett.* **B253** (1991) 439;
    T. Appelquist and G. Triantaphyllou, *Phys. Lett.* **B278** (1992) 345;
    T. Appelquist, Proceedings of the Rencontres de la Vallée d'Aoste, La Thuile, Italy, 1993.
38. J. Ellis, G.L. Fogli and E. Lisi, *Phys. Lett.* **B343** (1995) 282.
39. CHARM Collaboration, J.V. Allaby et al., *Phys. Lett.* **B177** (1986) 446; *Z. Phys.* **C36** (1987) 611;
    CDHS Collaboration, H. Abramowicz et al., *Phys. Rev. Lett.* **57** (1986) 298;
    A. Blondel et al., *Z. Phys.* **C45** (1990) 361;
    CCFR Collaboration, A. Bodek, Proceedings of the EPS Conference on High Energy Physics, Marseille, France, 1993.
40. M.C. Noecker et al., *Phys. Rev. Lett.* **61** (1988) 310;
    M. Bouchiat, Proceedings of the 12th International Atomic Physics Conference (1990).
41. CHARM II Collaboration, R. Berger, Proceedings of the EPS Conference on High Energy Physics, Marseille, France, 1993.
42. S. Dittmaier, D. Schildknecht and G. Weiglein, *Nucl. Phys.* **B465** (1996) 3.
43. See, for example, G.G. Ross, "Grand Unified Theories", Benjamin, New York, 1984;
    R. Mohapatra, *Prog. Part. Nucl. Phys.* **26** (1991) 1.
44. See, for example, M.B. Green, J.H. Schwarz and E. Witten, "Superstring Theory", University Press, Cambridge, 1987.
45. E. Gildener, *Phys. Rev.* **D14** (1976) 1667;
    E. Gildener and S. Weinberg, *Phys. Rev.* **D15** (1976) 3333.
46. H.P. Nilles, *Phys. Rep.* **C110** (1984) 1;
    H.E. Haber and G.L. Kane, *Phys. Rep.* **C117** (1985) 75;
    R. Barbieri, *Riv. Nuovo Cim.* **11** (1988) 1.
47. For a review, see, for example, C.T. Hill, in "Perspectives on Higgs Physics", ed. G. Kane, World Scientific, Singapore, 1993, and references therein.
48. W.A. Bardeen, T.E. Clark and S.T. Love, *Phys. Lett.* **B237** (1990) 235;
    M. Carena et al., *Nucl. Phys.* **B369** (1992) 33.
49. A. Hasenfratz et al., UCSD/PTH 91-06(1991).
50. A. Chamseddine, R. Arnowitt and P. Nath, *Phys. Rev. Lett.* **49** (1982) 970;
    R. Barbieri, S. Ferrara and C. Savoy, *Phys. Lett.* **110B** (1982) 343;
    E. Cremmer et al., *Phys. Lett.* **116B** (1982) 215.

51. S. Dimopoulos, S. Raby and F. Wilczek, *Phys. Rev.* **D24** (1981) 1681;
    S. Dimopoulos and H. Georgi, *Nucl. Phys.* **B193** (1981) 150;
    L.E. Ibáñez and G.G. Ross, *Phys. Lett.* **105B** (1981) 439.
52. P. Langacker, Proceedings of the Tennessee International Symposium on Radiative Corrections ed. B.F.L. Ward, Gatlinburg, USA, 1994, p. 415.
53. L.E. Ibáñez and G.G. Ross, *Phys. Lett.* **110B** (1982) 215;
    L. Alvarez-Gaumé, M. Claudson and M.B. Wise, *Nucl. Phys.* **B207** (1982) 96.
54. R. Barbieri and G.F. Giudice, *Nucl. Phys.* **B306** (1988) 63;
    G.G. Ross and R.G. Roberts, *Nucl. Phys.* **B377** (1992) 571.
55. B. de Carlos and J.A. Casas, CERN-TH.7024/93.
56. G. Altarelli et al., CERN-TH/95-151.
57. M.S. Chanowitz, J. Ellis and M.K. Gaillard, *Nucl. Phys.* **B128** (1977) 506;
    A.J. Buras et al., *Nucl. Phys.* **B135** (1978) 66.
58. V. Barger, M.S. Berger and P. Ohmann, *Phys. Rev.* **D47** (1993) 1093;
    P. Langacker and N. Polonsky, *Phys. Rev.* **D49** (1994) 1454;
    M. Carena, S. Pokorski and C. Wagner, *Nucl. Phys.* **B406** (1993) 59;
    G.L. Kane et al. *Phys. Rev.* **D49** (1994) 6173, **D50** (1994) 3498;
    W. de Boer et al., Karlsruhe preprint IEKP-KA/94-05 (1994).
59. M. Carena et al., *Nucl. Phys.* **B419** (1994) 213.
60. R.G. Roberts and L. Roszkowski, *Phys. Lett.* **B309** (1993) 329.
61. R. Barbieri, F. Caravaglios and M. Frigeni, *Phys. Lett.* **B279** (1992) 169.
62. R. Barbieri and L. Maiani, *Nucl. Phys.* **B224** (1983) 32;
    L. Alvarez-Gaumé, J. Polchinski and M. Wise, *Nucl. Phys.* **B221** (1983) 495.
63. W. Hollik, *Mod. Phys. Lett.* **A5** (1990) 1909.
64. A. Djouadi et al., *Nucl. Phys.* **B349** (1991) 48;
    M. Boulware and D. Finell, *Phys. Rev.* **D44** (1991) 2054. The sign discrepancy between these two papers now appears to be solved in favour of the second one.
65. D. Garcia and J. Sola, *Phys. Lett.* **B354** (1995) 335.
66. S. Pokorski, Proceedings of ICHEP '96, Warsaw, 1996.
67. G. Altarelli, R. Barbieri and F. Caravaglios, *Phys. Lett.* **B314** (1993) 357.
68. J.D. Wells, C. Kolda and G. L. Kane, *Phys. Lett.* **B338** (1994) 219.
69. X. Wang, J.L.Lopez and D.V. Nanopoulos, *Phys. Rev.* **D52** (1995) 4116.
70. G. Kane, R. Stuart and J. Wells, *Phys. Lett.* **B354** (1995) 350; see also hep-ph/9510372.
71. P. Chankowski and S. Pokorski, hep-ph/9505308.
72. Y. Yamada, K. Hagiwara and S. Matsumoto, hep-ph/9512227.
73. E. Ma and D. Ng, hep-ph/9508338.
74. A. Brignole, F. Feruglio and F. Zwirner, hep-ph9601293.
75. G.F. Giudice and A. Pomarol, CERN-TH/95-337.
76. J. Ellis, J.L. Lopez and D.V. Nanopoulos, CERN-TH/95-314 (hep-ph/9512288).
77. P. Chankowski and S. Pokorski, hep-ph9603310.
78. R.S. Chivukula, S.B. Selipsky and E.H. Simmons, *Phys. Rev. Lett.* **69** (1992) 575.
79. B. Holdom, *Phys. Lett.* **105** (1985) 301;
    K. Yamawaki, M. Bando and K. Matumoto, *Phys. Rev. Lett.* **56** (1986) 1335;
    V.A. Miransky, *Nuov. Cim.* **90A** (1985);
    T. Appelquist, D. Karabali and L.C.R. Wijewardhana, *Phys. Rev.* **D35** (1987) 389; 149;
    T. Appelquist and L.C.R. Wijewardhana, *Phys. Rev.* **D35** (1987) 774; *Phys. Rev.* **D36** (1987) 568.
80. R.S. Chivukula et al., Preprint BUHEP-93-11 (1993).
81. R.S. Chivukula, E.H. Simmons and J. Terning, Preprint BUHEP-94-08 (1994).
82. ALEPH Collaboration, D. Buskolic et al., CERN/PPE/96-52.
83. G. Farrar, hep-ph/9512306.
84. A.K. Grant et al., hep-ph/960139253.
85. CDF Collaboration, F. Abe et al., Preprint Fermilab-Pub-96/020-E(1996).
86. J. Huston et al., Michigan Preprint, MSU-HEP-50812;
    W.K. Tung, Proceedings of DIS'96, Rome, 1996.
87. E.W.N. Glover et al., hep-ph/9603327.
88. D0 Collaboration, Proceedings of the Rencontres de Moriond, Les Arcs, 1996.
89. P. Chiappetta, J. Layssac, F.M. Renard and C. Verzegnassi, hep-ph/9601306.

90. G. Altarelli, N. DiBartolomeo, F. Feruglio, R. Gatto and M. Mangano, *Phys. Lett.* **B375** (1996) 292.
91. CDF Collaboration ,S. Park, Proceedings of the Workshop on Proton-Antiproton Collider Physics, ed. R. Raja and J. Yoh, AIP Press, 1995.
92. M. Mangano et al., Proceedings of ICHEP '96, Warsaw, 1996.
93. S. Ambrosanio et al., *Phys. Rev. Lett.* **76** (1996) 3494;
    S. Dimopoulos et al., *Phys. Rev. Lett.* 76 (1996) 3498;
    S. Ambrosanio and B. Mele, hep-ph-9609212.
94. M. Dine and A.E. Nelson, *Phys. Rev.* **D48** (1993) 1277;
    M. Dine, A.E. Nelson and Y. Shirman, *Phys. Rev.* **D51** (1995) 1362;
    M. Dine, A.E. Nelson, Y. Nir and Y. Shirman, *Phys. Rev.* **D53** (1996)2658;
    T. Gherghetta, G. Jungman and E. Poppitz, hep-ph/9511317;
    G. Dvali, G.F. Giudice and A. Pomarol, hep-ph/9603238.
95. G. Altarelli, T. Sjöstrand and F. Zwirner (eds.), "Physics at LEP2", Report CERN 95-03.
96. M. Sher, *Phys. Rep.* **179** (1989) 273; *Phys. Lett.* **B317** (1993) 159.
97. G. Altarelli and G. Isidori, *Phys. Lett.* **B337** (1994) 141.
98. J.A. Casas, J.R. Espinosa and M. Quiros, *Phys. Lett.* **B342** (1995) 171.
99. J.A. Casas et al., *Nucl. Phys.* **B436** (1995) 3; E**B439** (1995) 466.
100. M. Carena and C.E.M. Wagner, *Nucl. Phys.* **B452** (1995) 45.
101. See, for example, M. Lindner, *Z. Phys.* **31** (1986) 295.
102. H. Haber and R. Hempfling, *Phys. Rev. Lett.* **66** (1991) 1815;
    J. Ellis, G. Ridolfi and F. Zwirner, *Phys. Lett.* **B257** (1991) 83;
    Y. Okado, M. Yamaguchi and T. Yanagida, *Progr. Theor. Phys. Lett.* **85** (1991) 1;
    R. Barbieri, F. Caravaglios and M. Frigeni, *Phys. Lett.* **B258** (1991) 167. For a 2-loop improvement, see also:
    R. Hempfling and A.H. Hoang, *Phys. Lett.* **B331** (1994) 99.
103. N. Watson, presented in a seminar at CERN on 8 Oct. 1996.

SUSY AND SUCH

S. Dawson

Physics Department
Brookhaven National Laboratory
Upton, NY, 11973

INTRODUCTION

The Standard Model of particle physics is in stupendous agreement with experimental measurements; in some cases it has been tested to a precision of greater than .1%. Why then expand our model? The reason, of course, is that the Standard Model contains several nagging theoretical problems which cannot be solved without the introduction of some new physics. Supersymmetry is, at present, many theorists' favorite candidate for such new physics.

The single aspect of the Standard Model which has not been verified experimentally is the Higgs sector. The Standard Model without the Higgs boson is incomplete, however, since it predicts massless fermions and gauge bosons. Furthermore, the electroweak radiative corrections would be infinite and longitudinal gauge boson scattering would grow with energy and violate unitarity at an energy scale around $3\,TeV$ if there were no Higgs boson.[1] The simplest mechanism to cure these defects is the introduction of a single $SU(2)_L$ doublet of Higgs bosons. When the neutral component of the Higgs boson gets a vacuum expectation value, the $SU(2)_L \times U(1)_Y$ gauge symmetry is broken, giving the W and Z gauge bosons their masses. The chiral symmetry forbidding fermion masses is broken at the same time allowing the fermions to become massive. Furthermore, in the Standard Model, the coupling of the Higgs boson to gauge bosons is just that required to cancel the infinities in electroweak radiative corrections and to cancel the unitarity violation in the gauge boson scattering sector. What then is the problem with this simple and economical picture?

The argument against the simplest version of the Standard Model with a single Higgs boson is purely theoretical and arises when radiative corrections to the Higgs boson mass are computed. The scalar potential for the Higgs boson, h, is given schematically by,

$$V \sim M_{h0}^2 h^2 + \lambda h^4. \qquad (2)$$

At one loop, the quartic self-interactions of the Higgs boson (proportional to λ) generate a quadratically divergent contribution to the Higgs boson mass which must be cancelled by the mass counterterm, δM_h^2,[2]

$$M_h^2 \sim M_{h0}^2 + \frac{\lambda}{4\pi^2}\Lambda^2 + \delta M_h^2. \tag{3}$$

The scale Λ is a cutoff which, in the Standard Model with no new physics between the electroweak scale and the Planck scale, must be of order the Planck scale. In order for the Higgs boson to do its job of preventing unitarity violation in the scattering of longitudinal gauge bosons, however, its mass must be less than around 800 GeV.[1] This leads to an unsatisfactory situation. The large quadratic contribution to the Higgs boson mass-squared, of $\mathcal{O}(10^{18} GeV)^2$, must be cancelled by the counterterm δM_h^2 such that the result is roughly less than $(800\ GeV)^2$. This requires a cancellation of one part in 10^{16}. This is of course formally possible, but regarded by most theorists as an unacceptable fine tuning of parameters. Additionally, this cancellation must occur at every order in perturbation theory and so the parameters must be fine tuned again and again. The quadratic growth of the Higgs boson mass beyond tree level in perturbation theory is one of the driving motivations behind the introduction of supersymmetry, which we will see cures this problem. It is interesting that the loop corrections to fermion masses do not exhibit this quadratic growth (and we therefore say that fermion masses are "natural"). It is only when we attempt to understand electroweak symmetry breaking by including a Higgs boson that we face the problem of quadratic divergences.

In these lectures, I discuss the theoretical motivation for supersymmetric theories and introduce the minimal low energy effective supersymmetric theory, (MSSM). I consider only the MSSM and its simplest grand unified extension here. Some of the other possible low-energy SUSY models are summarized in Ref. [3]. The particles and their interactions are examined in detail in the next sections and a grand unified SUSY model presented which gives additional motivation for pursuing supersymmetric theories.

Finally, I discuss indirect limits on the SUSY partners of ordinary matter coming from precision measurements at LEP and direct production searches at the Tevatron and discuss search strategies for SUSY at both future e^+e^- and hadron colliders. Only a sampling of existing limits are given in order to demonstrate some of the general features of these searches. Up-to-date limits on SUSY particle searches at hadron colliders[4] and e^+e^- colliders[5] were given at the 1996 DPF meeting and can be used to map out the allowed regions of SUSY parameter space. There exist numerous excellent reviews of both the more formal aspects of supersymmetric model building [6, 7] and the phenomenology of these models [8, 9] and the reader is referred to these for more details. I present here a workmanlike approach designed primarily for experimental graduate students.

WHAT IS SUSY?

Suppose we reconsider the one loop contributions to the Higgs boson mass in a theory which contains both massive scalars, ϕ, and fermions, ψ, in addition to the Higgs field, h. Then the Lagrangian is given by:

$$\mathcal{L} \sim -g_F \overline{\psi}\psi h - g_S^2 h^2 \phi^2 \ . \tag{4}$$

If we again calculate the one-loop contribution to M_h^2 we find[2]

$$M_h^2 \sim M_{h0}^2 + \frac{g_F^2}{4\pi^2}\left(\Lambda^2 + m_F^2\right) - \frac{g_S^2}{4\pi^2}\left(\Lambda^2 + m_S^2\right)$$
$$+\text{logarithmic divergences} + \text{uninteresting terms} \quad . \tag{5}$$

The relative minus sign between the fermion and scalar contributions to the Higgs boson mass-squared is the well-known result of Fermi statistics. We see that if $g_S = g_F$ the terms which grow with Λ^2 cancel and we are left with a well behaved contribution to the Higgs boson mass so long as the fermion and scalar masses are not too different,[10]

$$M_h^2 \sim M_{h0}^2 + \frac{g_F^2}{4\pi^2}\left(m_F^2 - m_S^2\right) \quad . \tag{6}$$

Attempts have been made to quantify *"not too different"*.[9] One can roughly assume that the cancellation is unnatural if the mass splitting between the fermion and the scalar is larger than about a TeV. Of course, in order for this cancellation to persist to all orders in perturbation theory it must be the result of a symmetry. This symmetry is **supersymmetry**.

Supersymmetry is a symmetry which relates particles of differing spin, (in the above example, fermions and scalars). The particles are combined into a *superfield*, which contains fields differing by one-half unit of spin.[11] The simplest example, the scalar superfield, contains a complex scalar, S, and a two-component Majorana fermion, ψ. (A Majorana fermion, ψ, is one which is equal to its charge conjugate, $\psi^c = \psi$. A familiar example is a Majorana neutrino.) The supersymmetry completely specifies the allowed interactions. In this simple case, the Lagrangian is

$$\mathcal{L} = -\partial_\mu S^* \partial^\mu S - i\overline{\psi}\overline{\sigma}^\mu \partial_\mu \psi - \frac{1}{2}m(\psi\psi + \overline{\psi\psi})$$
$$-cS\psi\psi - c^*S^*\overline{\psi\psi} - \mid mS + cS^2 \mid^2, \tag{7}$$

(where σ is a 2×2 Pauli matrix and c is an arbitrary coupling constant.) This Lagrangian is invariant (up to a total derivative) under transformations which take the scalar into the fermion and *vice versa*. Since the scalar and fermion interactions have the same coupling, the cancellation of quadratic divergences occurs automatically, as in Eq. 5. One thing that is immediately obvious is that this Lagrangian contains both a scalar and a fermion *of equal mass*. Supersymmetry connects particles of different spin, but with all other characteristics the same. That is, they have the same quantum numbers and the same mass.

- Particles in a superfield have the same masses and quantum numbers and differ by 1/2 unit of spin in a theory with unbroken supersymmetry.

It is clear, then, that **supersymmetry must be a broken symmetry**. There is no scalar particle, for example, with the mass and quantum numbers of the electron. In fact, there are no candidate supersymmetric scalar partners for any of the fermions in the experimentally observed spectrum. We will take a non-zero mass splitting between the particles of a superfield as a signal for supersymmetry breaking.

Supersymmetric theories are easily constructed according to the rules of supersymmetry. I present here a cookbook approach to constructing the minimal supersymmetric version of the Standard Model. The first step is to pick the particles in superfields. There are two types of superfields relevant for our purposes:*

1. *Chiral Superfields*: These consist of a complex scalar field, S, and a 2-component Majorana fermion field, ψ.

2. *Massless Vector Superfields*: These consist of a massless gauge field with field strength $F_{\mu\nu}^A$ and a 2-component Majorana fermion field, λ_A, termed a *gaugino*. The index A is the gauge index.

The Particles of the MSSM

The MSSM respects the same $SU(3) \times SU(2)_L \times U(1)$ gauge symmetries as does the Standard Model. The particles necessary to construct the supersymmetric version of the Standard Model are shown in Tables 1 and 2 in terms of the superfields, (which are denoted by the superscript "hat"). Since there are no candidates for supersymmetric partners of the observed particles, we must double the entire spectrum, placing the observed particles in superfields with new postulated superpartners. There are, of course, quark and lepton superfields for all 3 generations and we have listed in Table 1 only the members of the first generation. The superfield $\hat{Q}$ thus consists of an $SU(2)_L$ doublet of quarks:

$$Q = \begin{pmatrix} u \\ d \end{pmatrix}_L \tag{8}$$

and their scalar partners which are also in an $SU(2)_L$ doublet,

$$\tilde{Q} = \begin{pmatrix} \tilde{u}_L \\ \tilde{d}_L \end{pmatrix} . \tag{9}$$

Similarly, the superfield $\hat{U}^c$ ($\hat{D}^c$) contains the right-handed up (down) anti-quark, $\overline{u}_R$ ($\overline{d}_R$), and its scalar partner, $\tilde{u}_R^*$ ($\tilde{d}_R^*$). The scalar partners of the quarks are fancifully called squarks. We see that each quark has 2 scalar partners, one corresponding to each quark chirality. The leptons are contained in the $SU(2)_L$ doublet superfield $\hat{L}$ which contains the left-handed fermions,

$$L = \begin{pmatrix} \nu \\ e \end{pmatrix}_L \tag{10}$$

and their scalar partners,

$$\tilde{L} = \begin{pmatrix} \tilde{\nu}_L \\ \tilde{e}_L \end{pmatrix} . \tag{11}$$

Finally, the right-handed anti-electron, $\overline{e}_R$, is contained in the superfield $\hat{E}^c$ and has a scalar partner $\tilde{e}_R^*$. The scalar partners of the leptons are termed sleptons.

The $SU(3) \times SU(2)_L \times U(1)$ gauge fields all obtain Majorana fermion partners in a SUSY model. The $\hat{G}^a$ superfield contains the gluons, g^a, and their partners the gluinos, $\tilde{g}^a$; $\hat{W}_i$ contains the $SU(2)_L$ gauge bosons, W_i and their fermion partners, $\tilde{\omega}_i$ (winos); and $\hat{B}$ contains the $U(1)$ gauge field, B, and its fermion partner, $\tilde{b}$ (bino). The usual notation is to denote the supersymmetric partner of a fermion or gauge field with the same letter, but with a tilde over it.

Table 1: Chiral Superfields of the MSSM

Superfield	SU(3)	$SU(2)_L$	$U(1)_Y$	Particle Content
$\hat{Q}$	3	2	$\frac{1}{6}$	$(u_L, d_L), (\tilde{u}_L, \tilde{d}_L)$
$\hat{U}^c$	$\overline{3}$	1	$-\frac{2}{3}$	$\overline{u}_R, \tilde{u}_R^*$
$\hat{D}^c$	$\overline{3}$	1	$\frac{1}{3}$	$\overline{d}_R, \tilde{d}_R^*$
$\hat{L}$	1	2	$-\frac{1}{2}$	$(\nu_L, e_L), (\tilde{\nu}_L, \tilde{e}_L)$
$\hat{E}^c$	1	1	1	$\overline{e}_R, \tilde{e}_R^*$
$\hat{H}_1$	1	2	$-\frac{1}{2}$	$(H_1, \tilde{h}_1)$
$\hat{H}_2$	1	2	$\frac{1}{2}$	$(H_2, \tilde{h}_2)$

Table 2: Vector Superfields of the MSSM

Superfield	SU(3)	$SU(2)_L$	$U(1)_Y$	Particle Content
$\hat{G}^a$	8	1	0	$g, \tilde{g}$
$\hat{W}^i$	1	3	0	$W_i, \tilde{\omega}_i$
$\hat{B}$	1	1	0	$B, \tilde{b}$

One feature of Table 1 requires explanation. The Standard Model contains a single $SU(2)_L$ doublet of scalar particles, dubbed the "Higgs doublet". In the supersymmetric extension of the Standard Model, this scalar doublet acquires a SUSY partner which is an $SU(2)_L$ doublet of Majorana fermion fields, $\tilde{h}_1$ (the Higgsinos), which contribute to the triangle $SU(2)_L$ and $U(1)$ gauge anomalies. Since the fermions of the Standard Model have exactly the right quantum numbers to cancel these anomalies, it follows that the contribution from the fermionic partner of the Higgs doublet remains uncancelled.[12] Since gauge theories cannot have anomalies, these contributions must be cancelled somehow if the SUSY theory is to be sensible. The simplest way is to add a second Higgs doublet with precisely the opposite $U(1)$ quantum numbers from the first Higgs doublet. In a SUSY Model, this second Higgs doublet will also have fermionic partners, $\tilde{h}_2$, and the contributions of the fermion partners of the two Higgs doublets to gauge anomalies will precisely cancel each other, leaving an anomaly free theory. It is easy to check that the fermions of Table 1 satisfy the conditions for anomaly cancellation:

$$Tr(Y^3) = Tr(T_{3L}^2 Y) = 0 \quad . \tag{12}$$

We will see later that 2 Higgs doublets are also required in order to give both the up and down quarks masses in a SUSY theory. The requirement that there be at least 2 $SU(2)_L$ Higgs doublets is a feature of all models with weak scale supersymmetry.

- In general, supersymmetric extensions of the Standard Model have extended Higgs sectors leading to a rich phenomenology of scalars.

---
*The superfields also contain "auxiliary fields", which are fields with no kinetic energy terms in the Lagrangian.[11] These fields are not important for our purposes.

The Interactions of the MSSM

Having specified the superfields of the theory, the next step is to construct the supersymmetric Lagrangian.[13] There is very little freedom in the allowed interactions between the ordinary particles and their supersymmetric partners. It is this feature of a SUSY model which gives it predictive power (and makes it attractive to theorists!). It is important to note here, however, that there is nothing to stop us from adding more superfields to those shown in Tables 1 and 2 as long as we are careful to add them in such a way that any additional contributions to gauge anomalies cancel among themselves. Recent popular models add an additional gauge singlet superfield to the spectrum, which has interesting phenomenological consequences.[14] The MSSM which we concentrate on, however, contains only those fields given in the tables.

The supersymmetry associates each 2-component Majorana fermion with a complex scalar. The massive fermions of the Standard Model are, however, Dirac fermions. A Dirac fermion has 4 components which can be thought of as the left-and right-handed chiral projections of the fermion state. It is straightforward to translate back and forth between 2- and 4- component notation for the fermions and we will henceforth use the more familiar 4- component notation when writing the fermion interactions.[6] The fields of the MSSM all have canonical kinetic energies:[†]

$$\mathcal{L}_{KE} = \sum_i \left\{ (D_\mu S_i^*)(D^\mu S_i) + i\overline{\psi}_i D\psi_i \right\}$$
$$+ \sum_A \left\{ -\frac{1}{4} F_{\mu\nu}^A F^{\mu\nu A} + \frac{i}{2} \overline{\lambda}_A D\lambda_A \right\}, \tag{13}$$

where D is the $SU(3) \times SU(2)_L \times U(1)$ gauge invariant derivative. The $\sum_i$ is over all the fermion fields of the Standard Model, ψ_i, and their scalar partners, S_i, and also over the 2 Higgs doublets with their fermion partners. The $\sum_A$ is over the $SU(3)$, $SU(2)_L$ and $U(1)_Y$ gauge fields with their fermion partners, the gauginos.

The interactions between the chiral superfields of Table 1 and the gauginos and the gauge fields of Table 2 are completely specified by the gauge symmetries and by the supersymmetry, as are the quartic interactions of the scalars,

$$\mathcal{L}_{int} = -\sqrt{2} \sum_{i,A} g_A \left[S_i^* T^A \overline{\psi}_{iL} \lambda_A + \text{h.c.} \right] - \frac{1}{2} \sum_A \left(\sum_i g_A S_i^* T^A S_i \right)^2, \tag{14}$$

where $\psi_L \equiv \frac{1}{2}(1-\gamma_5)\psi$. In Eq. 14, g_A is the relevant gauge coupling constant and we see that the interaction strengths are fixed in terms of these constants. **There are no adjustable parameters here.** For example, the interaction between a quark, its scalar partner, the squark, and the gluino is governed by the strong coupling constant, g_s. A complete set of Feynman rules for the minimal SUSY model described here is given in the review by Haber and Kane.[6] A good rule of thumb is to take an interaction involving Standard Model particles and replace two of the particles by their SUSY partners to get an approximate strength for the interaction. (This naive picture is, of course, altered by $\sqrt{2}$'s, mixing angles, etc.).

The only freedom in constructing the supersymmetric Lagrangian (once the superfields and the gauge symmetries are chosen) is contained in a function called the

---
[†]Remember that both the right- and left- handed helicity state of a fermion has its own scalar partner.

superpotential,W. The superpotential is a function of the chiral superfields of Table 1 only (it is not allowed to contain their complex conjugates) and it contains terms with 2 and 3 chiral superfields. Terms in the superpotential with more than 3 chiral superfields would yield non-renormalizable interactions in the Lagrangian. The superpotential also is not allowed to contain derivative interactions and we say that it is an analytic function. From the superpotential can be found both the scalar potential and the Yukawa interactions of the fermions with the scalars:

$$\mathcal{L}_W = -\sum_i \left|\frac{\partial W}{\partial z_i}\right|^2 - \frac{1}{2}\sum_{ij}\left[\overline{\psi}_{iL}\frac{\partial^2 W}{\partial z_i \partial z_j}\psi_j + \text{h.c.}\right], \qquad (15)$$

where z is a chiral superfield. This form of the Lagrangian is dictated by the supersymmetry and by the requirement that it be renormalizable. An explicit derivation of Eq. 15 can be found in Ref. [11]. To obtain the interactions, we take the derivatives of W with respect to the superfields, z, and then evaluate the result in terms of the scalar component of z.

The usual approach is to write the most general $SU(3) \times SU(2)_L \times U(1)_Y$ invariant superpotential with arbitrary coefficients for the interactions,

$$\begin{aligned}W &= \epsilon_{ij}\mu \hat{H}_1^i \hat{H}_2^j + \epsilon_{ij}\left[\lambda_L \hat{H}_1^i \hat{L}^{cj}\hat{E}^c + \lambda_D \hat{H}_1^i \hat{Q}^j \hat{D}^c + \lambda_U \hat{H}_2^j \hat{Q}^i \hat{U}^c\right] \\ &+ \epsilon_{ij}\left[\lambda_1 \hat{L}^i \hat{L}^j \hat{E}^c + \lambda_2 \hat{L}^i \hat{Q}^j \hat{D}^c\right] + \lambda_3 \hat{U}^c \hat{D}^c \hat{D}^c,\end{aligned} \qquad (16)$$

(where i,j are $SU(2)$ indices). In principle, a bi-linear term $\epsilon_{ij}\hat{L}^i\hat{H}_2^j$ can also be included in the superpotential. It is possible, however, to rotate the lepton field, $\hat{L}$, such that this term vanishes so we will ignore it. We have written the superpotential in terms of the fields of the first generation. In principle, the λ_i could all be matrices which mix the interactions of the 3 generations.

The $\mu \hat{H}_1 \hat{H}_2$ term in the superpotential gives mass terms for the Higgs bosons when we apply $|\partial W/\partial z|^2$ and μ is often called the Higgs mass parameter. We shall see later that the physics is very sensitive to the sign of μ. The terms in the square brackets proportional to λ_L, λ_D, and λ_U give the usual Yukawa interactions of the fermions with the Higgs bosons from the term $\overline{\psi}_i(\partial^2 W/\partial z_i \partial z_j)\psi_j$. Hence these coefficients are determined in terms of the fermion masses and the vacuum expectation values of the neutral members of the scalar components of the Higgs doublets and are not free parameters at all.

The Lagrangian as we have written it cannot, however, be the whole story as all the particles (fermions, scalars, gauge fields) are massless at this point.

R Parity

The terms in the second line of Eq. 16 (proportional to λ_1, λ_2 and λ_3) are a problem. They contribute to lepton and baryon number violating interactions and can mediate proton decay at tree level through the exchange of the scalar partner of the down quark. If the SUSY partners of the Standard Model particles have masses on the TeV scale, then these interactions are severely restricted by experimental measurements.[13, 15]

There are several possible approaches to the problem of the lepton and baryon number violating interactions. The first is simply to make the coefficients, λ_1, λ_2, and

λ_3 small enough to avoid experimental limits.[16, 17] This artificial tuning of parameters is regarded as unacceptable by many theorists, but is certainly allowed experimentally. Another tactic is to make either the lepton number violating interactions, λ_1 and λ_2, or the baryon number violating interaction, λ_3, zero, (while allowing the others to be non-zero) which would forbid proton decay. There is, however, not much theoretical motivation for this approach.

The usual strategy is to require that all of these undesirable lepton and baryon number violating terms be forbidden by a symmetry. (If they are forbidden by a symmetry, they will not re-appear at higher orders of perturbation theory.) The symmetry which does the job is called **R parity**.[18] R parity can be defined as a multiplicative quantum number such that all particles of the Standard Model have R parity +1, while their SUSY partners have R parity -1. R parity can also be defined as,

$$R \equiv (-1)^{3(B-L)+s} , \qquad (17)$$

for a particle of spin s. It is then obvious that such a symmetry forbids the lepton and baryon number violating terms of Eq. 16. It is worth noting that in the Standard Model, the problem of baryon and lepton number violating interactions does not arise, since these interactions are forbidden by the gauge symmetries to contribute to dimension-4 operators and first arise in dimension-6 operators which are suppressed by factors of some heavy mass scale.

The assumption of R parity conservation has profound experimental consequences which go beyond the details of a specific model. Because R parity is a multiplicative quantum number, it implies that the number of SUSY partners in a given interaction is always conserved modulo 2.

- SUSY partners can only be pair produced from Standard Model particles.

Furthermore, a SUSY particle will decay in a chain until the lightest SUSY particle is produced (such a decay is called a *cascade decay*). This lightest SUSY particle, called the LSP, must be absolutely stable when R parity is conserved.

- A theory with R parity conservation will have a lightest SUSY particle (LSP) which is stable.

The LSP must be neutral since there are stringent cosmological bounds on light charged or colored particles which are stable.[19, 20] Hence the LSP is stable and neutral and is not seen in a detector (much like a neutrino) since it interacts only by the exchange of a heavy virtual SUSY particle.

- The LSP will interact very weakly with ordinary matter.

- A generic signal for R parity conserving SUSY theories is missing transverse energy from the non-observed LSP.

In theories without R parity conservation, there will not be a stable LSP, and the lightest SUSY particle will decay into ordinary particles (possibly within the detector). Missing transverse energy will no longer be a robust signature for SUSY particle production.[21]

Supersymmetry Breaking

The mechanism of supersymmetry breaking is not well understood. At this point we have constructed a SUSY theory containing all of the Standard Model particles, but the supersymmetry remains unbroken and the particles and their SUSY partners are massless. This is clearly unacceptable. It is typically assumed that the SUSY breaking occurs at a high scale, say M_{pl}, and perhaps results from some complete theory encompassing gravity. At the moment the usual approach is to assume that the MSSM, which is the theory at the electroweak scale, is an effective low energy theory.[22] The supersymmetry breaking is implemented by including explicit "soft" mass terms for the scalar members of the chiral multiplets and for the gaugino members of the vector supermultiplets in the Lagrangian. These interactions are termed soft because they do not re-introduce the quadratic divergences which motivated the introduction of the supersymmetry in the first place. The dimension of soft operators in the Lagrangian must be 3 or less, which means that the possible soft operators are mass terms, bi-linear mixing terms ("B" terms), and tri-linear scalar mixing terms (" A terms"). The origin of these supersymmetry breaking terms is left unspecified. The complete set of soft SUSY breaking terms (which respect R parity and the $SU(3) \times SU(2)_L \times U(1)$ gauge symmetry) for the first generation is given by the Lagrangian:[13, 23]

$$\begin{aligned}
-\mathcal{L}_{soft} = \; & m_1^2 \mid H_1 \mid^2 + m_2^2 \mid H_2 \mid^2 - B\mu\epsilon_{ij}(H_1^i H_2^j + \text{h.c.}) + \tilde{M}_Q^2(\tilde{u}_L^* \tilde{u}_L + \tilde{d}_L^* \tilde{d}_L) \\
& + \tilde{M}_u^2 \tilde{u}_R^* \tilde{u}_R + \tilde{M}_d^2 \tilde{d}_R^* \tilde{d}_R + \tilde{M}_L^2(\tilde{e}_L^* \tilde{e}_L + \tilde{\nu}_L^* \tilde{\nu}_L) + \tilde{M}_e^2 \tilde{e}_R^* \tilde{e}_R \\
& + \frac{1}{2}\left[M_3 \bar{\tilde{g}}\tilde{g} + M_2 \bar{\tilde{\omega}}_i \tilde{\omega}_i + M_1 \bar{\tilde{b}}\tilde{b}\right] + \frac{g}{\sqrt{2}M_W}\epsilon_{ij}\left[\frac{M_d}{\cos\beta}A_d H_1^i \tilde{Q}^j \tilde{d}_R^* \right. \\
& \left. + \frac{M_u}{\sin\beta}A_u H_2^j \tilde{Q}^i \tilde{u}_R^* + \frac{M_e}{\cos\beta}A_e H_1^i \tilde{L}^j \tilde{e}_R^* + \text{h.c.}\right] \; .
\end{aligned} \quad (18)$$

This Lagrangian has arbitrary masses for the scalars and gauginos and also arbitrary tri-linear and bi-linear mixing terms. The scalar and gaugino mass terms have the desired effect of breaking the degeneracy between the particles and their SUSY partners. The tri-linear A-terms have been defined with an explicit factor of mass and we will see later that they affect primarily the particles of the third generation.[‡] When the A_i terms are non-zero, the scalar partners of the left- and right-handed fermions can mix when the Higgs bosons get vacuum expectation values and so they are no longer mass eigenstates. The B term mixes the scalar components of the 2 Higgs doublets.

The philosophy is to add all of the mass and mixing terms which are allowed by the gauge symmetries. To further complicate matters, all of the mass and interaction terms of Eq. 18 may be matrices involving all three generations. $\mathcal{L}_{soft}$ has clearly broken the supersymmetry since the SUSY partners of the ordinary particles have been given arbitrary masses. This has come at the tremendous expense, however, of introducing a large number of unknown parameters (more than 50!). It is one of the wonderful features of supersymmetry that even with all these new parameters, the theory is still able to make some definitive predictions. This is, of course, because the gauge interactions of the SUSY particles are completely fixed. What is really needed, however, is a theory of how the soft SUSY breaking terms arise in order to reduce the parameter space.

[‡]We have also included an angle β in the normalization of the A terms. The factor β is related to the vacuum expectation values of the neutral components of the Higgs fields and is defined in the next section. The normalization is, of course, arbitrary.

We have now constructed the Lagrangian describing a softly broken supersymmetric theory which is assumed to be the effective theory at the weak scale. A more complete theory would predict the soft SUSY breaking terms. In the next section we will examine how the electroweak symmetry is broken in this model and study the mass spectrum and interactions of the new particles.

The Higgs Sector and Electroweak Symmetry Breaking

The Higgs sector of the MSSM is very similar to that of a general 2 Higgs doublet model.[24] The scalar potential involving the Higgs bosons is

$$V_H = \left(|\mu|^2 + m_1^2\right)|H_1|^2 + \left(|\mu|^2 + m_2^2\right)|H_2|^2 - \mu B \epsilon_{ij}\left(H_1^i H_2^j + \text{h.c.}\right)$$
$$+ \frac{g^2 + g'^2}{8}\left(|H_1|^2 - |H_2|^2\right)^2 + \frac{1}{2}g^2|H_1^* H_2|^2 \quad . \tag{19}$$

The Higgs potential of the SUSY model can be seen to depend on 3 independent parameters,

$$|\mu|^2 + m_1^2,$$
$$|\mu|^2 + m_2^2,$$
$$\mu B \quad , \tag{20}$$

where B is a new mass parameter. This is in contrast to the general 2 Higgs doublet model where there are 6 arbitrary coupling constants (and a phase) in the potential. From Eq. 14, it is clear that the quartic couplings are fixed in terms of the gauge couplings and so they are not free parameters. This leaves only the mass terms of Eq. 20 unspecified. Note that V_H automatically conserves CP since any complex phase in μB can be absorbed into the definitions of the Higgs fields.

Clearly, if $\mu B = 0$ then all the terms in the potential are positive and the minimum of the potential occurs with $V = 0$ and $\langle H_1^0 \rangle = \langle H_2^0 \rangle = 0$, leaving the electroweak symmetry unbroken.[§] Hence all 3 parameters must be non-zero in order for the electroweak symmetry to be broken. [¶]

In order for the electroweak symmetry to be broken and for the potential to be stable at large values of the fields, the parameters must satisfy the relations,

$$(\mu B)^2 > \left(|\mu|^2 + m_1^2\right)\left(|\mu|^2 + m_2^2\right)$$
$$|\mu|^2 + \frac{m_1^2 + m_2^2}{2} > |\mu B| \quad . \tag{21}$$

We will assume that these conditions are met. The symmetry is broken when the neutral components of the Higgs doublets get vacuum expectation values,[‖]

$$\langle H_1^0 \rangle \equiv v_1$$
$$\langle H_2^0 \rangle \equiv v_2 \quad . \tag{22}$$

[§]It also leaves the supersymmetry unbroken, since $\langle V \rangle > 0$ is required in order for the supersymmetry to be broken.[25]

[¶]We assume that the parameters are arranged in such a way that the scalar partners of the quarks and leptons do not obtain vacuum expectation values. Such vacuum expectation values would spontaneously break the $SU(3)$ color gauge symmetry or lepton number. This requirement gives a restriction on $A_i/\tilde{m}$, where $\tilde{m}$ is a generic squark or slepton mass.

[‖]Our conventions for factors of 2 in the Higgs sector, and for the definition of the sign(μ), are those of Ref. [26].

By redefining the Higgs fields, we can always choose v_1 and v_2 positive.

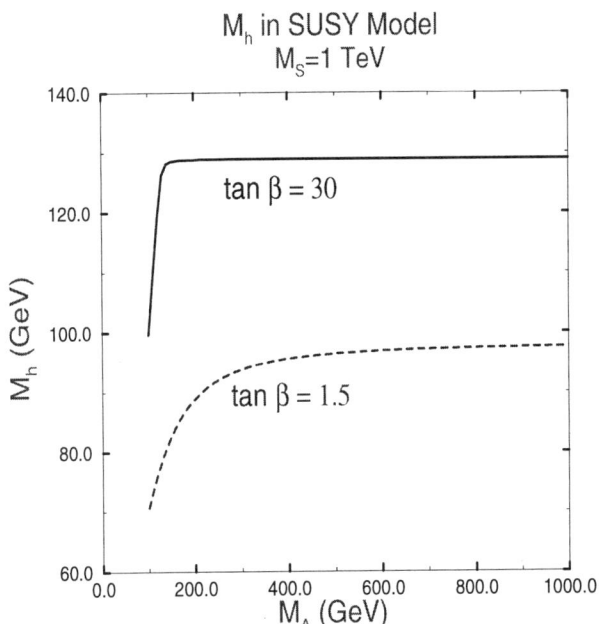

Figure 1: Mass of the lightest neutral Higgs boson as a function of the pseudoscalar mass, M_A, and $\tan\beta$. This figure includes radiative corrections to the Higgs mass[28], assumes a common scalar mass of $1\ TeV$, and neglects mixing effects, ($A_i = \mu = 0$).

When the electroweak symmetry is broken, the W gauge boson gets a mass which is fixed by v_1 and v_2,
$$M_W^2 = \frac{g^2}{2}(v_1^2 + v_2^2) \quad . \tag{23}$$
Before the symmetry was broken, the 2 complex $SU(2)_L$ Higgs doublets had 8 degrees of freedom. Three of these were absorbed to give the W and Z gauge bosons their masses, leaving 5 physical degrees of freedom. There is now a charged Higgs boson, $H^\pm$, a CP-odd neutral Higgs boson, A, and 2 CP-even neutral Higgs bosons, h and H. After fixing $v_1^2 + v_2^2$ such that the W gets the correct mass, the Higgs sector is then described by 2 additional parameters which can be chosen however you like. The usual choice is
$$\tan\beta \equiv \frac{v_2}{v_1} \tag{24}$$
and M_A, the mass of the pseudoscalar Higgs boson. Once these two parameters are

given, then the masses of the remaining Higgs bosons can be calculated in terms of M_A and $\tan\beta$. Note that we can chose $0 \leq \beta \leq \frac{\pi}{2}$ since we have chosen $v_1, v_2 > 0$.

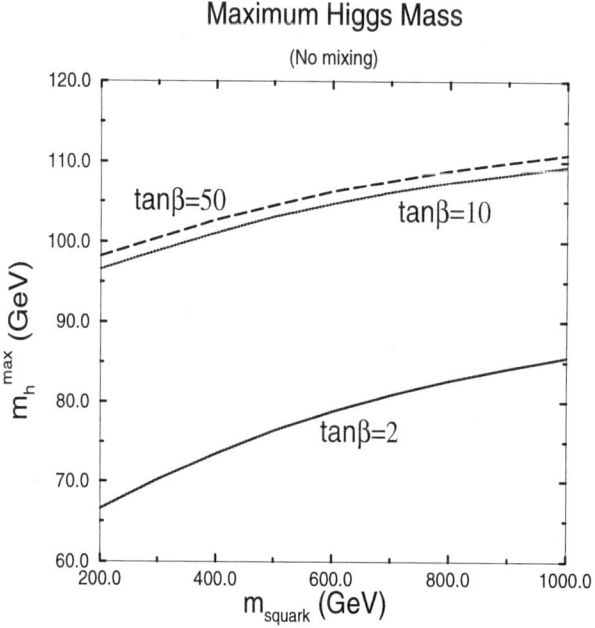

Figure 2: Maximum value of the lightest Higgs boson mass as a function of the squark mass including radiative corrections.[28] (We have assumed degenerate squarks and set the mixing parameters $A_i = \mu = 0$.)

It is straightforward to find the physical Higgs bosons and their masses in terms of the parameters of Eq. 19. Details can be found in Ref. [26]. The neutral Higgs masses are found by diagonalizing the 2×2 Higgs mass matrix and by convention, h is taken to be the lighter of the neutral Higgs. The pseudoscalar mass is given by,

$$M_A^2 = \frac{2\,|\,\mu B\,|}{\sin 2\beta}, \tag{25}$$

and the charged scalar mass is,

$$M_{H^\pm}^2 = M_W^2 + M_A^2 \ . \tag{26}$$

We see that at tree level[27], Eq. 19 gives important predictions about the relative masses of the Higgs bosons,

$$M_{H^+} > M_W$$

$$M_H > M_Z$$
$$M_h < M_A$$
$$M_h < M_Z |\cos 2\beta| \quad . \qquad (27)$$

These relations yield the desirable prediction that the lightest neutral Higgs boson is lighter than the Z boson and so must be observable at LEPII. Unfortunately (for experimentalists at least!) it was realized several years ago that loop corrections to the relations of Eq. 27 are large. In fact the corrections to M_h^2 grow like $G_F M_T^4$ and receive contributions from loops with both top quarks and squarks. In a model with unbroken supersymmetry, these contributions would cancel. Since the supersymmetry has been broken by splitting the masses of the fermions and their scalar partners, the neutral Higgs boson masses become at one- loop,[28]

$$M_{h,H}^2 = \frac{1}{2}\left\{M_A^2 + M_Z^2 + \frac{\epsilon_h}{\sin^2\beta} \pm \left[\left((M_A^2 - M_Z^2)\cos 2\beta + \frac{\epsilon_h}{\sin^2\beta}\right)^2 + \left(M_A^2 + M_Z^2\right)^2 \sin^2 2\beta\right]^{1/2}\right\} \qquad (28)$$

where ϵ_h is the contribution of the one-loop corrections,

$$\epsilon_h \equiv \frac{3 G_F}{\sqrt{2}\pi^2} M_T^4 \log\left(\frac{\tilde{m}^2}{M_T^2}\right) \quad . \qquad (29)$$

We have assumed that all of the squarks have equal masses, $\tilde{m}$, and have neglected the smaller effects from the mixing parameters, A_i and μ. In Fig. 1, we show the lightest Higgs boson mass as a function of the assumed common squark mass, $\tilde{m}$, and for two values of $\tan\beta$. For $\tan\beta > 1$, the mass eigenvalues increase monotonically with increasing M_A and give an upper bound to the mass of the lightest Higgs boson,

$$M_h^2 < M_Z^2 \cos^2 2\beta + \epsilon_h \quad . \qquad (30)$$

The corrections from ϵ_h are always positive and increase the mass of the lightest neutral Higgs boson with increasing top quark mass. From Fig. 1, we see that M_h obtains its maximal value for rather modest values of the pseudoscalar mass, $M_A > 300\ GeV$. The radiative corrections to the charged Higgs mass-squared are proportional to M_T^2 and so are much smaller than the corrections to the neutral masses.

There are many sophisticated analyses[28] which include a variety of two-loop effects, renormalization group effects, etc., but the important point is that for given values of $\tan\beta$ and the squark masses, there is an upper bound on the lightest neutral Higgs boson mass. The maximum value of the lightest Higgs mass is shown in Fig. 2 and we see that there is still a light Higgs boson even when radiative corrections are included.** For large values of $\tan\beta$ the limit is relatively insensitive to the value of $\tan\beta$ and with a squark mass less than about $1\ TeV$, the upper limit on the Higgs mass is about $110\ GeV$. Different approaches can raise this limit slightly to around $130\ GeV$.

- The minimal SUSY model predicts a neutral Higgs boson with a mass less than around $130\ GeV$.

**The leading logarithmic corrections are included in Fig. 2 and lower the result slightly from that obtained using Eq. 28.

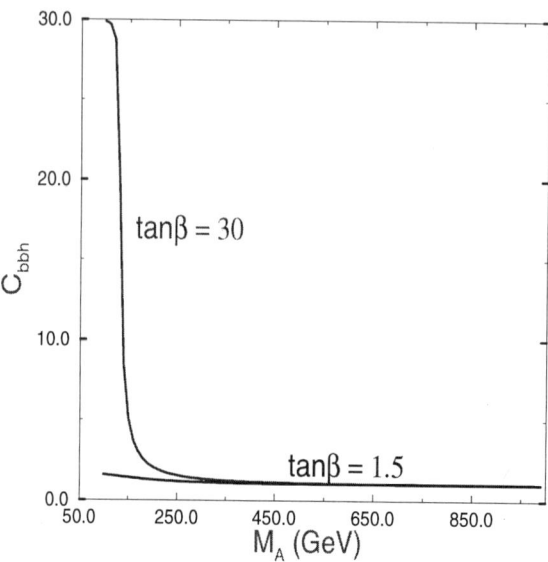

Figure 3: Coupling of the lightest Higgs boson to charge $-1/3$ quarks including radiative corrections [28] in terms of the couplings defined in Eq. 32. The value $C_{bbh} = 1$ corresponds to the Standard Model coupling of the Higgs boson to charge $-1/3$ quarks.

Such a mass scale will be accessible at LEPII or the LHC and provides a definitive test of the MSSM.

In a more complicated SUSY model with a richer Higgs structure, this bound will, of course, be changed. However, the requirement that the Higgs self coupling remain perturbative up to the Planck scale gives an upper bound on the lightest SUSY Higgs boson of around 150 GeV in all models.[29] This is a very strong statement. It implies that either there is a relatively light Higgs boson (which would be accessible experimentally at LEPII or the LHC) or else there is some new physics between the weak scale and the Planck scale which causes the Higgs couplings to become non-perturbative.

The Higgs boson couplings to fermions are dictated by the gauge invariance of the superpotential and at lowest order are completely specified in terms of the two parameters, M_A and $\tan\beta$. From Eq. 16, we see that the charge 2/3 quarks get their masses entirely from v_2, while the charge $-1/3$ quarks receive their masses from v_1. This is a consequence of the $U(1)$ hypercharge assignments for H_1 and H_2 given in Table 1. In the Standard Model, it is possible to give both the up and down quarks mass using a single Higgs doublet. This is because in the Standard Model the up quarks can get their masses from the charge conjugate of the Higgs doublet. Terms involving the charge conjugates of the superfields are not allowed in SUSY models, however, and so a second Higgs doublet with opposite $U(1)$ hypercharge from the first Higgs doublet is necessary in order to give the up quarks mass. Requiring that the fermions have their observed masses fixes the couplings in the superpotential of Eq. 16,[30]

$$\lambda_D = \frac{gM_d}{\sqrt{2}M_W\cos\beta}$$
$$\lambda_U = \frac{gM_u}{\sqrt{2}M_W\sin\beta}$$
$$\lambda_L = \frac{gM_l}{\sqrt{2}M_W\cos\beta}, \quad (31)$$

where g is the $SU(2)_L$ gauge coupling, $g^2 = 4\sqrt{2}G_F M_W^2$. We see that the only free parameter in the superpotential now is the Higgs mass parameter, μ, (along with the angle β in the λ_i couplings).

It is convenient to write the couplings for the neutral Higgs boson to the fermions in terms of the Standard Model Higgs couplings,

$$\mathcal{L} = -\frac{gm_i}{2M_W}\left[C_{ffh}\overline{f}_i f_i h + C_{ffH}\overline{f}_i f_i H + C_{ffA}\overline{f}_i \gamma_5 f_i A\right], \quad (32)$$

where C_{ffh} is 1 for a Standard Model Higgs boson. The C_{ffh} are given in Table 3 and plotted in Figs. 3 and 4 as a function of M_A. We see that for small M_A and large $\tan\beta$, the couplings of the neutral Higgs boson to fermions can be significantly different from the Standard Model couplings; the b-quark coupling becomes enhanced, while the t-quark coupling is suppressed. It is obvious from Figs. 3 and 4 that when M_A becomes large the Higgs-fermion couplings approach their standard model values, $C_{ffh} \to 1$. In fact even for $M_A \sim 300\ GeV$, the Higgs-fermion couplings are very close to their Standard Model values.

Table 3: Higgs Boson Couplings to fermions

f	C_{ffh}	C_{ffH}	C_{ffA}
u	$\frac{\cos\alpha}{\sin\beta}$	$\frac{\sin\alpha}{\sin\beta}$	$\cot\beta$
d	$-\frac{\sin\alpha}{\cos\beta}$	$\frac{\cos\alpha}{\cos\beta}$	$\tan\beta$

The Higgs boson couplings to gauge bosons are fixed by the $SU(2)_L \times U(1)$ gauge invariance. Some of the phenomenologically important couplings are:

$$Z^\mu Z^\nu h : \quad \frac{igM_Z}{\cos\theta_W}\sin(\beta-\alpha)g^{\mu\nu}$$
$$Z^\mu Z^\nu H : \quad \frac{igM_Z}{\cos\theta_W}\cos(\beta-\alpha)g^{\mu\nu}$$
$$W^\mu W^\nu h : \quad igM_W\sin(\beta-\alpha)g^{\mu\nu}$$
$$W^\mu W^\nu H : \quad igM_W\cos(\beta-\alpha)g^{\mu\nu}$$

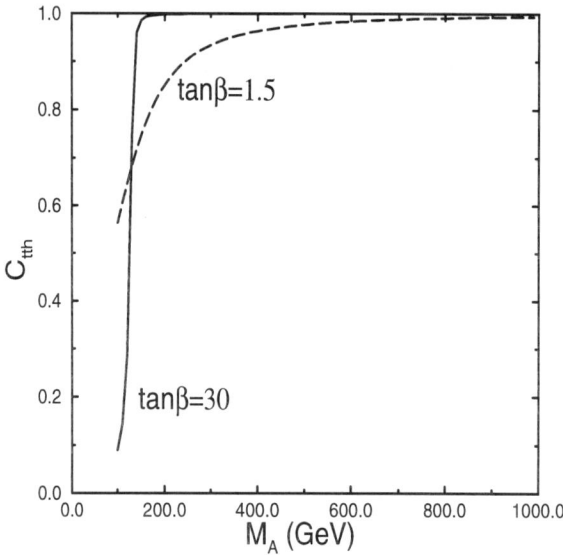

Figure 4: Coupling of the lightest Higgs boson to charge 2/3 quarks including radiative corrections [28] in terms of the couplings defined in Eq. 32. The value $C_{tth} = 1$ yields the Standard Model coupling of the Higgs boson to charge 2/3 quarks.

$$Z^\mu h(p)A(p'): \quad \frac{g\cos(\beta-\alpha)}{2\cos\theta_W}(p+p')^\mu$$
$$Z^\mu H(p)A(p'): \quad -\frac{g\sin(\beta-\alpha)}{2\cos\theta_W}(p+p')^\mu \quad . \tag{33}$$

We see that the couplings of the Higgs bosons to the gauge bosons all depend on the same angular factor, $\beta - \alpha$. The pseudoscalar, A, has no tree level coupling to pairs of gauge bosons. The angle β is a free parameter while the neutral Higgs mixing angle, α, which enters into many of the couplings, can be found in terms of the physical masses:

$$\tan 2\alpha = \frac{(M_A^2 + M_Z^2)\sin 2\beta}{(M_A^2 - M_Z^2)\cos 2\beta + \epsilon_h/\sin^2\beta} \quad . \tag{34}$$

With our conventions, $-\frac{\pi}{2} \leq \alpha \leq 0$. It is clear that the couplings of the SUSY Higgs to gauge bosons are always suppressed relative to those of the Standard Model. A complete set of couplings for the Higgs bosons (including the charged and pseudoscalar Higgs) at tree level can be found in Ref. [26]. These couplings completely determine the decay modes of the SUSY Higgs bosons and their experimental signatures. The important point is that (at lowest order) all of the couplings are completely determined in terms

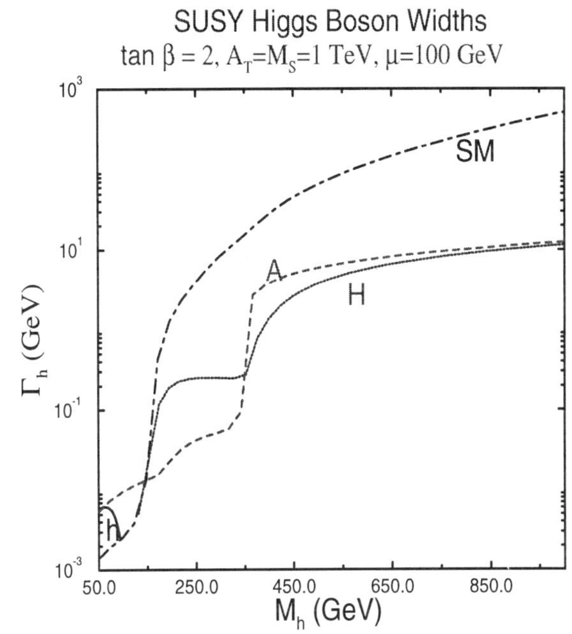

Figure 5: Total SUSY Higgs boson decay widths including two-loop radiative corrections as a function of the Higgs masses. The curve for the lightest Higgs boson is cut off at the maximum M_h. The program HDECAY [32] was used to obtain this plot.

of M_A and $\tan\beta$. When radiative corrections are included there is a dependence on the squark masses and the mixing parameters of Eq. 18. This dependence is explored in detail in Ref. [31].

It is an important feature of the MSSM that for large M_A, the Higgs sector looks like that of the Standard Model. As $M_A \to \infty$, the masses of the charged Higgs bosons, $H^\pm$, and the heavier neutral Higgs, H, also become large leaving only the lighter Higgs boson, h, in the spectrum. In this limit, the couplings of the lighter Higgs boson, h, to fermions and gauge bosons take on their Standard Model values. We have,

$$\begin{aligned}\sin(\beta-\alpha) &\to 1 \text{ for } M_A \to \infty \\ \cos(\beta-\alpha) &\to 0 \ .\end{aligned} \qquad (35)$$

From Eq. 33, we see that the heavier Higgs boson, H, decouples from the gauge bosons in the heavy M_A limit, while the lighter Higgs boson, h, has Standard Model couplings. Figs. 3 and 4 demonstrate that the Standard Model limit is also rapidly approached in the fermion-Higgs couplings for $M_A > 300 \ GeV$. In the limit of large M_A, it will thus be exceedingly difficult to differentiate a SUSY Higgs sector from the Standard Model Higgs boson.

- The SUSY Higgs sector with large M_A looks like the Standard Model Higgs sector.

The total width of the Higgs boson depends sensitively on $\tan\beta$ and is illustrated in Fig. 5 for $\tan\beta = 2$.[32] We see that the lightest Higgs boson has a width $\Gamma_h \sim$

$10 - 100$ MeV, while the heavier Higgs boson has a width $\Gamma_H \sim .1 - 1$ GeV, which is considerably narrower than the width of the Standard Model Higgs boson with the same mass. (The curve for the lighter Higgs boson is cut off at the kinematic upper limit.) The pseudoscalar, A, is also narrower than a Standard Model Higgs boson with the same mass.

The Squark and Slepton Sector

We turn now to a discussion of the scalar partners of the quarks and leptons. The left-handed $SU(2)_L$ quark doublet has scalar partners,

$$\tilde{Q} = \begin{pmatrix} \tilde{u}_L \\ \tilde{d}_L \end{pmatrix} . \tag{36}$$

The right-handed quarks also have scalar partners, $\tilde{u}_R$ and $\tilde{d}_R$. The L and R subscripts denote which helicity quark the scalars are partners of– **they are for identification purposes only.** These are ordinary complex scalars. Before SUSY is broken the fermions and scalars have the same masses and this mass degeneracy is split by the soft mass terms of Eq. 18. The tri-linear A terms allow the scalar partners of the left- and right-handed fermions to mix to form the mass eigenstates. In the top squark sector, the mixing between the scalar partners of the left- and right handed top (the stops), $\tilde{t}_L$ and $\tilde{t}_R$, is given by

$$M_{\tilde{t}}^2 = \begin{pmatrix} \tilde{M}_Q^2 + M_T^2 + M_Z^2(\frac{1}{2} - \frac{2}{3}\sin^2\theta_W)\cos 2\beta & M_T(A_T + \mu\cot\beta) \\ M_T(A_T + \mu\cot\beta) & \tilde{M}_U^2 + M_T^2 + \frac{2}{3}M_Z^2\sin^2\theta_W\cos 2\beta \end{pmatrix} . \tag{37}$$

For the scalars associated with the lighter quarks, the mixing effects will be negligible, since the mixing is proportional to the quark mass, (except if $\tan\beta >> 1$, when $\tilde{b}_L - \tilde{b}_R$ mixing may be large).

From Eq. 37, we see that there are two important cases to consider. If the soft breaking occurs at a large scale, much greater than M_Z, M_T, and A_T, then all the soft masses will be approximately equal, and we will have 12 degenerate squarks with mass $\tilde{m} \sim \tilde{M}_Q \sim \tilde{M}_U \sim \tilde{M}_D$. On the other hand, if the soft masses and the tri-linear mixing term, A_T, are on the order of the electroweak scale, then mixing effects become important.

If mixing effects are large, then one of the stop squarks will become the lightest squark, since the mixing effects are proportional to the relevant quark masses and hence will be largest in this sector. The case where the lightest squark is the stop is particularly interesting phenomenologically, and we discuss it in the section on squark mass limits.[33] In Fig. 6, we show the stop squark masses for $\tilde{M}_Q = \tilde{M}_t = \tilde{M}_b \equiv \tilde{m}$ and for several values of $\tan\beta$. Of course the mixing effects cannot be too large, or the stop squark mass-squared will be driven negative, leading to a breaking of the color $SU(3)$ gauge symmetry. Typically, the requirement that the correct vacuum be chosen leads to a restriction on the mixing parameter on the order of $\mid A_T \mid < \tilde{m}$.[7]

The couplings of the squarks to gauge bosons are completely fixed by gauge invariance, with no free parameters. A few examples of the couplings are:

$$\gamma^\mu \ \tilde{q}_{L,R}(p) \ \tilde{q}_{L,R}^*(p'): \quad -ieQ_q(p+p')^\mu$$

$$W^{\mu-} \ \tilde{u}_L(p) \ \tilde{d}_L^*(p'): \quad -\frac{ig}{\sqrt{2}}(p+p')^\mu$$

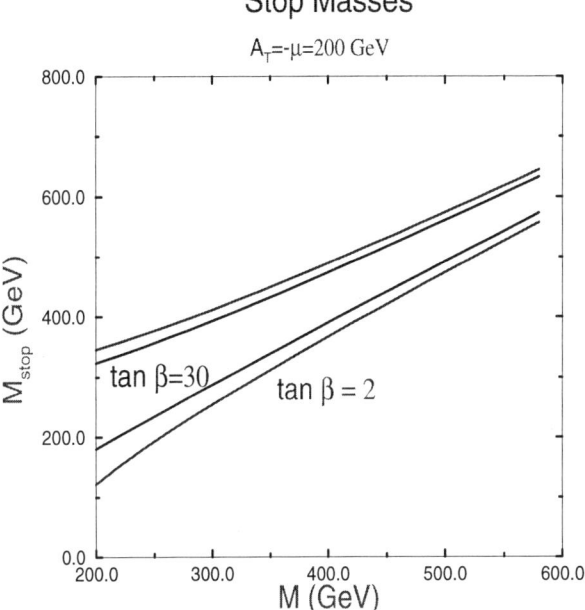

Figure 6: Stop squark masses for large mixing parameters, $A_T = \mu = 200\ GeV$, and for $\tan \beta = 2$ and $\tan \beta = 30$. $M \equiv \tilde{M}_Q = \tilde{M}_u$ are the squark mass parameters of Eq. 37.

$$Z^\mu\ \tilde{q}_{L,R}(p)\ \tilde{q}^*_{L,R}(p'): \quad -\frac{ig}{\cos\theta_W}\left[T_3 - Q_q \sin^2\theta_W\right](p+p')^\mu\ , \qquad (38)$$

where T_3 and Q_q are the quantum numbers of the corresponding quark. The strength of the interactions are clearly given by the relevant gauge coupling constants. A complete set of Feynman rules can be found in Ref. [6]

The mixing in the slepton sector is analogous to that in the squark sector and we will not pursue it further. From Table 1, we see that the scalar partner of the ν_L, $\tilde{\nu}_L$, has the same gauge quantum numbers as the H_2^0 Higgs boson. It is possible to give $\tilde{\nu}_L$ a vacuum expectation value and use it to break the electroweak symmetry. Such a vacuum expectation value would break lepton number (and R parity) thereby giving the neutrinos a mass and so its magnitude is severely restricted. [17]

The Chargino Sector

There are two charge 1, spin-$\frac{1}{2}$ Majorana fermions; $\tilde{\omega}^\pm$, the fermion partners of the $W^\pm$ bosons, and $\tilde{h}^\pm$, the charged fermion partners of the Higgs boson, termed the

Higgsinos. The physical mass states, $\tilde{\chi}^{\pm}_{1,2}$, are linear combinations formed by diagonalizing the mass matrix and are usually called charginos. In the $\tilde{\omega}^{\pm} - \tilde{h}^{\pm}$ basis the chargino mass matrix is,

$$M_{\tilde{\chi}^{\pm}} = \begin{pmatrix} M_2 & \sqrt{2} M_W \sin \beta \\ \sqrt{2} M_W \cos \beta & -\mu \end{pmatrix} . \tag{39}$$

The physics is extremely sensitive to M_2/μ. The mass eigenstates are then,

$$M^2_{\tilde{\chi}^{\pm}_{1,2}} = \frac{1}{2}\left\{ M_2^2 + 2M_W^2 + \mu^2 \mp \left[(M_2^2 - \mu^2)^2 + 4M_W^4 \cos^2 2\beta + 4M_W^2(M_2^2 + \mu^2 - 2M_2 \mu \sin^2 \beta) \right]^{1/2} \right\}. \tag{40}$$

By convention $M_{\tilde{\chi}^{\pm}_1}$ is the lighter chargino.

The Neutralino Sector

In the neutral fermion sector, the neutral fermion partners of the B and W^3 gauge bosons, $\tilde{b}$ and $\tilde{\omega}^3$, can mix with the neutral fermion partners of the Higgs bosons, $\tilde{h}^0_1, \tilde{h}^0_2$. Hence the physical states, $\tilde{\chi}^0_i$, are found by diagonalizing the 4×4 mass matrix,

$$M_{\tilde{\chi}^0_i} = \begin{pmatrix} M_1 & 0 & -M_Z \cos\beta \sin\theta_W & M_Z \sin\beta \sin\theta_W \\ 0 & M_2 & M_Z \cos\beta \cos\theta_W & -M_Z \sin\beta \cos\theta_W \\ -M_Z \cos\beta \sin\theta_W & M_Z \cos\beta \sin\theta_W & 0 & \mu \\ M_Z \sin\beta \sin\theta_W & -M_Z \sin\beta \cos\theta_W & \mu & 0 \end{pmatrix} \tag{41}$$

where θ_W is the electroweak mixing angle and we work in the $\tilde{b}, \tilde{\omega}^3, \tilde{h}^0_1, \tilde{h}^0_2$ basis. The physical masses can be defined to be positive and by convention, $M_{\tilde{\chi}^0_1} < M_{\tilde{\chi}^0_2} < M_{\tilde{\chi}^0_3} < M_{\tilde{\chi}^0_4}$. In general, the mass eigenstates do not correspond to a photino, (a fermion partner of the photon), or a zino, (a fermion partner of the Z), but are complicated mixtures of the states. The photino is only a mass eigenstate if $M_1 = M_2$. Physics involving the neutralinos therefore depends on M_1, M_2, μ, and $\tan \beta$. The lightest neutralino, $\tilde{\chi}^0_1$, is usually assumed to be the LSP.

WHY DO WE NEED SUSY?

Having introduced the MSSM as an effective theory at the electroweak scale and briefly discussed the various new particles and interactions of the model, I turn now to a discussion of the reasons for constructing a SUSY theory in the first place. We have already discussed the cancellation of the quadratic divergences, which is automatic in a supersymmetric model. There are, however, many other reasons why theorists are excited about supersymmetry. Theorists will often state that the mathematics of a supersymmetric model is *beautiful*. However, in my mind, the beauty of supersymmetry is largely obscured by the ugliness of the SUSY breaking sector which we have introduced, and it is therefore essential to have a solid motivation for studying SUSY theories.

Coupling constants run!

In a gauge theory, coupling constants scale with energy according to the relevant β-function. Hence, having measured a coupling constant at one energy scale, its value at any other energy can be predicted. At one loop,

$$\frac{1}{\alpha_i(Q)} = \frac{1}{\alpha_i(M)} + \frac{b_i}{2\pi}\log\left(\frac{M}{Q}\right) \ . \tag{42}$$

In the Standard (non-supersymmetric) Model, the coefficients b_i are given by,

$$\begin{aligned} b_1 &= \frac{4}{3}N_g + \frac{N_H}{10} \\ b_2 &= -\frac{22}{3} + \frac{4}{3}N_g + \frac{N_H}{6} \\ b_3 &= -11 + \frac{4}{3}N_g \ , \end{aligned} \tag{43}$$

where $N_g = 3$ is the number of generations and $N_H = 1$ is the number of Higgs doublets. The evolution of the coupling constants is seen to be sensitive to the particle content of the theory. We can take $M = M_Z$ in Eq. 42, input the measured values of the coupling constants at the Z-pole and evolve the couplings to high energy. The result is shown in Fig. 7. There is obviously no meeting of the coupling constants at high energy.

If the theory is supersymmetric, then the spectrum is different and the new particles contribute to the evolution of the coupling constants. In this case we have,[34]

$$\begin{aligned} b_1 &= 2N_g + \frac{3}{10}N_H \\ b_2 &= -6 + 2N_g + \frac{N_h}{2} \\ b_3 &= -9 + 2N_g \ . \end{aligned} \tag{44}$$

Because a SUSY model of necessity contains two Higgs doublets, we have $N_H = 2$. If we assume that the mass of all the SUSY particles is around $1\ TeV$, then the coupling constants scale as shown in Fig. 8. We see that the coupling constants meet at a scale around 10^{16} GeV.[13, 35, 36] This meeting of the coupling constants is a necessary feature of a Grand Unified Theory (GUT).

- SUSY theories can be naturally incorporated into Grand Unified Theories.

There are many variations on this theme including two loop beta functions, effects from passing through SUSY particle thresholds, etc., but they all allow us to take the picture of SUSY as resulting from a GUT theory seriously.[36, 37]

SUSY GUTS

The observation that the measured coupling constants tend to meet at a point when evolved to high energy assuming the β-function of a low energy SUSY model has led to widespread acceptance of a standard SUSY GUT model. We assume that the $SU(3) \times SU(2)_L \times U(1)$ gauge coupling constants are unified at a high scale $M_X \sim 10^{16}\ GeV$:[††]

$$\sqrt{\frac{5}{3}}g_1(M_X) = g_2(M_X) = g_3(M_X) \equiv g_X \ . \tag{45}$$

[††]This normalization of the $U(1)_Y$ coupling constant is canonical in Grand Unified Theories.

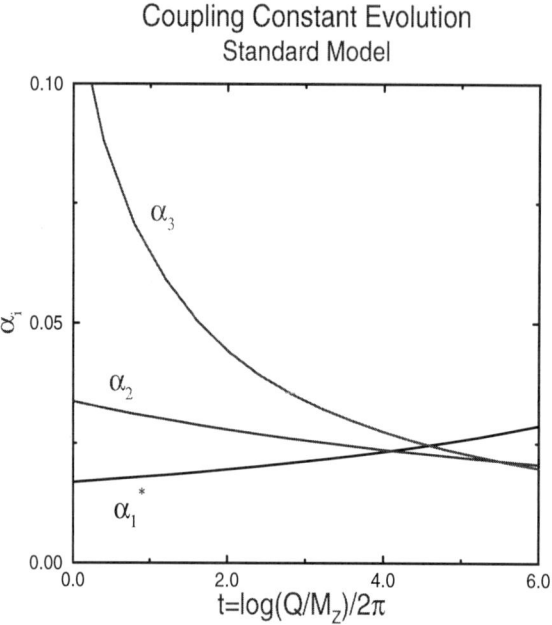

Figure 7: Evolution of the gauge coupling constants in the Standard Model from the experimentally measured values at the Z-pole. $\alpha_1^* \equiv 5/3\alpha_1$, since this is the relevant coupling in Grand Unified Theories.

The gaugino masses, M_i, are also assumed to unify,

$$M_i(M_X) \equiv m_{1/2} \ . \tag{46}$$

At lowest order, the gaugino masses then scale in the same way as the corresponding coupling constants,

$$M_i(M_W) = m_{1/2} \frac{g_i^2(M_W)}{g_X^2} \tag{47}$$

yielding

$$M_2 = \frac{\alpha}{\sin^2\theta_W} \frac{1}{\alpha_s} M_3$$
$$M_1 = \frac{5}{3} \tan^2\theta_W M_2 \ . \tag{48}$$

The gluino mass is always the heaviest of the gaugino masses. This relationship between the gaugino masses is a fairly robust prediction of SUSY GUTS and persists in models where the supersymmetry is broken dynamically.[3, 14]

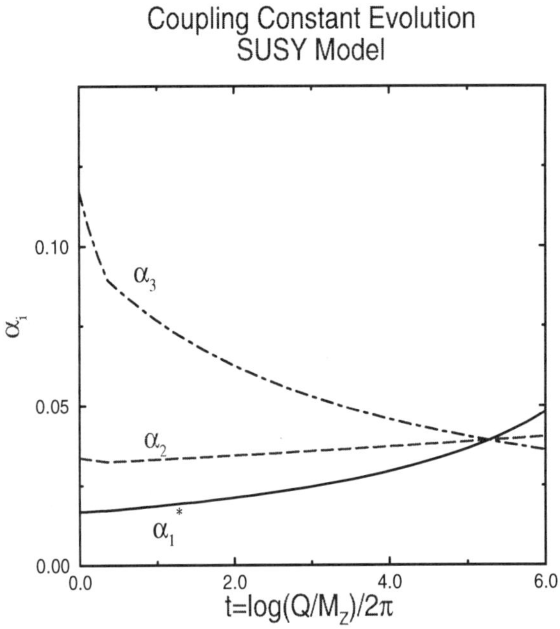

Figure 8: Evolution of the coupling constants in a low energy SUSY model from the experimentally measured values at the Z-pole. The SUSY thresholds are taken to be at $1\,TeV$. $\alpha_1^* \equiv 5/3\alpha_1$, since this is the relevant coupling in Grand Unified Theories.

Typical SUSY GUTS also assume that there is a common scalar mass at M_X,

$$m_1^2(M_X) = m_2^2(M_X) \equiv m_0^2$$
$$\tilde{M}_Q^2(M_X) = \tilde{M}_d^2(M_X) = \tilde{M}_u(M_X) = \tilde{M}_L^2(M_X) = \tilde{M}_e^2(M_X) \equiv m_0^2 \quad . \qquad (49)$$

The neutral Higgs boson masses at M_X are then $M_{h,H}^2 = m_0^2 + \mu^2$. As a final simplifying assumption, a common A parameter is assumed,

$$A_T(M_X) = A_b(M_X) = ... \equiv A_0 \quad . \qquad (50)$$

With these assumptions, the SUSY sector is completely described by 5 input parameters at the GUT scale,[38]

1. A common scalar mass, m_0.

2. A common gaugino mass, $m_{1/2}$.

3. A common trilinear coupling, A_0.

4. A Higgs mass parameter, μ.

5. A Higgs mixing parameter, B.

This set of assumptions is often called the "superstring inspired SUSY GUT" or SUGRA (although the connection with superstrings and/or supergravity is mostly wishful thinking) or the "constrained MSSM" (CMSSM). Although this framework is somewhat *ad hoc*, it does provide guidance to reduce the immense parameter space of a SUSY model. In actual practice, these relationships are satisfied only in the simplest models.

The strategy is now to input the 5 parameters given above at M_X and to use the renormalization group equations to evolve the parameters to M_W. In fact, the requirement that the Z boson obtain its measured value when the parameters are evaluated at low energy can be used to restrict $|\mu B|$, leaving the $sign(\mu)$ as a free parameter. We can also trade the parameter B for $\tan\beta$. In this way the parameters of the model become

$$m_0, m_{1/2}, A_0, \tan\beta, \text{sign}(\mu) \quad . \tag{51}$$

This form of a SUSY theory is extremely predictive, as the entire low energy spectrum is predicted in terms of a few input parameters. Within this scenario, contours for the various SUSY particle masses can be found as a function of m_0 and $m_{1/2}$ for given values of $\tan\beta$, A_0 and $\text{sign}(\mu)$.[37, 38]

It is instructive to study the scalar masses within this scenario. The evolution of the sleptons between M_X and M_W is small and we have the approximate result for the slepton masses,[8, 37]

$$\tilde{M}_L(M_W)^2 \sim \tilde{M}_e(M_W)^2 \sim m_0^2, \tag{52}$$

while the squark masses are roughly

$$\tilde{M}_q^2(M_W) \sim m_0^2 + 4m_{1/2}^2 \quad . \tag{53}$$

Since the squarks have strong interactions, (which drives the masses upwards), their masses at the weak scale tend to be larger than the sleptons. Once all the particle masses have been computed in this scheme, then their production cross sections and decay rates at any given accelerator can be computed unambiguously.

Changing the input parameters at M_X (for example, assuming non-universal scalar masses) of course changes the phenomenology at the weak scale. A preliminary investigation of the sensitivity of the low energy predictions to these assumptions has been made in Ref. [3]. For now, we will consider the Grand Unified Model described above as a starting point for phenomenological investigations into SUSY and hope that the general search strategies developed for this model will be applicable to other models.

Electroweak Symmetry Breaking

The simple SUSY model described above has the appealing feature that it explains the mechanism of electroweak symmetry breaking. Below, we sketch the argument.

In the Standard Model (non-supersymmetric) with a single Higgs field, ϕ, the scalar potential is given by:

$$V(\phi) = \mu^2 \phi^2 + \lambda \phi^4 \quad . \tag{54}$$

By convention, $\lambda > 0$. If $\mu^2 > 0$, then $V(\phi) > 0$ for all ϕ not equal to 0 and there is no electroweak symmetry breaking. If, however, $\mu^2 < 0$, then the minimum of the potential is not at $\phi = 0$ and the potential has the familiar Mexican hat shape. When the Lagrangian is expressed in terms of the physical field, $\phi' \equiv \phi - v$, which has zero vacuum expectation value, then the electroweak symmetry is broken and the W and Z gauge bosons acquire non-zero masses. We saw in the previous sections that this same mechanism gives the W and Z gauge bosons their masses in the MSSM. This simple picture leaves one looming question:

$$\text{Why is } \mu^2 < 0? \tag{55}$$

It is this question which the SUSY GUT models can answer.

In the minimal SUGRA model which we have described above, the neutral Higgs bosons both have masses, $M_{h,H}^2 = m_0^2 + \mu^2$, at M_X while the squarks and sleptons have mass m_0 at M_X. Clearly, at M_X, the electroweak symmetry is not broken since the Higgs bosons have positive mass-squared. The masses scale with energy according to the renormalization group equations.[39] If we neglect gauge couplings and consider only the scaling of the third generation scalars we have,[40]

$$\frac{d}{d\log(Q)} \begin{pmatrix} M_h^2 \\ \tilde{M}_{t_R}^2 \\ \tilde{M}_{Q_L^3}^2 \end{pmatrix} = -\frac{8\alpha_s}{3\pi} M_3^2 \begin{pmatrix} 0 \\ 1 \\ 1 \end{pmatrix} + \frac{\lambda_T^2}{8\pi^2} \left(\tilde{M}_{Q_L^3}^2 + \tilde{M}_{t_R}^2 + M_h^2 + A_T^2 \right) \begin{pmatrix} 3 \\ 2 \\ 1 \end{pmatrix}, \tag{56}$$

where $\tilde{Q}_L^3$ is the $SU(2)_L$ doublet containing $\tilde{t}_L$ and $\tilde{b}_L$, h is the lightest Higgs boson, λ_T is the top quark Yukawa coupling constant given in Eq. 31, and Q is the effective scale at which the masses are measured. The signs are such that the Yukawa interactions (proportional to M_T) decrease the masses, while the gaugino interactions increase the masses. Because of the $3 - 2 - 1$ structure of the last term in Eq. 56, the Higgs mass decreases faster than the squark masses and it is possible to drive $M_h^2 < 0$ at low energy, while keeping $\tilde{M}_{Q_L^3}^2$ and $\tilde{M}_{t_R}^2$ positive. A generic set of scalar masses in a typical SUSY GUT model is shown in Fig. 9. We can clearly see that the lightest Higgs boson mass becomes negative around the electroweak scale.[41]

For large λ_T, we have the approximate solution,

$$M_h^2(Q) = M_h^2(M_X) - \frac{3}{8\pi^2} \lambda_T^2 (\tilde{M}_{Q_L^3}^2 + \tilde{M}_{t_R}^2 + M_h^2 + A_T^2) \log\left(\frac{M_X}{Q}\right). \tag{57}$$

Hence the larger M_T is, the faster M_h^2 goes negative. This of course generates electroweak symmetry breaking. If M_T were light, M_h^2 would remain positive.[40] This observation was made ten years ago when we thought the top quark was light, ($\sim 40\ GeV$). At that time it was ignored as not being phenomenologically relevant. In fact, this mechanism only works for $M_T \sim 175\ GeV$!

- SUSY GUTS can explain electroweak symmetry breaking. The lightest Higgs boson mass is negative, $m_h^2 < 0$, because M_T is large.

The $3 - 2 - 1$ structure of Eq. 56 drives M_h^2 negative faster than the squark masses. This is important because driving the squark mass negative would have the undesired effect of breaking the color $SU(3)$ symmetry. The requirement that the electroweak symmetry breaking occur through the renormalization group scaling of the Higgs boson mass, (as given in Eq. 56) also restricts the allowed values of $\tan \beta$ to $\tan \beta > 1$. (Remember that λ_T depends on β through Eq. 31.)

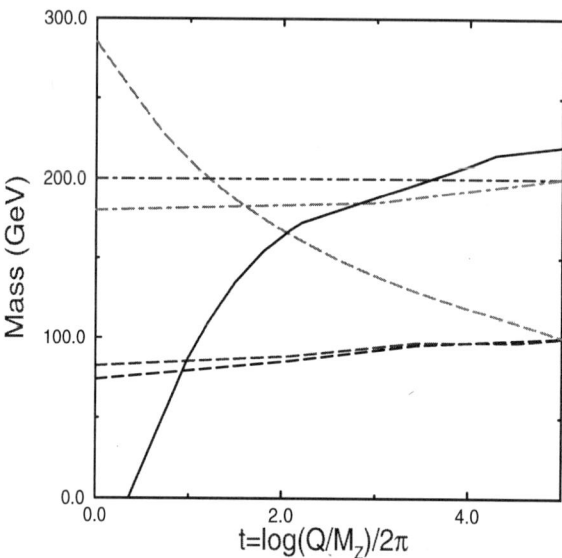

Figure 9: Sample masses of SUSY particles in a SUSY GUT. At the GUT scale M_X, we have taken $m_0 = 200\ GeV, m_{1/2} = 100\ GeV, \mu = 100\ GeV$ and $A_i = 0$. The solid line is the lightest neutral Higgs boson mass. The dashed lines are the gaugino masses (the largest is the gluino) and the dot-dashed lines are typical squark masses.

Fixed Point Interactions

In the previous subsection we saw that a large top quark mass could generate electroweak symmetry breaking in a SUSY GUT model. Here we show that the simplest SUSY GUT actually *predicts* a large top quark mass.

The top quark mass is determined in terms of its Yukawa coupling and scales with energy, Q,[42]

$$\lambda_T(Q) = \frac{M_T(Q)}{M_W} \frac{g}{\sqrt{2}\sin\beta} \quad . \tag{58}$$

Including both the gauge couplings and the Yukawa couplings to the t- and b- quarks, the scaling is:

$$\frac{d\lambda_T}{d\log(Q)} = \frac{\lambda_T}{16\pi^2}\left\{-\frac{13}{9}g_1^2 - 3g_2^2 - \frac{16}{3}g_3^2 + 6\lambda_T^2 + \lambda_B^2\right\} \quad . \tag{59}$$

To a good approximation, we can consider only the contributions from the strong coupling constant, g_3, and the top quark Yukawa coupling, λ_T. If we begin our scaling at M_X and evolve λ_T to lower energy, we will come to a point where the evolution of the Yukawa coupling stops,

$$\frac{d\lambda_T}{d\log(Q)} = 0 \quad . \tag{60}$$

At this point we have roughly,

$$-\frac{16}{3}g_3^2 + 6\lambda_T^2 = 0 \tag{61}$$

which gives,

$$\lambda_T \sim \frac{4}{3}\sqrt{2\pi\alpha_s} \sim 1, \tag{62}$$

or

$$M_T \sim (200\ GeV)\sin\beta \ . \tag{63}$$

This point where the top quark mass stops evolving is called a *fixed point*. What this means is that no matter what the initial condition for λ_T is at M_X, it will always evolve to give the same value at low energy. For $\tan\beta \sim 2$, the fixed point value for the top quark mass is close to the experimental value. More sophisticated analyses do not change this picture substantially.

- SUSY GUTS can naturally accommodate a large top quark mass for $\tan\beta \sim 1-3$.

$b - \tau$ Unification

The unification of the b- and τ- Yukawa coupling constants, λ_B and λ_τ, at the GUT scale is a concept much beloved by theorists since

$$\lambda_B(M_X) = \lambda_\tau(M_X) \tag{64}$$

occurs naturally in many GUT models. Requiring that the b quark have its experimental value at low energy leads to a prediction for the top quark mass in terms of $\tan\beta$. There are two solutions which yield $M_T = 175\ GeV$,[42]

$$\begin{aligned}\tan\beta &\sim 1 \\ \text{or}\quad \tan\beta &\sim \frac{M_T}{M_b}\ .\end{aligned} \tag{65}$$

The first solution roughly corresponds to the fixed point solution of the previous subsection. The second solution with $\tan\beta \sim 35$ has interesting phenomenological consequences, since for large $\tan\beta$ the coupling of the lightest Higgs boson to b quarks is enhanced relative to the Standard Model. (See Fig. 3). The values in the $\tan\beta - M_T$ plane allowed by $b - \tau$ unification depend sensitively on the exact value of the strong coupling constant, α_s, used in the evolution and so there is a significant uncertainty in the prediction.

- SUSY GUTs allow for the unification of the $b - \tau$ Yukawa coupling constants at the GUT scale along with the experimentally observed value for the top quark mass.

Similar relationships to Eq. 64 involving the first two generations do not work.

Comments

We see that SUSY plus grand unification has many desirable features and can explain a lot:

1. There are no troubling quadratic divergences requiring disagreeable cancellations.

2. M_T is large because λ_T evolves from the GUT scale to its fixed point.

3. Electroweak symmetry is broken, $m_h^2 < 0$, because M_T is large.

4. $b-\tau$ unification can be incorporated, leading to the experimentally observed value for the top quark mass.

Afficianados of SUSY can add many more items to this list.[43] For instance, the LSP is a leading candidate for cold, dark matter.[44] The conclusion is inescapable:
SUSY IS HERE TO STAY !

SEARCHING FOR SUSY

We begin this section with a description of the effects of SUSY particles on precision measurements and rare decays. We then turn to experimental limits on the various particles and search strategies at current and future machines. A more detailed expose can be found in the lectures of Tata[8] along with up to the minute limits in Refs. [4, 5].

Indirect Hints for SUSY

One might hope that the precision measurements at the Z-pole could be used to garner information on the SUSY particle spectrum. Since the precision electroweak measurements are overwhelmingly in good agreement with the predictions of the Standard Model, it would appear that stringent limits could be placed on the existence of SUSY particles at the weak scale. There are two reasons why this is not the case.

The first is that SUSY is a *decoupling theory*. With the exception of the Higgs particles, the effects of SUSY particles at the weak scale are suppressed by powers of M_W^2/M_{SUSY}^2, where M_{SUSY} is the relevant SUSY mass scale, and so for M_{SUSY} larger than a few hundred GeV, the SUSY particles give negligible contributions to electroweak processes. The second reason why there are not stringent limits from precision results at LEP has to do with the Higgs sector. The Higgs bosons are the only particles in the spectrum which do not decouple from the low energy physics when they are very massive. The fits to electroweak data tend to prefer a Higgs boson in the 100 GeV mass range.[45] Since the MSSM requires a light Higgs boson with a mass in this region anyways, the electroweak data is completely consistent with a SUSY model with a light Higgs boson and all other SUSY particles significantly heavier.

Attempts have been made to perform global fits to the electroweak data and to fix the SUSY spectrum this way.[46, 47] It is possible to obtain a fit where the χ^2/degree of freedom is roughly the same as in the Standard Model fit. Although the fits do not yield stringent limits on the SUSY particle masses, they do exhibit several interesting features. They tend to prefer either small $\tan\beta$, $\tan\beta \sim 2$, or relatively large values, $\tan\beta \sim 30$. In addition, the fitted values for the strong coupling constant at M_Z, $\alpha_s(M_Z)$, are slightly smaller in SUSY models than in the Standard Model. (For $\tan\beta =$

1.6, Ref. [46] finds $\alpha_s(M_Z) = .116 \pm .005$ and for $\tan\beta = 34$, they find $\alpha_s(M_Z) = .119 \pm .005$.) It is clear that all precision electroweak measurements can be accommodated within a SUSY model, but the data show no preference for these models.

There are also numerous indirect limits coming from the effects of SUSY particles on rare decays. Since the SUSY particles circulate in loops, they can affect rare B and K decays (among others). One of the most restrictive limits is from the CLEO measurement of the inclusive decay $B \to X_s\gamma$,[48]

$$BR(B \to X_s\gamma) = (2.32 \pm .67) \times 10^{-4}, \quad \text{CLEO} \qquad (66)$$

which is sensitive to loops containing the new particles of a SUSY model. The contribution from $tH^\pm$ loops always adds constructively to the Standard Model result and hence non-supersymmetric two-Higgs doublet models are severely restricted by the measurement of $b \to s\gamma$.

The situation is different in a SUSY model, however, since there are additional contributions from squark-chargino loops, squark-neutralino loops, and squark-gluino loops. The contributions from the squark-neutralino and squark-gluino loops are small and are typically neglected. The dominant contribution from the squark-chargino loops is proportional to $A_T\mu$ and thus can have either sign relative to the Standard Model and charged Higgs loop contributions. There will therefore be regions of SUSY parameter space which are excluded depending upon whether there is constructive or destructive interference between the Standard Model/ charged Higgs contributions and the squark-chargino contribution.[49] The limit which can be obtained is obviously very sensitive to the sign($A_T\mu$) and can be easily understood for large $\tan\beta$ where the squark-chargino contribution is completely dominant. Neglecting QCD corrections (which are significant) we have,[50]

$$\frac{BR(b \to s\gamma)}{BR(b \to ce\overline{\nu})} \sim \frac{|V_{ts}V_{tb}|^2}{|V_{cb}|^2} \frac{6\alpha}{\pi} \left\{ C + \frac{M_T^2 A_T\mu}{\tilde{m}_T^4} \tan\beta \right\}^2 , \qquad (67)$$

where C (positive) is the contribution from the Standard Model and charged Higgs loops and $\tilde{m}_T$ is the stop mass. For $A_T\mu$ positive, this leads to a larger branching ratio, $BR(b \to s\gamma)$, than in the Standard Model. Since the Standard Model prediction is already somewhat above the measured value, we require $A_T\mu < 0$ to avoid conflict with the experimental measurement if $\tilde{m}_T$ is at the electroweak scale and $\tan\beta$ is large. ‡‡ Detailed plots of the allowed regions for various assumptions about $\tan\beta$, μ, and A_T are given in Ref. [51]. Depending on $\tan\beta$ and the sign of $A_T\mu$, this process probes stop masses in the $100 - 300$ GeV region. For large $\tan\beta$, $B \to X_s\gamma$ may probe mass scales as large as a TeV.[52]

Another class of important indirect limits on SUSY models comes from flavor changing neutral current (FCNC) processes such as $K^0 - \overline{K}^0$ mixing. In general, the matrix which diagonalizes the squark mass matrix is different from that which diagonalizes the quark mass matrix and so there are off-diagonal interactions which can mediate FCNC's. The contributions from squarks to FCNC processes vanish if the squarks have degenerate masses and so the limits are typically of the form:

$$\frac{\Delta \tilde{m}^2}{\tilde{m}^2} < \mathcal{O}(10^{-3}) \quad , \qquad (68)$$

---

‡‡Reader beware: There are conflicting definitions of the sign of μ in the literature. The only way to be sure is to go back to the superpotential of Eq. 16 to see how μ is defined.

where $\Delta \tilde{m}^2$ is the mass-squared splitting between the different squarks and $\tilde{m}$ is the average squark mass. A detailed discussion of FCNCs in SUSY models and references to the literature is given in Ref. [53]. As a practical matter, the assumption is often made that there are 10 degenerate squarks, corresponding to the scalar partners of the u, d, c, s, and b quarks, while the stop squarks are allowed to have different masses from the others. This avoids phenomenological problems with FCNCs.

Experimental Limits and Search Strategies

We turn now to a discussion of some of the existing experimental limits on the various SUSY particles and also to the search strategies applicable at present and future accelerators. This section is intended only to give the flavor of how SUSY searches proceed and not as a comprehensive guide. We begin with the Higgs sector.

Observing SUSY Higgs Bosons

The goal in the Higgs sector is to observe the 5 physical Higgs particles, $h, H, A, H^{\pm}$, and to measure as many couplings as possible to verify that the couplings are those of a SUSY model. The lightest neutral Higgs boson in the minimal SUSY model is unique in the SUSY spectrum because there is an upper bound to its mass,

$$M_h < 130 \ GeV. \tag{69}$$

All other SUSY particles in the model can be made arbitrarily heavy just by adjusting the soft SUSY breaking parameters in the model and so be just out of reach of today's or tomorrow's accelerators (although if they are heavier than around $1 \ TeV$, much of the motivation for low energy SUSY disappears). The lightest SUSY Higgs boson, however, cannot be much outside the range of LEPII and can almost certainly be observed at the LHC . Hence an extraordinary theoretical effort has gone into the study of the reach of various accelerators in the SUSY Higgs parameter space since in this sector it will be possible to experimentally exclude the MSSM if a light Higgs boson is not observed.

If we find a light neutral Higgs boson, then we want to map out the parameter space to see if we can distinguish it from a Standard Model Higgs boson. The only way to do this is to measure a variety of production and decay modes and attempt to extract the various couplings of the Higgs bosons to fermions and gauge bosons. Since as $M_A \to \infty$, the h couplings approach those of the standard model, there will clearly be a region where the SUSY Higgs boson and the Standard Model Higgs boson are indistinguishable. This is obvious from Figs. 3 and 4.

The search strategies for the SUSY Higgs boson depend sensitively on the Higgs boson branching ratios, which in turn depend on $\tan \beta$. In Figs. 10 and 11, we show the branching ratios for the lightest SUSY Higgs boson, h, into some interesting decay modes assuming that there are no SUSY particles light enough for the h to decay into. (These figures include radiative corrections to the branching ratios, which can be important.[32]) For a Higgs boson below the WW threshold, the decay into $b\bar{b}$ is completely dominant. Unfortunately, there are large QCD backgrounds to this decay mode and so it is often necessary to look at rare decay modes. The branching ratios to $b\bar{b}$, $\tau^+\tau^-$, and $\mu^+\mu^-$ are relatively insensitive to $\tan \beta$, but the WW^*, ZZ^*, and $\gamma\gamma$ rates have strong dependences on $\tan \beta$ as we can see from Figs. 10 and 11.

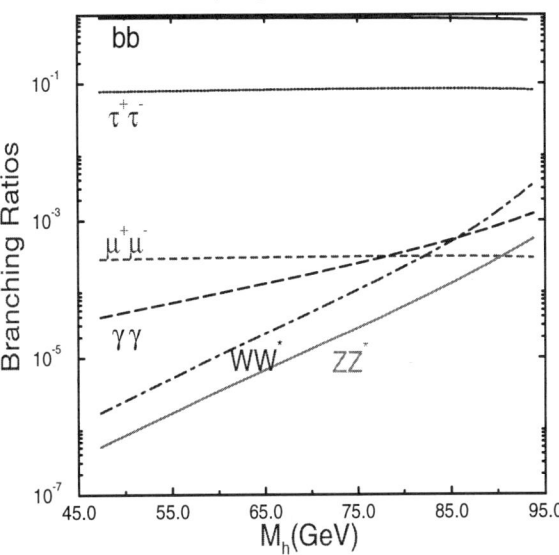

Figure 10: Branching ratios of the lightest Higgs boson assuming decays into other SUSY particles are kinematically forbidden. WW^* and ZZ^* denote decays with one off-shell gauge boson and M_S is a typical squark mass.[32]

Direct limits on SUSY Higgs production have been obtained at LEP by searching for the complementary processes,[31]

$$\begin{aligned} e^+e^- &\to Zh \\ e^+e^- &\to Ah \quad . \end{aligned} \qquad (70)$$

From the couplings of Eq. 33, we see that the process $e^+e^- \to Zh$ is suppressed by $\sin^2(\beta - \alpha)$ relative to the Standard Model Higgs boson production process, while $e^+e^- \to Ah$ is proportional to $\cos^2(\beta-\alpha)$. The moral is that it is impossible to suppress both processes simultaneously if both the h and the A are kinematically accessible! The experimental searches look for final states with b's and τ's since these have the largest branching ratios. Because the Higgs sector can be described by the two parameters, M_h and $\tan\beta$, searches exclude a region in this plane. (Remember that M_A can be expressed in terms of M_h and $\tan\beta$ at lowest order. When radiative corrections are included, there will be a dependence on the mixing parameters, A_i and μ, and on the squark masses). The LEP searches for Higgs bosons, $e^+e^- \to Zh$ and $e^+e^- \to AH$,

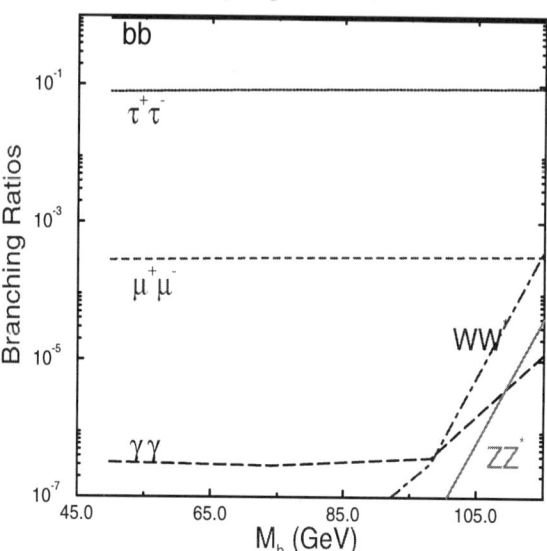

Figure 11: Branching ratios of the lightest Higgs boson assuming decays into other SUSY particles are kinematically forbidden.[32]

exclude the region, [31]

$$M_h > 44 \; GeV, \text{ for any } \tan\beta \; . \qquad (71)$$

For a given value of $\tan\beta$, there may be a stronger bound. It is important to note that the LEP searches do not leave any window for a very light (on the order of a few GeV) Higgs boson. The limit on a SUSY Higgs boson is weaker than the corresponding limit on the Standard Model Higgs boson, $M_h^{SM} > 65 \; GeV$, due to the suppression in the couplings of the Higgs boson to vector bosons.

At LEPII, the cross section for either Zh (small $\tan\beta$) or Ah (large $\tan\beta$) is roughly .5 pb. With a luminosity of $150/pb/yr$, this leads to 75 events/yr before the inclusion of branching ratios. Fig. 12 shows the cross sections for two different values of $\tan\beta$ and the complementarity of the two processes can be clearly observed. (The dependence on the top quark mass arises from the inclusion of radiative corrections.)

The limits on the Higgs boson mass could be substantially altered if there is a significant branching rate into invisible decay modes, such as the neutralinos,

$$h, A \to \tilde{\chi}_1^0 \tilde{\chi}_1^0 \; . \qquad (72)$$

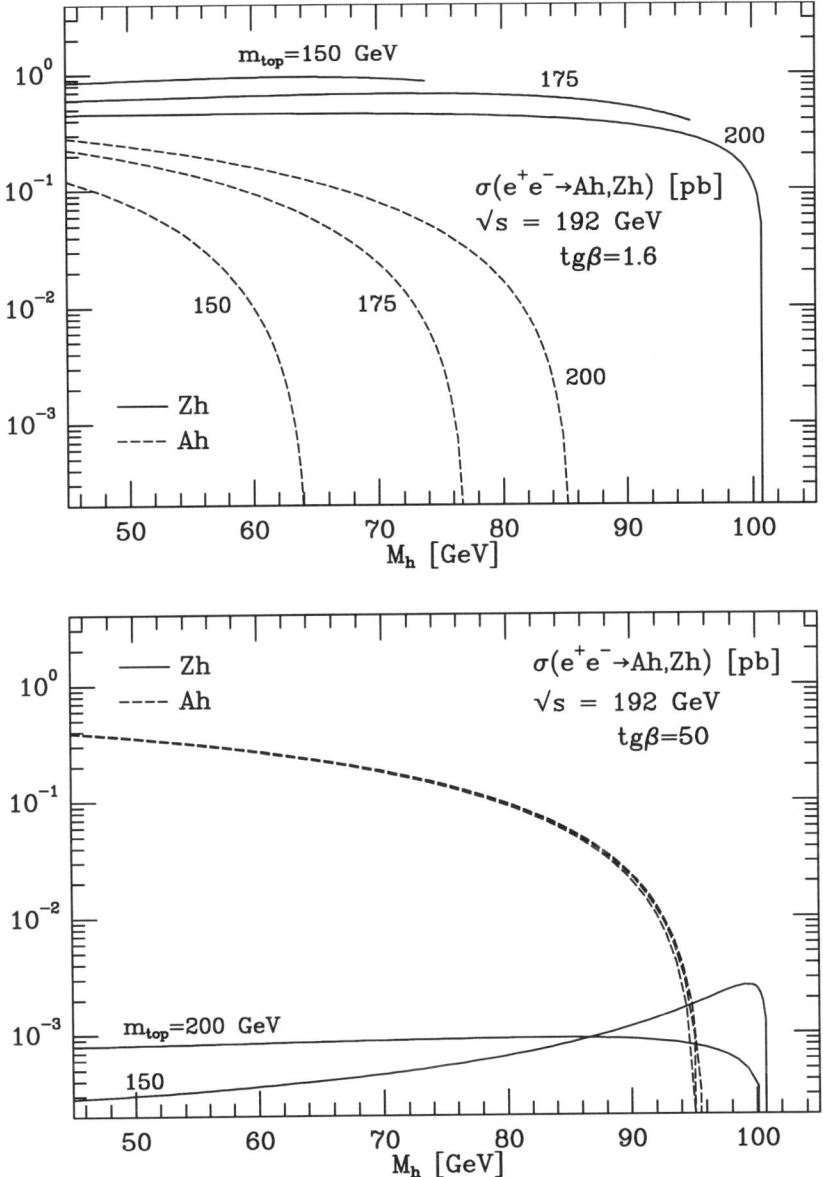

Figure 12: Cross sections for $e^+e^- \to Zh$ and $e^+e^- \to Ah$ at LEP. From Ref. [31].

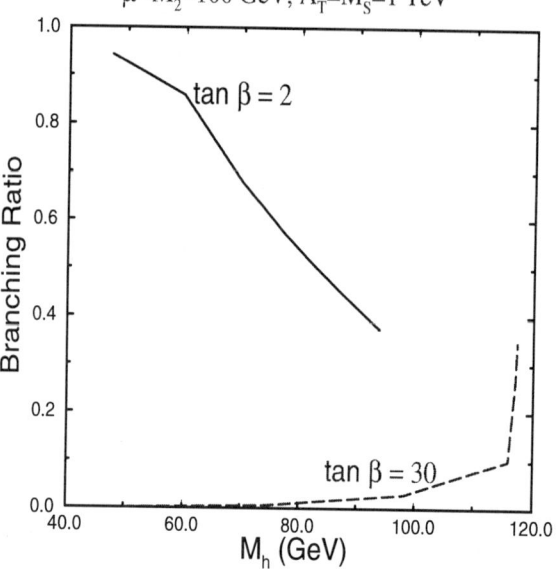

Figure 13: Branching ratio of the lightest Higgs boson to $\tilde{\chi}_1^0 \tilde{\chi}_1^0$. The curve with $\tan\beta = 30$ has $M_{\tilde{\chi}_1^0} = 33\ GeV$, while that with $\tan\beta = 2$ has $M_{\tilde{\chi}_1^0} = 7\ GeV$.[32]

These branching ratios could be as high as 80%, but are extremely model dependent since they depend sensitively on the parameters of the neutralino mixing matrix. In Fig. 13, we show the branching ratio of the lightest Higgs boson to $\tilde{\chi}_1^0 \tilde{\chi}_1^0$ for several choices of parameters. For $\tan\beta = 2$, with the set of parameters which we have chosen, the branching ratio is always greater than 40%. If the invisible decay modes are significant, a different search strategy for the Higgs boson must be utilized and LEPII can put a limit on the product of the Higgs boson mixing angles, $\beta - \alpha$, and the branching ratio to invisible modes:

$$\begin{aligned} R_1^2 &\equiv \sin^2(\beta - \alpha) BR(h \to \text{visible}) \\ R_2^2 &\equiv \sin^2(\beta - \alpha) BR(h \to \text{invisible}) \ . \end{aligned} \quad (73)$$

For $M_h = 40\ GeV$, the 95% confidence level excluded region from LEP is,[31]

$$\begin{aligned} R_1^2 &< .3 \\ R_2^2 &< .1 \ . \end{aligned} \quad (74)$$

These limits can be reinterpreted in terms of the parameters of the MSSM (A_i, μ, M_2, M_1, $\tilde{m}$, etc.) and will be greatly improved at LEPII. For $M_h = 80\ GeV$ and an integrated luminosity of $150/pb$ at $\sqrt{s} = 192\ GeV$, the 95% confidence level limit will be:

$$R_1^2 < .1$$
$$R_2^2 < .3\ . \tag{75}$$

These limits will significantly restrict the allowed SUSY parameter space.

A $\mu^+\mu^-$ collider could in principle obtain stringent bounds on a SUSY Higgs boson through its s-channel couplings to the Higgs.[54] Since these couplings are proportional to the lepton mass, the s-channel Higgs couplings will be much larger at a $\mu^+\mu^-$ collider than at an e^+e^- collider. For large $\tan\beta$, the lighter Higgs boson could be found in the process $e^+e^- \to Zh$ at LEPII or at an NLC.[31, 55] However, for large $\tan\beta$, the coupling of the heavier Higgs boson to gauge boson pairs is highly suppressed, (see Eq. 33), so the H can't be found through $e^+e^- \to ZH$. Instead the H can be found through $\mu^+\mu^- \to H \to b\bar{b}$, which is enhanced by the factor $\tan^2\beta$ relative to $\mu^+\mu^- \to h_{SM} \to b\bar{b}$.

A muon collider could also be very useful for obtaining precision measurements of the lighter Higgs boson mass. The idea is that the h has been discovered through either the process $e^+e^- \to Zh$ or $\mu^+\mu^- \to Zh$ and so we have a rough idea of the Higgs boson mass. A muon collider could be tuned to sit right on the resonance, $\mu^+\mu^- \to h$. By doing an energy scan around the region of the resonance, a precise value of the mass could be obtained due in large part to the narrowness of the muon beam as compared to the beam in an electron collider. (The narrowness of the beam is due to the suppression of synchrotron radiation in a muon collider.)

At the LHC, for most Higgs masses the dominant production mechanism is gluon fusion, $gg \to h, H$ or A. These processes proceed through triangle diagrams with internal b and t quarks and also through squark loops. In the limit in which the top quark is much heavier than the Higgs boson, the top quark contribution is a constant, while the b quark contribution is suppressed by $(M_b/v)^2 \log(M_h/M_b)$ and so only the top quark contribution is numerically important. For large $\tan\beta$, however, the dominance of the top quark loop is overtaken by the large $\bar{b}bh$ coupling and the bottom quark contribution becomes important, (as seen in Fig. 3). The production rate is therefore extremely sensitive to $\tan\beta$. Both QCD corrections and squark loops can also be numerically important.[56] In fact, the QCD corrections increase the rate by a factor between 1.5 and 2. The rate for $pp \to h$ at the LHC is shown in Fig. 14 as a function of $\tan\beta$ for $M_h = 80\ GeV$. We see that there are a relatively large number of events. For example, for $M_h \sim 80\ GeV$, the LHC cross section is roughly 100 pb. With a luminosity of $10^{33}/cm^2/sec$, this yields 10^6 events/year.

Unfortunately, there are large backgrounds to the dominant decay modes, ($b\bar{b}, \mu^+\mu^-$, and $\tau^+\tau^-$), for a Higgs boson in the 100 GeV region.[57] The decay $h \to ZZ^*$ will be useful, but its branching ratio decreases rapidly with decreasing Higgs mass. In order to cover the region around $M_h \sim 80 - 100\ GeV$, it will be necessary to look for the Higgs decay to $\gamma\gamma$,

$$gg \to h, H \to \gamma\gamma\ . \tag{76}$$

(From Figs. 10 and 11, we see that the $BR(h \to \gamma\gamma)$ is typically $< 10^{-3} - 10^{-5}$.) This process will be extremely difficult to observe at the LHC due to the small rate and the desire to observe the $h \to \gamma\gamma$ decay has been one of the driving forces behind the design

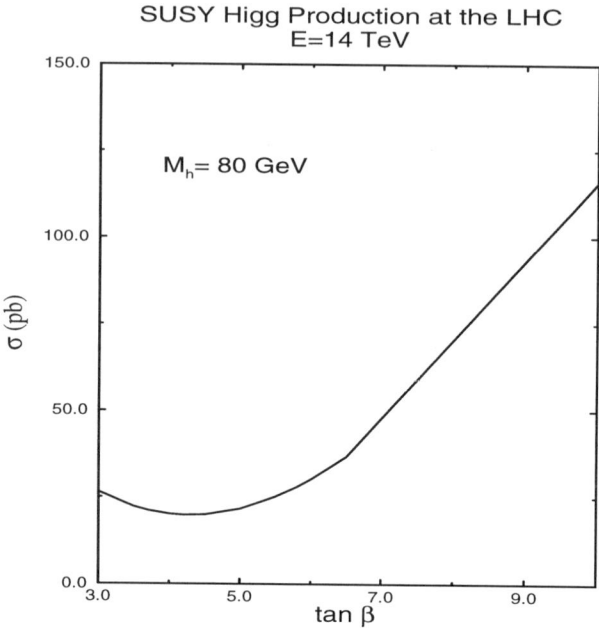

Figure 14: Cross section for production of the lightest SUSY Higgs boson at the LHC as a function of $\tan\beta$.

of both LHC detectors.[58] For large M_A, the rate is roughly independent of $\tan\beta$ for $\tan\beta > 3$ and can be used to exclude $M_A > 150 \ GeV$ with the full design luminosity of $3 \times 10^5/pb$. (With a smaller luminosity of $3 \times 10^4/pb$, the $h \to \gamma\gamma$ process is sensitive to roughly $M_A > 270 \ GeV$. See Fig. 15 for the exact region.)

In order to exclude the region with smaller $\tan\beta$, the process $pp \to Wh \to l\nu\bar{b}b$ can be used.[59] This process can exclude a region with $M_A > 100 \ GeV$ and $\tan\beta < 4$ (see Fig. 15) and demonstrates the crucial need for b-tagging at the LHC in order to cover all regions of SUSY parameter space. In Fig. 15, we see the excluded region formed by combining the LHC and LEP limits.[60] A variety of Higgs production and decay channels can be utilized in order to probe the entire $\tan\beta - M_A$ plane. The most striking feature of Fig. 15 is the region around $M_A \sim 100 \ GeV$ for $\tan\beta > 5$ where the lightest Higgs boson cannot be observed. In the region with $M_A \sim 100 - 200 \ GeV$, both the $h\bar{t}t$ coupling and the $h \to \gamma\gamma$ branching ratios are suppressed relative to the Standard Model rates. Furthermore, the dominant decays, $h \to b\bar{b}$ and $h \to \tau^+\tau^-$, have large backgrounds from Z decays. It will be necessary to look for the decays of the heavier neutral Higgs boson, H, or the pseudoscalar, A, to $\tau^+\tau^-$ pairs in order to

probe this region,
$$H, A \to \tau^+\tau^- \to l\nu\bar{q}q \quad . \qquad (77)$$
Detector studies by the ATLAS and CMS collaborations suggest that these decay modes may be accessible.

Finding the Zoo of SUSY Particles

In addition to the multiple Higgs particles associated with SUSY models, there is a whole zoo of other new particles. There are the squarks and gluinos which are produced through the strong interactions and the sleptons, charginos, and neutralinos which are produced weakly.

We begin by discussing some generic signals for supersymmetry. All SUSY particles in a theory with R parity conservation eventually decay to the LSP, which is typically taken to be the lightest neutralino, $\tilde{\chi}_1^0$, although in some models it could be the gravitino.[22] The LSP's interactions with matter are extremely weak and so it escapes detection leading to missing energy.

- A basic SUSY signature is missing energy, E_T^{miss}, from the undetected LSP.

A SUSY model typically produces a cascade of decays, until the final state consists of only the LSP plus jets and leptons. Hence typical final states are:

- $l^\pm + \text{jets} + E_T^{miss}$
- $l^\pm l^\pm + \text{jets} + E_T^{miss}$
- $l^\pm l^\mp + \text{jets} + E_T^{miss}$.

Because of the presence of the LSP in the final state, it is not possible to completely reconstruct the masses of the SUSY particles, although a significant amount of information about the masses can be obtained from the event structure.

- A combination of characteristic signatures may determine the SUSY model.

Because the gluinos are Majorana particles, they have some special characteristics which may be useful for their experimental detection. They have the property:
$$\Gamma(\tilde{g} \to l^+ X) = \Gamma(\tilde{g} \to l^- X) \quad . \qquad (78)$$
Hence gluino pair production can lead to final states with same sign $l^\pm l^\pm$ pairs.[38, 61] The standard model background for this type of signature is rather small.

- Same sign di-lepton pairs are a useful signature for gluino pair production.

Another generic signature for SUSY particles is tri-lepton production.[62] If we consider the process of chargino-neutralino production, then it is possible to have the process:
$$\tilde{\chi}_1^\pm \tilde{\chi}_2^0 \to l\nu \tilde{\chi}_1^0 + \bar{l}'l' \tilde{\chi}_1^0 \quad . \qquad (79)$$
Again this is a signature with a small standard model background.

In the following sections we will examine several of these signatures in detail. In order to predict the SUSY particle production rates, it is necessary to have an event generator which includes both the production and decays of the SUSY particles. A number of generators exist for both e^+e^- and hadronic colliders. The physics assumptions of two of the most commonly used event generators for SUSY (ISASUSY and SPYTHIA) are reviewed in Ref. [63].

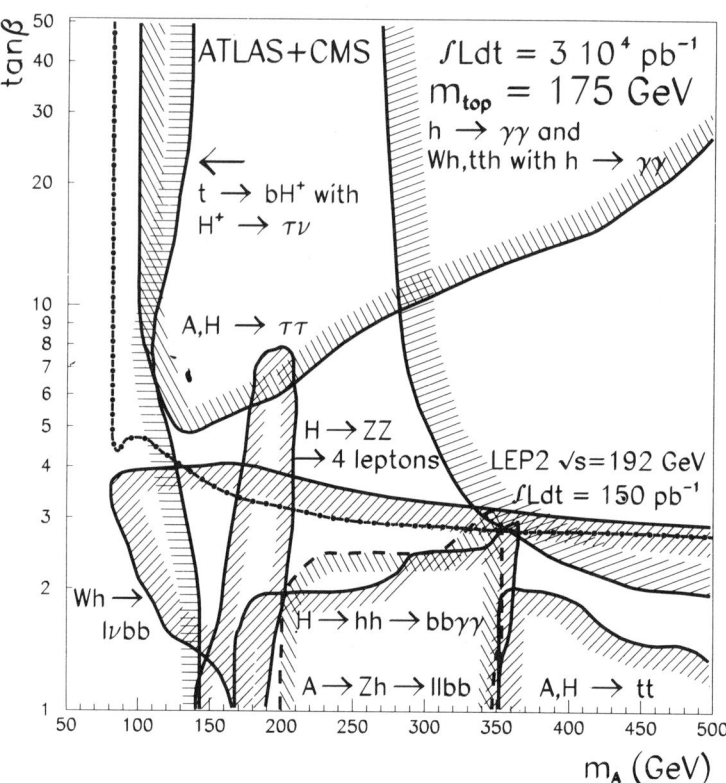

Figure 15: LHC (with low luminosity) and LEPII discovery limits for SUSY Higgs bosons. Figure from Ref. [60].

Chargino and Neutralino Production

As an example of SUSY particle searches, we consider the search for chargino pair production at an electron-positron collider,

$$e^+e^- \to \tilde{\chi}_1^+ \tilde{\chi}_1^- \quad , \tag{80}$$

(where $\tilde{\chi}_1^\pm$ are the lightest charginos.) The chargino mass matrix has a contribution from both the fermionic partner of the $W^\pm$, $\tilde{\omega}^\pm$, and from the fermionic partner of the charged Higgs, $\tilde{h}^\pm$, and so depends on the two unknown parameters in the mass matrix, μ and M_2. (See Eq. 40). If $|\mu| \ll M_2$, we say the chargino is "Higgsino-like", while if $|\mu| \gg M_2$, it is termed "gaugino-like". Results are usually presented in terms of the mass of the lightest chargino, $M_{\tilde{\chi}_1^+}$, and μ.

There are two types of Feynman diagrams contributing to chargino pair production: the first is an s-channel exchange of a γ or a Z, and the second is the t-channel exchange of the scalar partner of the neutrino, $\tilde{\nu}_L$. There is a destructive interference between the two types of diagrams. The largest interference occurs for light $\tilde{\nu}_L$ and $\tilde{\chi}_1^\pm$ "Gaugino-like". For light $\tilde{\nu}_L$, $\tilde{m}_{\nu_L} \lesssim 60~GeV$, the destructive interference can make the cross section significantly smaller, leading to a weaker limit. For a heavy $\tilde{\nu}_L$, the interference between the diagrams is small and the production cross section at LEP is $\sigma \sim 6 - 18~pb$ for $M_{\tilde{\chi}^+} \sim 60~GeV$. Hence any limits which may be obtained will depend on $\tilde{m}_{\nu_L}$, as well as μ and M_2.

The search proceeds by looking for the decay $\tilde{\chi}_1^\pm \to \tilde{\chi}_1^0 l^\pm \nu$. The assumption is made that the $\tilde{\chi}_1^0$ is stable and escapes the detector unseen. Using this technique, ALEPH obtains a limit,[64]

$$M_{\tilde{\chi}^\pm} > 67.8~GeV \quad @95\%CL \tag{81}$$

$$\text{For :} \quad \begin{cases} m_{\tilde{\nu}_L} > 200~GeV \\ M_{\tilde{\chi}^\pm}~\text{gaugino} - \text{like}, \quad |\mu| \gg M_2 \end{cases} . \tag{82}$$

This limit is not very sensitive to $\tan\beta$, but is considerably weaker when $|\mu| \lesssim 100~GeV$. It is clearly important to understand the input assumptions about the various SUSY parameters when interpreting this limit, as is the case with most limits on SUSY particles.

It is interesting to compare the search for charginos and neutralinos at LEP with what is possible at the LHC. At the LHC one clear signature will be,[65]

$$pp \to \tilde{\chi}_1^\pm \tilde{\chi}_2^0 \tag{83}$$

with,

$$\begin{aligned} \tilde{\chi}_1^\pm &\to l'^\pm \nu \tilde{\chi}_1^0 \\ \tilde{\chi}_2^0 &\to l\bar{l}\tilde{\chi}_1^0 \end{aligned} . \tag{84}$$

The cross section for this process is $\sigma \sim 1 - 100~pb$ for masses below $1~TeV$. This gives a "tri-lepton signature" with three hard, isolated leptons, significant E_T and little jet activity.[62] The dominant Standard Model backgrounds are from $t\bar{t}$ production (which can be eliminated by requiring that the 2 fastest leptons have the same sign) and $W^\pm Z$ production (which is eliminated by requiring that $M_{ll} \neq M_Z$).

To get reliable predictions at a hadron collider, it is not enough to use your Monte Carlo generator to simulate the process of interest (here chargino pair production). One must also simulate all the other SUSY production processes.[66] It is amusing to note that at the LHC the largest background to chargino and neutralino production is indeed from other SUSY particles, such as squark and gluino production, which also give events with leptons, multi-jets, and missing E_T. Since the squarks and gluinos are strongly interacting, they will generate more jets and a harder missing E_T spectrum than the charginos and neutralinos. This can be used to separate squark and gluino production from the chargino and neutralino production process of interest.[38]

- The biggest background to SUSY is SUSY itself.

As an example, we quote from a study of the tri-lepton signature at the LHC which assumes relatively light charginos and neutralinos,[38]

$$\begin{aligned} M_{\tilde{\chi}_1^+} &= 96 \; GeV \\ M_{\tilde{\chi}_2^0} &= 96 \; GeV \\ M_{\tilde{\chi}_1^0} &= 45 \; GeV \; . \end{aligned} \quad (85)$$

Once the SUSY particle masses are specified all the production rates can be computed unambiguously. After cuts, Ref. [38] finds (at the LHC):

$$\begin{aligned} \text{Signal} : &\quad 41 \; fb \\ t\bar{t} \; \text{bkdg} : &\quad 2.4 \; fb \\ WZ \; \text{bkgd} : &\quad .5 fb \\ \tilde{g}, \tilde{q} \; \text{bkgd} : &\quad 5.6 \; fb, \end{aligned} \quad (86)$$

demonstrating the viability of this signature at the LHC.

- The tri-lepton signal offers the possibility of untangling the $\tilde{\chi}^+\tilde{\chi}^0$ signal from the gluino and squark background.

CDF has searched for this decay chain and we see the results in Fig. 16.[67] Since the branching ratio to tri-leptons depends on the parameters of the chargino and neutralino mass matrix they also show the prediction from a specific Grand Unified Theory. Within this model, the limit translates to $M_{\tilde{\chi}_1^\pm} > 73 \; GeV$. This is roughly the same limit as that found at LEP, but involves different assumptions about the parameters of the model.

Aside from observing the process and verifying the existence of charginos and neutralinos, we would also like to obtain a handle on the masses of the SUSY particles. The kinematics are such that,

$$0 < M_{ll} < M_{\tilde{\chi}_2^0} - M_{\tilde{\chi}_1^0} \; , \quad (87)$$

and hence the distribution $d\sigma/dM_{ll}$ has a sharp cut-off at the kinematic boundary which can be used to obtain information on the masses. Recently, significant progress has been made in our understanding of the capabilities of a hadron collider for extracting values of the SUSY particle masses from different event distributions.[68]

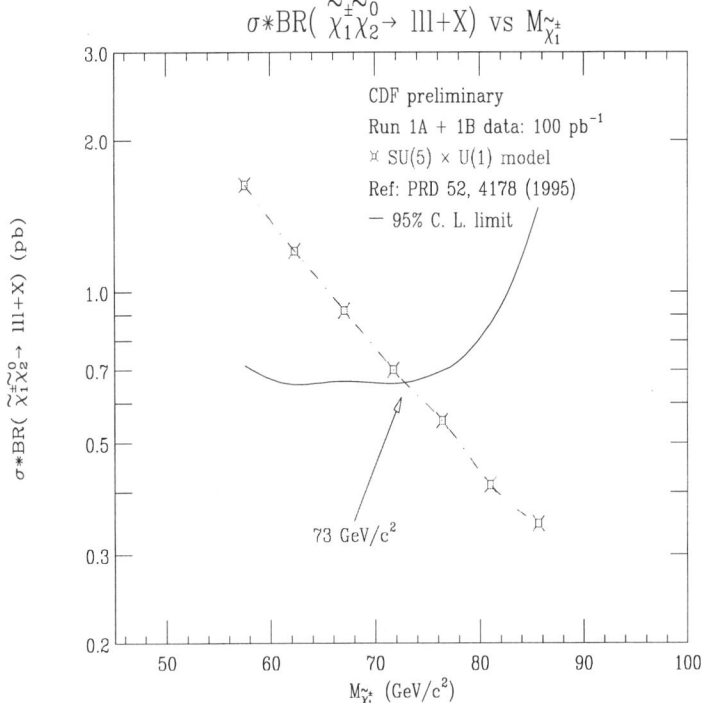

Figure 16: Limits on the tri-lepton signature in the reaction $p\bar{p} \to \tilde{\chi}_1^\pm \tilde{\chi}_2^0$ from CDF. This figure from Ref. [67].

Squarks, Gluinos, and Sleptons

Squarks and sleptons, $(\tilde{f}_i)$, can be produced at both e^+e^- and hadron colliders. At LEP, they would be pair produced via

$$e^+e^- \to \gamma, Z \to \tilde{f}_i \tilde{f}_i^*. \tag{88}$$

If there were a scalar with mass less than half the Z mass, it would increase the total width of the Z, Γ_Z. Since Γ_Z agrees quite precisely with the Standard Model prediction, the measurement of the Z lineshape gives

$$\tilde{m} > 35 - 40 \; GeV \tag{89}$$

for squarks and sleptons. The limit from the Z width is particularly important because it is independent of the squark or slepton decay mode and so applies for any model with low energy supersymmetry.

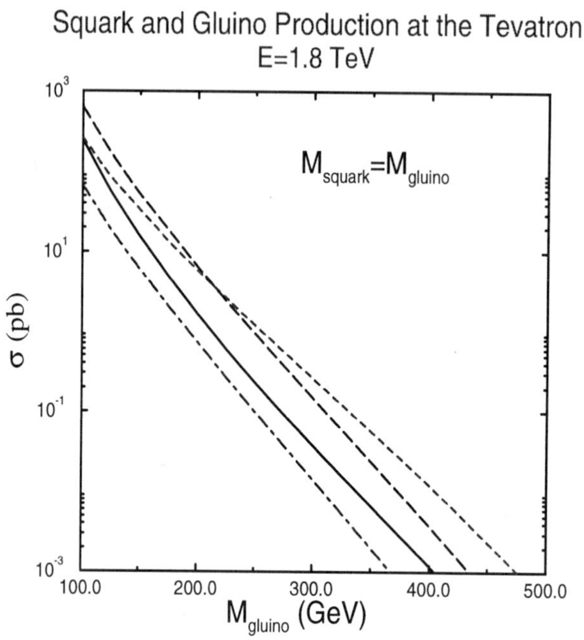

Figure 17: Squark and gluino production at the Tevatron assuming $M_{\tilde{q}} = M_{\tilde{g}}$. The solid line is $p\bar{p} \to \tilde{g}\tilde{g}$, the dot-dashed $\tilde{q}\tilde{q}$, the dotted $\tilde{q}\tilde{q}^*$, and the dashed is $\tilde{q}\tilde{g}$. This figure includes only the Born result and assumes 10 degenerate squarks.

There are limits on the direct production of squarks and gluinos from the Tevatron. The rates for squark and gluino production at both the Tevatron and the LHC are shown in Figs. 17 and 18 and analytic expressions for the Born cross sections can be found in Ref. [69]. The QCD radiative corrections to these process are large and increase the cross sections by up to a factor of two.[70] We neglect the mixing effects in the squark mass matrix and assume that there are 10 degenerate squarks associated with the light quarks. (The top squarks are assumed to be different since here mixing effects are clearly relevant.) The cross sections are significant, around 1 pb for squarks and gluinos in the few hundred GeV range.

The cleanest signatures for squark and gluino production are jets plus missing E_T from the undetected LSP, assumed to be $\tilde{\chi}_1^0$, and jets plus multi leptons plus missing E_T.[71] It will clearly be exceedingly difficult to separate the effects of squarks and gluino production, since they both contribute to the same experimental signature. The patterns of squark decays in various scenarios are examined in Ref. [72]. To obtain a limit on the gluino mass, we must therefore assume a limit on the squark mass. For 10 degenerate squarks, the limit from the Tevatron is, [4, 19]

$$M_{\tilde{q}} > 218 \; GeV \qquad \text{for} M_{\tilde{g}} = M_{\tilde{q}} \;. \tag{90}$$

This limit assumes a cascade decay, $\tilde{q} \to (....)\tilde{\chi}_1^0$. (There are similar limits for $M_{\tilde{g}} \ll M_{\tilde{q}}$ and $M_{\tilde{g}} \gg M_{\tilde{q}}$.)

Limits on the stop squark are particularly interesting since in many models it is the lightest squark. There are 2 types of stop squark decays which are relevant. The

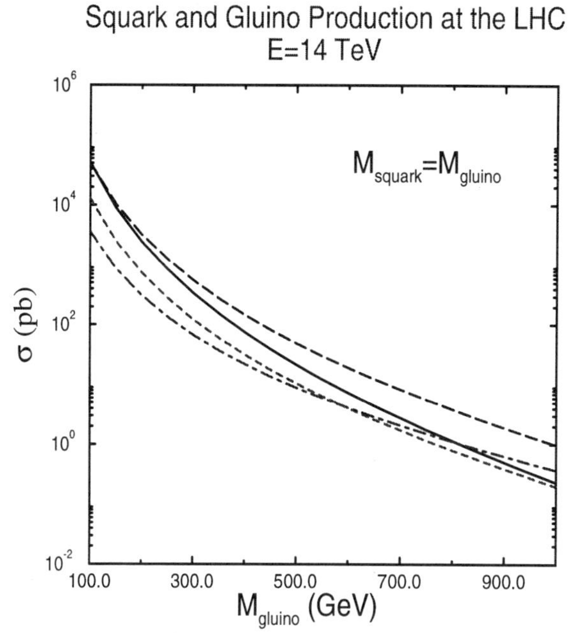

Figure 18: Squark and gluino production at the LHC assuming $M_{\tilde{q}} = M_{\tilde{g}}$. The solid line is $p\bar{p} \to \tilde{g}\tilde{g}$, the dot-dashed $\tilde{q}\tilde{q}$, the dotted $\tilde{q}\tilde{q}^*$, and the dashed is $\tilde{q}\tilde{g}$. This figure includes only the Born result and assumes 10 degenerate squarks.

first is,
$$\tilde{t} \to b\tilde{\chi}_1^+ \to bf\bar{f}'\tilde{\chi}_i^0 \quad . \tag{91}$$

The signal for this decay channel is jets plus missing energy. This signal shares many features with the dominant top quark decay, $t \to bW^+$, and in fact there have been suggestions in the literature that there may be some experimental confusion between the 2 processes.[73] Another possible decay chain for the stop squark is

$$\tilde{t} \to c\tilde{\chi}_1^0, \tag{92}$$

which also leads to jets plus missing energy. The 2 cases must be analyzed separately. The current limit on the stop squark mass from D0 is shown in Fig. 19.[74] We see that the limit depends sensitively on the mass of the LSP, $\tilde{\chi}_1^0$.

A spectacular signal for squark pair production which can result from the cascade decays is the production of same sign leptons,

$$pp \to \tilde{q}\tilde{q}^* \to (l^{\pm}l^{\pm}) + \text{jets} + E_T^{\text{miss}} \tag{93}$$

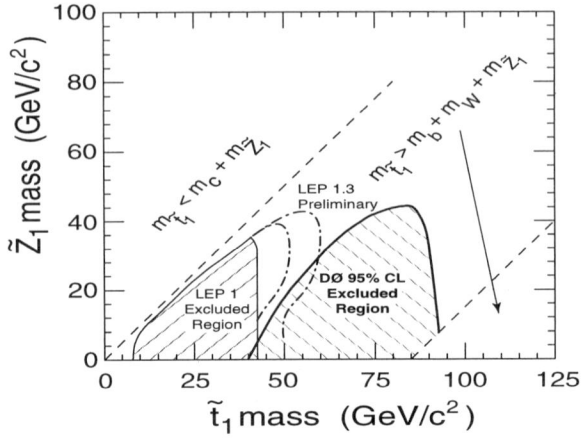

Figure 19: Limit on the lightest stop squark mass here labelled $\tilde{t}_1$, as a function of the lightest chargino mass (here labelled $\tilde{Z}_1$) from D0. This figure from Ref. [74].

At the Tevatron with $M_{\tilde{q}} \sim M_{\tilde{g}} \sim 100\ GeV$, the cross section for jets $+\ E_T^{\text{miss}}$ is $\sigma \sim 1\ pb$, while the rate for $l^\pm l^\pm +$ jets $+\ E_T^{\text{miss}}$ is $\sigma \sim .1pb$, which is still significant. The Standard Model background for this signal is quite small.

From the examples we have given, it is clear that searching for SUSY at a hadron collider is particularly challenging since there will typically be many SUSY particles which are kinematically accessible. Hadron colliders thus have a large discovery potential, but it is difficult to separate the various processes. To a large extent, one must trust the generic signatures of supersymmetry: E_T^{miss}, plus multi-jet and multi-lepton signatures. One will need to observe a signal in many channels in order to verify the consistency of the model.

A Case Study

It is instructive to consider an example of how the discovery of a SUSY particle might occur. Several years ago, CDF presented a single event,

$$p\bar{p} \to e^+ e^- \gamma\gamma + E_T^{miss}, \qquad (94)$$

for which it was difficult to find a Standard Model explanation.[75] By now, you all know that events with large missing energy are candidates for SUSY particle production. The scenario which we can construct is then,

$$p\bar{p} \to \tilde{e}\tilde{e}^* \quad , \qquad (95)$$

where $\tilde{e}$ is the scalar partner of either the right- or left-handed electron. The production cross section is then fixed unambiguously in terms of the selectron mass. The fact that only one event was seen fixes the selectron mass to be in the 100 GeV region. The selectron is then assumed to decay to an electron and a neutralino,

$$\tilde{e} \to e\tilde{\chi}^0 \quad . \tag{96}$$

The question which has engendered furious debate is how the neutralino might decay,

$$\tilde{\chi}^0 \to \tilde{X}\gamma, \tag{97}$$

where $\tilde{X}$ is either the lightest neutralino or a gravitino.[76] By examining the kinematics of the event, we could hope to learn about the underlying SUSY model. Unfortunately, examination of the 2 photon plus E_T^{miss} spectrum has produced no more SUSY candidates of the type of Eq. 94.[77]

CONCLUSIONS

Weak scale supersymmetry is a theory in desperate need of experimental input. The theoretical framework has evolved to a point where predictions for cross sections, branching ratios, and decay signatures can be reliably made. In many cases, calculations exist beyond the leading order in perturbation theory. However, without experimental observation of a SUSY particle or a precision measurement which disagrees with the Standard Model (which could be explained by SUSY particles in loops) there is no way of choosing between the many possible manifestations of low energy SUSY and thereby fixing the parameters in the soft SUSY breaking Lagrangian.

With the coming of LEPII, the Fermilab Main Injector, and the LHC, large regions of SUSY parameter space will be explored and we can only hope that some evidence for supersymmetry will be uncovered. The ball is definitely in the experimentalist's court !

Acknowledgments

I thank all the students at this school who asked such wonderful questions and really made me think about the experimental consequences of supersymmetry I also thank Tom Ferbel for his superb organization of all the lectures and other happenings at this school. I am grateful to Michael Spira for the use of his program HDECAY for the calculation of the radiatively corrected SUSY Higgs decay widths and also for discussions. Helpful discussions with Frank Paige are also gratefully acknowledged. I am indebted to Ken Kiers for a careful reading of the manuscript. This work has been supported by the DOE under contract number DE-AC02-76-CH-00016.

REFERENCES

1. B. Lee, C. Quigg, and H. Thacker, *Phys. Rev.* **D16** (1977) 1519; D. Dicus and V. Mathur, *Phys. Rev.* **D7** (1973) 3111.
2. G. t'Hooft in *Recent Developments in Gauge Theories*, eds. G. t'Hooft *et.al.* (Plenum, N.Y., 1980).
3. J. Amundson *et.al.*, *Report of the Supersymmetry Theory Subgroup*, Snowmass, 1996, hep-ph/9609374.

4. W. Merritt, *Proceedings of the 1996 DPF Meeting*, Minneapolis, MN.
5. M. Schmidt, *Proceedings of the 1996 DPF Meeting*, Minneapolis, MN.
6. H. Haber and G. Kane, *Phys. Rep.* **117C** (1985) 75.
7. J. Bagger, Lectures presented at the 1991 Theoretical Advanced Study Institute, Boulder, CO, June, 1991; Lectures presented at the 1995 Theoretical Advanced Study Institute, Boulder, CO, June, 1995, hep-ph/9604232; H.P. Nilles, *Phys. Rep.* **110** (1984) 1; H. Haber, Lectures presented at the 1986 Theoretical Advanced Study Institute, Santa Cruz, CA, June, 1986; R. Arnowitt, A. Chamseddine and P. Nath, *Applied N=1 Supergravity*, (World Scientific, 1984); V. Barger and R. Phillips, *Recent Advances in the Superworld*, J. Lopez and D. Nanopoulos, Ed. (World Scientific, 1994).
8. X. Tata, Lectures presented at the 1995 Theoretical Advanced Study Institute, *QCD and Beyond*, Boulder, CO, June, 1995, hep-ph/9510287.
9. H. Murayama and M. Peskin, *Ann. Rev. Nucl. Part. Sci.* 1996, hep-ex/9606003.
10. G. Anderson, D. Castano, and A. Riotto, hep-ph/9609463, 1996.
11. J. Wess and J. Bagger, *Supersymmetry and Supergravity*, (Princeton University Press, Princeton, N.J. 1983); P. Fayet and S. Ferrara, *Phys. Rep.* **32** (1977) 249.
12. D. Gross and R. Jackiw, *Phys.Rev.* **D6** (1972) 477; C. Bouchiat, J. Iliopoulos and P. Meyer, *Phys. Lett.* **B38** (1972) 519; H. Georgi and S. Glashow, *Phys. Rev.* **D6** (1972) 429; L. Alvarez-Gaume and E. Witten, *Nucl. Phys.* **B234** (1983) 269.
13. S. Dimopoulos and H. Georgi, *Nucl. Phys.* **B193** (1981) 150; N. Sakai, *Z. Phys.* **C11** (1981) 153; P. Fayet, *Phys. Lett.* **B69** (1977) 489; **B84** (1979) 416.
14. M. Dine, A. Nelson, Y. Shirman, *Phys. Rev.* **D51** (1995) 1362; M. Dine, A. Nelson, Y. Nir, and Y. Shirman, *Phys. Rev.* **D53** (1996) 2658.
15. S. Weinberg, *Phys. Rev.* **D26** (1982) 287; N. Sakai and T. Yanagida, *Nucl. Phys.* **B197** (1982) 533; S. Dimopoulos, S. Raby, and F. Wilczek, *Phys. Lett.* **B112** (1982) 133; J. Ellis, D. Nanopoulos, and S. Rudaz, *Nucl. Phys.* **B202** (1982) 43.
16. C. Carlson, P. Roy, and M. Sher, *Phys. Lett* **B357** (1995) 99.
17. G. Bhattacharyya, hep-ph/9608415 ,1996.
18. G. Farrar and P. Fayet, *Phys. Lett.* **B76** (1978) 575; F. Zwirner, *Phys. Lett.* **132B** (1983) 103; L. Hall and M. Suzuki, *Nucl. Phys.* **B231** (1984) 419; J. Ellis, G. Gelmini, C. Jarlskog, G. Ross, and J. Valle, *Phys. Lett.* **B150** (1985) 142; G. Ross and J. Valle, *Phys. Lett.* B151 (1985) 375; S. Dawson, *Nucl. Phys.* **B261**(1985) 297; S. Dimopoulos and L. Hall, *Phys. Lett.***B207** (1988) 210.
19. Particle Data Group, *Phys. Rev.* **D 54**, (1996) 1.
20. P. Smith *et.al.*, *Nucl. Phys.* **B144** (1979) 525; *Nucl Phys.* **B206** (1982) 333; E. Norman *et.al.*, *Phys. Rev. Lett.* **58** (1987) 1403; T. Hemmik *et.al.*, *Phys. Rev.* **D41** (1990) 2074; S. Wolfram, *Phys. Lett.* **B82** (1979) 65; C. Dover, T. Gaisser, and G. Steigman, *Phys. Rev. Lett.* **42** (1979) 1117.
21. H. Baer, C. Kao, and X. Tata, *Phys. Rev.* **D51** (1995) 2180; H. Baer, C. Chen, and X. Tata, hep-ph/9608221, 1996.
22. L. Hall, J. Lykken, and S. Weinberg, *Phys. Rev.* **D 27** (1973) 2359.
23. L. Giradello and M. Grisaru, *Nucl. Phys.* **B194** (1982) 65; K. Harada and N. Sakai, *Prog. Theor. Phys.* **67** (1982) 67.
24. H. Haber, G. Kane, and T. Sterling, *Nucl. Phys.* **B161** (1979) 493.
25. E. Witten, *Nucl. Phys.* **B185** (1981) 513; M. Dine, W. Fischler, and M. Srednicki, *Nucl. Phys.* **B189** (1981) 575; S. Dimopoulos and S. Raby, *Nucl. Phys.* **B192** (1981) 353; J. Polchinski and L. Susskind, *Phys. Rev.* **D26** (1982) 3661; L. Ibanez and G. Ross, *Phys. Lett.* **B105** (1981) 439.
26. J. Gunion, H. Haber, G. Kane, and S. Dawson, *The Higgs Hunter's Guide* (Addison Wesley, Menlo Park, CA) 1990.
27. S. Li and M. Sher, *Phys. Lett.* **B140** (1984) 339; H. Nilles and M. Nusbaumer, *Phys. Lett.* **B145** (1984) 73; J. Gunion and H. Haber, *Nucl. Phys.* **B272** (1986) 1.
28. P. Chankowski, S. Pokorski, and J. Rosiek, *Phys. Lett.* **B274** (1992) 191; **B281** (1992) 100; Y. Okada, M. Yamaguchi, and T. Yanagida, *Prog. Theor. Phys.* **85** (1991) ; *Phys. Lett.* **B262** (1991) 54; J. Espinosa and M. Quiros, *Phys. Lett.* **B267** (1991) 27; *Phys. Lett.* **B266** (1991) 389; H. Haber and R. Hempfling, *Phys. Rev.* **D48** (1993)4280; *Phys. Rev. Lett.* **66** (1991) 1815; J. Gunion and A. Turski, *Phys. Rev.* **D39** (1989) 2701; **D40** (1990) 2333; M. Berger, *Phys. Rev.* **D41** (1990) 225; K. Sasaki, M. Carena and C. Wagner, *Nucl. Phys.* **B381** (1992) 66; R. Barbieri and M. Frigeni, *Phys. Lett.* **B258** (1991) 395; J. Ellis, G. Ridolfi and F. Zwirner,

Phys. Lett. **B257** (1991) 83; **B262** (1991) 477; R. Hempfling and A. Hoang, *Phys. Lett.* **B331** (1994) 99; R. Barbieri, F. Caravaglios, and M. Frigeni, *Phys. Lett.* **B258** (1991)167; H.Haber, R. Hempfling, and H. Hoang, hep-ph/9609331,1996; M.Carena, M. Quiros, and C. Wagner, *Nucl. Phys.* **B461** (1996) 407; M. Carena, J. Espinosa, M. Quiros, and C. Wagner, *Phys. Lett.* **B355** (1995) 209.

29. M. Quiros, *XXIV International Meeting on Fundamental Physics: From Tevatron to LHC*, Gandia, Spain, 1996, hep-ph/9609392; T. Elliot, S. King, and P. White, *Phys. Lett.* **B305** (1993) 71; G. Kane, C. Kolda, and J. Wells, *Phys. Rev. Lett.* **70** (1993) 2686.

30. H. Haber and J. Gunion, *Nucl. Phys.* **B272** (1986) 1; *Nucl. Phys.* **B278** (1986) 449; erratum, **B402** (1993) 567.

31. M. Carena, P. Zerwas, *et.al.*, *Higgs Physics at LEPII*, hep-ph/9602250, 1996.

32. The FORTRAN program HDECAY is documented in M. Spira, CERN-TH-95-285, hep-ph/9610350 along with references to the original calculations.

33. S. Mrenna and C. Yuan, *Phys. Lett.* **B367** (1996) 188.

34. J. Bagger, S. Dimopoulos, and E. Masso, *Phys. Lett.* **B156** (1985) 357; *Phys. Rev. Lett.* **55** (1985) 920; M. Einhorn and D. Jones, *Nucl. Phys.* **B196** (1982) 475.

35. S. Dimopoulos, S. Raby, and F. Wilczek, *Phys. Rev.* **D24** (1981) 1681; U. Amaldi *et.al.*, *Phys. Rev.* **D36** 1987 1385; P. Langacker and M. Luo, *Phys. Rev.* **D44** (1991) 514; J. Ellis, S. Kelley, and D. Nanopoulos, *Phys. Lett.* **B260** (1991) 447; U. Amaldi, W. deBoer, and H. Furstenau, *Phys. Lett.* **B260** (1991) 447; N. Sakai, *Z. Phys* **C11** (1982) 153.

36. J. Ellis, S. Kelley, and D. Nanopoulos, *Phys. Lett.* **B260** (1991)131; P. Langacker and M. Luo, *Phys. Rev.* **D44** (1991) 817; U. Amaldi, W. deBoer, and H. Furstenau, *Phys. Lett.* **B260**(1991) 447; M. Carena, S. Pokorski, and C. Wagner, *Nucl. Phys.* **B406** (1993) 59; P. Langacker and N. Polonsky, *Phys. Rev.* **D47** (1993) 4028.

37. J. Bagger, K. Matchev, D. Pierce, and R. Zhang, hep-ph/9608444, 1996; J. Bagger, K. Matchev, and D. Pierce, *Phys. Lett.* **B348** (1995) 443; D. Pierce, J. Bagger, K. Matchev, and R. Zhang, hep-ph/9606211, 1996.

38. H. Baer, C. Chen, F. Paige, and X. Tata, *Phys. Rev.* **D54** (1996) 5866; *op. cit.* **D53** (1996) 6241; *op.cit.* **D52** (1995) 1565; **D52** (1995) 2746. 39. M. Machacek and M. Vaughn, *Nucl. Phys.* **B222** (1983) 83; C. Ford, D. Jones, P. Stephenson, and M. Einhorn, *Nucl. Phys.* **B395** (1993) 17.

40. L. Ibanez, *Nucl. Phys.* **B218** (1983) 514; *Phys. Lett.* **B118** (1982) 73; L. Ibanez and G. Ross, *Phys. Lett.* **B110** (1982) 215; J. Ellis, D. Nanopoulos, and K. Tamvakis, *Phys. Lett.* **B121** (1983) 123; L. Alvarez-Gaume, J. Polchinski, and M. Wise, *Nucl. Phys.* **B221** (1983) 495; B. Ananthanarayan, G. Lazarides, and Q. Shafi, *Nucl. Phys.***D44** (1991) 1613.

41. V. Barger, M. Berger, and P. Ohmann, *Phys. Rev.* **D49** (1994) 4908.

42. B. Pendleton and G. Ross, *Phys. Lett.* **B98** (1981)291; V. Barger, M. Berger, P. Ohmann, and R. Phillips, *Phys. Lett.* **B314** (1993) 351; S. Kelley, J. Lopez, and D. Nanopoulos, *Phys. Lett.* **B274** (1992) 387; M. Carena, M. Olechowski, S. Pokorski, and C. Wagner, *Nucl. Phys.* **B426** (1994) 269; N. Polonsky, *Phys. Rev.* **D54** (1996)4537; N. Polonsky, LMU-TPW-96-04, hep-ph/9602206, 1996..

43. G. Kane in *Proceedings of the 28th Rencontres de Moriond*, Les Arcs, France, 1993; H. Haber in *Workshop on Recent Advances in the Superworld*, Woodlands, TX, 1993.

44. H. Goldberg, *Phys. Rev. Lett.* **50** (1983) 1419; J. Ellis, K. Olive, D. Nanopoulos, J. Hagelin, and M. Srednicki, *Nucl. Phys.* **B328** (1984) 453; M. Drees and M. Nojiri, *Phys. Rev.***D47** (1993) 376; H. Baer, M. Brhlik, and D. Castano, hep-ph/9607465, 1996.

45. S. Dawson, Invited talk given at the 1996 DPF Meeting, Minneapolis, MN, hep-ph/9609340,1996.

46. W.deBoer *et.al.*, IEKP-KA/96-08, KA-TP-18-96, hep-ph/969209, 1996.

47. P. Chankowski and S. Pokorski, *Acta. Phys. Polon.* **27** (1996) 1719; G. Kane, R. Stuart, and J. Wells, *Phys. Lett.* **B354** (1995) 350; T. Blazek, M. Carena, S. Raby, and C. Wagner, OHSTPY-HEP-T-96-026, hep-ph/9611217, 1996.

48. M. Alam *et.al.*, (CLEO Collaboration), *Phys. Rev. Lett* **74** (1995) 2885.

49. S. Bertolini, F. Borzumati, A. Masiero and G. Ridolfi, *Nucl. Phys.* **B353** (1991) 591; R. Barbieri and G. Giudice, *Phys. Lett.* **309** (1993) 86; P. Nath and R. Arnowitt, *Phys. Lett.* **B336** (1994) 395; G. Kane, C. Kolda, L. Roszkowsi, and J. Wells, *Phys. Rev.* **D49** (1994) 6173; V. Barger, M. Berger, P. Ohmann, and R. Phillips, *Phys. Rev.* **D51** (1995) 2438; B. deCarlos and J. A. Casas, *Phys. Lett.* **B349** (1995) 300, *ibid* **B351** (1995) 604.

50. W. deBoer *et.al.*, IEKP-KA/96-04, hep-ph/9603350, 1996.

51. H. Baer and M. Brhlik, FSU-HEP-961001, hep-ph/9610224, 1996.
52. J. Hewett and J. Wells, SLAC-PUB-7290, hep-ph/9610323, 1996.
53. F. Gabbiani, E. Gabrielli, A. Masiero, and L. Silvestrini, ROM2F/96/21, hep-ph/9604387, 1996; L. Hall, A. Kostelecky, and S. Raby, *Nucl. Phys.* **B267** (1986) 415; L. Hall and L. Randall, *Phys. Rev. Lett.* **65** (1990) 2939; M. Dine, R. Leigh, and A. Kagan, *Phys. Rev.* **D48** (1993) 4269; Y. Nir and N. Seiberg, *Phys. Rev. Lett.* **B309** (1993) 337 .
54. V. Barger, M. Berger, J. Gunion, and T. Han, hep-ph/9606417, 1996; *Proceedings of the 3rd International Conference on Physics Potential and Development of $\mu^+\mu^-$ Colliders*, San Francisco, Dec. 1995, hep-ph/9604334.
55. A. Djouadi et. al., *Proceedings of the Workshop Physics with e^+e^- Linear Colliders*, (Annecy-Gran Sasso-Hamburg, 1995), Ed. P. Zerwas, hep-ph/9605437.
56. M. Spira, A. Djouadi, D. Graudenz, and P. Zerwas, *Nucl. Phys.* **B453** (1995) 17; *Phys. Lett.* **B318** (1993) 347; S. Dawson, A. Djouadi, and M. Spira, *Phys. Rev. Lett.* **77** (1996) 16.
57. J. Gunion, A. Stange, and S. Willenbrock, to appear in *Electroweak Symmetry Breaking and Physics at the TeV Scale*, Ed. T. Barklow, S. Dawson, H. Haber, and J. Siegrist, (World Scientific, 1996), hep-ph/9602238.
58. ATLAS Collaboration, Technical Proposal, LHCC/P2 (1994); CMS Collaboration, Technical Proposal, LHCC/P1 (1994).
59. A. Stange, W. Marciano, and S. Willenbrock, *Phys. Rev.* **D50** (1994) 4491; **D49** (1994) 1354.
60. D. Froidevaux *et.al.* ATLAS internal note, PHYS-No-74 (1995).
61. R.M. Barnett, J. Gunion, and H. Haber, *Phys. Lett.* **B315** (1993) 349; H. Baer, X. Tata, and J. Woodside, *Phys. Rev.* **D41** (1990) 906; M. Guchait and D.P. Roy, *Phys. Rev.* **D52** (1995) 133.
62. H. Baer, C. Chen, F. Paige, and X. Tata, *Phys. Rev.* **D50** (1994) 4516.
63. S. Mrenna, hep-ph/9609360, 1996; H. Baer, F. Paige, S. Protopopescu, and X. Tata, FSU-HEP-930329, hep-ph/9305342, 1993.
64. D. Buskulic *et.al.*, (ALEPH Collaboration), *Phys. Lett.* **B 373** (1996) 246.
65. H. Baer, C. Kao, and X. Tata, *Phys. Rev.* **D48** (1993) 5175.
66. H. Baer, C. Chen, R. Monroe, F. Paige, and X. Tata, *Phys. Rev.* **D51** (1995) 1046; S. Mrenna, G. Kane, G. Kribs, and J. Wells, *Phys. Rev.* **D53** (1996) 1168.
67. T. Kamon, *Proceedings of XXXI Rencontres de Moriond*, Les Arcs, France, 1996, hep-ex/9605006; F. Abe *et.al.*, (CDF Collaboration), *Phys. Rev. Lett.* **76** (1996) 4307.
68. I. Hinchliffe, F. Paige, M. Shapiro, J. Soderqvist, and W. Yao, hep-ph/9610544, 1996.
69. S. Dawson, E. Eichten, and C. Quigg, *Phys. Rev.* **D31** (1985) 1581; H. Baer, A. Bartl, D. Karatas, W. Majerotto, and X. Tata, *Int. Jour. Mod. Phys.* **A4** (1989) 4111.
70. W. Beenakker, R. Hoper, M. Spira, and P. Zerwas, *Z. Phys.* **C69** (1995) 163.
71. H. Baer, J. Ellis, G. Gelmini, D. Nanopoulos, and X. Tata, *Phys. Lett.* **B** (1985) 175; H. Baer, V. Barger, D. Karatas, and X. Tata, *Phys. Rev.* **D36** (1987) 96; R. Barnett, J. Gunion, and H. Haber, *Phys. Rev.* **D37** (1988) 1892; *Phys. Lett.* **B315** (1993) 349; H. Baer, C. Kao, and X. Tata, *Phys. Rev.* **D48** (1993) 2978.
72. R. Barnett, J. Gunion, and H. Haber, *Phys. Rev.* **D37** (1988) 1892; A. Bartl *et.al.*, *Z. Phys.* **C52** (1991) 477; H. Baer *et.al.*, *Phys. Lett.* **B161** (1985) 175; *Phys. Rev.* **D36** (1987) 96; G. Gamberini, *Z. Phys.* **C30** (1986) 605; G. Gamberini *et.al. Phys. Lett.* **B203** (1988) 453.
73. R. Barnett and L. Hall, hep-ph/9609313, 1996, *Proceedings of the 1996 DPF Meeting*, Minneapolis, MN.
74. W. Merritt, *Proceedings of the 4th International Conference on Supersymmetries in Physics*, College Park, MD, 1996, FERMILAB-CONF-96-242-E; A. Abachi *et.al.*, *Phys. Rev. Lett.* **76** (1976) 2222.
75. S. Park, *Proceedings of the 10th Topical Workshop on $p\bar{p}$ Collider Physics*, Ed. R. Raja and J. Yoh, AIP Press, 1995.
76. S. Ambrosanio, G. Kane, G. Kribs, S. Martin, and S. Martin, hep-ph/9607414; *Phys. Rev. Lett.* **75** (1996) 3498; S. Dimopoulos, M. Dine, S. Raby, and S. Thomas, *Phys. Rev. Lett.* **76** (1996) 3494; K. Babu, C. Kolda, and F. Wilczek, *Phys. Rev. Lett.* **77** (1996) 3070.
77. D. Toback, *Proceedings of the 1996 DPF Meeting*, Minneapolis, MN.

THE CHALLENGES OF LHC

Nick Ellis

PPE Division, CERN, 1211 Geneva 23, Switzerland

INTRODUCTION

My objective in these lectures is to highlight some of the experimental challenges at LHC and to illustrate how the general-purpose detectors have been adapted to meet these challenges. Some of the challenges result from the parameters of the LHC machine, others from the physics that one wants to do. I shall cover general aspects of the design of the detectors (e.g. why they are so big) and also technology aspects (e.g. why specific kinds of detectors have been chosen). Where relevant to the discussion, I shall touch on aspects of detector physics, although without going into much detail.

THE LHC MACHINE AND BENCHMARK PHYSICS PROCESSES
(Requirements on Detectors and the Trigger/DAQ System)

The LHC Machine

The Large Hadron Collider[1] (LHC) is a high-energy, high-luminosity proton–proton collider to be built in the existing LEP tunnel at CERN. We require high-energy because we are searching for massive particles. We also require high luminosity because we are searching for rare processes — cross-sections are small for producing massive particles and, in some cases, only rare decay modes will be detectable above the background (a disadvantage of a hadron machine compared to an e^+e^- one).

The LHC can also be used to collide beams of heavy ions, e.g. lead ions, with a centre-of-mass energy of about 6 TeV per nucleon. Here the luminosity will be relatively low; the interaction rate will be $\sim 10^4$ s^{-1} compared to $\sim 10^9$ s^{-1} for proton–proton collisions. In these lectures I shall mention only very briefly issues related to the heavy-ion programme.

It is also worth noting, although I shall not discuss it further here, that e–p collisions could be produced. Here one would use the fact that there is room in the tunnel for an e^+e^- machine such as LEP as well as the LHC beam elements. This would, however, only be considered at a later stage.

Physics Objectives of the LHC

The primary purpose of the LHC is to search for and to study new physics. In particular, one hopes to understand the origin of electroweak symmetry breaking (masses of W and Z bosons). Here one is looking for one or more Higgs bosons, for example. The LHC can also be used to search for Supersymmetric (SUSY) particles; note that the cross-sections for producing squarks and gluinos are large. Some other possibilities for new physics that could be studied at LHC are: quark compositeness, leptoquarks, and heavy vector bosons (W', Z'). Perhaps most important of all, one should be on the look-out for unpredicted new physics.

Experiments at LHC will be able to make many measurements relating to known physics. Examples are: top-quark mass and decay properties, B-physics and numerous cross-section measurements (W, Z, γ and jet production, for example). The B-physics programme will include studies of CP violation which will allow one to test the Standard Model in which CP violation is parametrized within the CKM matrix. It will also include the measurement of B_s^0 oscillations, the search for rare B decays such as $B_d^0 \to \mu^+\mu^-(X)$, and the study of doubly-heavy hadrons such as B_c.

The heavy-ion programme has as its primary objective the search for the quark-gluon plasma. One will try to investigate different stages in the production of the plasma and the return to normal hadronic matter: the initial conditions, the quark-gluon plasma, the phase transition and the hadronic matter.

It is important to remember when considering physics at LHC (and elsewhere) that the Standard Model may not be right. One has to be careful to be aware when ones conclusions depend on assuming the Standard Model. For LHC one tends to emphasise the direct search for new physics. However, indications for new physics may also come from inconsistencies between precision measurements that are related to each other by the Standard Model. Clearly, this motivates making precision measurements (e.g. at LEP — see lectures of A. Blondel).

As mentioned above, the new physics that we hope to see at LHC may not correspond to predictions — we must be prepared for the unexpected. Nevertheless, we use predictions for new physics to guide us in the detector design. The hope is that a detector that performs well for a wide variety of predicted possibilities for new-physics processes will also work well for unpredicted ones. The search for the unexpected must clearly be considered in the trigger where one will include 'inclusive' selection criteria, not focused on selecting particular processes.

Parameters of the LHC

In the following, I discuss proton–proton operation unless stated otherwise. The high-energy beams (7 TeV per beam) require powerful bending magnets given the limited size of the LEP ring (27 km circumference). Using $p = 0.3 \cdot B \cdot R$, where p is momentum in GeV, B is magnetic field in Tesla and R is radius in meters, one obtains an average field strength of 5.4 T for the given beam momentum and the radius of the LEP ring. Of course, one cannot fill the full circumference of the ring with bending magnets. The dipole magnets for LHC will have a field strength of 8.4 T (compared to 4.4 T at the Tevatron and 4.7 T at HERA).

A novel 'two-in-one' magnet design using superfluid-helium cooling has been adopted for the LHC dipoles — the lower temperature compared to 'conventional' superconducting magnets allows them to be operated at higher field strength. Each of the ~1300 dipole magnets will be 13 m long. Five prototype magnets have already been delivered and tested. A three-dipole 'string' has been operated for more than 1500 hours at a temperature of 1.9 K and more than 50 quenches have been induced without damage to the magnets.

Bunches of protons in counter-rotating beams will be made to collide at 'interaction points'. An important parameter of the machine is the 25 ns bunch-crossing period. The LHC will operate at very high luminosity — 10^{34} cm^{-2}s^{-1}. The total inelastic, non-diffractive cross-section at LHC energies will be about 100 mb, corresponding to an interaction rate of 10^9 Hz. By comparing this with the bunch-crossing rate of 40 MHz (period of 25 ns), we can see that each bunch crossing will contain on average about 25 'pile-up' interactions. Note that one must use detectors with a short response time, otherwise pile-up will get integrated over many bunch crossings. Note also that highly granular detectors must be used to minimise the probability for pile-up in a given detector element. We will consider both of these aspects later on in these lectures.

The particles from the pile-up interactions (and also from the 'underlying event') have a typical transverse momentum of about 500 MeV and a distribution approximately flat in pseudorapidity out to $\eta \sim 5$; the density of charged particles is about 7 per unit of pseudorapidity. Transverse momentum, p_T, is defined as $p \cdot sin(\theta)$, where θ is the polar angle measured from the beam axis (different from the definition in e$^+$e$^-$ machines where 'p_T' is measured relative to the jet axis). Pseudorapidity is defined as $\eta = -ln(tan(\theta/2))$; it corresponds to rapidity, $y = \frac{1}{2}ln\frac{E+p_z}{E-p_z}$, in the limit $E \gg m$ (E and m are the energy and mass of the particle; p_z is the component of momentum parallel to the beam direction). Changing variables to polar angle, one sees that there are many forward-going particles. Given that $p = p_T/sin(\theta)$ and that the average p_T is approximately independent of η, these forward-going particles have high energies.

Note that the choice of p_T and η as variables is motivated by the fact that p_T and differences in η are invariant under boosts in the beam direction. In pp collisions, the centre of mass of the colliding partons is not at rest in the lab frame, but the transverse momentum of the two-parton system is usually small.

The high flux of particles from proton–proton interactions (and to a lesser extent from beam losses) places the detectors and associated electronics in a high-radiation environment. Only radiation resistant detectors can be used and electronics must be 'radiation hard'. One must also take into account background signals induced in the detectors by such radiation.

The high interaction rate also has implications for the trigger and data-acquisition systems. Only a tiny fraction of the interactions can be recorded for off-line analysis, requiring a trigger selectivity of about one interaction in 10^7. Furthermore, as we will see later, massive amounts of data have to be transmitted to and stored in buffer memories while the trigger system performs its calculations. (Most of the data correspond to rejected events and will be discarded.) Note that, since it is impossible make a trigger decision within the 25 ns between bunch crossings, so-called 'pipelined readout' has to be used, typically with electronics mounted on the detector. The functionality of pipelined readout is illustrated in Fig. 1; the memory cells may be analogue (e.g. store charge), digital (store the result of an analogue-to-digital conversion) or binary (store a single bit of information, e.g. discriminator output). The data must be kept in the pipeline memory for a few μs, until the first-level trigger decision is available.

Two other machine parameters will be relevant in these lectures: the lateral and longitudinal size of the interaction region will be about 15 µm and 5.6 cm respectively (rms).

In summary, the following factors must be taken into account when designing the experiments:

- Bunches cross each 25 ns (40 MHz rate).

- Each bunch crossing produces on average about 25 interactions.

- The high intensity and high energy of the machine result in high radiation levels.

- The trigger system must be very selective and pipelined readout must be used.

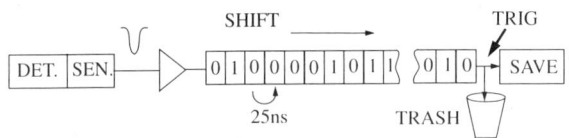

Figure 1. Illustration of pipelined readout.

Detectors for LHC

There will be at least four large-scale experiments at LHC. Two very big 'general-purpose' experiments have already been approved — ATLAS[2] and CMS[3]. A technical proposal has been submitted for a heavy-ion experiment — ALICE[4] — and a technical proposal is in preparation for a B-physics experiment — LHC-B[5]. There may also be some other, smaller, specialised experiments. I shall concentrate on the two general-purpose experiments in these lectures.

ATLAS uses a large (∼20 m diameter) air-core toroid system for its muon spectrometer, as shown in Fig. 2. The electromagnetic (EM) calorimetry, shown in Fig. 3, uses the liquid-argon technique. In the barrel, an iron–scintillator hadronic calorimeter is used; the endcap hadronic calorimeter is of the liquid-argon type. In front of the barrel EM calorimeter (integrated in the same cryostat) is a superconducting-solenoid coil that provides a 2 T field. The inner-tracking system consists of semiconductor detectors in the innermost part and straw-tubes in the outer part.

CMS uses a large (6 m diameter) solenoid magnet that provides the field for both the inner-tracking system and the muon spectrometer, as shown in Fig. 4. The coil surrounds the EM and hadronic calorimetry. The EM calorimeter is made of $PbWO_4$ crystals, while the hadronic calorimeter is made of copper plates and scintillator. The

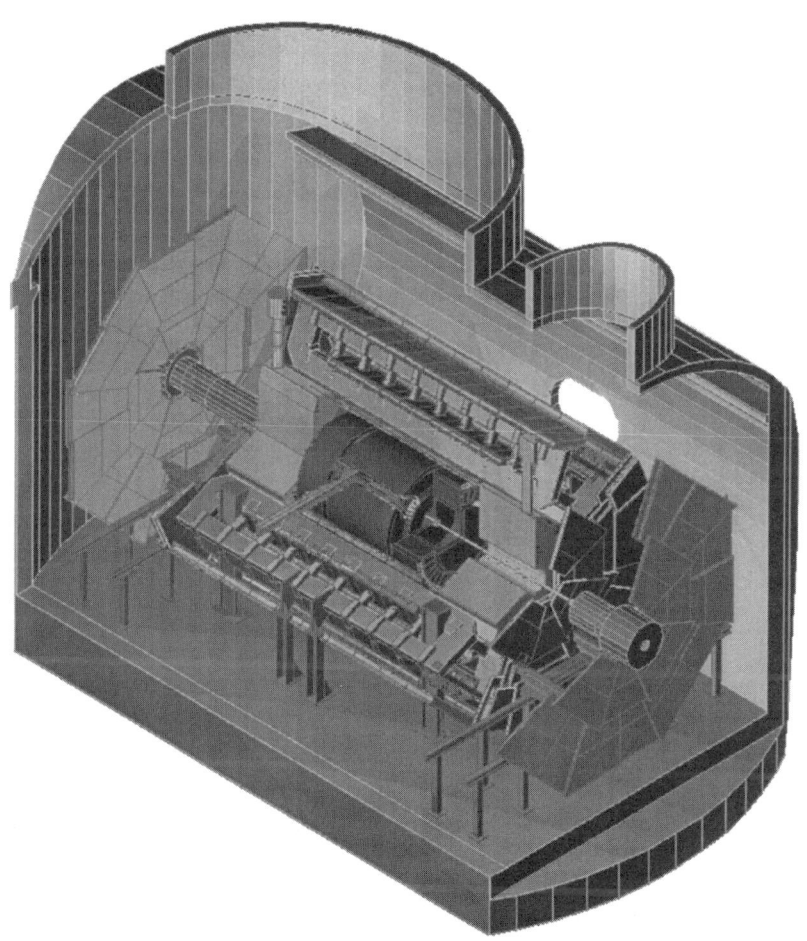

Figure 2. The ATLAS detector.

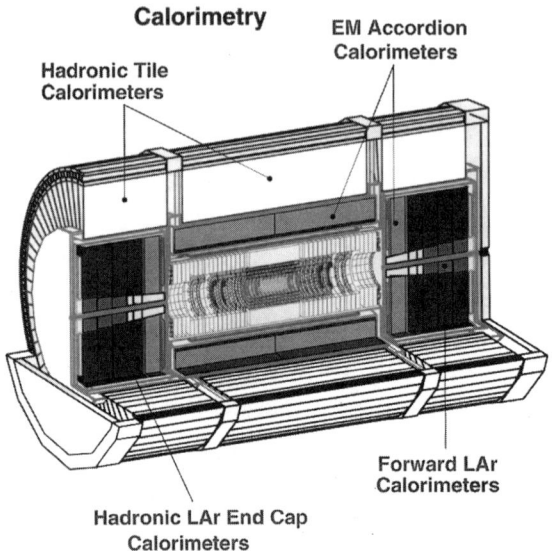

Figure 3. The ATLAS calorimeter system.

inner-tracking system, shown in Fig. 5, consists of semiconductor detectors in the innermost part and microstrip gas chambers in the outer part.

Benchmark Physics Processes

By considering some 'benchmark' processes we can gain an understanding of the required performance of LHC detectors. One should keep in mind that LHC, being a hadron collider, gives a 'dirty' environment compared to an e^+e^- machine. New particles may be produced by collisions between partons from the two protons, but the spectator partons also hadronize producing an 'underlying event'. Furthermore, as mentioned earlier, each bunch crossing produces about 25 collisions. Hence, the decay products of new particles have to be identified in the presence of a 'background' of hundreds of particles. At first sight this might appear hopeless as shown in Fig. 6a (from Ref. [6]). However, applying a cut on transverse momentum is very effective at reducing the background from the underlying event and pile-up, making visible the particles from the hard-scattering process (see Fig. 6b). Very detailed studies have shown that, provided detectors are used that can cope with the high-multiplicity environment, these backgrounds are manageable.

Even neglecting pile-up and the underlying event, it may be impossible to identify some decay modes of new particles. For example, it will be very hard to detect decays of new particles into pairs of jets because of the very large cross-section for high-p_T jet production due to Standard Model QCD processes. One is therefore often led to concentrate on channels containing clean signatures such as high-p_T electrons, muons and photons.

We consider some representative processes that influence the detector design. In the following 'low' luminosity refers to $L = 10^{33}$ cm^{-2} s^{-1} and 'high' (or 'design') luminosity refers to $L = 10^{34}$ cm^{-2} s^{-1}; the equivalent running time per year at these luminosities is taken as 10^7 seconds. Hence, a year's running at low luminosity gives an integrated luminosity of 10^4 pb^{-1} and a year's running at high luminosity corresponds to 10^5 pb^{-1}.

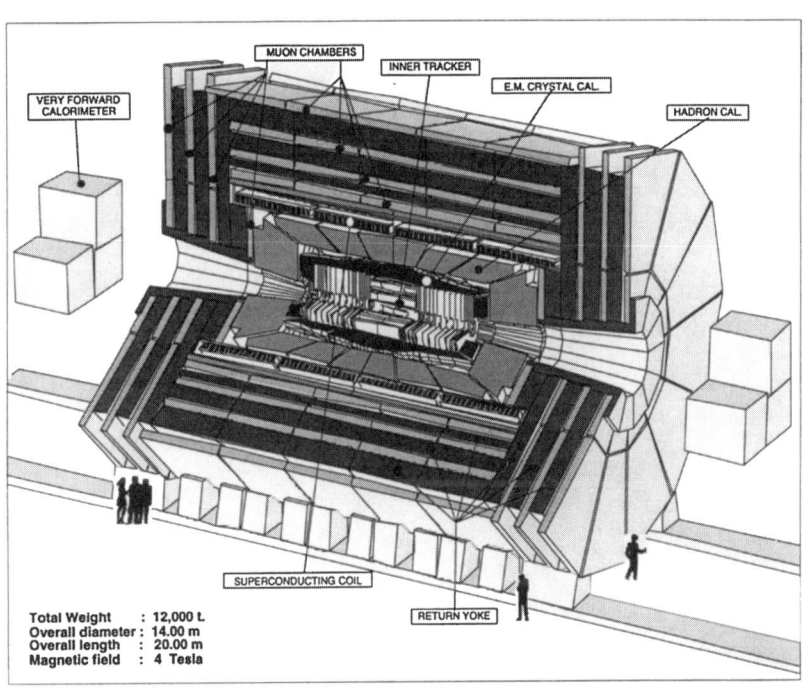

Figure 4. The CMS detector.

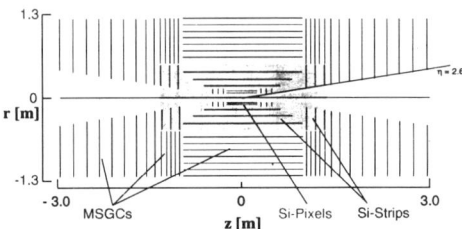

Figure 5. The CMS inner-tracking system.

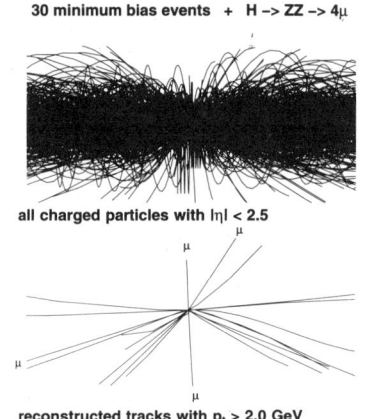

Figure 6. A simulated H → ZZ → 4μ event in the CMS inner tracker with pileup background for $L = 10^{34}\,\text{cm}^{-2}\,\text{s}^{-1}$: (a) all tracks, (b) tracks with $p_T > 2$ GeV.

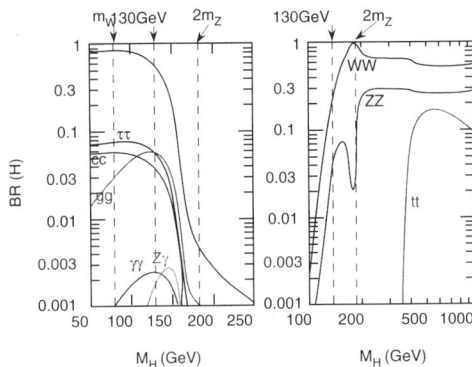

Figure 7. Branching ratios versus mass for SM Higgs decays.

Standard Model Higgs Boson Depending on the mass of the Standard Model (SM) Higgs boson, different decay modes will dominate as shown in Fig. 7. As already indicated above, not all decay modes would be visible above the background at LHC. Taking into account the backgrounds, the most promising channels for observing a Higgs boson are indicated in Fig. 8. Below a mass of ∼90 GeV, a SM Higgs boson should be detectable at LEP II. At LHC, the Higgs boson should be detectable via its decay into two Z bosons if its mass is larger than ∼130 GeV (one of the Zs may be virtual). For masses in the range 90–130 GeV, the two-photon decay is the most likely channel to give a significant signal. (The dominant $b\bar{b}$ final state is difficult to observe above the background from strong production of b-quark pairs.)

The dominant mechanism for Higgs-boson production for masses up to ∼700 GeV is gluon–gluon fusion (see Fig. 9). For higher masses, the W–W and Z–Z fusion mechanisms become important. Note that in the case of W–W and Z–Z fusion, the quarks that emit the W or Z bosons have typical transverse momenta $\sim m_W$. The corresponding jets may be detected in the forward directions ($2 < \eta < 5$) and used to 'tag' the reaction. In this way the signal-to-background ratio may be improved, allowing one to extend the mass range over which the experiments are sensitive up to ∼ 1 TeV. Note that this requires forward-jet detection.

Other processes that may be used to enhance the signal-to-background ratio are associated production of the Higgs with a W or with a $t\bar{t}$ pair. Consider the case $pp \to WH + X$, with $W \to \ell\nu$ and $H \to b\bar{b}$. The rate for associated production would only be a few percent of the total Higgs production rate. However, $BR(H \to b\bar{b}) \sim 100\%$ in contrast to $BR(H \to \gamma\gamma) \sim 10^{-3}$. The decay $W \to \ell\nu$ ($BR \sim 21\%$) would be used to provide a trigger and as a 'tag' for the Higgs production. Studies suggest that, while it will be difficult to establish a signal in the channel, it is not hopeless provided b-jets can be cleanly identified with good efficiency. This requires good b-jet tagging based on secondary-vertex reconstruction. As an example of what might be achieved, ATLAS estimates that a one-year run at $L = 10^{33}$ cm^{-2} s^{-1} would give a two-standard-deviation signal for $m_H = 80$ GeV, assuming 50% efficiency for tagging b jets and a factor 50 rejection against other jets.

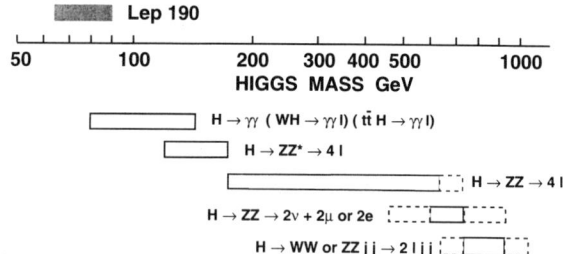

Figure 8. Decay modes for SM Higgs discovery at LHC (from Ref. [6]).

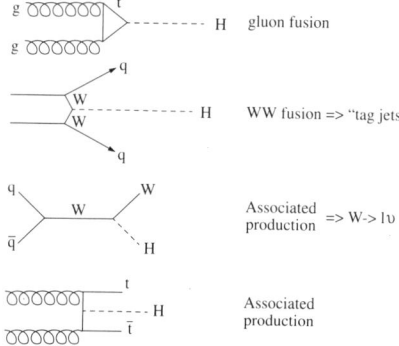

Figure 9. Diagrams of Higgs-boson production mechanisms.

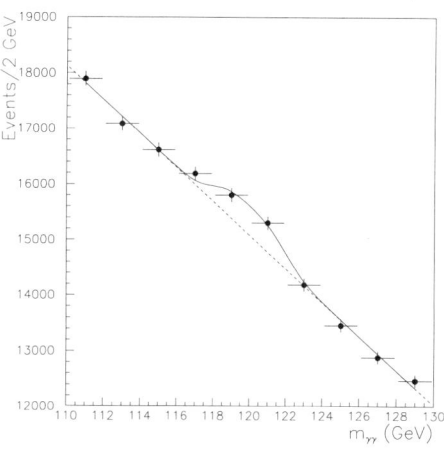

Figure 10. Simulated signal for H → γγ in ATLAS. The signal for $m_H = 120$ GeV is shown on top of the irreducible background from two-photon production, assuming an integrated luminosity of 10^5 pb^{-1}.

H → γγ The so-called 'intermediate-mass' Higgs boson ($m_Z < m_H < 2m_Z$) is narrow and the observed width of a signal would be determined by the experimental two-photon mass resolution. It is very important to have as good a mass resolution as possible because there is a large background from both real and fake photons. Note that good resolution in both energy and angle is required; the energy and angular resolution should be 'matched' in the sense that neither should dominate the mass resolution. The significance of the signal, defined as $N_S/\sqrt{N_B}$, is inversely proportional to $\sqrt{\sigma(m)}$ because the size of the search region, and hence N_B, is proportional to $\sigma(m)$. Note that the number of background events in the search region, N_B, is much larger than the number of signal events, N_S.

Each bunch crossing at the design luminosity will contain ∼25 interactions; the interaction region is about 5.6 cm long (rms). For the H → γγ search (at high luminosity), one will not be able to tell which of the many interaction vertices corresponds to Higgs production. Instead one requires precise measurements of the photon directions using the calorimeter system.

A number of backgrounds to the H → γγ signal have to be considered. The cross-section for jet production is enormously higher than that for photon production. It is therefore essential to have excellent photon–jet separation. The aim is to lower the 'reducible' (i.e. fake-photon) background well below the 'irreducible' one from genuine two-photon events. This requires a rejection factor of above ∼ 10^3 per jet; jets dominated by an isolated π^0 must also be rejected.

An example of what a Higgs signal might look like in this channel is shown in Fig. 10 (for $m_H = 120$ GeV with one year of running at design luminosity in ATLAS — note that this is the Higgs mass that gives the best significance for the H → γγ channel). Because of the large irreducible background, the signal-to-background ratio is only 1/30. Nevertheless, the significance of the signal is about six standard deviations.

Note that H → γγ events must be selected using a two-photon trigger (p_T threshold ∼ 20 GeV for each of the photons).

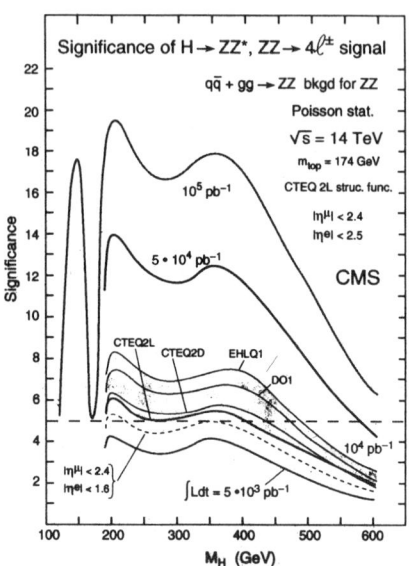

Figure 11. Expected significance for H → ZZ* → 4ℓ in CMS versus Higgs mass.

H → ZZ* → 4ℓ For Higgs bosons heavier than ∼ 130 GeV, the most promising channel is H → ZZ* → 4ℓ, where both the Zs decay into electron or muon pairs. If the mass of the Higgs is less than $2m_Z$, one of the Zs will be virtual. In this case, the leptons typically have quite low p_T (∼5–50 GeV). The natural width of such a Higgs boson would be small, as for H → γγ, and the observed width would be dominated by detector resolution; because of the small branching ratio to ZZ*, the rate would be low. For Higgs-boson masses above the ZZ threshold, the natural width becomes significant compared to the experimental resolution. Note that all four leptons should be isolated (i.e. they would not be inside hadronic jets, in contrast to leptons from b and c semileptonic decays, for example). Obviously, one has to cleanly identify real leptons and suppress fake ones; this is not expected to be a problem in ATLAS and CMS for high-p_T electrons and muons.

The major source of background to H → ZZ* → 4ℓ comes from $t\bar{t}$ production. Here, two of the leptons come from the Ws produced in the t-quark decays and are isolated; the other two come from semileptonic b decays (generally non-isolated). This background can be reduced by demanding that two of the leptons are consistent with the decay of a real Z. Hence, it is desirable to have a dilepton mass resolution comparable to or better than the natural width of the Z. Another source of background to H → ZZ* → 4ℓ is $Zb\bar{b}$ production where the b and $\bar{b}$ decay semileptonically. Both the above-mentioned backgrounds can be strongly reduced by requiring four isolated leptons. Further rejection against both backgrounds can be obtained by using vertex information to reject leptons from b decays; for Higgs decays, all of the leptons should be consistent with coming from the primary-interaction vertex.

Typically events in the H → ZZ* → 4ℓ channel are selected with a dilepton trigger (p_T threshold ∼ 20 GeV for each lepton).

The signal significance that can be expected in this channel as a function of m_H is shown in Fig. 11, for the example of CMS. A significance of more than five standard deviations is expected over almost the full mass range 130–600 GeV after one year of running at high luminosity (10^5 pb^{-1}).

In the case of a very heavy Higgs boson, the search in the H → ZZ → 4ℓ channel

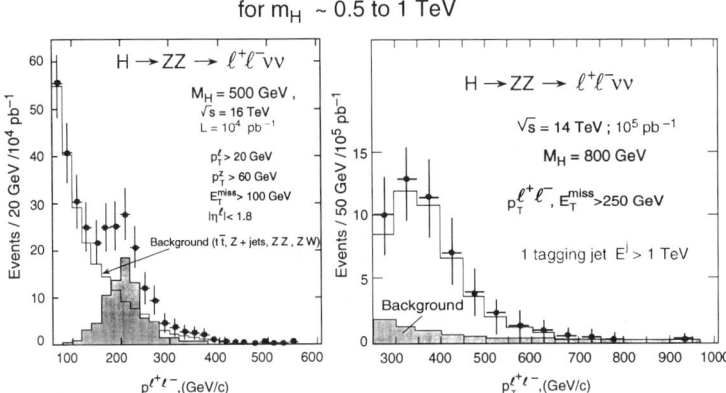

Figure 12. Simulated $H \to ZZ \to \ell\ell\nu\bar{\nu}$ signal for $m_H = 800$ GeV and 10^5pb^{-1}. One tagged jet with $E > 1$ TeV is assumed.

is limited by low statistics. One can extend the sensitivity to cover Higgs masses up to $m_H \sim 800$ GeV by using additional decay modes: $H \to ZZ \to \ell\ell\nu\bar{\nu}$, $H \to ZZ \to \ell\ell jj$ and $H \to WW \to \ell\nu jj$. Note that the branching ratios in these cases are significantly larger than for the $H \to ZZ \to 4\ell$ channel, by a factor of about six in the case of $H \to ZZ \to \ell\ell\nu\bar{\nu}$ and a factor of ~ 150 in the case of $H \to ZZ \to \ell\ell jj$ and $H \to WW \to \ell\nu jj$. Note that for such heavy Higgs masses, the decay products have very high p_T. Even heavier Higgs bosons (masses up to ~ 1 TeV) could be observed by using forward-jet tagging to enhance the signal-to-background ratio.

An example of a study by CMS using forward-jet tagging and $H \to ZZ \to \ell\ell\nu\bar{\nu}$ is shown in Fig. 12. Note that while the significance of the signal in terms of $N_S/\sqrt{N_B}$ is large, the shapes of the signal and background are similar, and establishing a signal would require a good understanding of the background normalisation.

SUSY 'SUSY and such' is covered in the lectures of S. Dawson. A large number of new particles are predicted, including SUSY partners for all of the known particles. The lightest SUSY particle (LSP) is often assumed to be stable and weakly interacting, in which case it will not be detected directly (neutrino-like particle). The LSP is expected to be produced in the decay chains of other SUSY particles. Hence its 'detection' due to the resulting large missing transverse energy is of crucial importance. The measurement of missing E_T requires (hadronic and EM) calorimeter coverage out to pseudo-rapidity $\eta \sim 5$. The calorimetry should be free of 'cracks' and be sufficiently deep to avoid non-gaussian tails on the missing E_T distribution for multi-jet events produced in QCD processes.

Jets and leptons may also be produced in the decays of SUSY particles. A combination of triggers will be required to select these events: jet trigger, missing-E_T trigger, lepton triggers and combinations thereof.

In SUSY models there are several Higgs bosons. For example, in the Minimal Supersymmetric Standard Model (MSSM), one has five Higgs bosons: H^0, h^0, A^0, H^+ and H^-. Here the masses and branching ratios depend on two parameters usually taken as $tan(\beta)$ and m_A. Depending on the values of these parameters which are poorly constrained by existing data, the experimentally easiest decay channels for the H^0 and h^0 search may have branching ratios smaller than in the case of the Standard Model Higgs. One may therefore have to search for decays such as $H \to \tau^+\tau^-$ that involve tau leptons. Another example is the search for top decays to charged Higgs ($t \to bH^+$ followed by $H^+ \to \tau^+\nu_\tau$. In view of these observations, the LHC detectors should be able to identify hadronic τ decays (using vertex information).

Top Physics The top quark has been observed at Fermilab (see lectures of M. Strovink) and its mass has been measured to be $m_t = 180 \pm 12$ GeV [8]. Very large numbers of top quarks will be produced at LHC, allowing a relatively precise measurement of the mass. One method considered for this measurement is to reconstruct the mass from the jets produced in the decay $t \to W b$; $W \to q\bar{q}$; here one would rely on a semi-leptonic decay of the other top quark in the event to provide the trigger. Reconstructing the mass in this way would require good jet resolution (energy and angle) and an excellent understanding of the calibration of the energy scale. In this analysis one would use b-jet tagging (based on vertex information) to identify the b jets from top decay.

B Physics An extensive programme of B-physics is planned for LHC, in a dedicated (optimised) experiment — LHC-B[5] — and also in the general-purpose ones — ATLAS and CMS. At $\sqrt{s} = 14$ TeV, the cross-section for B production is high ($\sigma(b\bar{b}) \sim 500$ μb) and the signal-to-background conditions are much more favourable than at lower energies: $\sigma(b\bar{b})/\sigma(tot) \sim 1\%$. Beauty events can be selected with high purity by requiring, for example, high-p_T leptons and displaced secondary vertices. High-p_T muons form the basis of the (first-level) B-physics triggers in ATLAS and CMS, while LHC-B uses also high-p_T hadrons.

The B-physics studies, which involve reconstructing multi-particle final states where the particles have modest transverse momenta, can best be done at relatively low luminosity where there is not much pileup. At $L = 10^{33}$ cm^{-2}s^{-1} the rate for $b\bar{b}$ production is ~ 500 kHz (500 μb$\times 10^{33}$ cm^{-2}s^{-1}); after requiring a muon from the semileptonic decay of one of the b quarks with $p_T > 6$ GeV (as in the ATLAS trigger, for example) the rate is still ~ 1 kHz.

A major aim of the B-physics programme is the measurement of CP violation (see lectures of S. Stone). The experimentally-easiest channel is the decay $B_d^0 \to J/\psi K_s^0$ which can be reconstructed exclusively (so-called 'gold-plated' channel for CP-violation studies). Another channel that is considered is $B_d^0 \to \pi^+\pi^-$. This is experimentally harder since one must rely heavily on secondary-vertex information to suppress the combinatorial background. Hadron identification is very important to separate the signal from decays such as $B_d^0 \to K\pi$ and $B_s^0 \to K^+K^-$ — this is foreseen in LHC-B but not in ATLAS or CMS which are optimised for 'high-p_T' physics. (Note that there are significant theoretical uncertainties, due to Penguin diagrams, in the interpretation of CP violation in the $B_d^0 \to \pi^+\pi^-$ channel.) Decays such as $B_s^0 \to \rho K_s^0$ will extremely difficult (if not impossible) to observe above the background at LHC.

Other aspects of B physics that one hopes to study at LHC include B_s^0 oscillations and rare-decay searches. Concerning B_s^0 oscillations, it is known from studies at LEP that the oscillation period is much less than the lifetime. Very good secondary-vertex resolution is therefore needed. In general, B-physics at LHC requires very good momentum resolution at $p_T \sim$ few GeV (needed to minimise background under mass peaks in exclusive decay modes).

Summary of Requirements on Detectors and the Trigger/DAQ System

- Fast detector response (25 ns bunch-crossing interval).

- Highly granular detectors (minimise effects of pile-up).

- Resistance to radiation damage (high radiation environment, mainly due to pp collisions).

- High trigger selectivity and also high efficiency for rare processes. The trigger must be based on photons, leptons, hadrons, jets, missing E_T and combinations thereof.

- Good photon identification, and photon energy and angle measurement (e.g. for H $\to \gamma\gamma$).

- Good electron and muon identification and momentum measurement (e.g. for H $\to$ ZZ* $\to 4\ell$).

- Isolation measurement of leptons and photons using calorimetry and tracking (e.g. to suppress background to H $\to$ ZZ* $\to 4\ell$).

- Good jet resolution (e.g. for top-mass measurement); note that linearity is also important (e.g. search for quark compositeness).

- Identification of b jets using vertex information for veto or tag (e.g. to suppress background to H $\to$ ZZ* $\to 4\ell$); decay-time measurements for B physics are also needed.

- Good missing E_T resolution — need calorimeter coverage to $\eta \sim 5$ (e.g. for SUSY studies).

- Forward jet detection (jet tagging of heavy Higgs production).

- Identification and measurement of hadronic tau decays (e.g. for Higgs decays in the MSSM).

- Good momentum resolution for hadrons (e.g. for reconstructing exclusive B-hadron decays such as $B_d^0 \to J/\psi K_s^0$).

Clearly, the benchmark channels discussed above represent only a small subset of physics processes that have been predicted for LHC energies. Very extensive studies have been performed in the context of the LHC collaborations and elsewhere. However, I hope that I have illustrated how one can gain an understanding of the required performance of the detector by considering the parameters of the LHC machine and by looking at which detector parameters limit the achievable precision of the measurements.

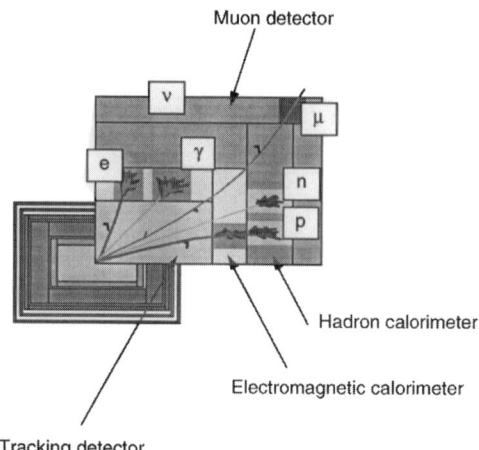

Figure 13. Illustration of identification of electrons, photons, hadrons and muons in a 'generic' detector.

Exercises

Why is there a dip in the branching ratio to ZZ* for $m_H \sim 160$ GeV (see Fig. 7)?

What are the detector requirements for studying the decays of new vector bosons, W' and Z', to lepton pairs?

GENERIC LHC DETECTOR (SUPERFICIAL VIEW)

The means by which different particle types are identified in a 'generic' LHC detector are shown in Fig. 13. Working outwards from the interaction point, the following functions can be identified:

- Momenta of charged particles are measured in the inner tracker (possibly also particle identification is performed, e.g. by Transition Radiation Detection in ATLAS).

- Primary and secondary vertices are reconstructed from track information, including precision 'vertexing' layers close to the beams.

- The energies and directions of electrons and photons are measured in EM calorimeters.

- The energies and directions of hadrons are measured in calorimeters (EM and hadronic).

- Muons are identified, and their momenta are measured, in the muon spectrometer.

- Forward calorimetry completes the missing transverse energy measurement.

In the following sections, each of the detector elements — inner tracking, calorimetry and muon detection — are discussed. More details on some of the detector technologies can be found in Refs. [9] and [10] and references therein.

INNER TRACKING

Inner tracking is required to measure the momenta of charged particles from the curvature of their tracks in a magnetic field. CMS uses a very large 4 T solenoid magnet (which also provides the bending for the muon spectrometer), whereas ATLAS uses a smaller 2 T solenoid (with a separate toroid magnet system for the muon spectrometer). In both cases the magnets use superconducting coils.

The choice of detectors for inner tracking at LHC is severely limited by the requirement to have a fast response time ~ 25 ns. For example, drift chambers used at LEP and elsewhere, such as the Time Projection Chamber, typically have a maximum drift time ~ 1–100 μs. Thus, although such detectors have good precision and give 3D information for complicated events, they are unsuitable for use in inner tracking for proton–proton experiments at LHC.

The requirements of radiation hardness, high granularity (to provide adequate pattern-recognition capability for the large number of tracks in each bunch crossing) and good momentum resolution further limit the choice of detectors. Note that one must be careful not to put too much material in the tracking volume, otherwise the resolution will be degraded unacceptably by multiple scattering. Other consequences of tracker material are electron bremsstrahlung and photon conversions. Electron bremsstrahlung reduces the efficiency for reconstructing electrons as well as degrading the resolution; this is important, for example, in the search for $H \to ZZ \to e^+e^-e^+e^-$. Photon conversions reduce the efficiency for reconstructing photons with obvious consequences for the $H \to \gamma\gamma$ search.

Two basic technologies are being used in the ATLAS and CMS inner tracking:

- Semiconductor detectors — used for the innermost part of the tracking in both experiments. Pixel detectors will be used in the part closest to the beam, with strip detectors at larger radii.

- Gaseous detectors with short charge collection times — MicroStrip Gas Chambers in CMS and straw tubes in ATLAS (MSGCs are discussed in much more detail in the lectures of R. Bellazzini).

In both cases, an enormous amount of work has taken place over the last few years to develop the new techniques required for LHC. One must consider radiation damage caused by ionising particles and that caused by neutrons (and other hadrons). The radiation doses are given in Gy for ionising radiation (absorbed dose 1 Gy = 1 joule/kg); for non-ionising radiation, the dose is often related to the equivalent dose for 1 MeV neutrons, expressed in neutrons per cm^2.

Semiconductor detectors

The principle of operation of a semiconductor detector is quite simple (see Fig. 14). When a charged particle traverses a reverse-biased p–n junction, ionisation is produced in the depletion region. This charge drifts in the presence of the electric field and is detected. Techniques developed in the micro-electronics industry allow structures to be implemented on semiconductor wafers, making finely-segmented strip or pixel detectors. The typical thickness of these detectors is about 300 μm, giving a charge-collection time < 25 ns.

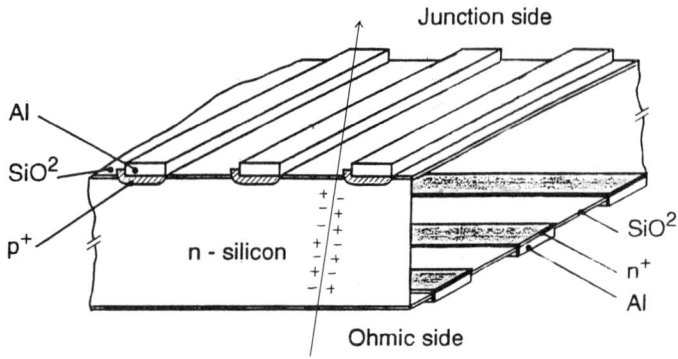

Figure 14. Principle of operation of a semiconductor detector.

By applying Gauss' theorem, it is easy to show that the voltage required for full depletion of a diode junction of thickness d and with charge density ρ (determined by the doping level) is given by $V_d = \rho d^2 / 2\epsilon_0 \epsilon_r$. For a typical detector thickness $d \sim 300 \mu m$ and currently-used doping levels, $V_d \sim 100$ V. Note that increasing ρ leads to an increase in V_d (as explained below, this is one of the effects of radiation). Also note that the signal increases with d, so one does not want to make the detectors too thin.

A very large effort has been made to understand the affect of radiation on semiconductor detectors. The most important effects stem from defects in the crystal lattice induced by hadrons. These result in an increase in leakage current for the detector and in the voltage required for full depletion. Consequences are a degradation in signal-to-noise ratio (increased shot noise) and higher operating voltages; the practical limit for the operating voltage is generally thought to be about 250 V. A further, dramatic consequence of irradiation at large doses is type inversion — the lightly-doped bulk material changes from n-type to p-type! This does not necessarily mean that the detector will stop working. However, when type inversion occurs, the junction moves from one side of the detector to the other, and this has to be taken into account in the detector design.

A detailed understanding of the fundamental processes involved in radiation damage of silicon detectors is not yet available. Nevertheless, based on phenomenological models, measures can be taken to reduce the effects of radiation damage. The amount of damage per unit dose depends on the temperature at which the detector is kept, both during operation and between exposures to radiation. If a detector irradiated at low temperature is warmed up, there is an initial improvement in the performance (annealing). However, this is followed by a more important degradation (reverse-annealing). By keeping the detectors at temperatures close to 0 C, the effects of radiation can be minimised. Operating the detectors at low temperature has the added advantage of reducing the leakage current which varies rapidly with temperature. On the other hand, novel cooling schemes are required — in ATLAS the use of 'binary ice' is considered, for example. (Binary ice is a suspension of water ice in an organic liquid.)

Semiconductor-strip detectors are typically made with a strip pitch $\sim$50–100 μm.

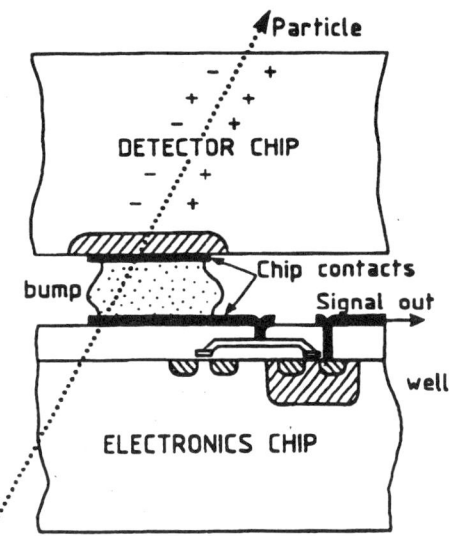

Figure 15. Illustration of bump bonding of pixel readout electronics to a pixel detector.

They may be read out using binary readout (strip hit or not hit), in which case the rms resolution is given approximately the pitch divided by $\sqrt{12}$. Alternatively, one may read out the pulse height on each strip, allowing reconstruction of the centre of gravity of the signal, improving the resolution for a given pitch. Note that the latter solution requires more complicated electronics than the former, with cost implications that may require a larger pitch to lower the channel count. Hence the choice is not an obvious one — ATLAS has adopted binary readout as its baseline, while CMS is considering an analogue solution. In practice, a resolution of ~ 15 μm is typical for silicon-strip detectors.

Semiconductor-strip detectors at LHC will have unprecedented numbers of readout channels ($\sim 10^7$ channels). It is not possible to move the corresponding amount of information off the detector each 25 ns; it is also not possible to make (or distribute) the first-level trigger decision in 25 ns. The solution is to place so-called 'pipeline memories' on the detector for each detector channel that store all of the data until the trigger decision is available (few microseconds). The information for rejected events is discarded, while that for retained events is read out for further analysis by multiplexing many channels onto a single data link, compressing the data by suppressing channels with no signal (the 'occupancy' is $\sim 1\%$). Of course, the electronics, like the detectors, must be radiation hard.

Semiconductor-pixel detectors at LHC will have even higher numbers of readout channels than strip ones ($\sim 10^8$ pixel channels). As an example, the CMS barrel pixel detector uses a pixel size of 125×125 μm^2. Here it will be necessary to integrate the detector and its corresponding electronics. This can be achieved by 'bump bonding' the electronics chip to the detector one. This is illustrated in Fig. 15.

The occupancy in pixel detectors is very small. For the readout, a technique called 'time stamping' is used, rather than storing all of the data in pipeline memories (almost all of the data would be zero). The basic principle is that each time a pixel is hit, the bunch-crossing number is recorded. When the trigger decides to retain an event for further analysis, pixels tagged with the corresponding bunch-crossing number are flagged as hit in the readout data. Practical implementations are more complicated than this, e.g. relying on the fact that, in a given bunch crossing, it is unlikely for more than one pixel in a column to be hit.

MicroStrip Gas Chambers (MSGCs)

MSGCs are covered in detail in the lectures of R. Bellazzini, so I shall only mention them very briefly here. MSGCs are gas ionisation detectors with small drift distances using electrodes deposited on, for example, glass plates. These will be used in CMS for the outer part of the inner tracker. Because of the short maximum drift distance ($\sim$ 2 mm), they have fast signal collection (collection time < 50 ns). The size of the electrodes can be made small (pitch $\sim$ 200 μm), so that good position resolution can be achieved ($\sim$ 50 μm for tracks at normal incidence).

An important issue for such detectors at LHC is radiation resistance; this depends on the choice of materials for the chamber construction and the gas mixture. A large amount of work has been invested to develop detectors that can survive the radiation levels at LHC.

Straw-Tube Detectors

Straw-tube detectors are cylindrical drift tubes with a small diameter ($\sim$4 mm). These will be used in ATLAS for the outer part of the inner tracker. The straws are arranged in layers, and each particle will traverse > 35 layers. Because of the small straw diameter, the maximum drift time is only 38 ns for the chosen gas mixture (Xe, CF_4, CO_2 in the ratio 70:20:10). By measuring the drift time, a position resolution per straw of about 150 μm can be achieved. While this is less precise than for the other detectors we have discussed, the worse precision per layer is compensated by the large number of measurements per track. The 'continuous tracking', i.e. many measurements along the length of the track, also provides good pattern-recognition performance. Tests have been made to check that these detectors can survive the radiation dose when operated at LHC over a 10-year period.

An interesting feature of the straw tracker in ATLAS is the inclusion of 'transition radiation' radiators, consisting of a plastic foam, between the straw layers. When ultra-relativistic particles (γ > 1000) pass between media of different refractive indices they may radiate 'transition radiation' (probability about 1% per boundary). Clearly, the only particles with such a high boost are electrons. The transition radiation is in the form of soft X-rays which may be detected in the straw tubes thanks to Xenon in the gas mixture which gives a high probability for conversion (photoelectric effect). The signature for transition radiation is a signal larger than that expected for a minimum-ionising particle and is detected by using a readout system with two thresholds. One can separate electrons from hadrons on the basis of the fraction of high-energy hits on the track. This is illustrated in Fig. 16 which shows results obtained with a detector prototype in a beam test.

For high-luminosity running, the ATLAS Transition-Radiation Tracker (TRT) will operate with a high fraction of hit tubes (more than 10% in some parts of the detector). Very extensive and detailed simulation studies have been made to check that track reconstruction will be possible under these conditions. Note that the drift-time information is used to reduce the probability for including spurious hits on the reconstructed tracks.

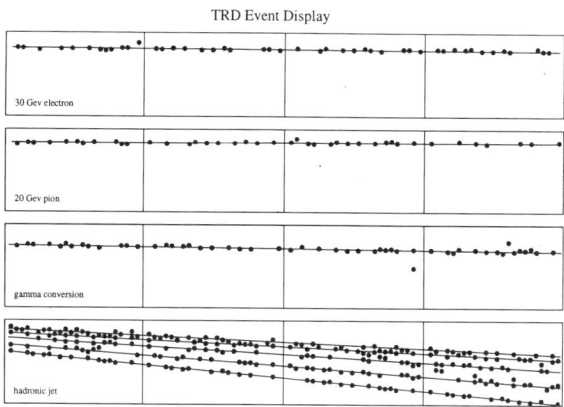

Figure 16. Electron and hadron tracks in a beam test of a prototype Transition Radiation Tracker module for ATLAS.

Tracker Performance

The performance of a tracker depends on the layout of the different detector layers, the magnetic field strength, the material distribution (including 'services' such as cooling), as well as on the spatial resolution in the individual layers. The following performance criteria must be considered:

- Resolution in p_T.

- Resolution in impact parameter (secondary-vertex resolution).

- Track-finding efficiency.

- Fake-track rate.

Of course, cost is also an important consideration.

At low p_T, the resolution will be dominated by multiple scattering which depends on the material distribution in the tracker volume; at high p_T the measurement will be limited by the detector resolution. Note that the multiple-scattering angle is given by $\theta \sim \frac{0.015}{\beta p}\sqrt{x/X_0}$. Here x/X_0 is the amount of material traversed, measured in radiation lengths, p is the momentum in GeV. Roughly speaking, if the extrapolation error from one plane to the next, $\sim \theta \cdot \delta r$ where δr is the difference in radius between the planes, is larger than the position resolution of the detector, the momentum resolution will be limited by multiple scattering. This behaviour can be seen in Figs. 17 and 18 for the example of the CMS tracker. One can see the transition from the multiple-scattering-dominated regime to the one dominated by detector resolution at $p_T \sim 20$ GeV.

Factors that affect the robustness of the pattern recognition are the number of points per track, the detector occupancy and the number of ambiguities, e.g. from stereo layers. Extensive studies have been made to check that track reconstruction will be possible up to the highest anticipated LHC luminosities. Similarly, it has been shown

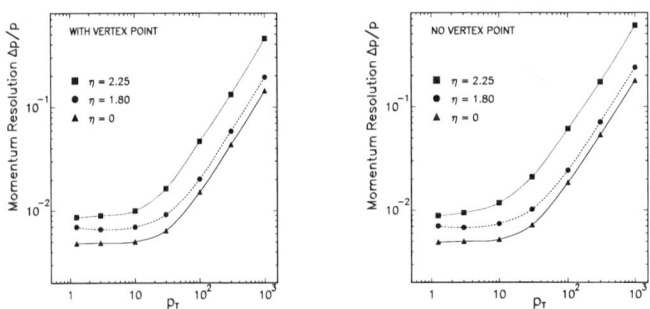

Figure 17. Simulation of CMS tracker: p_T resolution vs p_T.

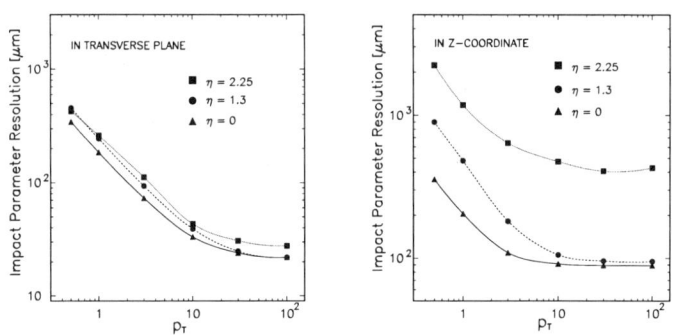

Figure 18. Simulation of CMS tracker: impact-parameter resolution vs p_T.

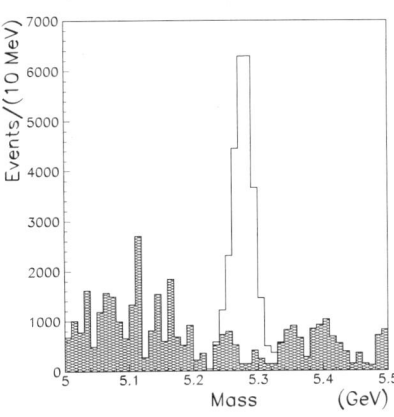

Figure 19. Simulated signal in ATLAS for $B_d^0 \to J/\psi K_s^0$ (for integrated luminosity of 10^4 pb^{-1}).

that b-quark jets can be tagged with good efficiency. For example, ATLAS estimates a factor 50 rejection against light-quark jets for a 50% efficiency for b-quark jets.

Let us review some of the requirements and see if they have been met:

- Fast detector response — achieved by the use of semiconductor, MSGC and straw-tube detectors.

- Highly granular detectors — achieved at the expense of very large numbers of detector channels.

- Resistance to radiation damage — achieved thanks to extensive R&D over the last few years.

- Lepton charge-sign determination up to very high p_T — achieved; the simulated resolution in the ATLAS and CMS detectors is $\delta p_T/p_T \sim 20\text{--}30\%$ at 1 TeV.

- Identification of b jets using vertex information to veto or tag the jets — achieved; the impact-parameter resolution is much smaller than the average impact-parameter value for tracks from B decays.

Note that inner tracking is essential for electron identification, as discussed later in these lectures. It is also used extensively in the second-level trigger.

In concluding this section, let us look at an example of a physics process that relies strongly on the inner detector. In Fig. 19 is shown the mass peak that one expects to see in ATLAS for the decay $B_d^0 \to J/\psi K_s^0$. By measuring the asymmetry between B_d^0 and $\overline{B}_d^0$ decays in this channel, it is hoped to study CP violation in the B system.

Exercise

The heavy-ion experiment ALICE plans to use a TPC (Time Projection Chamber). Why does this option work for ALICE and not for ATLAS or CMS?

CALORIMETRY

Calorimetry is required to measure the energy (and direction) of photons, electrons, isolated hadrons and jets, as well as the missing transverse energy (neutrinos, LSPs or other non-interacting particles). We distinguish between the electromagnetic (EM) calorimetry (electron and photon measurements) and the hadronic calorimetry. Combined with the inner tracking, calorimeter measurements are used for electron identification; here the hadronic calorimetry plays a role as a veto (for electrons, there should not be a significant signal in the hadronic calorimeter) and in isolation measurements. Calorimetry is also used for photon identification. Here the inner tracker is used as a veto.

Electron and photon identification are complicated by electron bremsstrahlung and photon conversions, respectively, in the beam pipe and tracker material. Algorithms have been developed to reconstruct these processes so as to minimise the loss of efficiency.

Electromagnetic Calorimetry

As discussed already in these lectures, the search for $H \to \gamma\gamma$ decays requires an EM calorimeter with excellent energy and angular resolution for photons. The choice of technology is limited by requirements of fast detector response and high granularity, essential if the performance is not to be degraded by pile up. Also, the calorimeter must be able to survive for the lifetime of the experiment in the high-radiation environment. The cost of the detector and its associated electronics (also photodetectors, etc.) must, of course, also be considered.

The basic principle of calorimetry is to measure the energy of an incident particle by total absorption. A fraction of the absorbed energy is converted into a measurable quantity such as light or charge. It is useful to recall how a shower develops when a particle interacts with matter. In the case of an incident EM particle, e.g. a photon, a shower of electrons, positrons and photons is produced; above ~ 1 GeV, photon conversion and bremsstrahlung are the dominant mechanisms. The development of an EM shower is illustrated in Fig. 20.

The characteristic length scale in EM shower development is the radiation length, X_0. This is defined as the distance over which an electron's energy is, on average, reduced to $1/e$ of its original value by bremsstrahlung. The radiation length of a material depends (amongst other parameters) on the inverse of the atomic number, Z, of the material — values of X_0 are 1.76 cm for iron ($Z=26$) and 0.56 cm for lead ($Z=82$). At low energies, the dominant energy-loss mechanism is ionisation — the critical energy, E_C, is defined as the energy below which energy loss by ionisation exceeds that from bremsstrahlung. The critical energy can be parametrized by $E_C \simeq (800 \text{ MeV})/(Z+1.2)$.

The average longitudinal shape of an EM shower can be described by $\frac{dE}{dt} = E_0 b \frac{(bt)^{(a-1)} e^{-bt}}{\Gamma(a)}$, where $t = x/X_0$ is the depth measured in radiation lengths, E_0 is the incident energy, and a and b are parameters that depend on the particle type (e or γ). Although the shower maximum typically occurs at $t \sim 5$, the calorimeter thickness must be greater than about 25 to avoid shower leakage that would spoil the energy resolution.

The lateral size of EM showers is an important parameter since it suggests a natural

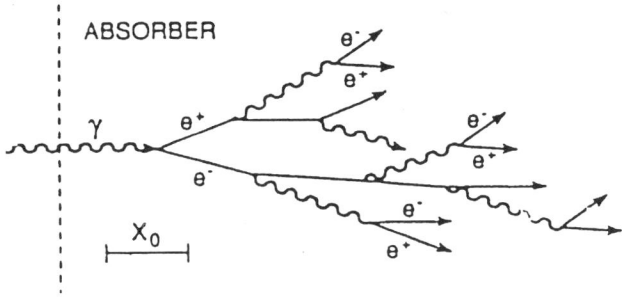

Figure 20. Illustration of EM shower development (from Ref. [10]).

limit to the calorimeter granularity — there is no point in making calorimeter 'cells' much smaller than the shower size. The finite width of the showers is due to multiple scattering of electrons which are deflected off the shower axis. The transverse size is generally expressed in terms of the Moliere radius given by $R_M \approx X_0(21 \text{ MeV})/E_C$ — typically an infinitely deep cylinder of radius $2R_M$ contains 95% of the shower energy. For an EM calorimeter, a typical value of the Moliere radius is $R_M \approx 2$ cm. In Fig. 21 is shown the average profile of EM showers in a calorimeter.

The most important parameter for calorimetry is the energy resolution which depends strongly on the technology used. The best resolution is obtained using 'homogeneous calorimeters' in which the particles release all of their energy in an active medium, for example scintillating crystals (CMS). Here the resolution is limited by statistical fluctuations in the way the shower develops and the way in which the electrons and positrons in the shower give rise to detected signals such as charge from a photodetector (shower $\rightarrow$ light $\rightarrow$ photoelectrons).

Good resolution can also be obtained using 'sampling calorimeters' in which the shower develops mainly in a passive medium, usually lead. The shower development is sampled at regular intervals with a detection medium, for example liquid-argon ionisation chambers (ATLAS). The resolution is worse than in a homogeneous calorimeter due to 'sampling fluctuations', i.e. fluctuations in the fraction of the shower seen in the active medium.

Because of the statistical nature of the processes described above, one finds that the resolution due to these effects scales with the square root of the energy: $\frac{\Delta E}{E} = \frac{a}{\sqrt{E}}$, where a is a parameter that depends on the calorimeter. Typically, one obtains $a \approx 10\%$ for a sampling calorimeter and $a \approx 2\%$ for a homogeneous one.

In an experiment at LHC, the so-called 'stochastic' term in the energy resolution described above is just one of three terms. We find $\frac{\Delta E}{E} = \frac{a}{\sqrt{E}} \oplus \frac{b}{E} \oplus c$. The second term results from 'noise' — it is due, for example, to electronic noise in the calorimeter readout; pileup (particles other than the electron or photon under consideration that deposit energy in the same calorimeter cells) also adds to this 'noise' term. The third term, called the 'constant term', results, for example, from non-uniformity in response

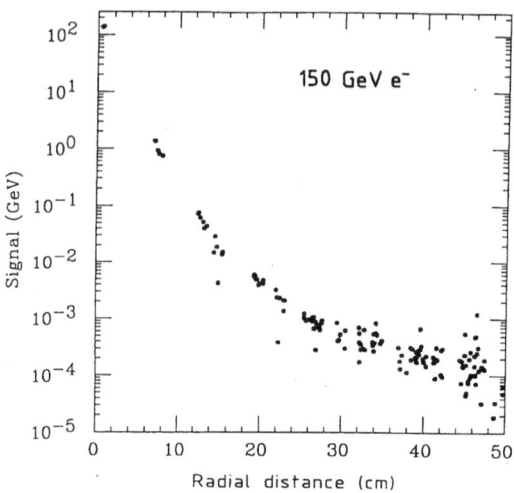

Figure 21. Lateral profile of EM shower (from Ref. [10]).

across calorimeter cells or from errors in the cell-to-cell calibration. The symbol '$\oplus$' indicates summation in quadrature.

The EM calorimeters at LHC have to measure the direction as well as the energy of photons, for example for the H $\to \gamma\gamma$ search. The principle for doing this is very simple: the shower centroid is determined as a function of depth, from which the direction of flight of the incident photon is inferred.

ATLAS EM Calorimeter ATLAS has chosen to use a lead–liquid-argon calorimeter. This has the advantages that it is radiation hard, has intrinsic stability of calibration (one simply measures the charge produced by ionisation in the liquid argon) and is easily segmented to the required cell size.

Conventional liquid-argon calorimeters have flat absorber plates oriented transverse to the direction of incidence of the particles. In ATLAS, a novel 'Accordion' geometry has been adopted in which corrugated absorber plates are aligned parallel to the incoming particles (see Fig. 22). This has the advantage that the electronic signals can be brought out on the front and back faces of the calorimeter, avoiding the need for dead regions occupied by electronics. It also allows one to make calorimeter cells with a truly projective geometry.

In my introduction I told you how important it is to use detectors with a fast response time. However, it is well known that the drift time in liquid-argon calorimeters is quite long (drift velocity ~ 5 mm/μs). Consider the case of ATLAS where the liquid-argon gap is about 2 mm — this gives a maximum drift time of $\sim$400 ns, much longer than the 25 ns bunch-crossing interval.

The current pulse produced in the calorimeter has a triangular shape (see Fig. 23) of the form $i = i_0(1 - t/t_d)$ where $t_d \sim 400$ ns is the maximum drift time. A 'trick' is used to derive a fast pulse starting from this pulse shape: the pulse shape is constant (neglecting noise), so the pulse amplitude can be inferred from the charge collected in the first few tens of ns. In practice, electronic shaping is used to transform the triangle-shape pulse to a bipolar pulse with rise time ≈ 40 ns (see Fig. 24). A disadvantage of fast shaping is that it adds electronic noise. This has to be balanced against the pileup

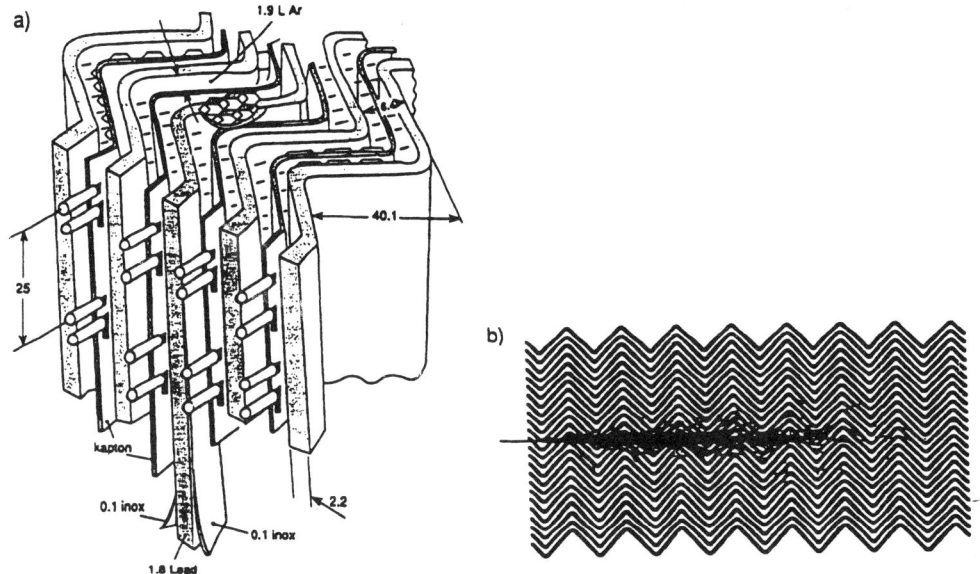

Figure 22. Illustration of Accordion calorimeter geometry (from Ref. [9]).

noise to find the optimum shaping time as shown in Fig. 25. Note that the optimum shaping time depends on the luminosity.

The validity of the techniques discussed above has been checked in test-beam measurements with the large prototype Accordion calorimeter which is illustrated in Fig. 26. Good energy resolution, with a stochastic term $9.8\%/\sqrt{E}$ and a constant term of 0.4% has been achieved. The noise term depends on luminosity due to the pileup; at low luminosity, where it is dominated by electronic noise, the constant term is 0.3 GeV.

The ATLAS EM calorimeter is divided into three samplings in depth, giving information about the longitudinal development of the shower and allowing a measurement to be made of the direction of photons. (In addition, there is a 'presampler' that makes measurements for showers that develop in the material upstream of the calorimeter.) The first depth compartment is segmented in strips of size $\Delta\eta \times \Delta\phi = 0.003 \times 0.1$, the second compartment is segmented in cells of size $\Delta\eta \times \Delta\phi = 0.025 \times 0.025$; $\Delta\eta \times \Delta\phi = 0.025 \times 0.05$ is used for the third compartment. Note that the inner radius of the calorimeter is at $R \approx 1.4$ m; the lateral size of the cells (in the middle compartment) is about 3.5 cm square, about twice the Moliere radius.

The direction of photons is obtained by measuring the shower centroid in the first compartment, and in the second and third compartments (combined). For 50 GeV photons as may be expected from $H \to \gamma\gamma$ decays, an angular resolution of about 5 mrad is obtained.

It is interesting to see how the different effects discussed above affect the physics, taking the example of $H \to \gamma\gamma$. The overall mass resolution is 1050 MeV at low luminosity and 1210 MeV at high luminosity. The various contributions to the resolution are shown in Table 1.

CMS EM Calorimeter CMS has decided to use scintillating-crystal calorimetry — this will give excellent resolution, with a stochasic term $\sim 2\%/\sqrt{E}$. A drawing of the CMS calorimeter is shown in Fig. 27. The barrel calorimeter alone weights 81 tonnes

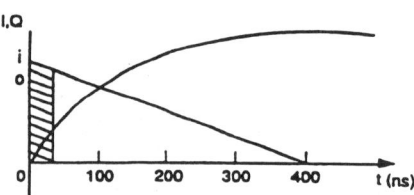

Figure 23. Pulse shape from liquid-argon calorimeter before shaping (from Ref. [9]).

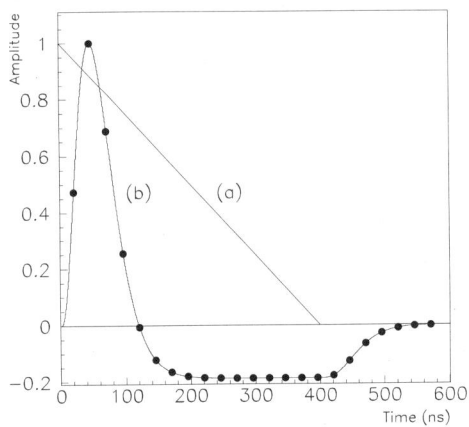

Figure 24. Fast pulse shaping of signals from a liquid-argon calorimeter.

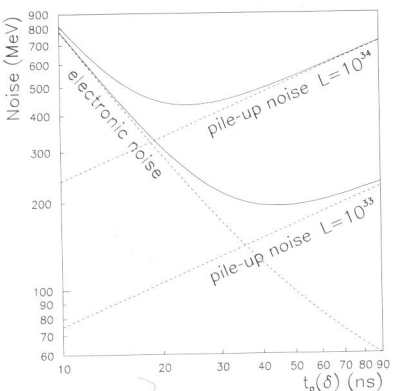

Figure 25. Electronic and pileup noise versus shaping time for the ATLAS liquid-argon calorimeter.

Figure 26. Photograph of prototype Accordion calorimeter.

Table 1. Contributions to the H → γγ mass resolution with the ATLAS detector. At low luminosity it is assumed that the photon directions are determined using the vertex position obtained with charged tracks from the underlying event. (From Ref. [10]).

	High luminosity	Low luminosity
Sampling term (MeV)	900	900
Constant term (MeV)	490	490
Total noise term (MeV)	500	210
Angular term (MeV)	400	70
Total (MeV)	1210	1050

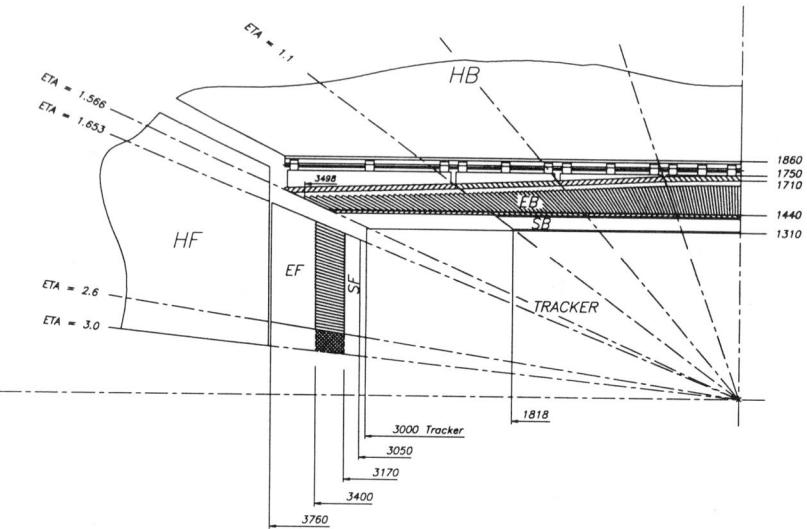

Figure 27. Drawing of CMS calorimeter.

and contains 93312 crystals. A silicon preshower detector is placed in front of the calorimeter for running at high luminosity. This is used in measuring the direction of photons.

The use of crystal calorimeters has already been demonstrated in large experiments, for example in L3 which has ~10000 BGO crystals. However, for LHC, it has been necessary to carry out an extensive programme of work to find new crystals that satisfy requirements such as:

- Fast response.
- Radiation hardness.
- Affordable cost.
- Sufficient light yield.

Many of the obvious candidates (thallium doped NaI and CsI, BGO, BaF$_2$) have too slow a decay time. Many are not sufficiently radiation hard for LHC, where the accumulated dose over the lifetime of the experiment will be up to 70 kGy (7 Mrad) at the shower maximum in the endcap calorimeter. CMS has decided to use PbWO$_4$; CeF$_3$ was also considered, but was rejected on the basis of practical considerations such as cost. Properties of a variety of candidate crystals are listed in Table 2.

Table 2. Properties of some scintillating crystals (from Ref. [10]).

	NaI(Tl)	CsI(Tl)	CsI	CeF$_3$	BGO	BaF$_2$	PbWO$_4$
Density (g/cm^{-3})	3.67	4.51	4.51	6.16	7.13	4.89	8.28
Radiation length (cm)	2.59	1.85	1.85	1.68	1.12	2.06	0.85
Moliere radius (cm)	4.8	3.5	3.5	2.63	2.33	3.39	2.2
Emission peak (nm)	410	560	420	340	480	300	450
			310	300		220	420
Decay time (ns)	230	1250	35	30	300	620	36
fast and slow if any			6	9		0.9	<10
Hygroscopic	yes	slightly	slightly	no	no	no	no
% of signal/K	0.22	0.1	0.1	0.15	−1.6	−2.0	−1.9
Light yield γ/MeV	4×10^4	5×10^4	4×10^4	2.10^3	8×10^3	10^4	1.5×10^2
Relative yield in pe	100	45	5.6	6.6	9	21	0.3
			2.3	2.0		2.7	0.2
Radiation hardness (rad)	10^2	10^3	10^5	10^7	10^2	10^3	10^7

An attractive feature of PbWO$_4$ is its very short radiation length (0.85 cm) which allows a calorimeter to be made with a relatively small volume of crystal. The decay time constant has a mean value of about 10 ns, and 85% of the light is collected within 25 ns. Drawbacks are a strong dependence of light yield on temperature (-1.9% per degree C) and a relatively small absolute light yield (compared to other crystals). Photomultipliers cannot be used to read out the calorimeter because of the high magnetic field (4 T), yet an efficient photodetector is needed. The solution is to use Avalanche PhotoDiodes (APDs) which have good photoefficiency, but, like the crystals, have a temperature-dependent response. The temperature of the calorimeter and the APDs will have be stabilised to a few tenths of a degree. Care has to be taken to use APDs that are not too sensitive to the passage of charged particles (the APDs will be mounted behind the crystals and hence exposed to a large flux of hadrons).

In Fig. 28 results are shown from a beam test of a 36-crystal prototype calorimeter read out with APDs. With a beam energy of 120 GeV, the resolution is 0.55%.

The crystal calorimeter has a single compartment in depth, yet at high luminosity one needs to measure the direction of photons in the calorimeter system. This is achieved using a silicon preshower detector placed after $3X_0$ of lead. Combining the information from the preshower detector and the calorimeter, the angular resolution is $\sigma(\theta) \approx 30$ mrad/$\sqrt{E} \oplus 4$ mrad. Note that the presence of the preshower material upstream of the crystal calorimeter significantly degrades the energy resolution.

As for ATLAS, it is interesting to see how the different effects discussed above affect the physics, taking the example of H $\to \gamma\gamma$. The overall mass resolution is 475 MeV at low luminosity and 775 MeV at high luminosity. The various contributions to the resolution are shown in Table 3.

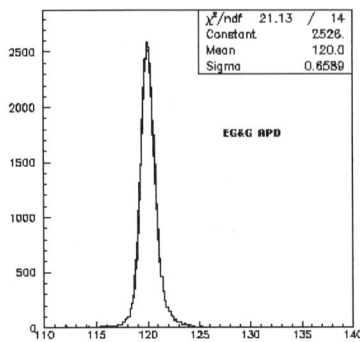

Figure 28. Results from a beam test of a prototype CMS PbWO$_4$ calorimeter (from Ref. [10]).

Table 3. Contributions to the CMS di-photon mass resolution for H $\to \gamma\gamma$ assuming $m_H = 100$ GeV. At low luminosity, the photon directions are obtained using the primary vertex position reconstructed with charged tracks.

	Low luminosity	High luminosity
Sampling term (MeV)	150	400
Constant term (MeV)	350	350
Noise term (MeV)	200	200
Pileup term (MeV)	–	200
Angular term (MeV)	200	500
Total (MeV)	475	775

Hadronic Calorimetry

Hadronic calorimetry is required to measure the energies and directions of jets and isolated hadrons, and for missing transverse energy determination. It also plays an important role in electron identification, providing a hadron veto and isolation measurement.

When a hadron interacts in matter it produces a shower of hadrons as indicated in Fig. 29. Part of the energy ($\sim$30% for 10 GeV incident energy) produces an EM component in the shower, mainly due to photons coming from the decays $\pi^0 \to \gamma\gamma$ and $\eta \to \gamma\gamma$. The interaction length, λ, defined as the mean free path between inelastic nuclear collisions, characterises the size of the hadronic shower. The depth of calorimeter required to contain a shower rises slowly with energy; for LHC, where hadrons may be produced with large energies, a depth of about 10λ is required. In the transverse direction, about 95% of the energy is contained within a radius of 1λ. Values of λ are given in Table 4 for some materials used in hadronic calorimeters.

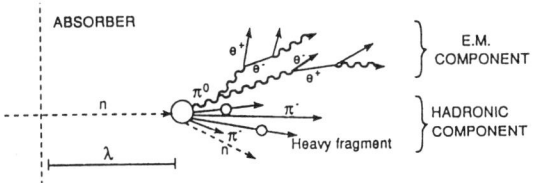

Figure 29. Illustration of hadronic shower development (from Ref. [10]).

Table 4. Interaction length for various materials.

Material	Density [g/cm^3]	λ [g/cm^2]	λ [cm]	X_0 [cm]
Fe	7.87	131.9	16.8	1.76
Cu	8.96	134.9	15.1	1.43
W	19.3	185	9.6	0.35
Pb	11.35	194	17.1	0.56

ATLAS Hadronic Calorimeters ATLAS will use an iron–scintillator sandwich calorimeter in the barrel region where radiation levels are relatively low, and (more expensive but more radiation hard) liquid-argon calorimetry in the endcap region. The barrel calorimeter, illustrated in Fig. 30, uses a novel geometry in which the scintillator tiles are oriented approximately parallel to the incident particles. The light is read out via wavelength-shifting fibres that pass along groves in the edges of the scintillator tiles. The calorimeter is divided into three compartments in depth with a cell size $\Delta\eta \times \Delta\phi$ = 0.1 × 0.1 (0.2 × 0.1 in the last compartment).

An energy resolution for pions of $\frac{\Delta E}{E} = \frac{45\%}{\sqrt{E}} + 1.3\%$ has been obtained in beam tests, including the energy measured in the liquid-argon EM calorimeter. The linearity of the hadron calorimeter has also been checked. The expected resolution for jets is $\frac{\Delta E}{E} = \frac{35\%}{\sqrt{E}} + 1.0\%$ (the EM component in jets is well measured in the EM calorimeter). Very recently a full-size prototype calorimeter module (see Fig. 31) has been successfully tested with beam.

The liquid-argon endcap hadronic calorimeter uses (non-magnetic) copper plates as the absorber with a conventional parallel-plate geometry.

CMS Hadronic Calorimeter CMS will use a copper–scintillator sandwich calorimeter in which wave-length shifting fibres are embedded in the scintillator plates.

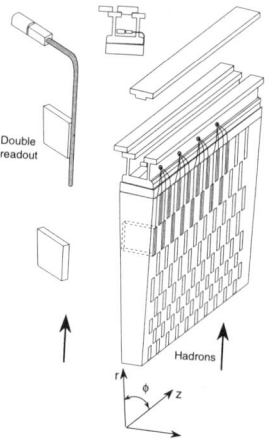

Figure 30. Illustration of ATLAS Tile calorimeter.

Figure 31. Photograph of 'Module-0' of the ATLAS Tile calorimeter.

Since the calorimeter is inside the coil, a non-magnetic absorber is required. Copper has the advantage of a shorter interaction length than steel, optimising shower containment for the limited available space within the coil. The expected performance for jets is about $\frac{\Delta E}{E} = 60\%/\sqrt{E}$ with a 3% constant term.

Comment on Resolution for Jets It should be noted that many factors other than the intrinsic calorimeter resolution affect the jet energy resolution. For the measurement of low-p_T jets at high luminosity, pileup falling within the jet cone has an important effect, for example.

Forward Calorimetry

Forward calorimetry is important for both missing transverse energy measurement and for forward jet tagging. In both ATLAS and CMS the calorimeter coverage extends to pseudorapidity $\eta \sim 5$. ATLAS uses a tungsten–liquid-argon calorimeter integrated in the endcap at about 5 m from the interaction point (the first section in depth, mainly sensitive to EM showers, is made of copper instead of tungsten). CMS uses quartz-fibre forward calorimeters.

Comments on Photon (and Electron) Identification

The performance of the ATLAS and CMS calorimeters in terms of energy resolution (also photon angular resolution) has been discussed above. In conjunction with the inner tracker, the calorimeters play an essential role in photon and electron identification. They are also used to study the isolation properties of electrons, photons and muons. (Isolation is a measure of the amount of activity around a particle; this can be used to distinguish, for example, leptons from decays of new particles from leptons produced in jets (e.g. from b-quark decays).)

As an example, consider photon identification in ATLAS. The EM cluster is measured in a $(\eta \times \phi)$ region of 3×7 calorimeter cells. Using the EM and hadronic calorimeters to make cuts on isolation and to veto on hadronic energy behind the EM cluster, photons can be selected with good efficiency while obtaining a rejection factor of ~ 1000 against jets. Further rejection can be obtained by using the profile of the cluster in the preshower detector. In particular, the two photons from $\pi^0 \to \gamma\gamma$ decays may be resolved. By demanding a single, narrow cluster in the preshower detector, an additional background-rejection factor of 3 is obtained. The inner tracker is used to separate photons from electrons, requiring the absence of an associated charged track for the former case. (For electrons it is relatively easy to reject the background. Here one demands consistency between the momentum measured in the inner tracker and the energy measured in the calorimeter. In ATLAS, the TRD signature is also used.)

Note that, in practice, the situation described above is complicated by bremsstrahlung of electrons and conversions of photons in the beam pipe and tracker material.

Concluding Remarks on Calorimetry

Let us check that the required performance has been met in the ATLAS and CMS calorimeters (not full list):

- Fast detector response — achieved by various means, e.g. fast shaping in the ATLAS liquid-argon calorimeters and choice of crystals in CMS.

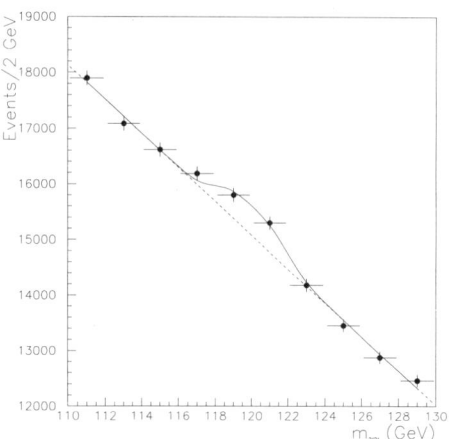

Figure 32. Simulated H → γγ signal in ATLAS.

- Highly granular detectors — achieved at expense of having > 100000 channels in the EM calorimeters.

- Resistance to radiation damage — achieved by, e.g., choice of radiation-hard crystal in CMS and liquid-argon technology in ATLAS.

- Good photon identification, and good photon energy and angle measurement — achieved.

- Good electron identification and energy measurement — achieved.

- Isolation measurement of leptons and photons using calorimetry — achieved.

- Good jet resolution and linearity — achieved.

- Good missing E_T resolution — achieved by including forward calorimeter coverage to $\eta \sim 5$.

- Forward jet detection — achieved thanks to forward-calorimeter coverage to $\eta \sim 5$.

In concluding this section on calorimetry, let us consider the example of the (Standard Model) H → γγ signal that might be observed in ATLAS for $m_H = 120$ GeV, assuming an integrated luminosity of 10^5 pb^{-1} (roughly one year at $L = 10^{34}$ cm^{-2} s^{-1}). While the signal-to-background ratio is small, the significance is 5.9 standard deviations (see Fig. 32).

CMS have a somewhat better sensitivity than ATLAS in the H → γγ channel thanks to their superior energy resolution, although this is may be partially offset by a lower signal efficiency due to the difficulty of reconstructing photon conversions in the 4 T magnetic field. Note that $m_H = 120$ GeV is the easiest case for the H → γγ search — towards lower masses the background increases while the acceptance falls, reducing the significance to 3.9 for $m_H = 100$ GeV and 2.5 for $m_H = 90$ GeV. Several years of running might be required to obtain a convincing signal in this very difficult channel.

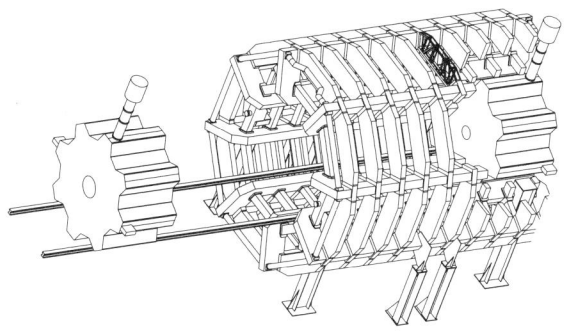

Figure 33. ATLAS muon spectrometer.

Exercise

ATLAS uses tungsten as the absorber material for its forward calorimeter. What are the advantages of using this (relatively expensive) material?

MUON-DETECTION SYSTEM

The muon detection systems have to identify muons and measure their momenta; precise momentum determination is required up to very high p_T. The systems must also provide a means to trigger on muons. It is considered important for the muon systems to be able to make measurements independently of the inner tracking. ATLAS and CMS have adopted very different approaches for the spectrometer magnets, although they use similar detector technologies.

ATLAS Spectrometer

ATLAS uses a very large system of air-core toroid magnets for its muon spectrometer. Note that ATLAS stands for 'Advanced Toroidal LHC ApparatuS'. The ATLAS spectrometer is illustrated in Fig. 33. The basis of the momentum measurement in ATLAS (in the barrel part) is to measure a position on the muon trajectory before and after the magnet, and a third point between the other two. One can understand the basis of the method in a simplified example, assuming constant magnetic field (in a toroid, the field falls as $1/R$ where R is radius).

It is easy to show that the sagitta, s, for a track of momentum p in a uniform magnetic field is given by $s \approx \frac{B\ell^2}{27p}$ (B in Tesla, s and ℓ (the track length) in meters, p in GeV). Very approximately, taking $\ell = 4$ m and $B = 0.8$ T for the ATLAS toroid, we see that the sagitta for a 1 TeV muon is about 0.5 mm. This immediately shows us that if we wish to measure with reasonable precision the momenta of multi-TeV muons, the sagitta must be measured with a precision < 100 μm. This is not easy given the

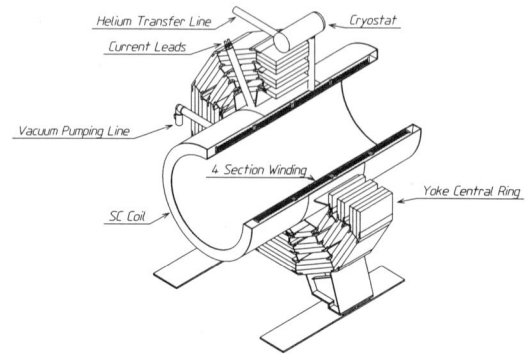

Figure 34. CMS magnet.

physical size of the system and the need to achieve alignment between chambers at this level of accuracy.

One can see from the simple formula $s \approx \frac{B\ell^2}{27p}$ why a very large magnet is required — the precision of the measurement improves as $B\ell^2$. An average field of $B = 0.8$ T over a large volume is not easy to achieve. By considering Ampere's theorem, one can easily see that about 2.5×10^6 Ampere-turns are required per coil for eight coils. This can be achieved using superconducting coils, but the ATLAS magnet represents an enormous engineering challenge.

An advantage of using an air-core magnet is that the amount of material seen by the muons between measurement stations is small, so there is relatively little multiple scattering. Rather precise momentum measurement is therefore possible using only the external muon system — this is called 'stand-alone' measurement. Note that the muon system is well shielded from the interaction region, so it should be relatively immune to unforeseen background conditions that might affect the inner tracking.

For low muon momenta, the momentum measurement in the external spectrometer is not as good as the one in the inner tracker. One of the reasons for this is that one must correct for the energy loss of the muon as it passes through the material of the calorimeters. Fluctuations compared to the average energy loss limit the precision of the estimate of the muon momentum at production. In practice, the inner-tracker and external measurements will be combined to give the best possible resolution.

CMS Spectrometer

CMS uses a very large, high-field, superconducting solenoid that provides bending power for the inner tracker measurement (electrons and hadrons as well as muons) and for the muon spectrometer. Note that CMS stands for 'Compact Muon Solenoid'. The CMS magnet is illustrated in Fig. 34. The field strength of the solenoid is 4 T, the radius is about 3.5 m and the full length of the magnet is about 13 m.

The trajectory of a muon is measured in the inner tracker and as it passes through the magnet return yoke (see Fig. 35). Note that, neglecting multiple scattering and

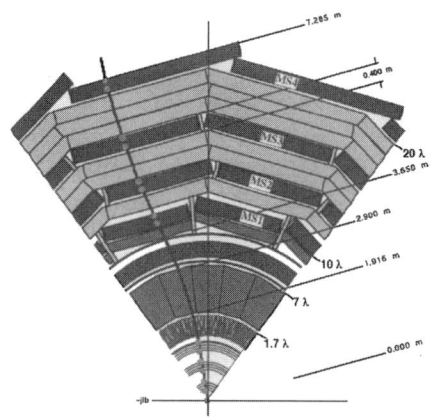

Figure 35. Cross-section of CMS spectrometer.

energy loss, the muon trajectory beyond the return yoke extrapolates back to the beam line due to compensation of the bending before and after the coil (straight-line extrapolation).

A 'stand-alone' measurement of momentum can be made by using information from the muon detectors outside the coil, together with the last layer of the inner tracking (in which comparatively low background levels are expected) and the known position of the beam line.

The stand-alone muon-momentum resolution of the ATLAS and CMS spectrometers is compared in Fig. 36. At moderate p_T (~100 GeV), the air-core toroid of ATLAS gives superior resolution (little multiple scattering). At higher p_T (~ 1 TeV) the resolution of the detectors dominates the error on p_T, and the advantage of the higher field strength in CMS can be seen.)

In order to understand in a very rough way the muon-momentum resolution of CMS, it is useful to consider a simplified example in which the change in angle is measured for a muon traversing a region of length ℓ and uniform field strength, B, filled with material of radiation length X_0. We have for the average multiple-scattering angle $\theta_{MS} \sim \frac{0.015}{\beta p}\sqrt{\ell/X_0}$ and for the bending angle $\theta_B = \ell/R$ where $R = \frac{p}{0.3B}$, i.e. $\theta_B = 0.3B\ell/p$. Here, the units are meters, Tesla and GeV, and angles are measured in radians. Combining these formulae, and assuming that multiple scattering dominates over measurement errors, we obtain $\frac{\delta p}{p} = \frac{\theta_{MS}}{\theta_B} = \frac{0.05}{B\beta\sqrt{X_0 \ell}}$, independent of p. Taking $B \approx 2$ T (saturated iron in the return yoke), $X_0 = 0.018$ m and $\ell \approx 5$ m, we obtain 8%, comparable with the value ($\approx 6\%$) obtained by CMS from detailed simulation which allows for the detector layout, the use of the beam spot, etc.

Muon Detectors

Two kinds of muon detectors are required at LHC, serving complementary functions: drift chambers that provide very accurate position measurements for use in the momentum measurement, and 'trigger chambers' that have less precise position resolution, but have a short response time (< 25 ns). The latter are required to uniquely identify the bunch crossing that gave rise to the muon and are used in the first level of triggering. The former have a response time much longer than the bunch-crossing period of LHC, but provide high-precision measurements over a large surface area.

Although the muon detectors are rather well shielded from the interaction region by the calorimeters, many sources of background to the prompt-muon signal have to be considered. These include muons from pion and kaon decays and shower leakage.

An important source of 'noise' in the muon detectors is that caused by neutrons in the cavern. These low-energy neutrons may produce hits in the chambers when they collide with protons in the gas. They may also produced photons if they are captured by nuclei — these photons may then induce hits in the gas. In ATLAS, the rate of neutron- and photon-induced hits (obtained by convolving the fluxes of neutrons and photons with their probabilities to produce hits) may be as high as 100 kHz/cm² in the forward region. Clearly this limits the choice of detectors to ones that can run with good efficiency when operated at very high rate. A consequence of such considerations is a change in chamber technology between the barrel, where rates are relatively low, and the endcap where rates are highest.

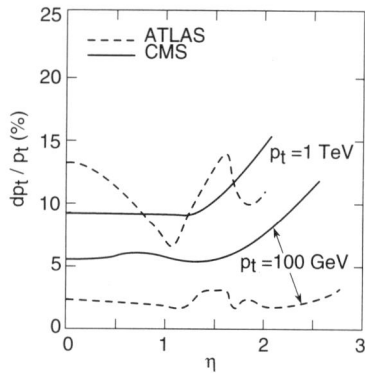

Figure 36. Comparison of ATLAS and CMS stand-alone muon momentum resolution (from Ref. [6]).

It is interesting to see where the neutrons are produced and how they reach the cavern. This is studied in great detail by Monte Carlo simulation in order to understand the source of the background and to find ways to improve the shielding. An example of a plot from such a study is shown in Fig. 37.

Note that, in order to avoid fake tracks due to accidental coincidences of noise hits, each measurement station has several layers of detectors and a local coincidence is required between layers.

There was not time in these lectures to go into the details of the different muon-chamber technologies being used at LHC. Very briefly, in ATLAS, the precision chambers are Monitored Drift Tubes of cylindrical geometry, except in the forward regions where Cathode Strip Chambers are used to cope with the high background rates; for the trigger chambers, Resistive Plate Chambers are used in the barrel and Thin Gap Chambers are used in the forward region. The technologies used in CMS are rectangu-

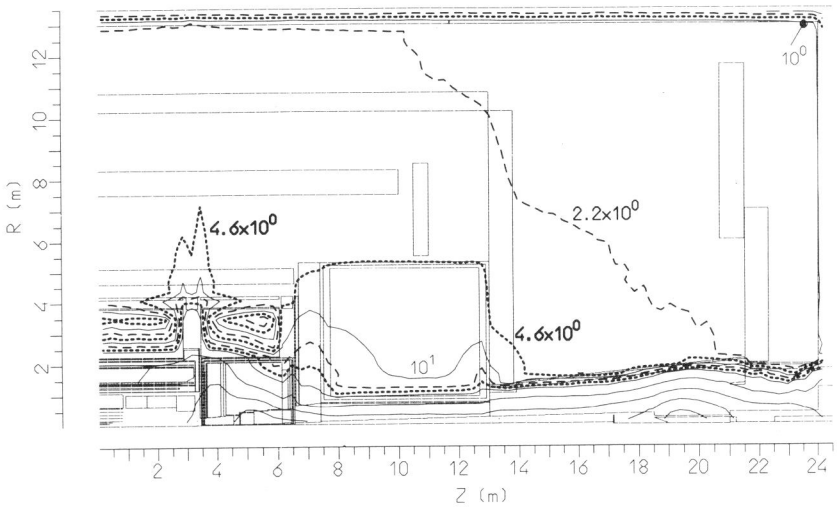

Figure 37. Map showing neutron flux in ATLAS (kHz/cm²).

lar drift tubes (so-called 'DTBX' chambers) and Cathode Strip Chambers; RPCs are used in the trigger.

Let us check that the requirements have been met in ATLAS and CMS (not full list):

- Fast detector response — achieved for trigger chambers. (The precision chambers have drift time ∼ few hundred ns, but this is acceptable given that they are well shielded from the interaction region.)

- Highly granular detectors — achieved at expense of $> 10^6$ channels in the ATLAS muon system, for example.

- Operation in high-radiation environment — achieved for the selected technologies which work at high rates.

- Good muon identification and momentum measurement — achieved $\frac{\delta p_T}{p_T} \sim$ few % at 100 GeV.

- Lepton charge-sign determination up to very high p_T — achieved $\frac{\delta p_T}{p_T} \sim$10% at 1 TeV.

In concluding this section on muon detection and measurement, let us look at an example of a physics process that is sensitive to the performance of the muon system: H → ZZ* → 4μ. The mass peak that might be observed is shown in Fig. 38 for the example of ATLAS.

Exercise

What is the stored energy in the CMS solenoid and how does it compare with the kinetic energy of an aircraft flying at cruising speed?

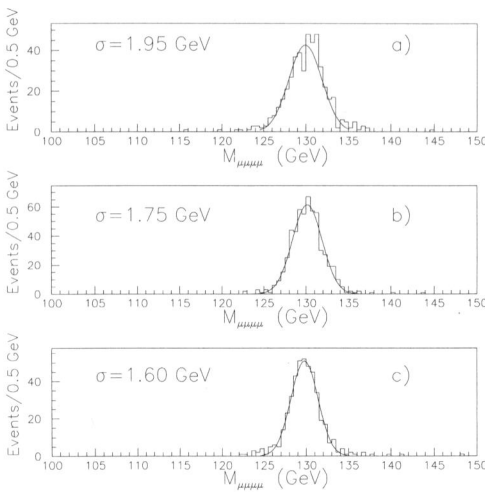

Figure 38. Simulated H → ZZ* → 4μ signal for $m_H = 130$ GeV in ATLAS: (a) stand-alone measurement, (b) measurement using only the inner tracker, (c) combined measurement.

TRIGGER AND DATA-ACQUISITION SYSTEMS

The role of the trigger and data-acquisition system is to select bunch crossings containing interesting interactions and to record the corresponding data on permanent storage. This is an extremely challenging task at LHC because of the following:

- The short bunch-crossing period (25 ns) — this is much shorter than the time required to make the first-level trigger decision.

- The fact that each bunch crossing contains ~25 interactions. The pile-up events add to the volume of data to be read out and complicate the task of recognising signatures of interesting interactions.

- The very high required trigger selectivity — the interaction rate of $\sim 10^9$ Hz (bunch-crossing rate 40 MHz) has to be reduced to about 100 Hz to be recorded on permanent storage.

- The need to select with high efficiency the events associated with rare physics processes.

- The need to select events with complicated signatures, based on high-p_T muons, electrons/photons, jets and large missing transverse energy at the first level, and more complex criteria in the higher-level triggers.

ATLAS and CMS both rely on multi-level trigger systems — two levels in the case of CMS and three in the case of ATLAS. In the following, we discuss the ATLAS trigger and data-acquisition (T/DAQ) system first as an example, and then describe some of the differences for CMS.

A functional view of the ATLAS T/DAQ system is shown in Fig. 39. The first-level (LVL1) trigger works on a subset of information from the calorimeter and muon detectors (trigger chambers) only. It requires about 2 μs to reach its decision, including

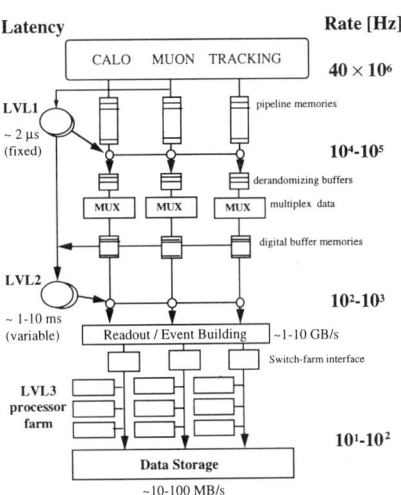

Figure 39. Functional view of the ATLAS T/DAQ system.

the propagation delays on cables between the detector and the underground counting room where the trigger logic is housed. All of the information from the detector must be stored in 'pipeline memories' until the LVL1 decision is available. The LVL1 trigger reduces the rate from 10^9 interactions per second to ~ 100 kHz. The LVL1 trigger searches for candidate high-p_T muons, electrons/photons, hadrons (e.g. from hadronic decays of τ leptons) and jets, and large missing transverse energy. Events (i.e. bunch crossings) are selected on the basis of combinations of these signatures. The LVL1 trigger has to select unambiguously the bunch crossing containing the interaction of interest.

For events selected by the LVL1 trigger, the information from the detector must be retained for further analysis. Data from the pipeline memories are transferred first to 'derandomizing' memories that can accept the very high instantaneous input rate (LVL1 can accept several events within the space of a few bunch crossings, even if the average rate is much lower than this). The derandomizing memories are then emptied at the lower average rate, via readout drivers (RODs) to readout buffers (ROBs); in this process, many individual channels are multiplexed over a single readout link into the ROBs, in many cases applying zero-suppression or data compression. The data for LVL1-selected events remain in the ROBs during the 'latency' of the second-level (LVL2) trigger (i.e. until the LVL2 decision is available).

The data in the ROBs can be accessed selectively by the LVL2 trigger which uses regions of interest defined by the LVL1 trigger. The LVL1 system identifies the locations in η–ϕ space of candidate muons, electrons/photons, jets and hadrons, including candidates of low p_T not actually used in making the LVL1 accept/reject decision. The LVL2 trigger refines the selection of candidate 'objects' compared to LVL1, using full-granularity information from all detectors, including the inner tracker which is not used at LVL1. In this way, the rate can be reduced to ~ 1 kHz. Many events are analysed concurrently by the LVL2 trigger system using processor farms and an average latency of up to ~ 10 ms is considered reasonable.

For LVL2-selected events, 'event building' is performed. Each ROB contains fragments of many events for a small part of one subdetector. The event builder collects

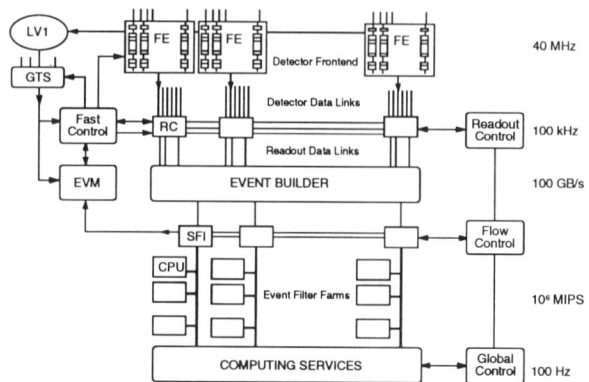

Figure 40. Functional view of CMS T/DAQ system.

all the fragments from one event into a single memory — the memory of a third-level (LVL3) trigger processor. The event building is performed using a data switch, somewhat similar to a telephone exchange; in fact, one may make use of switching technologies from the telecommunications industry.

LVL3 processing is performed using farms of processors acting on the full-event data. Because the input rate is only about 1 kHz, one can envisage using algorithms similar to those used offline, with complicated event-selection criteria. The amount of processing time per event could be about 1 second on a 1000 MIPS (Million Instructions Per Second) processor (today's processors are typically 100–200 MIPS).

The trigger and data-acquisition system of CMS has a few differences compared to ATLAS; a functional view of the CMS system is shown in Fig. 40. Examples of differences compared to ATLAS are:

- The LVL1 trigger latency is somewhat longer. This is due to a number of factors — the most important ones are longer cable routings due to the detector design and location of the counting room, and the use of muon drift chambers in the LVL1 trigger (drift time ∼400 ns).

- There are not separate LVL2 and LVL3 trigger systems and the region-of-interest mechanism is not used — instead there is a 'virtual LVL2 trigger' that accesses initially only the calorimeter and muon-detector data, rejecting most of the events without having to use data from the inner tracker.

First-Level Calorimeter Trigger

In both ATLAS and CMS, the LVL1 calorimeter trigger processors are digital and act on reduced-granularity information. Trigger towers are formed with a granularity of about $\Delta\eta \times \Delta\phi = 0.1 \times 0.1$, separately in the EM and hadronic calorimeters. The total number of trigger towers is about 8000, each giving an 8–10 bit E_T value every 25 ns. This represents a massive amount of data that has to be transmitted to the

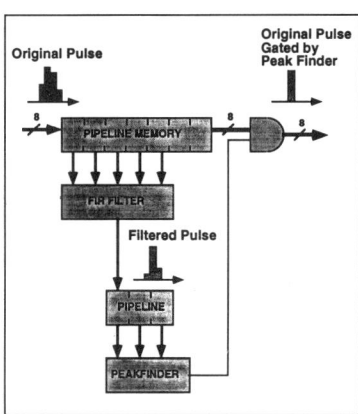

Figure 41. Bunch-crossing identification algorithm.

trigger processor — in total $\sim$ 3000 Gbits/sec. This is achieved by using many high-speed serial data links in parallel (e.g. $\sim$4000 links, each carrying data for two trigger towers). Work is in progress to achieve the density of inputs required for a reasonably compact trigger-processor system.

Although the pulses from the ATLAS and CMS calorimeters are quite fast, they extend over several bunch crossings. It is, of course, essential that the trigger always selects the right bunch crossing to be read out, not one too early or too late. This is achieved through bunch-crossing identification logic associated with each trigger tower.

Digitisation may be performed at 40 MHz rate on the full calorimeter granularity in which case a digital sum can be made to form the trigger towers — this is the case for CMS. In ATLAS it is under discussion whether to form the trigger towers by analogue or by digital summation.

An example of bunch-crossing identification is illustrated in Fig. 41. A digital filter is applied to the data from the trigger tower, modifying the pulse shape; a peak-finding algorithm is applied to the resulting pulse; non-zero data are sent to the trigger processor only for 'in-time' pulses, corresponding to the peaks of the filtered pulses.

The calorimeter trigger processors of ATLAS and CMS perform relatively simple algorithms to identify candidate high-p_T electrons/photons, jets, hadrons and large missing transverse energy. Given that all of the data for all detector channels have to be retained until the LVL1 trigger decision is available, it is important to make the processing as fast as possible. This is achieved by using custom-made hardware processors made with application-specific integrated circuits (ASICs).

As an example of a LVL1 calorimeter trigger algorithm, consider the ATLAS electron/photon trigger, illustrated in Fig. 42. Within a 4×4 trigger-tower window, one requires a pair of EM towers containing a high-E_T cluster, with little E_T in the surrounding ring of EM cells or in the 16 hadronic cells contained in the window; the cluster, isolation and hadron-veto thresholds are all programmable. CMS uses a similar algorithm based on a 3×3 window (Fig. 43).

The above algorithms are quite simple, but they have to be performed every 25 ns for each of $\sim$ 4000 window positions; this represents a massive computing task. The

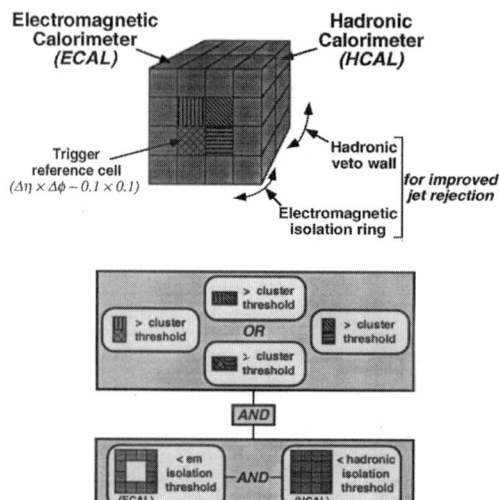

Figure 42. ATLAS first-level electron/photon trigger algorithm.

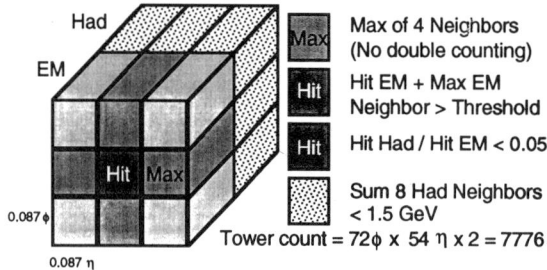

Figure 43. CMS first-level electron/photon trigger algorithm.

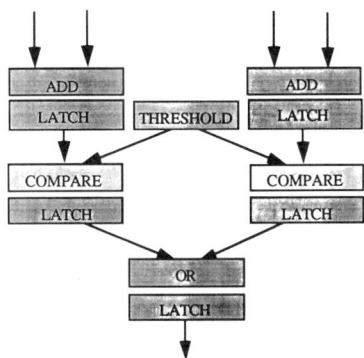

Figure 44. Illustration of pipelined and parallel processing.

solution is to use pipelined and parallel processing, as illustrated in Fig. 44. Pipelined processing means that one organises the processing logic in a chain of operations to be performed one after another for each event, with no processing loops. Each processing element in the chain performs its function in 25 ns and passes its results to the next element in the chain. The processing latency is then the number of steps in the chain multiplied by 25 ns. Data corresponding to successive bunch crossings follow each other down the processing 'pipe'.

Parallel processing means that many processing elements act in parallel, for example performing the same operations on different data. In the ATLAS and CMS calorimeter trigger processors, the processing of different EM cluster windows is carried out in parallel (there are ~ 4000 such windows); parallel processing is also employed within the processing elements associated with each window.

It is worth noting that the amount of data to be processed in parallel decreases as one works through the processing chain. Initially one has ~ 8000 bytes of data (one byte per trigger tower); finally, after combining results from the calorimeter and muon processing one has just one bit — yes, accept the event or no, reject it.

First-Level Muon Trigger

The first-level muon triggers in ATLAS and CMS are based on dedicated trigger chambers, as described in the section on the muon detection system. In CMS, but not in ATLAS, information from the drift chambers is used in addition.

The concept of the muon trigger is to define a 'road' from one station of muon chambers to the next. The width of the road depends on the desired p_T threshold; it is calculated allowing for magnetic deflection and multiple scattering.

The use of drift chambers is possible in CMS, at the expense of increasing the trigger latency, by exploiting the geometry of the 'DTBX' (Drift Time / Bunch Crossing) drift chambers. These rectangular drift tubes, arranged with alternate layers offset by half the tube width, provide an easy way to extract the bunch crossing of interest, as illustrated in Fig. 45. For constant drift velocity, v_d, the drift times in tubes 1 and

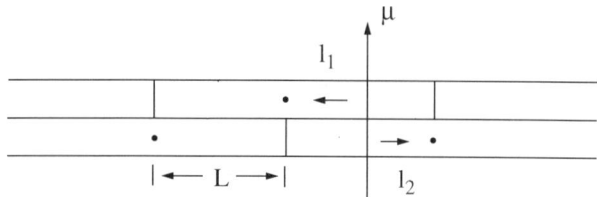

Figure 45. Illustration of the use of DTBX chambers to identify the bunch crossing containing the muon track.

2 are $t_1 = \ell_1/v_d$ and $t_2 = \ell_2/v_d$ respectively. Assuming tracks at normal incidence, $\ell_1 + \ell_2 = L$ and $t_1 + t_2 = L/v_d = t_{max}$ (constant).

The above concept can be extended to handle inclined tracks by combining information from three layers of drift tubes. In CMS, four layers are used; hits are required in at least three of the four layers, allowing some room for chamber inefficiency. An advantage of this method is that it provides relatively precise position (and angular) information for use in the subsequent muon-trigger processing. However, given the ~ 400 ns maximum drift time in the DTBX detectors, it is more sensitive to noise hits than methods based on fast detectors such as RPCs; multiple hits may give rise to ambiguities.

In both ATLAS and CMS most of the muon-trigger logic will be mounted on or near the detector, inaccessible during running. Given the huge detector area and the significant delays due to time of flight, etc., it will be a challenging task to set up the timing of the system.

Higher-Level Triggers and Data Acquisition

The LVL1 trigger systems of ATLAS and CMS will reduce the event rate to ~ 100 kHz. This rate is still far too high for recording on permanent storage given that the event size is expected to be ~ 1 MByte. Further selection is therefore needed. It is estimated that ~ 100 events per second can be recorded, so an additional rejection factor of ~ 1000 is required after LVL1.

It is thought that much (maybe all) of the trigger processing after LVL1 can be done using 'farms' of standard processors of the type used in workstations or PCs, assuming that current trends in performance versus price continue over the next few years. For example, ~ 1000 processors could be used to handle the ~ 100 kHz input rate, allowing 10 ms of processing time per event. It is reasonable to expect an order-of-magnitude increase in computing power per processor between now and the time when the processors will have to be purchased. Hence, relatively complicated algorithms can be considered.

The challenge of high-level triggering for LHC is not just the need for massive computing power. Perhaps more difficult are the tasks of selecting and moving the required data to the relevant processors, and of controlling such a large multi-processor system.

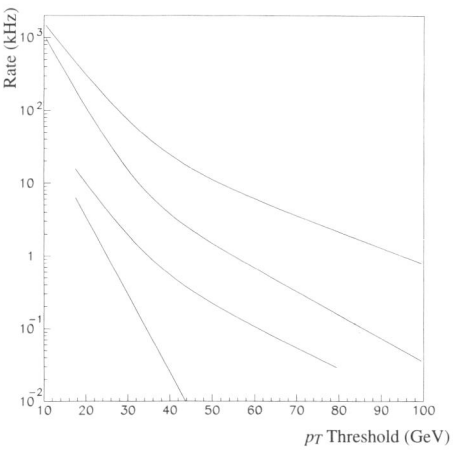

Figure 46. The electron trigger rate versus E_T threshold for $L = 10^{34}$ cm^{-2} s^{-1}. The uppermost curve is the LVL1 rate without using isolation. The next two curves are the LVL1 rate after isolation cut and after the LVL2 calorimeter trigger, respectively. The bottommost curve shows the LVL2 electron-trigger rate after using inner-tracker information in addition.

The full event data represent about 1 Mbyte; following a LVL1 trigger, these data are scattered in a large number of buffer memories (called ROBs in ATLAS). Neither ATLAS nor CMS plan to do full event building at 100 kHz rate which would require a network bandwidth of 100 GBytes/sec ($\sim$ 1000 Gbits/sec). Instead, the LVL2 trigger is performed using a subset of the data. Even so, very high performance switches will be needed.

As an example, consider one of the candidate technologies for the switch: ATM. This is a switching system developed for the telecommunications industry. Packets of data, carrying an address, can be transmitted by the network at a speed of 155 Mbits/sec (622 Mbits/sec also becoming available). One sees immediately why it is necessary to reduce the total bandwidth significantly below 1000 Gbits/sec in order to avoid the high cost of a switch with many thousands of ports.

A critical issue in the high-level trigger and data-acquisition systems will be error detection and recovery. Of course, we will design systems as best we can to avoid errors, but with such large systems operating at such high speeds, error conditions will be a fact of life. Error recovery must be designed into the systems from the start, and deadtime and loss of data due to error conditions must be minimised.

Trigger Performance

Studies have been made to check that acceptably low rates can be achieved while retaining high efficiency for the (predicted) physics processes that one hopes to search for and study at LHC. In fact, the trigger cuts are usually less restrictive than the ones that give optimal significance in the analysis.

As an example, consider how the single-electron trigger rate can be reduced between the LVL1 and LVL2 trigger in ATLAS (see Fig. 46). The basis of the LVL1 trigger is to require E_T in a pair of trigger towers in excess of a threshold. Isolation can also be required at LVL1 reducing the rate by an order of magnitude. The LVL2 trigger has access to full-granularity calorimeter information using which the rate can be reduced by another order of magnitude. Finally, guided by the position of the calorimeter cluster, a track search can be performed in the inner tracking. Requiring a track consistent with the calorimeter cluster and the electron hypothesis brings an additional rejection factor of ~ 10.

Exercise

How does the data rate into the ATLAS or CMS first-level calorimeter trigger processor compare with the total world telephone traffic (i.e. voice traffic)?

REFERENCES

1. The Large Hadron Collier, CERN/AC/95-05.
2. ATLAS Technical Proposal, CERN/LHCC/94-43, 15 December 1994.
3. CMS Technical Proposal, CERN/LHCC/94-38, 15 December 1994.
4. ALICE Technical Proposal, CERN/LHCC/95-71, 15 December 1995.
5. LHC-B Letter of Intent, CERN/LHCC/ 95-5, 25 August 1995.
6. N. Ellis and T.S. Virdee, Annu. Rev. Nucl. Part. Sci. 1994, 44:609-53.
7. M. Della Negra, Open Presentation to Large Hadron Collider Committee, CERN, January 1995.
8. Review of Particle Properties, Phys. Rev. D 50:1173-1826.
9. C. Fabjan, Detectors for Elementary Particle Physics, Proc. 1993 European School of High-Energy Physics, CERN 94-04, pp. 77-116.
10. D. Fournier and L. Serin, Experimental Techniques, Proc. 1995 European School of High-Energy Physics, CERN 96-04, pp. 291-363.

ACKNOWLEDGEMENTS

I would like to thank Tom Ferbel for giving me the opportunity to lecture at the 1996 ASI. I would also like to thank Daniel Fournier and Chris Fabjan for allowing me to reproduce figures from their lecture notes.

STATISTICAL ISSUES IN DATA ANALYSIS

Harrison B. Prosper

Department of Physics, Florida State University
Tallahassee, FL, 32306

INTRODUCTION

"We don't know all about the world to start with; our knowledge by experience consists simply of a rather scattered lot of sensations, and we cannot get any further without some *a priori* postulates. My problem is to get these stated as clearly as possible."

— Sir Harold Jeffreys, in a letter to Fisher dated 1 March 1934

My first encounter with statistical analysis was probably typical of the experience of many of us: it was a cursory exposure to a subject that seemed opaque, arid and dull. A necessary evil to be learned quickly then confined to a backwater of the mind only to be dredged up from time to time to compute a standard deviation or two. I don't blame my teachers, who for the most part were excellent. They too probably suffered in their youth. Then one day I found myself confronted with data for which I could find nothing sensible in any textbook I looked at[1]. I even delved into the mammoth tomes of Kendall and Stuart[2]. I scoured the bit of the statistical literature that physicists know about. But I had no success. I was forced, therefore, to go back to first principles—and the more I learned the more I came to realize that very little in statistical analysis is obvious or inevitable; there was no such thing as the "right" answer unless, first, it was specified what "right" meant. The subject was married to controversy and its handmaiden, confusion. In short, the subject was dynamic and interesting.

And then I was invited to deliver these lectures. Here was a wonderful opportunity to try to convey to a vigorous new generation of physicists the idea that statistical analysis is an important thing that is worth the effort to understand well. It can also be a lot of fun! I certainly had a lot of fun telling you about it. And though I came to St. Croix with the brazen ambition to "corrupt the youth" I'm happy to report that the youth got the upper hand!

The problem of statistical inference has been solved in two radically different ways. The first approach dates back to the time of the English cleric Thomas Bayes (1702

– 1761) and the great mathematical physicists of the 18th and 19th centuries Daniel and Jacob Bernoulli, Laplace, Gauss and others. During the past forty years or so their approach, called **Bayesian probability theory**, has undergone an impressive renaissance that, in the 1930s, was spearheaded by Sir Harold Jeffreys[3] (1891 – 1989). His important ideas lay dormant for many years until they were vigorously revived by many mathematicians and physicists, most notably Cox, de Finetti, Lindley, Savage and Jaynes.

In Bayesian probability theory probability is regarded as a measure of the *plausibility* of some assertion. Therefore, statements like "There is a 20% chance of rain tomorrow" make sense. Bayes and Laplace, were they around today, would interpret the 20% as a measure of the plausibility that the statement about rain tomorrow is true.

Towards the middle of the last century, this notion of probability and its associated methods fell into disfavor because of the apparent arbitrary nature of the assignment of prior probabilities and the perceived subjectivity of the Bayesian mode of reasoning. To correct these alleged defects an entirely different approach to inference was developed (largely by geneticists and biologists) at the start of this century, referred to variously as **classical**, **orthodox** or **frequentist** statistics. The Bayesian view of things was all but abandoned. The newer approach comprises the body of statistical ideas with which most of us are familiar. Many of the ideas were developed by Sir Ronald Aylmer Fisher (1890 – 1962). Important contributions were made by Neyman, Pearson, Cramer, Rao, von Mises, Kolmogorov and others.

Although it is common practice to lump these thinkers together it should be recognized that they did not all share the same point of view. There were severe disagreements between Jerzy Neyman and Egon Pearson on the one hand and Ronald Fisher on the other. Sir Ronald, in particular, pounded Bayesians and fellow frequentists alike with great vigor. While Neyman conceded, albeit reluctantly, the limited nature of his own approach to inference, Fisher believed he had solved the problem in its full generality and refused to accept that his approach raised as many problems as it purported to solve. Neyman and Pearson stressed the importance of specifying alternative hypotheses, when testing a particular hypothesis; Fisher never accepted this. And there were other disagreements. But the frequentists did agree, more or less, on one thing: that probability is to interpreted not as a measure of plausibility or degree of belief but rather as the **relative frequency** with which something happens. Statements like "There is a 20% chance of rain tomorrow" did not make sense to Fisher. Why? Because it is not possible, he would argue, to repeat that particular day lots of times and count how often it rained. However, the statement "There is a 20% chance of rain on days labelled by the date July 14th" is considered meaningful because July 14th happens often and we can, therefore, assess the relative frequency with which it rains on days so labelled. We have inherited this kind of reasoning in experimental physics. The following is a typical example: were we to repeat our experiment many times a certain statement would be true 90 % of the time, therefore, the probability is 90% that for the experiment we actually performed the statement is true. This kind of reasoning is so deeply ingrained in us that at least one physicist[4] has declared, in effect, that any self-respecting physicist will not knowingly use a statistical method that does not conform to this reasoning. I beg to differ; we physicists are more open-minded than this author suggests.

Sir Ronald was a great scientist who was very influential in shaping the debate about statistical inference. This is partly the reason why many universities, including my alma mater, do not teach the ideas of Bayes and Laplace even though, as physicists, we have a closer affinity with them than we do perhaps with Sir Ronald who was, after all, a geneticist. Some of us have been trying to correct this. And what better way to do so than to talk to a captive audience of some of our brightest young physicists from around the world!

I would, however, have been remiss had I ignored Neyman, Pearson and Sir Ronald, since their point of view currently prevails. More to the point, since their point of view prevails it is important to understand what precisely they have said. I decided, therefore, to divide the lectures equally between frequentist and Bayesian ideas. The first two lectures cover frequentist statistics and in the remaining lectures I give an introduction to Bayesian probability theory and its application. These notes have been re-arranged slightly with respect to the lectures delivered at St. Croix and reflects some of the interesting feedback I got from some students.

I have tried in these notes to offer clear precise explanations of some of the key statistical concepts, rather than give yet another data analysis manual. There are plenty of those already. In the last lecture, I describe two examples of Bayesian analysis that I hope you will find instructive.

At the end of these lecture notes there is a short bibliography of books and articles that you might find helpful to deepen your understanding of the topics discussed in the lectures. As a very good introduction to orthodox statistics I recommend highly Roger Barlow's book[5]. Although it is aimed primarily at (British) undergraduates his book covers many advanced topics in just enough detail to be useful. Also, please peruse the wonderful book by Prof. E. T. Jaynes who, displaying great generosity, has made it available on the Internet[6].

If sometimes, in these notes, I seem a tad provocative, be assured that it is just a useful pedagogical device to stimulate discussion! So, without further ado, let's begin.

BASIC NOTIONS

"Thou shalt not sit with statisticians nor commit a social science."

— W. H. Auden

One goal of science is to infer facts about the world through a combination of theory and experiment. The facts about the world are usually expressed as the parameters of a mathematical model. A model may have one parameter, or many parameters. But, even when a model has many parameters, our interest is generally focused on only one or two of them, for example the top quark mass. The background rate or the jet energy scale in our experiment may be part of the model but they are not usually thought to be of deep scientific interest.

We can never be certain that our model is correct nor are we privy to the exact values of its parameters. Instead, what we have is direct knowledge of a vector $\mathbf{x} = (x_1, \ldots, x_N)$, comprising N data, which we have collected in the hope that it will help us decide two things: 1) which of several competing models is best and 2) the values of its parameters. The first task is called **model selection** or hypothesis testing; the second is called **parameter estimation**. To execute these tasks sensibly we should first agree on some self-consistent rules and then agree to apply them consistently.

Unfortunately, the choice of rules has occasioned much controversy during the past two hundred years, and it is as well to recognize that there is no single right way of doing things. Nonetheless, some rules are better founded than others, as I shall try to make clear in these lecture notes.

One of the rules, which happily everyone accepts, is that the connection between the parameters Θ of the model being considered and our data $\mathbf{x}$ is a **probability**, $P(\mathbf{x}|\Theta)$. That symbol is to be read as "the probability of $\mathbf{x}$ given Θ". This is a conditional probability, as indeed are *all* probabilities. In general, the probability can be written as

$$P(\mathbf{x}|\Theta) = \sum_{\mathbf{z} \in \Omega} f(\mathbf{z}|\Theta), \tag{1}$$

where the sum is over a neighborhood Ω of the point $\mathbf{x}$. If the point $\mathbf{x}$ is a member of a continuous set the sum should be replaced by an integral. We often choose the neighborhoods to be very small in which case we can write

$$P(\mathbf{x}|\Theta) = f(\mathbf{x}|\Theta)\, d\mathbf{x}, \tag{2}$$

if $\mathbf{x}$ forms a continuous set, or

$$P(\mathbf{x}|\Theta) = f(\mathbf{x}|\Theta), \tag{3}$$

if $\mathbf{x}$ is discrete. If $\mathbf{x}$ is continuous the function $f(\mathbf{x}|\Theta)$ is called a **probability density function** (pdf).

The number $P(\mathbf{x}|\Theta)$ is the probability that we assign to our data given specified values of the parameters and assuming a particular model. In going from the parameters of a model to data we are performing **deductive** reasoning. That's what theorists do.

Statistical analysis solves the inverse problem: we have data and we want to infer something about the correctness or otherwise of the model and the values of its parameters. That is, when we do statistics we are engaged in **inductive** reasoning. That's what experimentalists do and even some theorists. But how do we assign these probabilities and, having assigned them, what are we to do with them? The only things that are undeniably objective are the data we have acquired; they exist, after all, on our computers. Everything else in statistical analysis is subjective. It is so because everything else depends upon which set of rules we consider to be reasonable. What is considered reasonable by one person may be considered contrived or absurd by another. There are no laws of nature to which we can appeal. It is precisely this aspect of statistical analysis that causes controversy.

Given a **sample** of data $\mathbf{x} = (x_1, \ldots, x_N)$, or to use the term we use in our field—data-set, a mathematical model M and the associated probability $P(\mathbf{x}|\Theta)$ we use statistical analysis to decide the best values to assign to the parameters Θ. If we have several models $M_1, M_2 \ldots$ then we will, in addition, want to decide which one is best. But what do we mean by "best"?

The mapping $(x_1, \ldots, x_N) \longrightarrow (\theta_1, \ldots, \theta_M)$ from our data-set to the parameters, or to our set of models, is called a **decision function**; it will be denoted by the symbol $d(x_1, \ldots, x_N)$. For the time being we shall limit our discussion to models that depend upon a single parameter θ. We shall denote by $\hat{\theta}$ any **estimate** of θ. If the decision function is chosen so that $\hat{\theta} = d(x_1, \ldots, x_N)$ then d is called an **estimator** for θ. So when we plug into it an actual data-set, $\mathbf{x}_0$ say, the function d gives a definite estimate, $\hat{\theta}_0$, of θ. Okay, fine; but how do we choose a decision function?

Loss and Risk Functions

Some decisions, of course, are better than others. For example, suppose you're outside on a sunny day and someone asks you for the time. Unfortunately, you left your atomic watch at home, but you observe that the sun is high in the sky. That's your datum. Here are two possible decision functions

- The sun is high in the sky; therefore, say it is not midnight.

- The sun is high in the sky; therefore, say it is noon.

The first time estimate is surely safe, but not too useful perhaps. On the other hand the second one is likely to be true only approximately. So which is the better estimate?

To choose a decision function we need a way to quantify the quality of its estimates. Bad decisions entail loss. We therefore introduce a **loss function**, $\mathcal{L}(\theta, d)$, to measure the loss arising from a decision. The loss function depends both on our decision function, which we know, and on the parameter, whose value we don't know. Notice that the loss function, through its dependence on the decision function, is a function of the data-set.

At this point we face a profound difference between frequentist and Bayesian ideas. A failure to recognize the importance of this difference can, and does, lead to lots of rather useless discussion and arguments. This difference is not mere nit-picking; it lies at the heart of the controversies. Here it is:

- **Frequentist:** In making inferences, data we could have observed are as relevant as data observed.

- **Bayesian:** In making inferences, only data observed are relevant.

Accordingly, in the frequentist approach one asserts that the loss function is only part of the story. To get a complete measure of loss we need to consider the loss pertaining to every data-set which could have been observed, as well as the loss pertaining to the data actually obtained. This is the same reasoning that requires that we compute the probability to obtain a χ^2 value equal to or greater than the value actually obtained—and to call that the "significance" of the data we have. In the Bayesian theory it is asserted that only the data we have are relevant; the data we could have obtained are deemed irrelevant. But because we do not know which hypothesis about the parameter is correct we must consider all possible hypotheses; so we average the loss function over all possible hypotheses, rather than over all possible data-sets.

In either case, this motivates the definition of a new function

$$\mathcal{R}_d = \mathrm{E}[\mathcal{L}(\theta, d)]_*, \tag{4}$$

called the **risk function**. For the frequentists, the risk function is an ordinary function of the parameter θ, and a *functional* of the decision function d, that is, it depends on the set of all possible values of d. For the Bayesian, the risk function is an ordinary function of the single data-set we have. It is also a functional of d except that now each value of d corresponds to a different value of θ, rather than to a different data-set. The symbol $\mathrm{E}[\cdot]_*$ is the averaging or **expectation operator**. The averaging is done either with respect to all possible data-sets **x** (frequentist) for fixed θ or with respect to all possible θ for fixed **x** (Bayesian).

FREQUENTIST STATISTICS

Sampling Distribution, Bias and Variance

Our starting point shall be the loss function that was described, in general terms, in the previous section. But what function should we choose? Well, a whole theory for that exists, about which I'm blissfully ignorant. So let's just pick the loss function that most of us use:
$$\mathcal{L}(\theta, d) = (\theta - d)^2. \tag{5}$$
Its associated risk function
$$\mathcal{R}_d(\theta) = \mathrm{E}[(\theta - d)^2]_{\mathbf{x}}, \tag{6}$$
is called the **mean squared error**. To write it in a more useful form let us apply the expectation operator to the function d itself. Remember, in Eq. (6), we are averaging over all possible data-sets, of which our particular data-set is but a singular sample. The probability to draw our data-set is assumed to be $P(\mathbf{x}|\theta)$. When this probability is so interpreted it is called the **sampling distribution** of $\mathbf{x}$. This is the fundamental quantity in frequentist statistics. The expectation value of d can be written as
$$\mathrm{E}[d(x_1, \ldots, x_N)]_{\mathbf{x}} \equiv \theta + b(\theta). \tag{7}$$
The function $b(\theta)$ is called the **bias** and is, in general, a function of the parameter to be estimated. Now, we can re-write Eq. (6) as
$$\mathcal{R}_d(\theta) = \mathrm{Var}[d]_{\mathbf{x}} + b^2(\theta), \tag{8}$$
where
$$\mathrm{Var}[d]_{\mathbf{x}} \equiv \mathrm{E}[d^2]_{\mathbf{x}} - \mathrm{E}[d]_{\mathbf{x}}^2, \tag{9}$$
defines the **variance operator**, $\mathrm{Var}[\cdot]_{\mathbf{x}}$. The **mean** and the **variance**—or its square-root the **standard deviation**—are important characteristics of a sampling distribution.

If our estimates $\hat{\theta} = d(x_1, \ldots, x_N)$ are such that $b(\theta) = 0$ for every possible value of θ then our estimates $\hat{\theta}$ and our estimator d are said to be **unbiased**. In that happy circumstance the mean squared error is equal to the variance of the sampling distribution. We can now answer the question: What is the best estimator?

- **Proposition**: The best estimator is unbiased and has the smallest variance.

These estimators are a kind of "Holy Grail" for frequentist statistics.

Discussion. As noted earlier, Sir Ronald was a ferocious advocate of frequentist methods. But even he noted, pointedly, that the mere specification of a data-set does not uniquely determine to which ensemble it belongs. This observation has far-reaching consequences. Why? Because of the fact that the sampling distribution associated with our data-set depends on which ensemble we consider our data-set to inhabit. This may seem patently obvious. It is! Yet how often have we witnessed arguments that arise simply because of a failure to acknowledge the obvious: that the choice of ensemble in statistical data analysis is just that, a choice. We often argue as if the ensemble we believe our experiment to inhabit has an objective reality, when in fact the ensemble exists only in the abstract. Run II of the CDF experiment has come and gone, never to be repeated. We have no objective way to determine to which ensemble the data-set of that experiment belongs, as no such ensemble physically exists. The reason we say that a counting experiment, like CDF, follows a Poisson distribution is because of what we *believe* we know about the characteristics of that experiment. But, unless we actually repeated the CDF experiment, we cannot *know* whether or not the assumption of a Poisson distribution is in fact correct. Indeed, it is much more likely, given the changing conditions during an experiment, that the counts are *not* exactly Poisson distributed. Let's pursue this a little further.

Consider an experiment that has run for one year and has found 100 single electron events. To which ensemble does this data-set belong? If we assert that the experiment ran for a fixed amount of time, one year say, then it would be reasonable to assume that the data-set belongs to an ensemble characterized by experiments each of which ran for one year, in which case it would be just fine to use a Poisson ensemble. If, however, some colleagues asserted that the experiment ran until we found 100 events, then they are at liberty to choose an ensemble consisting of experiments each of which contains exactly 100 events. Or perhaps the experiment ran until funding stopped. These three ensembles are different and each will give different physics results, because each corresponds to different sampling distributions. By simply changing our mind about which ensemble we think is reasonable, a subjective act, we can change our answers. Yet the data have not changed: we have 100 events that were collected in one year with a given amount of funding.

Why am I making such a big deal of this? Well, because it has real-world consequences: it affects what is published. Suppose that both CDF and DØ each have a data-set consisting of 100 electron plus jet events. The CDF collaboration chooses to use a binomial ensemble in their analysis, while DØ prefers a Poisson ensemble. Even if the data-sets of both experiments were identical in all respects the CDF collaboration would publish a result, say on the top quark mass, that would have a smaller uncertainty than that published by DØ simply because of the different choices of ensembles! So CDF's result would have greater weight.

Some ensembles, of course, are real—for example: a balloon full of helium molecules where each molecule is considered an element of the ensemble. The ones we use in statistical data analysis, typically, are not. Which ensemble we use in an analysis is governed largely by tradition and collective agreement. Do not be swayed by specious arguments to the effect that a Poisson ensemble is obviously the correct thing to use because "counts are known to Poisson distributed", or "we've found 100 events so we must use a binomial ensemble". Neither is "obviously correct", nor can it be said that either is wrong.

Efficiency, Likelihood and the Minimum Variance Bound

Should we be so lucky as to stumble across the best estimator for our particular inference problem, assuming that one exists, how would we recognize it? First some more definitions. The **efficiency** of an estimator d is defined by

$$e(\theta) \equiv \text{Var}[d^*]/\text{Var}[d], \tag{10}$$

where d^* is the best estimator, as defined above. The **likelihood function** $L(\mathbf{x}|\theta)$ is just another name for the probability density function $f(\mathbf{x}|\theta)$. Note that the probability is dimensionless, as any decent probability should be, but the likelihood function has the dimensions of $1/\mathbf{x}$. However, if the probability distribution is discrete then the likelihood and the probability are identical.

None of this seems terribly helpful, does it? Well, it wouldn't be were it not for a theorem proved by Aitken, Silverstone, Fréchet, Cramér and Rao. Sadly for the others it is usually called the **Cramér-Rao inequality**; a non-partisan name is the **minimum variance bound**. The theorem states that

$$\text{Var}[d] \geq \frac{(1 + D_\theta b(\theta))^2}{E[(D_\theta \ln L(\mathbf{x}|\theta))^2]}, \qquad (11)$$

with the corollary that

$$\text{Var}[d^*] = \frac{1}{E[(D_\theta \ln L(\mathbf{x}|\theta))^2]} \qquad (12)$$

is the variance of the best estimator for θ, if it exists. The symbol D_θ is the derivative operator with respect to θ. The main requirement on $L(\cdot|\cdot)$ so that the theorem holds true is that the boundary of the domain, over which the likelihood function is defined, be independent of the parameter θ.

So, if you have an estimator and you want to know if it's any good you simply calculate its efficiency. If the efficiency turns out to be unity you're in great shape because you then know that you have the best estimator for your parameter. This is all very well, but we still do not know how to find the best estimator for θ and even whether or not one exists.

The Principle of Maximum Likelihood and Consistency

Amongst the many ideas introduced by Sir Ronald Fisher his principle of **maximum likelihood** is justly famous. It provides a very general and systematic way to construct estimators. The principle is:

- Choose as an estimate the value of θ, call it $\hat{\theta}$—which will be equal to some function $d(\mathbf{x})$, that maximizes the likelihood function for a given data-set $\mathbf{x} = (x_1, \ldots, x_N)$. That is, we solve $D_\theta L(\mathbf{x}|d) = 0$, for d.

In the context of frequentist statistics this principle is not something to be proved. It is just a good general rule to find estimators $d(\mathbf{x})$. Like all good mathematical principles this one boasts many important theorems. Here is the "crown jewel":

- If a *maximally efficient unbiased* estimator exists for a parameter θ then the maximum likelihood principle will find it.

Therefore, if you make it a habit always to derive estimators by maximum likelihood then you are guaranteed to find the best estimator, if one exists, for the parameter whose value you're trying to estimate.

A lack of bias is held by some to be important, and by others to be irrelevant. Unfortunately, we cannot appeal to the laws of nature to decide which opinion is "correct", if any. But here are some points to ponder. In high energy physics rarely do we perform repeated sampling with a fixed sample size, the circumstance to which the concept of bias pertains. What we do, typically, is to append a new data-set to an older one. We add Run II's data-set to that of Run I. Nevertheless, bias it is something that the Particle Data Group[7] does worry about when they combine the results from different experiments, and it is important to understand why. We shall return to this point later.

The one requirement that everyone agrees is crucial is that an estimator be **consistent**. I find it ironic, therefore, that while we often fret about bias, and bias correction, we seldom check whether or not our estimators are consistent! An estimator is said to be consistent if, as the size of the data-set grows, its estimates converge to the true value of the parameter θ. More precisely, the probability that an estimate deviates by a stated amount from the true value goes to zero as the sample size grows to infinity. This implies that the bias $b(\theta)$ of a consistent estimator goes to zero as more and more data are accumulated. The great thing about maximum likelihood estimators is that they are, in general (although not always), consistent. For large samples, the distribution of the estimates $\hat{\theta}$ has a mean that gets ever closer to the true value of θ and a variance that attains that prescribed by the minimum variance bound. Consistency is a very desirable property, indeed. We like to believe that, as we accumulate more and more data, we will be able to report better and better results.

Another splendid property of estimators obtained from the maximum likelihood principle is invariance of the estimators under re-parametrization. What this means is if we had found $\hat{\theta}$ by solving

$$D_\theta L(\mathbf{x}|\hat{\theta}) = 0, \tag{13}$$

or what is usually more convenient

$$D_\theta \ln L(\mathbf{x}|\hat{\theta}) = 0, \tag{14}$$

and then we changed our minds and solved the above equation using instead the parameter $\alpha = g(\theta)$ then we would find that $\hat{\alpha} = g(\hat{\theta})$. (Try to figure out why this is true.) So you might as well solve the problem using the most convenient parameter and, at the end, transform to the parameter you really care about.

There is, however, a bit of unpleasantness. For a given likelihood function, if a best estimator exists it is unique; consequently, all other estimators will necessarily be biased! The easiest way to see this is to do a Taylor series expansion of $\hat{\alpha} = g(\hat{\theta})$ about the parameter θ:

$$\hat{\alpha} \approx g(\theta) + (\hat{\theta} - \theta) D_\theta g(\theta) + \frac{1}{2}(\hat{\theta} - \theta)^2 D_\theta^2 g(\theta), \tag{15}$$

and then apply the expectation operator to both sides, where the averaging is done with respect to all possible data-sets $\mathbf{x}$. Assuming $\hat{\theta}$ to be unbiased we get

$$E[\hat{\alpha}] \approx \alpha + \frac{1}{2}\text{Var}[\hat{\theta}]D_\theta^2 g(\theta), \tag{16}$$

which shows that the estimate $\hat{\alpha}$ is biased, in general, even if $\hat{\theta}$ is not.

The Sampling Distribution of an Estimator

As noted above the probability $P(\mathbf{x}|\theta) = L(\mathbf{x}|\theta)\,d\mathbf{x}$ is interpreted as the sampling distribution of the data-set $\mathbf{x}$. From the maximum likelihood principle we can obtain an estimate $\hat{\theta} = d(x_1, \ldots, x_N)$. It too has a sampling distribution that is determined by the sampling distribution of the underlying data-set $\mathbf{x}$. For the frequentist the sampling distribution of estimates is a fundamental quantity. It is the basis for quantifying uncertainty. Here is the point made plainly.

- The error on an estimate is, by definition, a measure of the width of the sampling distribution of the estimates.

Therefore, for a frequentist, the error depends not only on the data-set we have, but also on all the data-sets we could have obtained. It depends, therefore, on the *ensemble* to which we consider our data-set to belong. For those who advocate this measure of uncertainty, it does not matter whether our particular data-set is very spread out or tightly clustered. In high energy physics we sometimes refer to this as the **ensemble error**. In fact, in frequentist theory, it is *the* error!

Fisher was extremely good at computing the sampling distributions of estimators; this is partly the reason why he became so influential. However, as we shall see later in Bayesian probability theory, the sampling distribution of estimators does not play a fundamental role. It is used mainly as a way to investigate how Bayesian results behave on the average, in so-called **ensemble tests**.

To compute sampling distributions often demands a considerable amount of mathematical ingenuity. It certainly helps to be a Fisher! The basic method, however, is simple. Here it is:

$$P(\hat{\theta}|\theta) = d\hat{\theta} \int_{\mathbf{x}} \delta(\hat{\theta} - d(\mathbf{x})) P(\mathbf{x}|\theta). \tag{17}$$

The integral is over all possible data-sets $\mathbf{x} = (x_1, \ldots, x_N)$ subject to the constraint that the estimate $\hat{\theta}$ is equal to the value of the estimator $d(\mathbf{x})$. The δ-function is the simplest way to impose that constraint [8]. Equation (17) is an elegant formula, but it is rare that we can actually use it. Most of the time we must resort to Monte Carlo methods to calculate the sampling distribution, or as we high energy physicists would say the **ensemble distribution**.

Now suppose we've done all that; what next? Well, in accordance with the frequentist definition of error, we must provide some measure of the width of the sampling distribution. One such measure is $\sqrt{\text{Var}[\hat{\theta}]}$—that is, the standard deviation of the sampling distribution of the estimates. It measures the expected spread of our estimates were we to repeat our experiment many times in some agreed upon manner. Again, note that we are at liberty to *imagine* repeating our experiment in any manner we choose. It is then entirely possible that we would end up with a different error, even though the data we have acquired have not changed! The error we quote is subject to our collective whim regarding the ensemble to be used in our analysis, an unsettling point that is often overlooked.

From the function $P(\hat{\theta}|\theta) = L(\hat{\theta}|\theta) d\hat{\theta}$ it is straightforward, in principle, to compute the standard deviation. If the calculations can be done analytically it is often useful first to calculate the function

$$\begin{aligned} F(\omega) &= \text{E}[e^{i\omega\hat{\theta}}], \\ &= \int_{\hat{\theta}} e^{i\omega\hat{\theta}} P(\hat{\theta}|\theta), \end{aligned} \tag{18}$$

which we recognize as just the Fourier transform of the likelihood function $L(\hat{\theta}|\theta)$. The quantity $F(\omega)$ is called the *characteristic function* and from it a wealth of mathematical goodies flow. One of them is an elegant way to compute the moments of the sampling distributions:

$$\begin{aligned} m_n &\equiv \int_{\hat{\theta}} \hat{\theta}^n \, P(\hat{\theta}|\theta) \\ &= i^{-n} \text{D}_\omega^n \, F(\omega)_{\omega=0}. \end{aligned} \tag{19}$$

The standard deviation is then simply $\sigma = \sqrt{(m_2 - m_1^2)}$.

A Toy Example: Estimating the Mean Lifetime

Let's now try to apply some of the above frequentist theory to a simple example. We have N lifetime measurements, $\mathbf{x} = (x_1, \ldots, x_N)$, and we want to estimate the mean lifetime of an ensemble of systems that decay spontaneously. Let's suppose that the measurement errors can be neglected and that the system—some quantum state—decays according to an exponential law: $P(x|\theta) = \exp(-x/\theta)dx/\theta$. We shall do the calculation in several steps.

- **Step 1**

First we write down the likelihood function for our data-set $\mathbf{x}$:

$$L(\mathbf{x}|\theta) = \prod_{i=1}^{N} \exp(-x_i/\theta)/\theta$$

$$= \theta^{-N} \exp\left(-\theta^{-1} \sum_{i=1}^{N} x_i\right). \qquad (20)$$

Recall that this is just the joint probability density for our data-set. Be sure to include *all* terms that depend upon the parameter; but feel free to drop any constant multiplier that is independent of θ.

- **Step 2**

We invoke the maximum likelihood principle

$$D_\theta L(\mathbf{x}|\theta) = 0, \qquad (21)$$

and derive the estimator

$$d(\mathbf{x}) = \frac{1}{N} \sum_{i=1}^{N} x_i, \qquad (22)$$

which is just what our intuition told us it had to be!

- **Step 3**

The next step is to calculate the sampling distribution of $\hat{\theta} = d(\mathbf{x})$ using Eq. (17):

$$P(\hat{\theta}|\theta) = d\hat{\theta} \int_{\mathbf{x}} \delta\left(\hat{\theta} - \frac{1}{N}\sum_{i=1}^{N} x_i\right) \theta^{-N} \prod_{i=1}^{N} e^{-x_i/\theta},$$

$$= d\hat{\theta}\, \theta^{-N} \int_0^\infty dx_1\, e^{-x_1/\theta} \cdots \int_0^\infty dx_N\, e^{-x_N/\theta}\, \delta(\cdots). \qquad (23)$$

The easiest way to handle the δ-function is to write it as

$$\delta(x) = \frac{1}{2\pi} \int_{-\infty}^{\infty} d\omega\, e^{i\omega x}, \qquad (24)$$

whereupon we can write

$$P(\hat{\theta}|\theta) = d\hat{\theta}\, \theta^{-N} \frac{1}{2\pi i} \int_{-\infty}^{\infty} d\omega\, e^{i\omega\hat{\theta}} G(\omega), \qquad (25)$$

where the function $G(\omega)$ is given by

$$G(\omega) = i\theta^{-N} \prod_{j=1}^{N} \int_0^\infty dx_j \, e^{-(i\omega/N + 1/\theta)x_j},$$
$$= i\left(\frac{N}{i\theta}\right)^N \left(\frac{1}{\omega - N/i\theta}\right)^N. \tag{26}$$

The integral is most easily done by continuing ω into the complex plane and then applying a spot of residue theory. The function $G(\omega)$ has a simple pole of order N at $\omega = N/i\theta$; so the integral in Eq. (25) is easy to do. We get

$$P(\hat{\theta}|\theta) = \frac{\left(N\hat{\theta}/\theta\right)^{N-1} e^{-N\hat{\theta}/\theta}}{(N-1)!} d(N\hat{\theta}/\theta). \tag{27}$$

The above is a special case of the gamma distribution:

$$g(x) = x^{p-1} e^{-x}/\Gamma(p). \tag{28}$$

As far as the frequentist approach goes we are almost done; the rest is just a matter of computing the moments of the sampling distribution.

In practice, lifetimes are measured with some error; so the sampling distribution we have calculated is generally a poor model. A more realistic model is to approximate the distribution of measurement errors by a Gaussian distribution with known variance. It turns out that even for this more complicated case the sampling distribution of $\hat{\theta} = d(\mathbf{x})$ can be worked out exactly[9].

One question we haven't answered is: is the lifetime a parameter for which a best estimator exists; that is, one that is unbiased and has the smallest variance? If it does exist then we are guaranteed to have found it because we have derived our estimator by maximum likelihood. The answer is yes. Try to prove it.

A WORD OR TWO ABOUT ERRORS

Correlation

Imagine that we have two data-sets $\mathbf{x}$ and $\mathbf{y}$, each with N elements. They could be the same quantity sampled N times or N different quantities, each sampled once. Whatever they are think of these data-sets as column matrices. The **covariance matrix** of the data-sets (sometimes called the error matrix) is defined by

$$\text{Cov}[\mathbf{x}, \mathbf{y}] = E[(\mathbf{x} - E[\mathbf{x}])^T (\mathbf{y} - E[\mathbf{y}])],$$
$$= E[\mathbf{x}^T \mathbf{y}] - E[\mathbf{x}^T] E[\mathbf{y}] \tag{29}$$

where ^{T} is the transpose operator. The averaging is done over all possible data-sets with N elements. From the definition of the variance we note that $\text{Var}[\mathbf{x}] = \text{Cov}[\mathbf{x}, \mathbf{x}]$. The variance matrix is square and diagonal. The covariance matrix is square and symmetric. The two are related as follows:

$$\text{Var}[\mathbf{x} + \mathbf{y}] = \text{Var}[\mathbf{x}] + \text{Var}[\mathbf{y}] + 2\,\text{Cov}[\mathbf{x}, \mathbf{y}]. \tag{30}$$

(Try to show this.) This shows that, unlike the expectation operator, the variance operator is non-linear. The covariance matrix has the disadvantage that it has dimensions, so sometimes we prefer to use a dimensionless version of it called the **correlation matrix** ρ, whose matrix elements are given by

$$\rho_{ij} = \frac{\text{Cov}[\mathbf{x}, \mathbf{y}]_{ij}}{\sigma_i \sigma_j} \quad (31)$$

where $\sigma_i = \sqrt{\text{Var}[x_i]}$ are the standard deviations. Note that $-1 \leq \rho_{ij} \leq 1$. If a matrix element is close to 1 then we say that the quantities x_i and y_i are strongly **correlated**. If it is close to -1 then the quantities are said to be strongly **anti-correlated**.

Independence

Consider, for simplicity, two data x and y only. Call the sampling distribution of x and y, $P(x, y|\theta)$. We can write the sampling distribution as

$$P(x, y|\theta) = f(x|y\theta) g(y|\theta) \, dx \, dy, \quad (32)$$

where $f(x|y\theta)$ is the probability *density* of x given y and θ and $g(y|\theta)$ is the probability density of y given θ. In general, the data x and y are not independent in the sense that the probability to obtain x depends upon the value of y and vice versa. But if we can write

$$P(x, y|\theta) = f(x|\theta) g(y|\theta) \, dx \, dy, \quad (33)$$

then the data x and y are said to be **independent**. That is, the probability to draw x does not depend on having drawn y or vice versa. The following points are important to remember.

- If data are independent then they will be uncorrelated.

- If data are uncorrelated this does *not* imply that they are independent.

For example, suppose we define a uniform probability on the circle $x^2 + y^2 = a^2$. Obviously, x and y are not independent. Nonetheless, the correlation between these data will be zero! You've been warned.

Combining Results

In the last lecture we learned that the frequentist procedure to solve an inference problem is:

- Given a model M, derive the sampling distribution $P(\mathbf{x}|\theta)$ for all the possible data-sets $\mathbf{x} = (x_1, \ldots, x_N)$ that the model predicts.

- Solve the likelihood equation $D_\theta L(\mathbf{x}|d) = 0$ to obtain the estimate $\hat{\theta} = d(\mathbf{x})$.

- Compute the sampling distribution $P(\hat{\theta}|\theta)$ of the estimates $\hat{\theta}$ from the underlying sampling distribution $P(\mathbf{x}|\theta)$.

- Compute the moments of $P(\hat{\theta}|\theta)$.

There are at least two things that we usually report, after following this procedure: 1) the estimate $\hat{\theta}$ itself and 2) the standard deviation of its sampling distribution. The latter is important because it provides us with a way to judge the reliability of the estimate, that is, the **uncertainty** in the estimate. Recall that, according to the frequentists, the uncertainty, or error, is measured by the average spread of the estimates over an ensemble of estimates. For Bayesians, the error is determined by the spread in the particular data-set obtained. I repeat this mantra, at the risk of boring you to death, because sooner or later you will be engaged in an argument with an elder about whether the error to be published should be based upon the data-set obtained or upon the sampling distribution. The only sensible resolution of the argument is to note that both uncertainty measures are correct within their respective approaches. The real issue is deciding which approach to inference is, on the whole, the more satisfactory: frequentist or Bayesian. But, we are getting ahead of ourselves.

Let's agree for the moment to accept the frequentists' definition of error, namely the standard deviation. There is another important reason why it should be reported: it is needed to combine the results from different experiments.

Suppose several of us have tried to measure the mass of the top quark. The Particle Data Group now has the job of combining these results to produce a world average, $\hat{\theta}_W$, of the top quark mass. What do they do? They compute the weighted average

$$\hat{\theta}_W = \sum_i w_i \hat{\theta}_i. \qquad (34)$$

What is the error on the weighted average? The frequentist answer is: first find the sampling distribution of the world average $P(\hat{\theta}_W|\theta)$, and then compute its standard deviation. Usually, the Particle Data Group assumes that the results from the different experiments are uncorrelated in which case they obtain

$$\text{Var}[\hat{\theta}_W] = \sum_i w_i^2 \, \text{Var}[\hat{\theta}_i], \qquad (35)$$

as the variance of the world average sampling distribution. Actually, Eq. (35) can be derived directly from Eq. (34); it is not necessary to know the explicit form of the sampling distribution $P(\hat{\theta}_W|\theta)$. But how are they to choose the weights?

Their aim is to combine the results so that the world average sampling distribution has the smallest possible variance, that is, the smallest error. This they do by solving $D_{w_i}\text{Var}[\hat{\theta}_W] = 0$ subject to the constraint $\sum_i w_i = 1$. The problem is solved easily (try it) using the method of Lagrange multipliers. The answer is

$$w_i = \frac{1/\text{Var}[\hat{\theta}_i]}{\sum_j 1/\text{Var}[\hat{\theta}_j]}. \qquad (36)$$

The optimal weights are determined by the variances.

Notice that we have placed no restrictions on the form of the sampling distributions $P(\hat{\theta}_i|\theta)$, other than that their variances be finite; in particular, we have not assumed them to be Gaussian distributions. Therefore, Eq. (36) is valid for all sensible distributions, and not just Gaussians.

Let's apply the expectation operator to $\hat{\theta}_W$. We get

$$\text{E}[\hat{\theta}_W] = \theta + \sum_i w_i b_i(\theta), \qquad (37)$$

which is just the mean of the sampling distribution $P(\hat{\theta}_W|\theta)$. The Particle Data Group could be lucky, or very well informed, and choose just the right combination of experiments so that the weighted sum of biases $b_i(\theta)$ is zero. Alas, in the real world this seldom happens. Indeed, it may be that the experiments are all biased in the same way. In this case, no matter how many experiments were combined, the world average would be biased. But, if each experiment reports an unbiased estimate then no bias will arise in the world average.

However, forming a weighted average is not the only way to combine results. We shall discover another way when we discuss Bayesian statistics. Moreover, as I suggested earlier, our insistence upon a lack of bias in high energy physics experiments is a somewhat artificial requirement because rarely do we perform the averaging implied in Eq. (37): we do not ask experimenters to perform umpteen identical repetitions of their experiments and average their results. In practice, experiments acquire more data, report new results and the Particle Data Group updates its world average. Since this is what we do it seems to me that this is the procedure that should be modeled. Specifically, we should verify, by Monte Carlo simulation, that it leads to a world average that converges to the true value as each experiment accumulates more and more data. That is, we should verify that the procedure is *consistent*.

Confidence Intervals: Exact

So, reporting the variance is important. Unfortunately, there is a problem. The variance usually depends upon the parameter whose value we are trying to estimate. Since we have only an estimate of this parameter we can compute only an estimate of the variance. That in itself is not necessarily a problem except for the difficulty that sometimes the estimate of the variance can be very poor. If we use an unreliable estimate of the variance we risk either over-weighting or under-weighting an experimental result when forming the weighted average, Eq. (34).

There is another difficulty with Eq. (34); a conceptual one. The averaging of Eq. (34) to obtain Eq. (35) and Eq. (37) assumed that the weights remain constant during the averaging. That is, each repetition of an experiment leads to the same reported variance and hence is given the same weight ω_i. But, we would expect that, since each experiment provides only an estimate of the variance, that estimate would vary from one repetition of the experiment to another. Therefore, the weights would not be constant over the ensembles of experiments. In fact, the weights ω_i would be correlated with the estimates $\hat{\theta}_i$, with the possible consequence that the weighted average could be biased even if each estimate $\hat{\theta}_i$ were not. And that bias, like the error itself, depends on the choice of ensemble.

Neyman, Clopper and Pearson sought to sidestep these difficulties with their invention of the **confidence interval**, or **error interval** as we high energy physicists call it. These intervals are another measure of uncertainty. The confidence interval $\left[l(\hat{\theta}), u(\hat{\theta})\right]$ is two numbers, with $l(\hat{\theta}) < u(\hat{\theta})$, that depend on the estimate $\hat{\theta}$ so that

$$\text{Prob}\{l(\hat{\theta}) < \theta < u(\hat{\theta})\} \geq \beta, \tag{38}$$

whatever the true value of θ. The quantity β is a pre-assigned probability called the **confidence level**. The Bayesians have a similar concept but their interpretation is totally different, a point that is sometimes missed. The confidence interval, as described by Neyman and Pearson, is very much a frequentist notion which, curiously—in spite of his frequentist credentials, Sir Ronald disliked intensely. Perhaps, Sir Ronald disliked the indirectness of the reasoning behind Neyman's intervals. By sharp contrast, the equivalent Bayesian interval is interpreted rather more directly. But Sir Ronald didn't like those either, even though he himself introduced a similar notion in his theory of fiducial probability.

What Eq. (38) means is this. The parameter θ is an unknown, but *fixed*, number; for example, the mean lifetime of a quantum state. You do an experiment and get the estimate $\hat{\theta}$ and you calculate the corresponding interval $\left[l(\hat{\theta}), u(\hat{\theta})\right]$. Imagine the experiment repeated. In general, you would get a different estimate and, therefore, a different interval. Sometimes the interval you obtain will contain the true value of θ and sometimes it will not. Suppose you could count how many times the interval contained the true value, then you would be able to divide that count by the number of repetitions of your experiment. The fraction, so obtained, is called the **coverage probability**. It is the fraction of intervals, over the ensemble, that contain the true value of the parameter. If the coverage probability is found *never* to fall below β and if this is true for *all* possible *fixed* values of θ we say that $\left[l(\hat{\theta}), u(\hat{\theta})\right]$ is a confidence interval with a confidence level of $100 \times \beta\%$. The confidence level, then, is the minimum coverage probability over the parameter space.

So, in what sense are we to understand the word "confidence"? Consider a 90% interval. The basic idea is that if 90% of intervals, over an ensemble, contain the true value then it is very likely that, by chance, your experiment will yield an interval containing the true value.

According to Neyman and Pearson, the confidence level is a property of a given interval only in so far as that interval belongs to an ensemble. If you change your mind about the ensemble being used then you will, in general, change the confidence level. So, in frequentist theory, a 90% confidence level is devoid of meaning unless the ensemble is specified, just as the standard deviation is undefined until we give the set of data-sets that constitute the ensemble. This observation has practical ramifications.

Suppose two different experimental teams have set a limit on the squark mass, but do so using different ensembles to define their confidence intervals and levels. It is quite possible that one team's 90% limit may actually be more stringent than another's 95% limit! The only way to avoid such problems would be for both teams to agree to use the same ensemble to compute their limits.

But, wait a minute you say; since we don't know the true value of θ—that's why we are doing the experiment—how can we possibly construct such intervals? It is a remarkable fact that intervals with such properties can be constructed, and rather easily it turns out, for any sampling distribution that depends upon a *single* parameter. If, however, we have a sampling distribution that depends on many parameters of which only one is of interest, then the frequentist theory becomes rapidly unstuck. In fact, the near impossibility of constructing confidence intervals, with a pre-determined confidence level, in the face of unwanted parameters is the chief stumbling block of the frequentist theory.

When the sampling distribution depends only on the parameter of interest, or a function thereof, it is relatively easy to construct confidence intervals with a pre-determined confidence level; that is, with a coverage probability that does not fall below a specified value. To do so define the probabilities

$$L(\hat{\theta}|\theta) = \int_{z \leq \hat{\theta}} P(z|\theta), \tag{39}$$

and
$$R(\hat{\theta}|\theta) = \int_{z \geq \hat{\theta}} P(z|\theta), \qquad (40)$$

which could be called the left and right **cumulative distribution functions** (cdf), respectively. But because this is an awful mouthful, let's just refer to them as the left and right probabilities. The functions $u(\hat{\theta})$ and $l(\hat{\theta})$ are obtained by solving

$$L(\hat{\theta}|u) = \alpha_L, \text{ (Upper Limit)} \qquad (41)$$

and

$$R(\hat{\theta}|l) = \alpha_R, \text{ (Lower Limit)} \qquad (42)$$

where

$$\beta = 1 - \alpha_L - \alpha_R. \qquad (43)$$

Remember the rule: LEFT is UP and RIGHT is LOW! Notice there are infinitely many sets of intervals that could be computed, corresponding to our freedom to choose the value of α_L, or α_R. To make the choice unique we appeal to simplicity and convention: usually, we set $\alpha_L = \alpha_R = (1-\beta)/2$. This choice leads to intervals called **central confidence intervals**. By being exceptionally clever we my be able to construct intervals that have the smallest width on average for a given confidence level; but this is very tricky. I have written the Eqs. (39) and (40) in a form that is valid both for continuous and discrete distributions.

It may not be immediately obvious why this construction leads to intervals with the stated property. To understand that they do consider Fig 1 which shows a plot of θ versus $\hat{\theta}$. That figure depicts a mapping between the parameter space on the vertical axis and the estimate space on the horizontal axis. The two curves $\theta = u(\hat{\theta})$ and $\theta = l(\hat{\theta})$, which are solutions to Eq. (41) and Eq. (42), respectively, map an estimate $\hat{\theta}$ from the estimate space to an interval in the parameter space. The horizontal line $\theta = \theta_0$ represents the true value of θ, θ_0. That line crosses the upper limit curve $\theta = u(\hat{\theta})$ at the estimate $\hat{\theta}_1$ and the lower limit curve $\theta = l(\hat{\theta})$ at the estimate $\hat{\theta}_2$. The estimates $\hat{\theta}_1$ and $\hat{\theta}_2$ partition the estimate space into three regions, labelled I, II and III. Now that we have the geometry straight, let's do some thought experiments.

Suppose that an estimate lands in region I. Its associated interval will fail to include the true value. Should an estimate land in region III it too will fail to bracket the true value θ_0. On the other hand, any estimate landing in region II will contain θ_0. The question is how often does that happen? Well, by construction, the probability to obtain an estimate in region I is given by the left probability $L(\hat{\theta}_1|u) \equiv \text{Prob}\{\hat{\theta} \leq \hat{\theta}_1\} = \alpha_L$; likewise, the probability to obtain an estimate in region III is just $R(\hat{\theta}_2|l) \equiv \text{Prob}\{\hat{\theta} \geq \hat{\theta}_2\} = \alpha_R$. Consequently, the probability to obtain an estimate in region II—and, therefore, an interval that brackets the true value θ_0—is just $\beta = 1 - \alpha_L - \alpha_R$. Since we've made no assumption about θ_0 other than that it is a fixed number, then the above reasoning holds for all *a priori* possible values of θ_0.

Example: Exact intervals for the Poisson distribution. Let us use these rules to compute the exact 68.3% central confidence intervals for the Poisson distribution. The Poisson distribution is the probability

$$p(N|\theta) = \exp(-\theta)\theta^N/N!, \qquad (44)$$

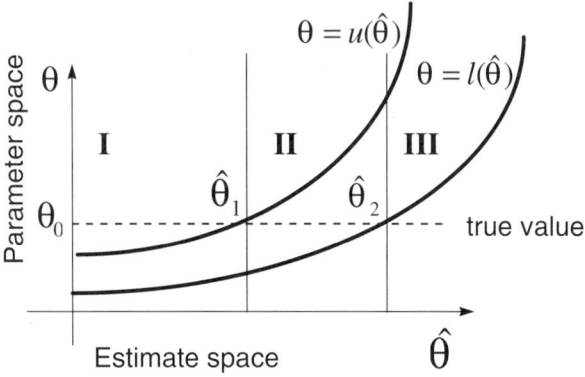

1. The plot shows how an estimate is mapped into an interval in the parameter space. As the estimates vary, so do the intervals. Estimates that land in region II lead to intervals that bracket the true value θ_0, while estimates that land in either region I or region II exclude θ_0.

to count N things in a fixed amount of time. It is the high energy physicist's favorite distribution. Applying the "left is up" and "right is low" rules we get

$$L(N|u) = \sum_{k=0}^{N} p(k|u) = (1-\beta)/2, \tag{45}$$

and

$$R(N|l) = \sum_{k=N}^{\infty} p(k|l) = (1-\beta)/2. \tag{46}$$

Note that the "right is low" equation can be re-written as

$$\sum_{k=0}^{N-1} p(k|l) = (1+\beta)/2. \tag{47}$$

In Table 1 we give the first 26 intervals obtained by solving these equations.

Everyone knows the rule-of-thumb $l(N) \approx N - \sqrt{N}$ and $u(N) \approx N + \sqrt{N}$ for computing approximate 68.3% confidence intervals for the Poisson distribution. But not everyone knows how bad an approximation this is even for values of N as large as 25! This is shown in Fig. 2 which is a plot of the ratio of the width of the exact central interval $u(N) - l(N)$ to the width, $2\sqrt{N}$, of the $\sqrt{N}$ interval. Another noteworthy observation is the peculiar behavior of the coverage probability as θ varies over the parameter space. This is shown in Fig. 3. The complex structure is due to the discreteness of the Poisson distribution. As expected, at no point does the coverage probability fall below a probability of 0.683. This is in contrast to the behavior of the $\sqrt{N}$ intervals which, as shown in Fig. 4, have a confidence level that is much lower than 0.683. Remember the confidence level is the smallest coverage probability over the parameter space.

1. Exact 68% Poisson Central Intervals.

N	$l(N)$	$u(N)$	$[u(N)-l(N)]/2\sqrt{N}$
0	0.00000	1.84200	—
1	0.17257	3.30080	1.56412
2	0.70774	4.63935	1.39003
3	1.36664	5.91986	1.31440
4	2.08483	7.16458	1.26994
5	2.83933	8.38444	1.23992
6	3.61895	9.58574	1.21797
7	4.41728	10.77250	1.20102
8	5.23025	11.94747	1.18745
9	6.05506	13.11264	1.17626
10	6.88972	14.26948	1.16684
11	7.73274	15.41915	1.15877
12	8.58296	16.56254	1.15175
13	9.43947	17.70038	1.14558
14	10.30155	18.83327	1.14010
15	11.16858	19.96172	1.13519
16	12.04008	21.08612	1.13076
17	12.91562	22.20685	1.12673
18	13.79484	23.32421	1.12305
19	14.67743	24.43848	1.11967
20	15.56312	25.54987	1.11655
21	16.45170	26.65861	1.11367
22	17.34293	27.76486	1.11098
23	18.23665	28.86879	1.10848
24	19.13268	29.97055	1.10614
25	20.03090	31.07028	1.10394

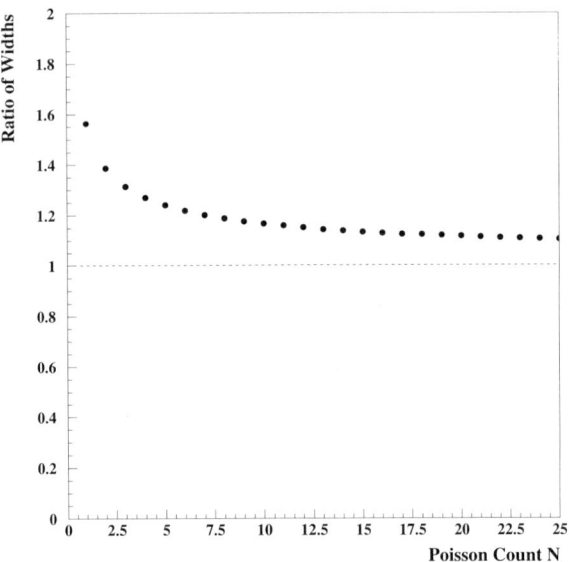

2. We plot the ratio of the width of the exact 68% central interval, for the Poisson distribution, to the that of the $\sqrt{N}$ interval. We see that for small counts, N, the $\sqrt{N}$ rule underestimates the width.

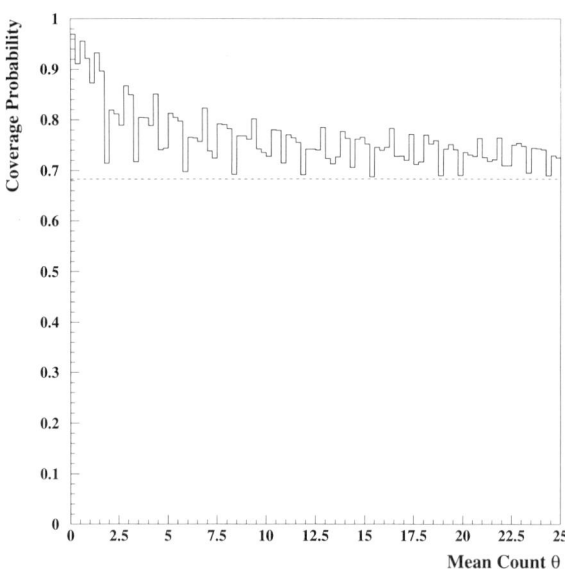

3. The coverage probability for the 68% central intervals is plotted as a function of the Poisson parameter θ. The strange behavior is due to the discrete nature of the Poisson distribution. The confidence level, which by definition is the minimum coverage probability over the parameter space, is 0.683.

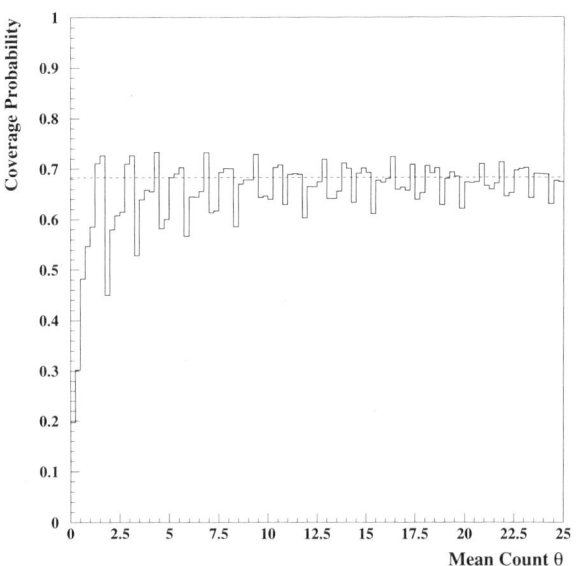

4. The coverage probability for the $\sqrt{N}$ intervals is plotted as a function of the Poisson parameter θ. The confidence level is much less than the nominal value of 0.683.

Example: Exact intervals for the exponential distribution. We observe a single unstable system decay after a time interval t and from that we wish to infer something about its mean lifetime θ. We shall assume an exponential decay model:

$$P(t|\theta) = e^{-t/\theta}\, dt/\theta. \tag{48}$$

The maximum likelihood estimate for θ is obviously $\hat{\theta} = t$. But how accurate is this estimate? Let's use a $100 \times \beta\%$ confidence interval to quantify the accuracy. All we have to do is calculate the left and right probabilities, Eq. (39) and Eq. (40), and set them equal to $\alpha = (1-\beta)/2$. The answer is

$$l(\hat{\theta}) = \hat{\theta}/\ln\left[2/(1-\beta)\right], \tag{49}$$

and

$$u(\hat{\theta}) = \hat{\theta}/\ln\left[2/(1+\beta)\right]. \tag{50}$$

Try to show this. The interval behaves as we would expect intuitively, namely: as we demand a higher confidence level, the wider the interval becomes; specifically, as $\beta \to 1$, $l(\hat{\theta}) \to 0$ and $u(\hat{\theta}) \to \infty$.

Example: Exact intervals for the Gaussian distribution Along with the Poisson distribution the Gaussian is the high energy physicist's most important distribution. Statisticians are positively besotted by it! Ironically, because of its beautiful symmetries it is the worst distribution to use to explain the pitfalls of statistical analysis. It is easy to muddle concepts that are distinct. For example, for a Gaussian there is no numerical distinction between a 68.3% confidence interval and the interval based upon the standard deviation, even though conceptually those intervals are very different.

For a Gaussian distribution,

$$P(\hat{\theta}|\theta) = \exp(-\chi^2/2)/\sigma\sqrt{2\pi}\, d\hat{\theta}, \qquad (51)$$

where $\chi^2 = [(\hat{\theta} - \theta)/\sigma]^2$. The boundaries of its 68.3% confidence interval are given by $l(\hat{\theta}) = \hat{\theta} - \sigma$ and $u(\hat{\theta}) = \hat{\theta} + \sigma$ where σ is the standard deviation. The Gaussian distribution owes its importance to the fact that many likelihood functions assume a Gaussian form when the data-sets become sufficiently large. We may then avail ourselves of the Gaussian's many useful mathematical properties to simplify analyses that might otherwise be extremely difficult, if not impossible.

Confidence Intervals: Approximate

Most of the time the exact procedure for computing confidence intervals is simply impossible to put into practice because of insurmountable mathematical difficulties. For example: Let N and B be Poisson distributed integers with mean values λ and μ, respectively, and let $\theta = \lambda - \mu$ be the parameter to be estimated; μ could be the mean background count and θ the mean signal count in a counting experiment. To the best of my knowledge, no one has been able to find a method to construct exact confidence limits for this basic counting experiment. This is a typical impasse in the frequentist theory.

Since the exact approach is usually impossible we must resort to approximations, the most widely used of which is based upon the following theorem: let

$$y = -2\ln\frac{L(x_1,\ldots,x_N|\theta)}{L(x_1,\ldots,x_N|\hat{\theta})}, \qquad (52)$$

then, under fairly mild assumptions, as $N \to \infty$, $y \to \chi^2$ variable with one degree of freedom. The theorem implies that in the large data-set limit y will become *independent* of the parameters of the likelihood function. This seems miraculous; that it is true, for sufficiently large data-sets, is crucial for computing approximate confidence intervals because a probability statement about the quantity y can be re-cast as one for the quantity θ, by inverting Eq. (52). The theorem states, in effect, that in the limit of large data-sets all reasonable likelihood functions approach a Gaussian form, as alluded to above.

More generally, if there are K un-constrained parameters in the numerator and we divide it by the absolute maximum of the likelihood function then the quantity y will follow a χ^2 distribution with K degrees of freedom, provided that the data-set is large enough.

Well, this is a great theorem. Its utility is best illustrated by an example. We have just completed a counting experiment; we watched American TV for one minute and we've counted the number of times president Bill Clinton uttered the phrase "bridge to the 21st century" and found that he did so 4 times. Now, we would like to estimate the average number of times he utters the phrase per minute and we wish to quantify the accuracy of our estimate by computing a 68.3% confidence interval for the estimated mean count. Let the mean count, in one minute, be denoted by θ and let the measured count be denoted by N. We'll model this problem with a Poisson distribution: $P(N|\theta) = \theta^N \exp(-\theta)/N!$. The maximum likelihood principle gives $\hat{\theta} = N$ as our estimate of the mean count; no surprise here! As shown in the previous section, the confidence intervals can be computed exactly because this is a single parameter problem. But let's pretend that the problem is a mathematical quagmire, so we must resort to approximations. Even for Mr. Clinton N isn't quite infinite but let us nonetheless apply the theorem stated above, assuming that $N = 4$ is close enough to infinity, and write

$$y = 2(N \ln N - N - N \ln \theta + \theta) \approx \chi^2. \tag{53}$$

Notice that $y = 0$ at $\theta = \hat{\theta} = N$. For $\theta \neq N$, $y \to +\infty$. The function y has a roughly parabolic form, becoming more so as $N \to +\infty$. Accordingly, in that limit, we should be able to write Eq. (53) as

$$y \approx \chi^2 = (\theta - \hat{\theta})^2/\sigma^2, \tag{54}$$

and find that for every value of y we get two values of θ: $l(\hat{\theta}) = \hat{\theta} - \sigma$ and $u(\hat{\theta}) = \hat{\theta} + \sigma$. But we do not have to re-write Eq. (53) in that way to obtain the lower and upper limits; we just solve Eq. (53) directly.

From the properties of the χ^2 distribution we know that the 68.3% boundary occurs at $\chi^2 = 1$, and that

$$\text{Prob}\{\chi^2 < 1\} = 0.683. \tag{55}$$

So setting $y = 1$, corresponding to $\chi^2 = 1$, in Eq. (53) we find $l(4) = 2.3$ and $u(4) = 6.3$. We, therefore, report the results of our little counting experiment as

$$\hat{\theta} = 4^{+2.3}_{-1.7} \text{ annoying utterances per minute.} \tag{56}$$

If, in addition to the above, we wanted to report an estimate of the standard deviation we could calculate it from the formula $\sigma = [u(\hat{\theta}) - l(\hat{\theta})]/2$, which becomes ever more accurate as $N \to \infty$.

We note that the interval (2.3,6.3) is close to that provided by the $\sqrt{N}$ rule: $l(4) = 2$ and $u(2) = 6$. But, the $\sqrt{N}$ rule itself is only an approximation to the exact 68.3% central confidence interval. The exact 68.3% central confidence interval for $N = 4$ is (2.1,7.2), according to the results given in Table 1.

The log-likelihood interval (2.3,6.3) is an approximate 68.3% confidence interval in the sense that it belongs to an ensemble in which the minimum coverage probability falls below 0.683. Therefore, we can state only that

$$\text{Prob}\{l(\hat{\theta}) < \theta < u(\hat{\theta})\} \approx 0.683. \tag{57}$$

Discussion. The method just described is how "errors" are estimated from a negative log-likelihood curve: you find the minimum, go up by 0.5 and read off the error interval. This is a procedure that is used routinely, and sometimes blindly. Hence a word of caution is in order. The log-likelihood method of estimating an error is an approximation that saves you the trouble of having first to compute the sampling distribution of the estimates in order to compute a confidence interval. It is a short-cut that computes an approximate confidence interval that could otherwise be computed exactly from the sampling distribution, provided that the latter contains no unwanted parameters. Arguments rage about whether we should quote an error derived from the log-likelihood curve or from the sampling distribution, as if these were two alternative definitions of error. They are not. In frequentist theory the error is, by definition, either a confidence interval computed from the sampling distribution, or its standard deviation. The "up-0.5" rule is just a handy way to get at that interval, approximately, with a minimum of fuss.

This well-known rule is valid for sufficiently large data-sets. What about small data-sets? Within the frequentist fold, I know of no theorem that can shed light on how to proceed in that case. Nevertheless, we scientists, being free spirits, feel free to use the "up-0.5" rule regardless of its provenance—and then argue interminably about what kind of function to fit to the log-likelihood curve. But, if we insist on using this rule for small data-sets, we have no right to claim that we have computed a "one-sigma"—that is, a 68.3% interval. More to the point: we have no idea, *a priori*, what confidence level to associate with the interval we've calculated!

One reason, I suspect, that we cling so stubbornly to the 0.5-rule is that it is easy to use. Another is our desire to have a definition of error that reflects the intrinsic accuracy of the data-set we actually have, rather than one that reflects the average accuracy of an ensemble of hypothetical data-sets. For a justification of this commendable bit of intuition you must wait for our discussion of Bayesian probability theory.

I think the safest policy, when data-sets are small, is to go back to first principles: compute, usually by Monte Carlo means, the sampling distribution of the estimates; and then apply the theory for computing exact intervals. This policy is valid whatever the size of the data-set.

But, even if we follow this policy, it may not help. Suppose we have gone through all the trouble of computing the sampling distribution of the estimates only to find that the latter depends upon several parameters, of which but one is of interest; that is, we find that our sampling distribution contains **nuisance** parameters. This dependence on nuisance parameters, alas, is the norm rather than the exception. In the DØ and CDF top quark mass analyses, the sampling distributions of the mass estimates depend not only on the top quark mass but also on the mean signal and mean background counts. Why is that a problem: it's a problem because we don't know their true values. What is done in these analyses is to replace the true values with estimates.

Nuisance parameters bedevil all experiments. Another example is the Grenoble neutron-antineutron experiment, discussed below, in which the exact sampling distribution of the estimated signal is a function both of the mean signal count, which is the parameter of interest in that problem, as well as of the mean background.

The general strategy to eliminate nuisance parameters is to find a function, called a **statistic**, of the parameter of interest and the data-set such that the sampling distribution of the statistic is *independent of all parameters*. Needless to say this is a tall order. As yet, no general method exists for implementing it. The closest thing to a general method is the likelihood ratio statistic discussed above which, alas, works only for large data-sets.

But isn't all of this just so much nit-picking? What possible relevance can this have to the down-to-earth world of experimental physics? Well, to the degree that we wish to publish results that are well founded this stuff is relevant. If, on the other hand, our wish is simply to follow established recipes irrespective of their validity then these issues indeed are irrelevant.

Let us grant, for a moment, that these subtleties are relevant. Then, why have we been able to make headway in spite of them? The reason is simple: we basically just ignore them in most of our analyses. Our working rule is to replace all the nuisance parameters in our problem by their estimates and then proceed as if these estimates were infinitely precise. We physicists are a pragmatic lot!

Example: The Grenoble Experiment

In the heady days of SO(10) Grand Unified Theories some physicists suggested that the neutron and the antineutron could form a two-state system[10]. A neutron would then be able to change into an antineutron, and vice versa, with a time-dependent probability given by

$$P(t|\epsilon) = \left[\epsilon^2/(\epsilon^2 + \Delta E^2)\right] \sin^2\left[(\epsilon^2 + \Delta E^2)^{1/2} t\right], \tag{58}$$

where ϵ is a fundamental interaction energy that characterizes the oscillation effect. The neutron's normal interaction energy is ΔE. The main interaction of a free neutron is with the ambient magnetic field. That is both a blessing and a curse: a blessing because it can be used to suppress, at will, the hypothesized oscillation effect, and a curse because it is difficult to achieve a "field-free" ($\Delta E = 0$) environment to look for the effect. The saving grace is that at a research reactor the flight time t of free neutrons from the core to an experiment is small enough (about 25 ms) so that to a good approximation

$$P(t|\epsilon) \approx (\epsilon t)^2, \tag{59}$$

and is independent of ΔE. This is called the "quasi-free" limit. The parameter ϵ is smaller than 10^{-22} eV.

So, more than a decade ago, a bunch of us trooped off to the Institut Laue Langevin in Grenoble, France, to test this intriguing idea. We got hold of a beam of low energy neutrons and caused them to collide with a graphite foil some tens of microns thick. The idea was to detect any antineutron component in the beam by recording the annihilations in the foil. Should that happen convincingly we thought then bingo, next stop for us would be Stockholm! This was one of those beautiful experiments where we could measure directly the background rate, and then measure the rate of background plus signal. Subtract, and you're done. Well, it was not quite so simple.

In the grenoble experiment we obtained a background count of 7 events[11]; call this count B. We measured the signal plus background, independently, and we got a count of 3; call this count N. This is a perfectly plausible result, unsettling perhaps, but statistically unremarkable given that the two counts were measured independently. However, as I discovered then, the analysis of these data leaves the frequentist theory severely challenged!

Suppose that the mean background count is μ and let θ be the mean number of annihilation events, and let's assume further that the observation time for the background and signal plus background experiments were identical, as in fact they were, then our likelihood function—that is, the sampling distribution for our data-set (N, B), can be written as

$$P(N, B|\theta, \mu) = \frac{(\theta + \mu)^N e^{-(\theta + \mu)}}{N!} \frac{\mu^B e^{-\mu}}{B!}. \tag{60}$$

This is a 2-parameter problem. But we really only care about θ. The mean background count μ is a nuisance parameter. And in the frequentist theory it really is! Unfortunately, no progress can be made until we get rid of it. Let's be pragmatic and try to do so by replacing the nuisance parameter with some reasonable estimate. It is hardly news to report that the maximum likelihood estimates for θ and μ are

$$\hat{\theta} = N - B, \tag{61}$$

and

$$\hat{\mu} = B. \tag{62}$$

Casting caution to the wind, and ignoring our own advice, we shall ignore the fact that N and B are a bit short of infinity and apply the likelihood ratio theorem to compute approximate confidence intervals $[l(N,B), u(N,B)]$ by solving

$$-2\ln\frac{P(N,B|\theta,\mu=B)}{P(N,B|\theta=N-B,\mu=B)} \approx \chi^2. \tag{63}$$

Since we are interested only in an upper limit the appropriate interval is $(-\infty, u(N,B)]$. To calculate an approximate 90% upper limit we use the fact that $\text{Prob}\{\chi^2 < 2.7\} = \beta$ with $\beta = 0.9$. For the Grenoble data-set the answer is $u(3,7) = -0.192$! Needless to say, we did not regard this as satisfactory. A decade later it still does not seem satisfactory to me, so let's indulge in some tinkering.

One possibility is to modify the estimate $\hat{\theta}$ so that whenever $B > N$ we set $\hat{\theta} = 0$. After all, maximum likelihood is a heuristic principle so, surely, we are at liberty to make heuristic changes to the estimates so derived? Besides, we know that the signal *cannot* be negative. The modification seems eminently reasonable. However, a simple Monte Carlo simulation shows that if we replace the negative estimates by zero (and incidentally construct a pathological likelihood function) the distribution of the log-likelihood ratio will no longer be χ^2. By itself this isn't a disaster. The disaster is that this non-χ^2 distribution depends on θ and μ, and thus violates the log-likelihood ratio theorem. It would appear that the only way to comply with the theorem is to keep the unphysical negative estimates and negative upper limits.

Things are looking a little bleak, so let's try another tack. We could try to compute the sampling distribution of $N - B$. It turns out that this can be done exactly[12] and it is fun to do; try it. Unfortunately, nothing is gained because the distribution also depends on θ and μ. Try as we might we seem unable to excise the mean background count μ from our problem.

How can we salvage this? Well, given an ensemble of intervals, whose confidence level we do not know, there is one more thing we can do. We can perform an exhaustive search of the 2-dimensional parameter space $\theta \otimes \mu$ to find the minimum coverage probability, which by definition is the confidence level associated with the set of intervals, however they may be calculated. We shall use Eq. (63) to compute nominal 90% upper limits and then search the parameter sub-space $(\theta, \mu) \in [0, 20] \otimes [0, 20]$ to find the minimum coverage probability. Figure 5 shows how the coverage probability varies over the 2-d parameter space. We find that our putative 90% upper limits are really only 78% limits. Therefore, to be strictly correct, we should report these intervals as 78% intervals. Yet, looking at Fig. 5, it seems that for the sake of making Neyman happy, we are selling ourselves a bit short. The minimum coverage probability of 78% occurs only in a tiny region of parameter space. It does not adequately represent the fact that over most of the parameter space the coverage probability is much greater than 78%. But at least Neyman's ghost is smiling!

When I encountered this problem more than a decade ago I was astonished by the amount of difficulty it caused. As far as I know, no one has found a wholly satisfactory solution to this most basic of 2-parameter problems using frequentist ideas.

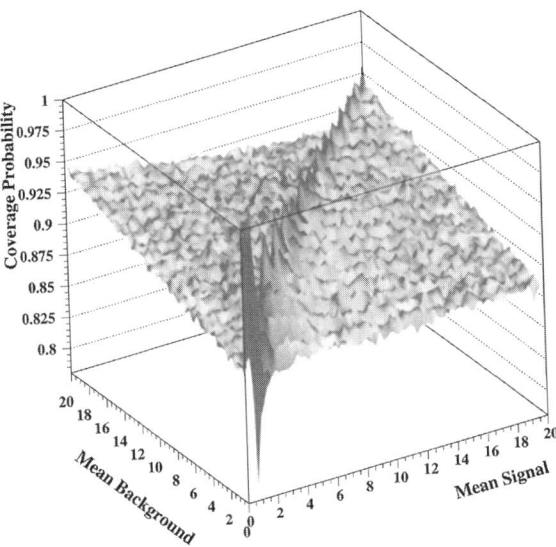

5. The coverage probability for the double Poisson counting experiment is plotted as a function of the Poisson parameters θ and μ. The confidence level is nominally 0.90. But because the minimum coverage probability is about 0.78, that is the number that should be reported as the confidence level.

PROBABILITY THEORY: AN APPRECIATION

"Probability theory is nothing but common sense reduced to calculation."

— Laplace, 1819

Probability theory, according to Laplace, is human reasoning abstracted. Not everyone agrees with him. In the first couple of lectures we learned about a method of statistical inference that was invented in the early part of this century, by Fisher, Neyman, Pearson and others. These people thought Laplace was wrong and decried his position as a metaphysical aberration that ought to have no role to play in science. Anything that smacked of subjectivity—a slippery word—was to be avoided at all cost; even at the cost of throwing the baby out with the bathwater, which, in effect, they did by abandoning the mathematical methods of Bayes and Laplace. The demands of science required a statistical theory that was wholly objective—another slippery word. The frequentists believed that they had created such a theory; one that let the data "speak for themselves". But as we saw in the previous lectures, contrary to their claim, that theory has its share of subjective elements, also.

I think Bayes and Laplace made a profound observation: no theory of inference can be wholly objective. Yet, in spite of our affinity with Laplace, most of us believe that probability has nary a thing to do with common sense and everything to do with relative frequency. This is not surprising. In a series of debates in the 1930s between Sir Ronald Fisher and Sir Harold Jeffreys, Fisher was particularly aggressive and, consequently, his view prevailed. Actually, that's a bit unfair! The Bayes-Laplace point of view has its own crop of difficulties of which many scientists are leery. We shall address those difficulties in due course.

Yet it is undoubtedly true that if we maintain a strict frequentist point of view then, as Jaynes[6] has noted, the domain of application of probability theory is so severely restricted that the majority of real-world inference problems that we wish to solve cannot be, except by heuristic means. The virtue of Bayes' and Laplace's point of view, and that of Sir Harold Jeffreys, Jaynes and many others, is that it leads to a theory of probability that can solve the kind of complicated statistical problems that occur in the real world. For that reason alone it is worth learning about. It is impossible to do justice to the theory of probability in these brief notes. My aim here is simply to give you a feel for what it is all about.

Deductive Reasoning

Way back around 350 BC, give or take, Aristotle—described by his teacher Plato as the *Nous* (intelligence personified) of his Academy—noticed that when we reasoned well we did so according to some definite rules all of which can be reduced to the **syllogisms**:

	modus ponens (ponere=affirm)	*modus tollens* (tollere=deny)
Major premise	If A is TRUE, then B is TRUE	If A is TRUE, then B is TRUE
Minor premise	A is TRUE	B is FALSE
Conclusion	Therefore, B is TRUE	Therefore, A is FALSE

In addition, if A is TRUE then its negation, written as $\overline{A}$, is FALSE. We say that A **contradicts** $\overline{A}$. A useful mnemonic for the syllogisms are the set of symbolic expressions:

	modus ponens	*modus tollens*
Major premise	$AB = A$	$AB = A$
Minor premise	$A = 1$	$B = 0$
Conclusion	$B = 1$	$A = 0$

The symbols A, B, 1, 0, and their **negations**, $\overline{A}$, $\overline{B}$, $\overline{1}$ and $\overline{0}$, are called **propositions**. The proposition 1 has a **truth value** which is always TRUE; 0 is always FALSE.

Here is an example. Let A = (You finished school) and let B = (You do well). Our major premise is: If (You finished school), then (You do well). Our minor premise is: (You finished school) is TRUE. Our logical conclusion is: Therefore, (You do well) is TRUE. Note, however, that if B is TRUE then it does *not* follow that A is TRUE. If (You do well) is TRUE it does *not* follow that (You finished school) is TRUE. Conversely, if A is FALSE we cannot conclude that B is FALSE. However, if (You do well) is FALSE then we conclude that (You finished school) is FALSE.

The simplest way to see this is to use the symbolic expressions and note that if B is set to 1 (that is, to the proposition that is always TRUE) in $AB = A$ we get $A = A$ and we are non the wiser about the truth value of A; likewise, if $A = 0$ then the truth value of B is undefined. This is as far as you can get with purely deductive reasoning. We say a bit more about these rules below.

An important point to notice is that from the nature of the propositions it is not clear that A and B are causally related. In fact, whether they are or not doesn't matter: the syllogisms deal only with the *logical* relationship between two propositions A and B. The syllogisms say nothing about whether A causes B or B causes A. Logical implication and causal implication are not the same thing.

Inductive Reasoning

Deductive reasoning can only get us so far. Most of the time we are not well enough informed to reason deductively. Actually, most of the time we do not even reason; witness some of the goings-on in the US Congress or the British Houses of Parliament! In the example given above no one doubts that if B is false then A is false. But what if B is true? We could be wrong, of course, but we feel intuitively that if B is true then this makes the proposition A more likely. Bayes and Laplace saw this as the essence of common sense. The question is: can such a vague thing be made rigorous?

It is extraordinary that the answer is an unequivocal yes. Common sense reasoning, called **plausible reasoning** by Jaynes, follows definite rules. One can view the Bayes-Laplace theory and its subsequent developments by Sir Harold Jeffreys, R. T. Cox, E. T. Jaynes and others as an extension of logic to include truth values that lie between FALSE and TRUE. These truth values, which are represented by real numbers in the interval $[0,1]$, are called **probabilities**; they are a quantitative measure of the plausibility of propositions. This theory assigns a quantitative meaning to the weaker syllogisms:

Major premise	If A is TRUE, then B is TRUE	If A is TRUE, then B is TRUE
Minor premise	B is TRUE	A is FALSE
Conclusion	Therefore, A is more likely	Therefore, B is less likely.

The Basic Rules

Probability theory, viewed as extended logic, is based upon two sets of rules. The first set is the algebra of propositions, called **Boolean algebra** after George Boole (1854). If A, B, C, 1, 0 and their negations are propositions, and $+$ and $\cdot$ are binary operations then:

	$A + 0 = A$	$A \cdot 1 = A$
	$A + \overline{A} = 1$	$A \cdot \overline{A} = 0$
Commutativity law	$A \cdot B = B \cdot A$	$A + B = B + A$
Distributivity law	$A \cdot (B + C) = A \cdot B + A \cdot C$	$A + B \cdot C = (A + B) \cdot (A + C)$

The above are called the **Huntington axioms**. Notice the duality between the operations ".". and "+". To simplify the notation we usually don't write the "." operation explicitly. From these rules you can derive thousands of theorems; here are a few:

$$\begin{array}{r|rcl|rcl}
& A+1 & = & 1 & A0 & = & 0 \\
& \overline{0} & = & 1 & \overline{1} & = & 0 \\
& A+AB & = & A & A(A+B) & = & A \\
\text{Idempotency law} & AA & = & A & A+A & = & A \\
\text{Associativity law} & A(BC) & = & (AB)C & A+(B+C) & = & (A+B)+C \\
\text{de Morgan's laws} & \overline{AB} & = & \overline{A}+\overline{B} & \overline{A+B} & = & \overline{A}\,\overline{B}.
\end{array}$$

See if you can prove them.

The second set of rules are the **product rule**

$$P(AB|C) = P(B|AC)P(A|C), \qquad (64)$$
$$= P(A|BC)P(B|C), \qquad (65)$$

the **sum rule**

$$P(A|B) + P(\overline{A}|B) = 1, \qquad (66)$$

and the conventional numerical assignments for certainty and its converse

$$P(1|B) = 1, \qquad (67)$$
$$P(0|B) = 0. \qquad (68)$$

From just the two sets of rules, plus the rules of arithmetic, a truly impressive mathematical theory has been built, some of whose applications we shall learn about in the last lecture.

Most of us would probably agree that the Huntington axioms seem clear enough intuitively. What of the product and sum rules, Eqs. (65) and (66)? They seem less intuitively obvious.

After the geneticist onslaught from the 1930s it was time for physicists to go on the offensive. In 1946 a major intellectual breakthrough occurred when, fittingly, a physicist R. T. Cox published a beautiful article[13] * in which he derived the product and sum rules from more primitive and intuitive foundations:

- **Axiom 1)** If we specify the plausibility of proposition A and then specify the plausibility of proposition B given proposition A, then we shall have specified, implicitly, the plausibility of propositions A and B together.

- **Axiom 2)** If we specify the plausibility of proposition A then we shall have specified, implicitly, the plausibility of its negation $\overline{A}$.

The empire has tried to strike back, but as far as I know no one has found fault with these axioms.

What Cox showed was that the seemingly vague Bayes-Laplace notion of probability as a measure of plausibility or "degree of belief" can be securely anchored, in stark constrast to the frequency interpretation that has, so far, resisted all attempts to secure its foundations, axiomatically. Ironically, it is the plausibility interpretation of probability that provides, via Jacob Bernoulli's theorem, the most satisfactory understanding of relative frequency and its relationship to probability.

---

*R. T. Cox, Probability, Frequency, and Reasonable Expectation, *Am. J. Phys.*, 14:1 (1946). The article is declared by Jaynes as "the most important advance in the conceptual (as opposed to the purely mathematical) formulation of probability theory since Laplace".

To illustrate the use of the rules let's try to prove a theorem. We want to relate $P(A+B|C)$ to $P(A|C)$ and $P(B|C)$. All we need do is apply the above rules repeatedly in more or less the right order. As you "walk through" the proof, check off the rule that is being used:

$$\begin{aligned}
P(A+B|C) &= 1 - P(\overline{A+B}|C) \\
&= 1 - P(\overline{A}\,\overline{B}|C) \\
&= 1 - P(\overline{B}|\overline{A}C)P(\overline{A}|C) \\
&= 1 - \left[1 - P(B|\overline{A}C)\right]P(\overline{A}|C) \\
&= 1 - P(\overline{A}|C) + P(B|\overline{A}C)P(\overline{A}|C) \\
&= P(A|C) + P(B|\overline{A}C)P(\overline{A}|C) \\
&= P(A|C) + P(\overline{A}B|C) \\
&= P(A|C) + P(\overline{A}|BC)P(B|C) \\
&= P(A|C) + [1 - P(A|BC)]P(B|C) \\
&= P(A|C) + P(B|C) - P(A|BC)P(B|C) \\
P(A+B|C) &= P(A|C) + P(B|C) - P(AB|C). \quad (69)
\end{aligned}$$

Of course, this is something we could have guessed from drawing Venn diagrams and thinking about sets. And that's fine, because that's how many theorems are found: you guess a result and then you try to prove it. However, the beauty of this version of probability theory is that it is *not* tied to sets, measures and other esoterica, but to something much more profound: plausible human reasoning.

Bayes' Theorem

In 1763, Thomas Bayes published a paper [†] in which he gave a special case of a theorem that has proven to be immensely fruitful and controversial. Bayes' theorem

$$P(B_k|AC) = \frac{P(A|B_kC)P(B_k|C)}{\sum_i P(A|B_iC)P(B_i|C)}, \quad (70)$$

where A, B_k and C are propositions is a direct consequence of the product rule of probability theory. We shall see later that it enjoys a ubiquitous presence in data analysis.

Here is a simple non-controversial example of its use. You're being hounded by a pack of dogs of which 25% are rabid. You have a good nose for rabid dogs, so if a dog is in fact rabid then 80% of the time you would recognize it as such. If you like, this is your rabid-dog identification efficiency. However, you have a 10% chance to mis-identify a poor mutt as rabid when in fact it is not. But, in our politically correct world, you must refrain from opening fire on a dog if there is more than a 30% chance of shooting one that is healthy, otherwise you risk a grisly encounter with the Animal Liberation Front. A dog pounces at you. Do you shoot it? Ask the Reverend. We summarize the probabilities as follows

$$P(\text{Ok}|C) = 0.75,$$

[†] Posthumously as it turned out—he died in 1761, even though there is some evidence that he had figured out his result as early as 1748. Maybe the Reverend was so worried about how his ideas might be received that he decided to die before publishing them!

$$P(\text{Bad}|C) = 0.25,$$
$$P(\text{rabid}|\text{Bad} \cdot C) = 0.80,$$
$$P(\text{rabid}|\text{Ok} \cdot C) = 0.10, \tag{71}$$

where C could be some proposition like C =(The dog is a pit-bull terrier). So we read $P(\text{Bad}|C)$ as the probability that the dog is bad, that is, rabid, given that it is a pit-bull terrier; $P(\text{rabid}|\text{Ok} \cdot C)$ is the probability to classify as rabid a dog that is, in fact, healthy. Applying Bayes' theorem we get

$$\begin{aligned} P(\text{Ok}|\text{rabid}) &= \frac{P(\text{rabid}|\text{Ok} \cdot C)P(\text{Ok}|C)}{P(\text{rabid}|\text{Ok} \cdot C)P(\text{Ok}|C) + P(\text{rabid}|\text{Bad} \cdot C)P(\text{Bad}|C)}, \\ &= \frac{0.1 \times 0.75}{0.1 \times 0.75 + 0.80 \times 0.25}, \\ &= 0.27. \end{aligned} \tag{72}$$

So according to the Reverend, the chance that a dog, having been classified as rabid, is actually healthy is 27%. Being less than your cut of 30%, you open fire secure in the knowledge that you've approached the issue with due rigor!

THE WORLD ACCORDING TO THOMAS BAYES

In the first two lectures we surveyed the body of statistical ideas called orthodox or frequentist statistics. This is the stuff that most of us know and use in our everyday scientific work. The essential notion is of an ensemble of possible outcomes, or data-sets. The ensembles can be real; for example: all the forty-year olds in St. Croix with a bad haircut. But often they are not. This, however, does not make them necessarily less useful. On the contrary, ensembles are a powerful and ubiquitous tool in physics. So useful are they, however, that sometimes we forget to distinguish between those that have an objective reality and those that are mental constructs. Conceptual difficulties can arise. If we have performed an experiment only once, how do we know to which ensemble it belongs? By a mere act of will we can put our multi-million dollar experiment into a different ensemble and perhaps get a different experimental result, without altering a single datum! It seems futile to argue about something that exists only in the imagination, but we do. The arguments arise because we cannot agree on whose imaginary ensemble is "the correct one" to analyze real data. The issue is to agree on what aspects of the experiment are to be considered held constant, and thereby define the imagined repetitions.

The notion of an ensemble of data-sets leads, naturally, to the idea of the sampling distribution of estimates. We have seen how the error on an estimate can be quantified using either the variance of the sampling distribution or a confidence interval derived from the same. Frequentist statistics deals with much more than this, of course. Of the many topics we have not covered one is of particular interest to everyone, namely: **model selection** or hypothesis testing.

My omission was deliberate. Hypothesis testing is the least satisfactory bit of the frequentist theory. Even the frequentists themselves could not, and do not, agree about how hypothesis testing should be done. Fisher was adamant that an hypothesis could be tested in isolation. Neyman, echoing the Bayesians, thought this was unsatisfactory and argued that alternatives must be specified to render the problem well-posed. These disagreements are bad enough, but are minor in comparison to a much more serious practical problem. The frequentist theory of hypothesis testing, which is well developed for testing simple hypotheses, becomes essentially unworkable in the presence of nuisance parameters.

In sharp contrast, in Bayesian probability theory, hypothesis testing or model selection is well-founded and is intuitively straightforward. For this reason I shall deal with it from a Bayesian viewpoint. There are many other issues that arise in data analysis. Far too many to be covered in these notes. I have, therefore, elected to cover, in this lecture, topics that I hope will command broad interest. Again, all done from the Bayesian viewpoint. Before I do that, however, I would like to say a bit more about the clash between the two schools of thought on the meaning of probability.

Relative Frequency Interpretation

Frequentists interpret probability as the relative frequency with which something happens. Let T be the number of experiments or **trials**; for example, the number of times protons have been bashed against antiprotons in a particle accelerator. Let S be the number of events of interest or **successes**; this could be the number of times a top-quark antitop-quark pair has been created in a particular volume of phase space. The relative frequency of a success is

$$\frac{S}{T}. \tag{73}$$

Our expectation, borne of experience, is that as $T \to \infty$ the relative frequency S/T settles down to a number, call it p, which we interpret as the probability of a success. Many people have tried to build a theory of probability on this basis, but have met with failure. The reason is easy to see.

Any such theory of probability must deal with the following annoying possibility. It is possible, though exceedingly unlikely, that on every trial we get a success, or a failure. We, therefore, have to be more careful about what we mean by the "settling down" of the fraction S/T. The correct statement, first noted by Jacob Bernoulli, is the **law of large numbers** which states that

$$\text{Prob}\{|\frac{S}{T} - p| > \epsilon\} \to 0, \text{ as } T \to \infty, \tag{74}$$

for any number $\epsilon > 0$. That is, as the number of trials grows to infinity, the *probability* that the fraction S/T differs from the *probability* p by more than ϵ goes to zero. But notice that the last sentence contains a potential difficulty: a recursive definition of probability. Now, there is nothing wrong, in principle, with recursive definitions. They are often extremely useful. In this case, however, the implied recursion fails. It fails because the same word is used in two fundamentally different ways. The second probability in that sentence can be interpreted loosely as the "limit" of the relative frequency S/T; loose because no limit, in a mathematical sense, exists for the fraction S/T. A necessary, though not sufficient, condition for a limit to exist is the existence of a rule that tells us how to compute the next outcome from the previous one. But no such rule exists because the outcomes are, by hypothesis, random!

What of the first probability? A moment's reflection will reveal that, if we try to take the implied recursion seriously and try to interpret the first probability also as the limit of a relative frequency, we shall be ensnared in an infinite hierarchy of infinite sets of trials.

Plausibility Interpretation

Perhaps, one day, a mathematician with a Cantorian sensibility will be able to make sense of the infinite tower of infinite trials. But until that day arrives we are obliged to try something else. We can avoid the infinite regression by interpretating the first probability as a measure of the *plausibility* of the proposition: $A = (S/T \to p$ as $T \to \infty)$. Bernoulli's law of large numbers is then a declaration that it is plausible to the point of certainty that

$$S/T \to p,$$

for an infinite number of trials. I find this interpretation of Bernoulli's theorem to be intuitively cogent. Its corollary is that the relative frequency interpretation is a *derived* notion pertaining to a special class of circumstances, namely, those in which one can entertain, in principle, performing *identically* repeated trials in which the probability of success at each trial is the same.

As we saw in the last lecture, the apparently vague idea of the "plausibility of a proposition" can be rendered mathematically sound. Once this is realized we have cleared the path to the use of probability theory, to perform inductive reasoning, in its full generality.

BAYESIAN PROBABILITY THEORY

Proof of Bayes' Theorem

The product rule

$$P(AB|C) = P(B|AC)P(A|C), \tag{75}$$
$$= P(A|BC)P(B|C), \tag{76}$$

and the sum rule

$$P(A|B) + P(\overline{A}|B) = 1, \tag{77}$$

are the basis of Bayesian inference. They are a precise statement of inductive reasoning. Two propositions A and B are said to be **mutually exclusive** if the truth of one denies the truth of the other, that is: $P(AB|C) = 0$. In that case, from the theorem we proved in the last lecture, we get

$$P(A + B|C) = P(A|C) + P(B|C), \tag{78}$$

which is easily generalized to any number of mutually exclusive propositions. A set of mutually exclusive propositions B_k is said to be **exhaustive** if their probabilities sum to unity:

$$\sum_k P(B_k|C) = 1. \tag{79}$$

If $B_k D_j$ are a set of mutually exclusive and exhaustive **joint propositions**, then we can write Bayes' theorem as

$$P(B_k D_j|AC) = \frac{P(A|B_k D_j C)P(B_k D_j|C)}{\sum_{i,l} P(A|B_i D_l C)P(B_i D_l|C)}, \tag{80}$$

which is a generalized form of that given in the last lecture. To prove the theorem start by replacing B, in the product rule Eq. (76), by the joint proposition $B_k D_j$

$$P(A|B_k D_j C)P(B_k D_j|C) = P(B_k D_j|AC)P(A|C), \qquad (81)$$

and then sum over the set of joint propositions:

$$\sum_{k,j} P(A|B_k D_j C)P(B_k D_j|C) = P(A|C) \sum_{k,j} P(B_k D_j|AC)$$
$$= P(A|C). \qquad (82)$$

In the last step we have used the assumption that the joint propositions $B_k D_j$ form an exhaustive set. We now go back to the product rule Eq. (81), substitute in our formula for $P(A|C)$ and re-arrange to get Bayes' theorem, Eq. (80). At this point let us note that, to reduce Bayes' theorem to a statement about the proposition B_k only, we need merely to sum Eq. (80) over the propositions D_j. Bayes' theorem can be extended easily to encompass larger and larger sets of joint propositions.

The Bayesian Procedure

We are almost ready to state how Bayesian probability theory solves the inference problem. But first some more definitions. The probability $P(B_k D_j|C)$ is called the **prior** probability of the joint proposition $B_k D_j$; $P(A|B_k D_j C)$ is the **likelihood** assigned to proposition A given the joint proposition $B_k D_j C$ and $P(B_k D_j|AC)$ is the **posterior** probability of proposition $B_k D_j$. We shall suppose that our wish is to assess the probability of the propositions B_k regardless of propositions D_j, which are of no interest to us. The Bayesian inference procedure is:

- Assign a prior probability $P(B_k D_j|C)$ to each proposition $B_k D_j$ in light of what we know about them. This **prior knowledge** is represented by the proposition C.

- Acquire some pertinent evidence, represented by proposition A, and assign to these data a probability $P(A|B_k D_j C)$.

- Compute the posterior probability $P(B_k D_j|AC)$ from Bayes' theorem.

- Since we are interested in the set of propositions $\{B_k\}$ only, we excise the uninteresting ones, $\{D_j\}$, from the problem by summing over them: $P(B_k|AC) = \sum_j P(B_k D_j|AC)$. This operation is called **marginalization**. Now that we have the posterior probability $P(B_k|AC)$ we can answer all rational questions about the proposition B_k.

An example might make this less abstract.

Suppose A is the proposition that several meteorological variables have certain values. Then perhaps D_j is the proposition that a hurricane of strength j will batter St. Croix; B_k could be the proposition that Professor Ferbel will have to pay the WCCC (World Center for Chicken Cuisine) 10^k dollars of lost revenue due to the cancellation of the school. It may be that the dear professor could care less about the strength of the hurricane, but cares dearly about his potential financial liability. So, being a born-again Bayesian, he sums over the uninteresting propositions, D_j—about the potential strength of the hurricane, to leave himself with a probability, $P(B_k|AC)$, specific to his current concern.

The above procedure is a mathematical model of inductive reasoning. As such it captures only some aspect of human reasoning, which surely is vastly more complex than the model implies. Yet even this simplified model of reasoning has proven to be extraordinarily powerful.

This is impressive stuff! So why aren't we impressed? A majority of high energy physicists are still of the opinion that Bayesian reasoning is somehow ill-founded. Even a cursory reading of the Bayesian literature will attest that it is this opinion that is ill-founded. The real reason for our reluctance to embrace these powerful ideas must be sought elsewhere. I would say that the chief reason we are leery of Mr. Bayes and Mr. Laplace is a serious practical one: the rules say only *how to reason*; they do not say *how to assign the probabilities*. Most of us are happy with the standard arguments for assigning likelihoods. All of these arguments reduce ultimately to (sometimes difficult) combinatoric reasoning. Few of us, however, agree on how to assign prior probabilities. Many of us are not persuaded that prior probabilities are even assignable. A plethora of methods have been advanced to assign them; none is compelling[14]. With such pervasive uneasiness about prior probabilities it seems much easier to "do a Fisher": throw the water overboard, even at the risk of throwing out the baby, and thereby sidestep the problem. But we have seen that these attempts did not meet with complete success either. And they are decidedly less well-founded.

So what are we to do? I would suggest we do what any good scientist does: ask a few pertinent questions. We should ask which theory is the more general, which can solve the most problems, which theory works better in practice and what, in practice, do working scientists do? On the last point the evidence is clear: experimental physicists, having long ago sensed the inadequacy of a purely frequentist approach, have been happy to patch up the statistical theory they learned at college with Bayesian grafts. I would hazard a guess that, for the most part, we did so blithely unaware that these fixes had anything to do with Mr. Bayes. We got reasonable answers; so who cared that we weren't purists. Certainly not the reviewers of our most prestigious journals. Such pragmatism is laudable. Had we been mired in the Bayesian-frequentist controversy we would have gone nowhere.

But pragmatism, like any good thing, can be overdone. Sooner of later the *ad hoc* edifice that we have built to do our data analysis comes crashing down and degenerates into numerology. We can do better; we should do better. In spite of the difficulty with prior probabilities, it is worth taking a closer look at Bayesian probability theory. If we find it lacking then by all means we should re-build our fallen edifice and muddle on. However, as I shall try to argue, the whole controversy about prior probabilities, while one should acknowledge its existence, is somewhat overblown. Spectacular progress can be, and has been, made in spite of it. So, let's press on!

Continous Sets of Propositions

In many applications in high energy physics the propositions we make are just declarations that a set of parameters have a set of values, which we assume form a continuous set. Suppose that

$$P(\mathbf{x}|\theta, \lambda, I) = \int_\Omega L(\mathbf{z}|\theta, \lambda, I) d\mathbf{z}, \tag{83}$$

is the probability assigned to the data-set **x**, contained in a neighborhood Ω of **x**, and that θ and λ are the parameters of the model being tested. Perhaps, θ is the parameter that is of interest to us while λ is not—that is, λ represents one or more nuisance parameters. If $P(\theta, \lambda | I) = h(\theta, \lambda) d\theta d\lambda$ is the prior probability we have assigned to the proposition that θ and λ have certain values, we can write Bayes' theorem as

$$P(\theta, \lambda | \mathbf{x}, I) = \frac{P(\mathbf{x} | \theta, \lambda, I) P(\theta, \lambda | I)}{\int_{\theta, \lambda} P(\mathbf{x} | \theta, \lambda, I) P(\theta, \lambda | I)}, \quad (84)$$

which is valid for a continuous set of propositions. In the denominator we are summing over all possible propositions about the value of θ and λ. The I in Bayes' theorem is the proposition that earlier we called C. It reminds us that the probabilities are conditional on whatever prior assumptions we have made or prior information we have about the inference problem. Every probability is conditional. It follows, therefore, that there is no such thing as an absolute probability. Sometimes, for simplicity, we shall drop the I; but we should not be misled by this bit of sloppiness.

In the frequentist theory we saw how hard it was to rid ourselves of nuisance parameters; try as we might they kept re-appearing all over the place like wayward mushrooms. How do we get rid of them in Bayesian theory? The answer was given above. We excise the nuisance parameters λ from the problem by marginalization, that is, by integrating over the posterior probability:

$$P(\theta | \mathbf{x}, I) = \int_{\lambda} P(\theta, \lambda | \mathbf{x}, I). \quad (85)$$

This is very beautiful, for now we have a probability that summarizes all that we know about the parameter θ, given the evidence we have acquired, represented by the symbol **x**. This is the general method of performing inferences using Bayesian probability theory and of dealing with nuisance parameters.

The posterior probability displays a very important philosophical, and practical, difference between frequentist and Bayesian statistics:

- The posterior probability depends only on the data observed.

This fundamental principle, namely that inferences are to be based upon the data observed and whatever prior information we have, is called the **likelihood principle**. (Not to be confused with the principle of maximum likelihood!)

The use of an ensemble of outcomes is sharply at odds with the likelihood principle; consequently it is at odds with Neyman and Pearson's notion of confidence intervals and a host of standard frequentist practice. So is the likelihood principle sensible? Certainly, Sir Harold Jeffreys thought so. Ironically, even his nemesis, Fisher—a forceful critic of all things Bayesian, was an advocate of the likelihood principle. Indeed, Fisher's development of his theory of fiducial inference was his enigmatic attempt to make good on this principle, without conceding an inch—so he thought—to Bayesians. As for me, I'm inclined to agree with both of them: why is it more objective to base an inference upon an ensemble of hypothetical data-sets, rather than on the single data-set I got from one real experiment? Yet this mode of reasoning is precisely what we are asked to accept when we perform the following, typical, frequentist test: given some χ^2 value we ask for the probability that a χ^2 value greater than or equal to that observed could have been observed. Or we observe a top quark signal and we calculate the probability to observe a count equal to or greater than that actually observed. We are so accustomed to this mode of reasoning that we fail to recognize how utterly subjective it is. How can we possibly *know* what we could have observed? All we can do is to *imagine* what could have been observed, which is not the same thing as knowing. It is logically possible that, were we to actually repeat our experiment, we might observe that which we failed to imagine. Indeed, that's often how progress is made.

Parameter Estimation

Once we have the posterior probability for the parameter θ, having integrated out all nuisance parameters, we can derive an estimator for θ by applying a modicum of decision theory. Furthermore, in accordance with the likelihood principle, we can quantify the uncertainty of our estimate directly from the posterior probability. So here, as promised, is the justification for quoting an error based on the particular data-set we have, rather than on an ensemble thereof. The justification is the likelihood principle, which follows from Bayes' theorem, which in turn is ultimately a consequence of the fundamental axioms of rationality, stated by the physicist R. T. Cox. In short, the justification is that probability does not require an ensemble of repeated trials for its interpretation. This is not to say that ensembles are a bad thing. It simply means that if an ensemble does not exist it does not matter.

In the first lecture we were introduced to the idea of a loss function as a way to measure the quality of a decision. A typical decision is: Given a data-set $\mathbf{x} = (x_1, \ldots, x_N)$ decide that the estimate of θ is $\hat{\theta} = d(\mathbf{x})$, where $d(\mathbf{x})$ is a special kind of decision function called an estimator. Let's use the same loss function as used in the frequentist theory, namely:

$$\mathcal{L}(\theta, d) = (\theta - d)^2 . \tag{86}$$

We also introduced the risk function as some kind of average of the loss function; we'll do likewise in the Bayesian theory. But whereas in the frequentist theory the averaging was done with respect to all possible data-sets $\mathbf{x}$, in the Bayesian theory the likelihood principle tells us that we should consider only the data-set we have. Instead of averaging over data-sets, we average over all possible propositions about the value of θ—that is, over all **hypotheses**, constrained by the fact that we have a definite data-set. The probability of a particular hypothesis about the value of θ, given the data-set $\mathbf{x}$, is precisely what we have called the posterior probability for θ: $P(\theta|\mathbf{x})$. Accordingly, we are led to consider the risk function

$$\begin{aligned}\mathcal{R}_d(\mathbf{x}) &= \mathrm{E}[\mathcal{L}(\theta, d)]_\theta, \\ &= \int_\theta \mathcal{L}(\theta, d) P(\theta|\mathbf{x}),\end{aligned} \tag{87}$$

that is,

$$\mathcal{R}_d(\mathbf{x}) = \int_\theta (\theta - d)^2 P(\theta|\mathbf{x}), \tag{88}$$

when we use our favorite loss function.

As in the frequentist theory, the best estimator, for a specified loss function, is declared to be that which minimizes the risk:

$$\mathrm{D}_d \mathcal{R}_d(\mathbf{x}) = \mathrm{D}_d \int_\theta \mathcal{L}(\theta, d) P(\theta|\mathbf{x}),$$

$$= \int_\theta D_d \mathcal{L}(\theta, d) P(\theta|\mathbf{x}),$$
$$= 0. \tag{89}$$

The symbol D_d is the derivative operator with respect to $d(\mathbf{x})$. We have made the usual assumption that the derivative and integral operators commute. When we perform the minimization using the quadratic loss function we obtain as the best Bayesian estimator for θ the intuitively pleasing result

$$\hat{\theta} = d(\mathbf{x}) = \int_\theta \theta P(\theta|\mathbf{x}). \tag{90}$$

In words:

- The best estimator for a parameter is the mean of the posterior probability.

Of course, this is true only for the quadratic loss function. Another loss function will, in general, lead to another estimator. Here is a problem for you to ponder: What kind of loss function would you need so that the best estimator is the value of θ at the peak of the posterior probability? This would be the Bayesian analog of a maximum likelihood estimator. Incidentally, it is now clear that maximum likelihood is nothing more than one of many ways to extract an estimate from the posterior probability. Indeed, there is no mathematical distinction between a posterior probability and the likelihood function when we choose a flat prior probability for the parameter of interest. Hidden behind the maximum likelihood principle is an implicit assumption of a flat prior!

Do we now need to compute the sampling distribution of $\hat{\theta}$? No! Some people do; but we don't have to. According to the likelihood principle, the posterior probability contains all the information we have about the parameter—so, nothing else is needed. As alluded to earlier, we can measure the accuracy of the estimator from the posterior probability. One measure of the accuracy of the estimator is the variance

$$\text{Var}[\mathbf{x}] = \int_\theta \theta^2 P(\theta|\mathbf{x}) - d^2. \tag{91}$$

Another measure is the Bayesian analog of a confidence interval, $\left[l(\hat{\theta}), u(\hat{\theta})\right]$, obtained from the formulae:

$$\int_{\theta \leq l(\hat{\theta})} P(\theta|\mathbf{x}) = \alpha_L \tag{92}$$

and

$$\int_{\theta \geq u(\hat{\theta})} P(\theta|\mathbf{x}) = \alpha_R. \tag{93}$$

The numbers α_L and α_R as chosen so that $\beta = 1 - \alpha_L - \alpha_R$, that is, so that

$$\int_{\theta=l(\hat{\theta})}^{\theta=u(\hat{\theta})} P(\theta|\mathbf{x}) = \beta. \tag{94}$$

These formulae appear to be rather like those used to compute intervals in the frequentist theory. However, the similarity is superficial and misleading. Here the confidence level β is not defined by reference to an ensemble of intervals. The interpretation of a $100 \times \beta\%$ Bayesian confidence interval is direct: β is the probability that the proposition $A=$(The true value of θ is bounded by $l(\hat{\theta})$ and $u(\hat{\theta})$) is true. Contrast the intuitive simplicity of this statement with the convoluted reasoning required to interpret Neyman's intervals, which even Fisher did not like.

Optimal Event Selection

Before we can measure something, we must have events. A basic task of data analysis, therefore, is to separate signal from background. It is one of the more enjoyable tasks because we get a chance to show off our cleverness. The most important work, of course, is using one's physical intuition to devise fiendishly clever variables to dig out the signal. But, having found our fantastic variables, the drudgery begins: finding cuts.

The traditional method to find cuts combines a judicious use of common sense with a dose of trial and error. We all have experienced, I'm sure, how extremely time consuming that can be; especially if, along the way, we have to refine the odd variable or two. Surely, there is a better way. Well, of course there is; and it should come as no surprise that it has something to do, again, with Mr. Bayes!

It helps to think about the problem geometrically. Suppose we have found n variables that we consider useful for separating signal from background. The n variables can be thought of as a point in an n-dimensional space, sometimes referred to as **feature space**. Presumably, by construction, the signal tends to cluster in one part of this space while the background occupies a different region. However, inevitably, there will be some overlap between the signal and background distributions. The problem to be solved is to find the boundary that separates optimally signal from background. Usually we do the simplest thing: we construct a boundary from planes that are perpendicular to the axes, where each plane corresponds to a cut on a specific variable. However, in general, the optimal boundary cannot be built from such intersecting planes; in general, it will be a curved surface.

The problem of finding this surface, however, is indeterminate until we have specified what we mean by optimal. This is what most people mean by optimal: a boundary is optimal if it minimizes the probability to misclassify events. So all we have to do is to write down a function of this probability, minimize it and we're done. For the moment, let's suppose that we know the signal and background distributions: $p(\mathbf{x}|S)$ and $p(\mathbf{x}|B)$, respectively. Let us further assume that we know the signal and background prior probabilities $p(S)$ and $p(B)$. These prior probabilities are not controversial: $p(S)$ is just the chance to pick a signal event without regard to its **feature vector x**, and likewise for $p(B)$. Since the event must be either signal or background we must have that $p(S) + p(B) = 1$. What's the chance to misclassify a signal event, with feature vector **x**? It is just the chance for the signal event to land on the background side of the boundary. It's easiest to see what's going on if we consider a one dimensional problem, with the boundary, say, at $x = x_0$. The probability E_S to misclassify a signal event is

$$E_S(x_0) = p(S) \int_x p(x|S) h(x_0 - x). \tag{95}$$

Here $h(z)$ is the step function, which is defined by $h(z) = 1$, if $z > 0$, but zero otherwise. The probability to misclassify the background is likewise the probability for the background to land on the signal side:

$$E_B(x_0) = p(B) \int_x p(x|B) h(x - x_0). \tag{96}$$

Hence, the probability to misclassify events, regardless of whether they are signal or background, is just

$$E(x_0) = E_S(x_0) + r\, E_B(x_0), \tag{97}$$

where we have introduced the weight r to allow for the possibility that we may wish to weight the background more (or less) than the signal. We now minimize $E(x_0)$ with respect to the choice of boundary, that is, we set $D_{x_0} E = 0$ and obtain

$$p(S) \int_x p(x|S)\delta(x_0 - x) + p(B) \int_x p(x|B)\delta(x - x_0) = 0. \tag{98}$$

The derivative operator has conveniently converted the step functions into delta functions, thereby rendering the integrals trivial. After almost no algebra we can savor our prize; the optimal boundary is obtained as the solution to the equation

$$r(x_0) = \frac{p(x_0|S)p(S)}{p(x_0|B)p(B)}. \tag{99}$$

The function $r(\cdot)$ is called the **Bayes discriminant**, because of its intimate connection with Bayes' theorem:

$$p(S|x) = \frac{r}{1+r} = \frac{p(x|S)p(S)}{p(x|S)p(S) + p(x|B)p(B)}. \tag{100}$$

The n-dimensional generalization of this has the same Bayesian form. The posterior probability $p(S|\mathbf{x})$ is precisely that needed for event classification. It is the probability that an event characterized by the vector $\mathbf{x}$ is of the signal class. By using this probability we have succeeded in mapping the original n-dimensional problem into a more tractable one-dimensional one.

This is all very pretty, but there is a serious practical problem. Rarely do we have analytical expressions for the signal and background distributions $p(\mathbf{x}|S)$ and $p(\mathbf{x}|B)$. We seem, alas, to have achieved a pyrrhic victory! Well, this would be the sad conclusion were it not for the existence of some impressive mathematical results obtained in the last few years, regarding artificial neural networks. A neural network is a highly non-linear function that maps n inputs into one or more outputs. It has been shown[15] that, under suitable circumstances, its output is a direct approximation to the probability $p(S|\mathbf{x})$.

This is a significant conceptual breakthrough: we now know what the network output means. Before this understanding was gained we could enjoy the parody of neural networks as mysterious black boxes that perform miracles in unfathomable ways. The reality is more prosaic: a neural network is an example of a special kind of mathematical function[16] that was first studied by Kolmogorov in the 1950s. He described a general way to build functions that could approximate, with arbitrary accuracy, any map of the unit n-cube into the unit interval. And that's exactly what we need to do to perform an optimal demarcation between signal and background, which demarcation can be calculated straightforwardly with a neural network, often in a matter of minutes on today's average desktop computer.

Combining Results

At some point we must sit down with our rivals and combine our results. In the frequentist theory the results from different experiments are combined using a weighted average. That approach requires, ideally, the results to be unbiased. There is, however, a more general way to combine results: use Bayes' theorem. The posterior probability obtained from one experiment serves as the prior probability for another. Actually, the order in which experiments are combined is completely immaterial. By repeatedly applying Bayes' theorem it is easy to show that for K independently obtained data-sets $\mathbf{x}_1, \ldots, \mathbf{x}_K$ the overall posterior probability is

$$P(\theta, \lambda | \mathbf{x}_1, \ldots, \mathbf{x}_K) = \frac{P(\mathbf{x}_1 | \theta, \lambda) \cdots P(\mathbf{x}_K | \theta, \lambda) P(\theta, \lambda)}{\int_{\theta, \lambda} P(\mathbf{x}_1 | \theta, \lambda) \cdots P(\mathbf{x}_K | \theta, \lambda) P(\theta, \lambda)}. \qquad (101)$$

Try to prove it! This is essentially just the joint likelihood function for all the experiments times the prior probability. Unlike the weighted average, this method will converge to the true value as more and more experiments are combined, even if each experiment is biased the same way, provided that the result from each experiment is consistent.

Model Selection

Let's say we have a set of competing models M, which may depend upon different sets of parameters θ_M. Given some prior information I and some data-set $\mathbf{x}$, how do we decide between the various models? This is the problem of hypothesis testing or model selection.

The first thing we do is assign a probability, $P(\mathbf{x} | \theta_M, M, I)$, to our data-set given a model M and hypotheses about the value of the corresponding parameters θ_M. We must also assign a prior probability $P(\theta_M, M | I)$. Then we write down Bayes' theorem

$$P(\theta_M, M | \mathbf{x}, I) = \frac{P(\mathbf{x} | \theta_M, M, I) P(\theta_M, M | I)}{\sum_M \int_{\theta_M} P(\mathbf{x} | \theta_M, M, I) P(\theta_M, M | I)}. \qquad (102)$$

The symbol $P(\theta_M, M | \mathbf{x}, I)$ represents the probability of the proposition that M is the "correct" model with parameter values θ_M, given the evidence $\mathbf{x}$ and I.

It is very important to understand that the probabilities $P(\theta_M, M | \mathbf{x}, I)$ are conditioned on the set of models considered. That's why the word correct is in quotes. Correct, in this context, means the best of the current bunch. If someone came up with yet another model, then the probabilities would, in general, change were we to include the new model in the set under consideration. Therefore, $P(\theta_M, M | \mathbf{x}, I)$ cannot be construed as an absolute measure of the validity of a model. But it *is* a measure of the conditional validity of a model: it provides a way to *compare* models within a given set in light of what we know. If a rational thinker had to choose a single model she would opt for the model with the highest posterior probability. Of course, should she acquire further pertinent information that information, via Bayes' theorem, could cause our rational thinker to change her mind. This is not a defect, but a virtue.

One last thing. We can marginalize $P(\theta_M, M | \mathbf{x}, I)$ with respect to θ_M to obtain $P(M | \mathbf{x}, I)$, the probability of model M. This is potentially very useful if each model within the set are identical, except for a single parameter α. For example, the M could label models that differ by the assumed top quark mass. We then have a way to estimate that parameter:

$$\hat{\alpha} = \sum_M \alpha_M P(M | \mathbf{x}, I), \qquad (103)$$

and its associated uncertainty

$$\sigma_\alpha^2 = \sum_M (\alpha_M - \hat{\alpha})^2 P(M | \mathbf{x}, I). \qquad (104)$$

DATA ANALYSIS

We now have enough of the Bayesian theory in hand to see how it works in practice. To this end we shall conclude these notes with a look at two realistic examples of Bayesian analysis. They are:

- How to analyze a counting experiment.

- How to analyze the solar neutrino problem.

But first we need to discuss the thorny issue of prior probabilities.

Prior Probabilities

To solve an inference problem we must assign two probabilities: a prior and a likelihood. Many of the techniques of sampling theory can be used to determine the latter and there is little disagreement about the results. Everyone is happy to use the Poisson, binomial, gamma, χ^2, Gaussian and other standard distributions as appropriate. However, even when people agree that prior probabilities are necessary, they often disagree about how to assign them. The problem is particularly acute when we have minimal prior information about the parameters to be estimated.

The critics of the Bayesian use of Bayes' theorem charge that the choice of prior probabilities is subjective and arbitrary and, therefore, so too are the inferences derived from the same. It may be that the only prior information we have about a parameter θ is that it is positive. What prior probability should we assign to various hypotheses about its value? Laplace argued that if we know nothing about the value of a parameter then we should assign a flat prior probability: $P(\theta|I) \propto d\theta$. This seems reasonable, until we realize that *any* choice of prior probability for a given parameter implies that we have specified, implicitly, the prior probability for the infinity of parameters that are functions of θ. Clearly, we have done a lot more than we bargained for!

For example, suppose we transform from θ to the parameter $\alpha = 1/\theta$ then mathematical consistency appears to demand that its prior probability distribution be $P(\alpha|I) \propto d\alpha/\alpha^2$; a form that looks, at best, non-intuitive.

This prior probability would be fine were it not for the following question: what reason do we have to suppose that the prior probability is uniform in θ rather than in the parameter α or some other parameter, like $\tau = \theta^{18}$? It seems that the assignment of prior probabilities for a parameter about which we are almost totally ignorant is, indeed, arbitrary. This is the core of the controversy about prior probabilities that has raged unabated for more than 200 years.

The problem of how to assign prior probabilities to represent a state of "almost complete ignorance" about the value of a parameter was re-examined early this century by Sir Harold Jeffreys[3]. Later, Jaynes[17] suggested that the prior probability, representing a state of "almost complete ignorance", should respect the symmetries of the problem. An example will make this clear.

Let θ be the rate at which accidents occur at a particular intersection. The probability that exactly n accidents occur in a fixed time t is

$$P(n|\theta) = e^{-\theta t} \frac{(\theta t)^n}{n!}. \tag{105}$$

We wish to infer something about the parameter θ. All that we know beforehand is that the accident rate is a positive number. Actually, we know a bit more than that: we know that the probability to observe n accidents would remain unchanged whether we measured the rate in accidents per day or in accidents per month, or for that matter whatever unit of time we used. That is, the probability to get n accidents does not change if we apply the transformation $t \to qt'$ and $\theta \to \theta'/q$ to the Poisson distribution, where q is an arbitrary positive scale change.

You could choose to measure things in accidents per day, while I might choose to use accidents per week. But, because we have the *same state of prior knowledge* about the accident rate, namely, we are completely ignorant of its value apart from knowing that it is positive—and because we think rationally, we *must*, Jaynes suggests, assign the same prior probabilities, just as we assign the same probability to observe n accidents, irrespective of the unit used. Another way to state this is that if two rational people assign the same probability for a given data-set, then for equivalent hypotheses about the accident rate (θ' in one case and θ in the other), being rational, they will assign the same prior probabilities to them. The only reason why they might not do so is if they had different prior information; but, by assumption, their prior information is the same.

Okay, let's for the moment buy this argument. If $P(\theta|I) = h(\theta)d\theta$ is your prior probability and $P(\theta'|I) = f(\theta')d\theta'$ is mine then, according to Jaynes, consistency demands that

$$P(\theta'|I) = P(\theta|I). \tag{106}$$

Here the I is absolutely essential, because it must be the same I on both sides of the equation. We get

$$qf(q\theta) = h(\theta). \tag{107}$$

The above is true for all $q > 0$; it is true, in particular, for $q = 1$. Setting $q = 1$, in Eq. (107), we find that $f(\cdot)$ and $h(\cdot)$ must be the same function. We can, therefore, write

$$qh(q\theta) = h(\theta). \tag{108}$$

This is a simple example of a **functional equation**. Its general solution, to within an arbitrary constant, is $h(\theta) = 1/\theta$. We conclude, therefore, that

$$P(\theta|I) \propto d\theta/\theta, \tag{109}$$

is the prior probability that encodes the minimal information we have about the accident rate θ. Such prior probabilities are called **non-informative priors**. This particular one is called the **Jeffreys prior**.

One might object that the Jeffreys prior and the uniform prior favored by Laplace do not make sense because they are non-normalizable. This objection is dismissed easily by noting that in the real world we always work within a finite parameter interval; we never go quite to zero nor do we ever go to infinity. Those limits are mathematical idealizations that we use to make our lives simpler. On any realistic interval the prior probabilities can always be normalized. Indeed, they have to be lest they fail to satisfy the laws of probability theory.

This is pretty slick reasoning; perhaps, a bit too slick! Why? Because other equally cogent arguments[18] lead to the prior probability

$$P(\theta|I) \propto d\theta/\sqrt{\theta}. \tag{110}$$

And there is something intuitively attractive about the flat prior. So, how is one to choose which prior probability to use? This is the Bayesian conundrum; it is the analog of the frequentist problem of choosing an ensemble. The problem of which prior to use is considered by many to be so severe as to negate any advantage that might be gained by the use of Bayesian theory. However, I shall now argue that, in practice, the problem is less severe than it seems.

The basic thesis of the advocates of these various non-informative priors is that, for any given prior information, there must be a *unique* prior probability. But why should that be true? If it is true, then there is only one prior probability for any given problem and all others are wrong. Yet, the many arguments that have been advanced for particular prior probabilities are not all obviously crazy, but give nonetheless different answers. Perhaps, the real problem is the assumption that there must be a unique prior probability. Perhaps, when we have next to no information about something, it should not be expected that even rational beings would agree about how to encode such information, or rather a lack thereof, into a prior probability. It could be that there is an irreducible element of uncertainty when we are faced with reasoning based on almost no information. There is some fuzziness that no amount of mathematical wizardry can "de-fuzzify", as the fuzzy set enthusiasts would say.

What I'm suggesting is this: our search for uniqueness in our inferences may be misguided, because non-uniqueness of outcome may be an inherent property of inductive reasoning. There is uncertainty arising from the data-set as well as from the prior information. Even in the frequentist theory both kinds of uncertainty exist, except that the second source is more deeply hidden. It is hidden in the implicit choice of a flat prior probability when using the maximum likelihood principle. It is hidden in the infinite number of ways we can choose an ensemble.

What matters is that we make reasonable inferences. And, as a practical matter, experience shows that any one of the priors discussed above leads to approximately the same inferences. Furthermore, these inferences merge rapidly as we acquire more data and the likelihood overwhelms the prior probability. My advice is this: use the prior probability that seems most reasonable to you or, better, the one that has been agreed upon by the community, for the given problem. Then check the robustness of your answers by trying different reasonable priors. If your answers are unduly sensitive to the choice of prior probability, then the honest conclusion should be that the data are inadequate and more should be acquired.

No one has succeeded in creating a theory of inference that is devoid of the influence of the inference-maker, in spite of some claims to the contrary. Perhaps such a thing is not possible, even in principle.

How To Analyze A Counting Experiment

Every Bayesian analysis contains at least four ingredients:

- A model
- A data-set
- A likelihood
- A prior probability

For a counting experiment, of which the Grenoble experiment is an example, the model is

$$n = \theta + \mu, \tag{111}$$

where n is the mean number of events, θ the mean signal count and μ the mean background count. Let N be the total number of events observed. What probability should we assign to the data-set $\{N\}$?

In a typical high energy physics experiment the number of reactions produced is very large. For example, at the Fermilab Tevatron accelerator more than a trillion collisions were needed to find the top quark. Call that the number of trials T. Suppose that out of the trillions of trials we succeed in recording N interesting events. What probability should we assign to the data-set? Straightforward combinatoric reasoning, combined with the assumption that at each trial the probability of success q is the same and that the order of events is immaterial, leads *uniquely* to the binomial distribution

$$P(N|q,T,I) = \binom{T}{N} q^N (1-q)^{T-N}. \tag{112}$$

While rigorously correct, it would be absurd to use the probability in this form because of the fact that T is usually so much greater than N. Noting that $n = qT$, by definition, we can write

$$P(N|q,T,I) = P(N|n,T,I) = \frac{T!}{(T-N)!N!} \frac{n^N}{T^N}(1-\frac{n}{T})^{T-N}, \tag{113}$$

which, for $T \gg N$, becomes the Poisson distribution

$$P(N|n,I) = n^N \exp(-n)/N!, \tag{114}$$

which is the standard likelihood for a counting experiment.

Recall that the Grenoble experiment yielded a total of $N = 3$ events. However, that experiment was slightly more complicated in that it also yielded an independent background count of $B = 7$ events, a result that caused some grief. So what does the Reverend have to offer?

We begin, as always, with Bayes' theorem:

$$P(\theta,\mu|\mathbf{x},I) = \frac{P(\mathbf{x}|\theta,\mu,I)P(\theta,\mu|I)}{\int_{\theta,\mu} P(\mathbf{x}|\theta,\mu|I)P(\theta,\mu|I)}, \tag{115}$$

and eliminate the nuisance parameter μ by marginalization:

$$P(\theta|N,B,I) = \int_\mu P(\theta,\mu|N,B,I). \tag{116}$$

The likelihood for the data-set (N,B) is

$$P(N,B|\theta,\mu,I) = \frac{(\theta+\mu)^N e^{-(\theta+\mu)}}{N!} \frac{\mu^B e^{-\mu}}{B!}, \tag{117}$$

which is the product of two Poisson distributions, one for the count N and the other for the count B. What about the prior probability $P(\theta,\mu|I)$?

Using Jaynes' invariance argument one can show[19] that

$$P(\theta, \mu | I) = \frac{d\theta \, d\mu}{(\theta + \mu)\mu}, \quad (118)$$

is the appropriate prior probability; appropriate, that is, for this argument. A striking feature of this strange looking prior probability is that it doesn't factorize: it cannot be written as a product of a function of θ times a function of μ. So here is another thing to think about: Is this prior reasonable and, if so, what does the non-factorizability mean? For this prior and likelihood the posterior probability can be worked out. The answer is

$$P(\theta | N, B, I) = e^{-\theta} \sum_{i=0}^{N-1} \omega_i \frac{\theta^{N-1-i}}{(N-1-i)!}, \quad (119)$$

where

$$\omega_i = \frac{(B-1+i)!/(2^i i!)}{\sum_{j=0}^{N-1}(B-1+j)!/(2^j j!)}. \quad (120)$$

Finally, by calculating the mean of the posterior probability we derive

$$\hat{\theta} = d(N, B) = N - \sum_{i=0}^{N-1} i \omega_i, \quad (121)$$

as our best estimate (for a quadratic loss function) of the mean signal count θ. To compute an upper limit u on the mean signal count θ we need merely to solve

$$\beta = \int_{\theta=0}^{u} P(\theta | N, B, I). \quad (122)$$

For $\beta = 0.9$ and $(N, B) = (3, 7)$ we obtain $u(N, B) = 3.3$ events, a perfectly plausible answer. It is possible to re-work the problem using a flat prior probability. Why don't you try it, compute the 90% upper limit and see by how much it differs from the value given here.

How To Analyze The Solar Neutrino Problem

The sun shines by destroying, in its core, 4.4 billion kilograms of hydrogen every second. Ironically, this prodigious giver of light is opaque to it. One consequence is that the light energy takes about a million years to reach the Sun's surface. The Sun also is a copious emitter of electron neutrinos, which travel from the core to the surface in about a couple of seconds. Every second, every square centimeter of your body is being bathed with about 60 billion neutrinos that were created deep within the stellar interior a mere 8 and half minutes ago! Unfortunately, theory predicts a higher flux than observed. This is the solar neutrino problem.

For about 25 years the Chlorine solar neutrino experiment at the Homestake[20] mine was the only one to report a discrepancy. However, today, three other experiments record fewer neutrinos than predicted by standard solar models[21]: the Gallium experiments, SAGE[22] and GALLEX[23] and the water experiment Kamiokande[24]. It is, therefore, unlikely that the problem is a symptom of an unknown experimental effect. Moreover, recent progress on helioseismology indicates that the solar models are

2. Solar neutrino reaction rates.

Experiment	Measured Rates	BP	TCL
(Cl) Homestake (SNU)	2.55 ± 0.25	8 ± 1	6.4 ± 1.4
(Ga) SAGE and GALLEX (SNU)	74 ± 9.5	132 ± 7	123 ± 7
(H_2O) Kamiokande (BP)	0.51 ± 0.07	1 ± 0.14	0.77 ± 0.19

able to account for the acoustic modes, thousands of which have been measured with great precision. The agreement between theory and experiment is impressive[25].

The Homestake experiment detects neutrinos through the reaction

$$\nu_e + {}^{37}Cl \to e^- + {}^{37}Ar. \qquad (123)$$

The Argon atoms are extracted chemically, and counted. The SAGE and GALLEX experiments use the reaction

$$\nu_e + {}^{71}Ga \to e^- + {}^{71}Ge. \qquad (124)$$

The Kamiokande experiment uses neutrino electron scattering:

$$\nu_e + e^- \to \nu_e + e^-. \qquad (125)$$

It is standard practice to quote the results of the radiochemical experiments, Homestake, SAGE and GALLEX in solar neutrino units (SNU): one SNU is 10^{-36} neutrino reactions per atom per second. The water experiments quote their measured neutrino fluxes as a fraction of that predicted by a standard solar model, for example, that of Bahcall and Pinsonneault[26].

In Table 2 I give the predictions of two standard solar models, those of Bahcall and Pinsonneault (BP) and those of Turck-Chièze and Lopes (TCL), along with the experimental results. The results for SAGE and GALLEX have been combined.

The solar neutrino problem is the discrepancy between the measured and predicted rates in *all* experiments.

But is there really a problem, and if so how big is it? To try to answer these questions we need to extract information about the individual neutrino fluxes. This has been done by many authors[27], who all come to the same broad conclusions. But in detail their conclusions differ; not because of different assumptions about the physics—almost everyone agrees on the physics, but because of differences in the definitions of the χ^2 functions, confidence levels and limits used by different authors. The morass is bewildering.

The key problem, as noted recently [28], is that a χ^2 analysis is ill-suited to answer these kinds of questions. So can Mr. Bayes help? Why, of course!

Standard solar models provide the connection between the neutrino fluxes ϕ and the measured rates **x**. Given **x** we can infer information about ϕ by using Bayes' theorem:

$$P(\phi|\mathbf{x}, I) = \frac{P(\mathbf{x}|\phi, I)P(\phi|I)}{\int_\phi P(\mathbf{x}|\phi, I)P(\phi|I)}, \qquad (126)$$

where $P(\mathbf{x}|\phi, I)$ is the likelihood assigned to the measured rates **x**, $P(\phi|I) \equiv f(\phi)d\phi$ is the prior probability assigned to hypotheses about the fluxes ϕ and $P(\phi|\mathbf{x}, I)$ is the posterior probability.

We shall consider first the likelihood function $P(\mathbf{x}|\phi, I)$. I shall use the model

$$\mathbf{x} = R \cdot \phi, \tag{127}$$

where $x^T = (x_{Cl}, x_{Ga}, x_{H_2O})$, $\phi^T = (\phi_B, \phi_{Be}, \phi_{pp}, \phi_O)$, R is the matrix

$$R = \begin{pmatrix} 6.2 & 1.2 & 0.0 & 0.6 \\ 13.8 & 35.2 & 70.8 & 12.2 \\ 1.0 & 0.0 & 0.0 & 0.0 \end{pmatrix}, \tag{128}$$

and the fluxes ϕ are given in units of those predicted by Bahcall and Pinsonneault, in their 1992 paper. The elements of the matrix R have been pieced together from Bahcall's book[21]. The matrix shows how the different neutrino fluxes $\phi_B, \phi_{Be}, \phi_{pp}$ and ϕ_O, from the ^8B, ^7Be, pp and all other nuclear reactions, respectively, contribute to the measured rates x_{Cl}, x_{Ga} and x_{H_2O}. The basic assumption is that the neutrino spectra are unaffected by the solar plasma—which is true to a high degree of accuracy—and that nothing untoward happens to the neutrinos from their point of creation to their point of detection.

The water experiment measures exclusively the flux of neutrinos from the ^8B reactions; the Chlorine experiment samples the neutrinos from the ^8B and ^7Be reactions and the Gallium experiments sample all neutrinos including, in particular, those from the proton-proton (pp) chain, which account for the bulk of the neutrinos emitted by the sun.

If we assume the measured rates to be uncorrelated the error matrix σ—associated with the rates $\mathbf{x}^T = (2.55, 74, 0.51)$—is diagonal, with diag-$\sigma = (0.25, 9.5, 0.072)$. A reasonable probability assignment for these data is a Gaussian,

$$P(\mathbf{x}|\phi, I) = \exp(-\frac{1}{2} z^T z). \tag{129}$$

The vector $z = \sigma^{-1} \cdot (\mathbf{x} - R \cdot \phi)$.

Now let's turn to the prior probability $P(\phi|I)$. We know that $\phi > 0$, and that ϕ isn't infinite! But that's about all we know. Let's try the simplest option: $P(\phi|I) \propto$ constant. As explained above other choices are possible, and the result will depend upon which choice is made. But it turns out the answers are essentially the same for a reasonable variety of prior probabilities. You're welcome to check this out!

For a flat prior we obtain

$$P(\phi_B, \phi_{Be}, \phi_{pp}, \phi_O|\mathbf{x}, I) \propto P(\mathbf{x}|\phi, I). \tag{130}$$

Here is an important observation: we have four unknowns, but only three data. This would be a problem in the frequentist theory. But Bayes' theorem doesn't care. What happens is this: as the number of unknowns increases relative to the data-set the information that can be extracted, about any given parameter, degrades gradually. By marginalizing with respect to the fluxes ϕ_B, ϕ_{Be} and ϕ_O

$$P(\phi_{pp}|\mathbf{x}, I) = \int_{\phi_B, \phi_{Be}, \phi_O} P(\phi_B, \phi_{Be}, \phi_{pp}, \phi_O|\mathbf{x}, I), \tag{131}$$

we obtain a probability $P(\phi_{pp}|\mathbf{x}, I)$ for the pp flux alone. We can do likewise for ϕ_B and ϕ_{Be}. The posterior probabilities are shown in Fig. 6. The results indicate a modest

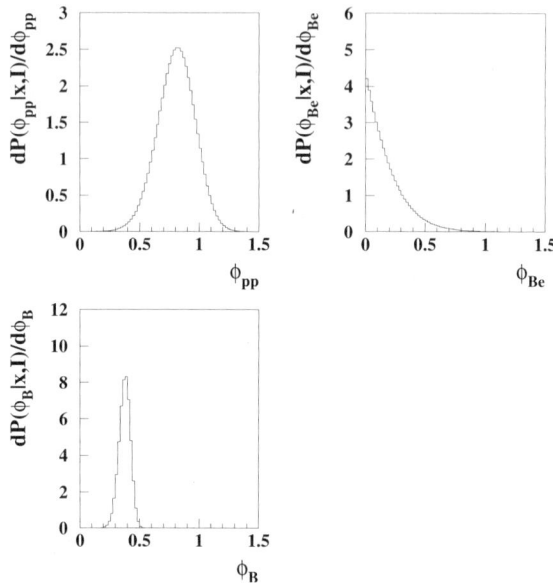

6. Posterior probabilities for the pp, ^7Be and ^8B neutrino fluxes, relative to the 1992 Bahcall-Pinsonneault predictions. Unit flux represents perfect agreement between theory and experiment. We observe significant disagreements for the fluxes from the ^7Be and ^8B reactions.

suppression of the pp neutrino flux, a reduction by about 2 of the ^8B neutrino flux and there is evidence of a marked reduction in the flux from the ^7Be reaction.

These conclusions are not new. What is interesting here is the directness and simplicity with which they flow from Bayes' theorem. Moreover, given the posterior probability for the fluxes there is a great deal that can be studied in a rigorous, but intuitively compelling manner.

Acknowledgements

Over the years I have benefited greatly from many discussions, and arguments, with colleagues and friends. I would like to thank in particular Roger Barlow, Maria Fidecaro, Daniel Drijard, Pushpa Bhat, Chip Stewart, Jim Linnemann, Tom Trippe, Rajendran Raja, Marc Paterno and Mark Strovink. And I cannot forget my many discussions with Fred James, who helped me understand that statistical analysis is a bit more than just "propagating errors". Edwin Jaynes, whom I have never met, continues to be particularly inspiring. His ideas are interesting and he writes about them beautifully, avoiding the disease that plagues contemporary technical writing, which strives mightily to be as impersonal as possible and succeeds to a fault. I would like, also, to thank my dear companion-in-life, Marie-France Prosper-Chartier, for her excellent proofreading. The responsibility for any remaining errors is, of course, mine!

And last but not least I should like to thank both Tom Ferbel, for giving me the chance to do this, and the wonderful young people who made St. Croix so enjoyable.

****

I dedicate these lectures to the memory of George Michail, who left us much too soon.

REFERENCES

1. W. T. Eadie, D. Drijard, F. E. James, M. Roos and B. Sadoulet, "Statistical Methods in Experimental Physics," North-Holland, Amsterdam (1982).
2. M.G. Kendall and A. Stuart. "The Advanced Theory Of Statistics," Griffin and Co. Limited, London (1973).
3. H. Jeffreys. "The Theory of Probability," Oxford University Press, Oxford (1939).
4. R. D. Cousins, Why Isn't Every Physicist A Bayesian?, *Am. J. Phys.* 63:398 (1995).
5. R. J. Barlow, "Statistics: A Guide To The Use Of Statistical Methods In The Physical Sciences," The Manchester Physics Series, John Wiley and Sons, New York (1989).
6. E. T. Jaynes, "Probability Theory—The Logic Of Science," http://www.math.albany.edu:8008/JaynesBook.html (1995).
7. L. Montanet et al., Review Of Particle Properties. Particle Data Group, *Phys. Rev.* D50:1173 (1994).
8. D. T. Gillespie, A Theorem For Physicists In The Theory Of Random Variables, *Am. J. Phys.* 51:520 (1983).
9. H. B. Prosper, On Estimating Mean Lifetimes By A Weighted Sum Of Lifetime Measurements, *Phys. Rev.* D36:2047 (1987).
10. D. Chang, et al., *Phys. Rev.*, D31:1718 (1985).
11. CRISP Collaboration, G. Fidecaro et al., Experimental Search For Neutron-Antineutron Transitions With Free Neutrons, *Phys. Lett.* 156B:122 (1985).
12. H.B. Prosper, The Distribution Of The Difference Of Two Poisson Variates, *Nucl. Instrum. Methods* A238:500 (1985).
13. R. T. Cox, Probability, Frequency, and Reasonable Expectation, *Am. J. Phys.* 14:1 (1946).
14. R. E. Kass and L. Wasserman, The Selection Of Prior Distributions By Formal Rules, *Journal of the American Statistical Association*, 91:1343 (1996).
15. D. W. Ruck et al., *IEEE Trans. Neural Networks*, 4:296 (1990). E. A. Wan, *IEEE Trans. Neural Networks*, 4:303 (1990).
16. E. K. Blum and L. K. Li, *Neural Networks*, 4:511 (1991).
17. E.T. Jaynes, Prior Probabilities, *IEEE Trans. Syst. Sci. Cybern.*, 227:SSC-4 (1968).
18. G. E. P. Box and G. C. Tiao. "Bayesian Inference In Statistical Analysis," John Wiley and Sons, New York (1992).
19. H. B. Prosper, Small Signal Analysis In High-Energy Physics: A Bayesian Approach, *Phys. Rev.* D37:1153 (1988).
20. B. T. Cleveland et al., *Nucl. Phys. B (Proc. Suppl.)*, 47:38 (1995).
21. J.N. Bahcall. "Neutrino Astrophysics," Cambridge University Press, Cambridge (1989).
22. P. Anselmann et al., *Phys. Lett.*, B327:377 (1995).
23. P. Nico et al., in "Proceedings of the XXVII International Conference On High Energy Physics," Glasgow, 1994, ed. by P.J. Bussey and I.H. Knowles, IOP, Bristol, p. 965 (1995).
24. Y. Suzuki et al., *Nucl. Phys. B (Proc. Suppl.)*, 38:54 (1995).
25. J. N. Bahcall, http://www.sns.ias.edu/~jnb.
26. J.N. Bahcall and M.H. Pinsonneault, *Rev. Mod. Phys.* 64:885 (1992). *Rev. Mod. Phys.*, 67:781 (1995).
27. N. Hata and P. Langacker, *Phys. Rev.* D50:632 (1994). S. Parke, *Phys. Rev. Lett.* 74:839 (1995).
28. M. Roos and L. A. Khalfin, Dangers of Unphysical Regions, HU-TFT-96-13, e-Print Archive: hep-ex/9605008, May (1996).

MUON-MUON AND OTHER HIGH ENERGY COLLIDERS

R. B. Palmer, J. C. Gallardo

Center for Accelerator Physics
Brookhaven National Laboratory
Upton, NY 11973-5000, USA

COMPARISON OF COLLIDER TYPES
Introduction

Before we discuss the muon collider in detail, it is useful to look at the other types of colliders for comparison. In this chapter we consider the high energy physics advantages, disadvantages and luminosity requirements of hadron (pp, $p\bar{p}$), of lepton (e^+e^-, $\mu^+\mu^-$) and photon-photon colliders. Technical problems in obtaining increased energy in each type of machine are presented. Their relative size, and probable relative costs are discussed.

Physics Considerations

General. Hadron-hadron colliders (pp or $p\bar{p}$) generate interactions between the many constituents of the hadrons (gluons, quarks and antiquarks); the initial states are not defined and most interactions occur at relatively low energy, generating a very large background of uninteresting events. The rate of the highest energy events is higher for antiproton-proton machines, because the antiproton contains valence antiquarks that can annihilate on the quarks in the proton. But this is a small effect for colliders above a few TeV, when the interactions are dominated by interactions between quarks and antiquarks in their seas, and between the gluons. In either case the individual parton-parton interaction energies (the energies used for physics) are a relatively small fraction of the total center of mass energy. This is a disadvantage when compared with lepton machines. An advantage, however, is that all final states are accessible. Many, if not most, initial discoveries in Elementary Particle Physics have been made with these machines.

In contrast, lepton-antilepton collider generate interactions between the fundamental point-like constituents in their beams, the reactions generated are relatively simple to understand, the full machine energies are available for "physics", and there is negligible background of low energy events. If the center of mass energy is set equal to the mass of a suitable state of interest, then there can be a large cross section in the s-channel, in which a single state is generated by the interaction. In this case, the mass and quantum numbers of the state are constrained by the initial beams. If the energy spread of the beams is sufficiently narrow, then precision determination of masses and widths are possible.

A gamma-gamma collider, like the lepton-antilepton machines, would also have all the machine energy available for physics, and would have well defined initial states, but these states would be different from those with the lepton machines, and thus be complementary to them.

For most purposes (technical considerations aside) e^+e^- and $\mu^+\mu^-$ colliders would be equivalent. But in the particular case of s-channel Higgs boson production, the cross section, being proportional to the mass squared, is more than 40,000 times greater for muons than electrons. When technical considerations are included, the situation is more complicated. Muon beams are harder to polarize and muon colliders will have much higher backgrounds from decay products of the muons. On the other hand muon collider interactions will require less radiative correction and will have less energy spread from beamstrahlung.

Each type of collider has its own advantages and disadvantages for High Energy Physics: they would be complementary.

Required Luminosity for Lepton Colliders. In lepton machines the full center of mass of the leptons is available for the final state of interest and a "physics energy" E_{phy} can be defined that is equal to the total center of mass energy.

$$E_{\text{phy}} = E_{c \; of \; m} \qquad (2)$$

Since fundamental cross sections fall as the square of the center of mass energies involved, so, for a given rate of events, the luminosity of a collider must rise as the square of its energy. A reasonable target luminosity is one that would give 10,000 events per unit of R per year (the cross section for lepton pair production is one R, the total cross section is about 20 R, and somewhat energy dependent as new channels open up):

$$\mathcal{L}_{\text{req.}} \approx 10^{34} \; (\text{cm}^{-2}\text{s}^{-1}) \left(\frac{E_{\text{phy}}}{1 \; (\text{TeV})} \right)^2 \qquad (3)$$

Fig. 1 shows this required luminosity, together with crosses at the approximate achieved luminosities of some lepton colliders. Target luminosities of possible future colliders are also given as circles.

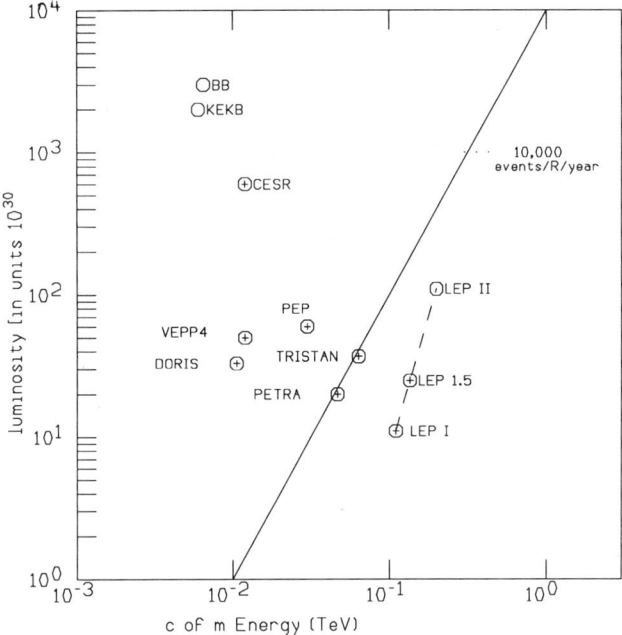

Figure 1. Luminosity of lepton colliders as a function of Energy

The Effective Physics Energies of Hadron Colliders. Hadrons, being composite, have their energy divided between their various constituents. A typical collision of constituents will thus have significantly less energy than that of the initial hadrons. Studies done in Snowmass 82 and 96 suggest that, for a range of studies, and given the required luminosity (as defined in Eq. 3), then the hadron machine's effective "physics" energy is between about 1/3 and 1/10 of its total. We will take a value of 1/7:

$$E_{\text{phy}}(\mathcal{L} = \mathcal{L}_{\text{req.}}) \approx \frac{E_{\text{c of m}}}{7}$$

The same studies have also concluded that a factor of 10 in luminosity is worth about a factor of 2 in effective physics energy, this being approximately equivalent to:

$$E_{\text{phy}}(\mathcal{L}) = E_{\text{phy}}(\mathcal{L} = \mathcal{L}_{\text{req.}}) \left(\frac{\mathcal{L}}{\mathcal{L}_{\text{req}}}\right)^{0.3}$$

From which, with Eq. 3, one obtains:

$$E_{\text{phy}} \approx \left(\frac{E_{\text{c of m}}}{7(TeV)}\right)^{0.6} \left(\frac{\mathcal{L}}{10^{34}(\text{cm}^{-2}\text{s}^{-1})}\right)^{0.2} (TeV) \qquad (4)$$

Table 1. Effective Physics Energy of Some Hadron Machines

Machine	C of M Energy TeV	Luminosity $cm^{-2}s^{-1}$	Physics Energy TeV
ISR	.056	10^{32}	0.02
TeVatron	1.8	7×10^{31}	0.16
LHC	14	10^{34}	1.5
VLHC	60	10^{34}	3.6

Tb. 1 gives some examples of this approximate "physics" energy. It must be emphasized that this effective physics energy is not a well defined quantity. It should depend on the physics being studied. The initial discovery of a new quark, like the top, can be made with a significantly lower "physics" energy than that given here. And the capabilities of different types of machines have intrinsic differences. The above analysis is useful only in making very broad comparisons between machine types.

Hadron-Hadron Machines

Luminosity. An antiproton-proton collider requires only one ring, compared with the two needed for a proton-proton machine (the antiproton has the opposite charge to the proton and can thus rotate in the same magnet ring in the opposite direction - protons going in opposite directions require two rings with bending fields of the opposite sign), but the luminosity of an antiproton- proton collider is limited by the constraints in antiproton production. A luminosity of at least 10^{32} $cm^{-2}s^{-1}$ is expected at the antiproton-proton Tevatron; and a luminosity of 10^{33} $cm^{-2}s^{-1}$ may be achievable, but LHC, a proton-proton machine, is planned to have a luminosity of 10^{34} $cm^{-2}s^{-1}$. Since the required luminosity rises with energy, proton-proton machines seem to be favored for future hadron colliders.

The LHC and other future proton-proton machines might even[1] be upgradable to 10^{35} $cm^{-2}s^{-1}$, but radiation damage to a detector would then be a severe problem. The 60 TeV Really Large Hadron Colliders (RLHC: high and low field versions) discussed at Snowmass are being designed as proton-proton machines with luminosities of 10^{34} $cm^{-2}s^{-1}$ and it seems reasonable to assume that this is the highest practical value.

Size and Cost. The size of hadron-hadron machines is limited by the field of the magnets used in their arcs. A cost minimum is obtained when a balance is achieved between costs that are linear in length, and those that rise with magnetic field. The optimum field will depend on the technologies used both for the the linear components (tunnel, access, distribution, survey, position monitors, mountings, magnet ends, etc) and those of the magnets themselves, including the type of superconductor used.

The first hadron collider, the 60 GeV ISR at CERN, used conventional iron pole magnets at a field less than 2 T. The only current hadron collider, the 2 TeV Tevatron, at FNAL, uses NbTi superconducting magnets at approximately $4\,°K$ giving a bending field of about 4.5 T. The 14 TeV Large Hadron Collider (LHC), under construction at CERN, plans to use the same material at $1.8\,°K$ yielding bending fields of about 8.5 T.

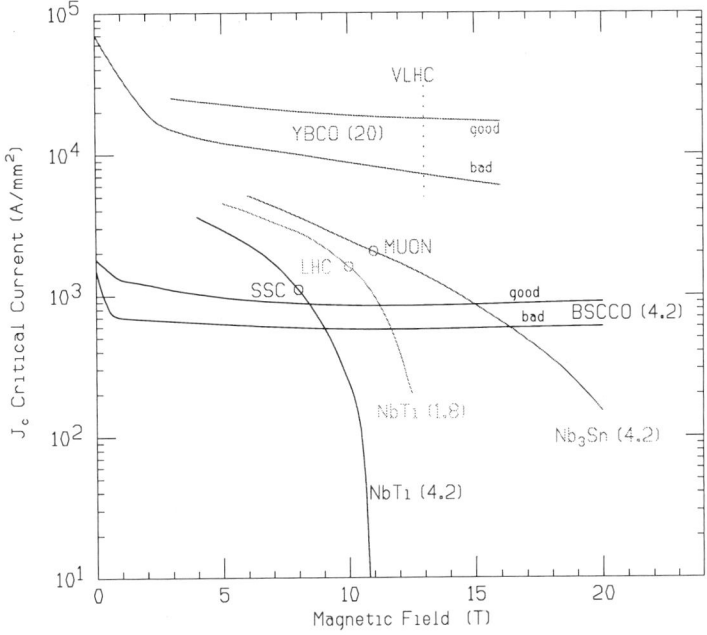

Figure 2. Critical current densities of superconductors as a function of magnetic field.

Future colliders may use new materials allowing even higher magnetic fields. Fig. 2 shows the critical current densities of various superconductors as a function of magnetic field. The numbers in parenthesis refer to the temperatures in $°$ K. *good* and *bad* refer to the best and worst performance according to the orientation of the tape with respect to the direction of the magnetic field. Model magnets have been made with Nb_3Sn, and studies are underway on the use of high T_c superconductor. $Bi_2Sr_2Ca_1Cu_2O_8$ (BSCCO) material is currently available in useful lengths as powder-in-Ag tube processed tape. It has a higher critical temperature and field than conventional superconductors, but, even at $4\,°K$, its current density is less than Nb_3Sn at all fields below 15 T. It is thus unsuitable for most accelerator magnets. In contrast $YBa_2Cu_3O_7$ (YBCO) material has a current density above that for Nb_3Sn $(4\,°K)$, at all fields and temperatures below $20\,°K$. But this material must be deposited on specially treated metallic substrates and is not yet available in lengths greater than 1 m. It is reasonable to assume, however, that it will be available in useful lengths in the not too distant future.

A parametric study was undertaken to learn what the use of such materials might do for the cost of colliders. 2-in-1 cosine theta superconducting magnet cross sections

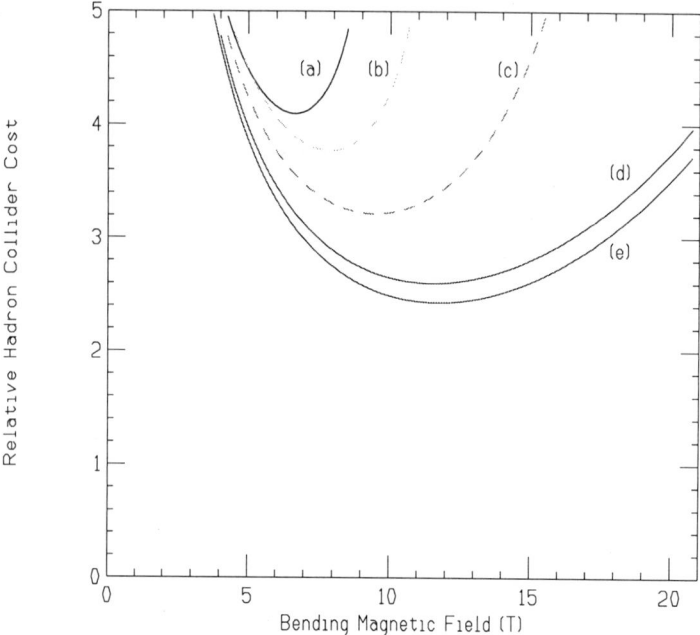

Figure 3. Relative costs of a collider as a function of its bending magnetic field, for different superconductors and operating temperatures. Costs are given for NbTi at (a) $4\,°K$, and (b) $1.8\,°K$, (c) Nb_3 Sn at $4.3\,°K$, and (d) and (e) YBCO High T_c at $20\,°K$. NbTi and Nb_3 Sn costs per unit weight were taken to be the same; YBCO was taken to be either equal to NbTi (in (d)), or 4 times NbTi (in (e)).

(in which the two magnet coils are circular in cross section, have a cosine theta current distributions and are both enclosed in a single iron yoke) were calculated using fixed criteria for margin, packing fraction, quench protection, support and field return. Material costs were taken to be linear in the weights of superconductor, copper stabilizer, aluminum collars, iron yoke and stainless steel support tube. The cryogenic costs were taken to be inversely proportional to the operating temperature, and linear in the outer surface area of the cold mass. The values of the cost dependencies were scaled from LHC estimates.

Results are shown in Fig. 3. Costs were calculated assuming NbTi at (a) $4\,°K$, and (b) $1.8\,°K$, Nb_3 Sn at (c) $4.3\,°K$ and YBCO High T_c at $20\,°K$ (d) and (e). NbTi and Nb_3 Sn costs per unit weight were taken to be the same; YBCO was taken to be either equal to NbTi (in (d)), or 4 times NbTi (in (e)). It is seen that the optimum field moves from about 6 T for NbTi at $4\,°K$ to about 12 T for YBCO at $20\,°K$; while the total cost falls by almost a factor of 2.

One may note that the optimized cost per unit length remains approximately constant. This might have been expected: at the cost minimum, the cost of linear and field dependent terms are matched, and the total remains about twice that of the linear terms.

The above study assumes this particular type of magnet and may not be indicative of the optimization for radically different designs. A group at FNAL[2] is considering an iron dominated, alternating gradient, continuous, single turn collider magnet design (Low field RLHC). Its field would be only 2 T and circumference very large (350 km for 60 TeV), but with its simplicity and with tunneling innovations, it is hoped to make its cost lower than the smaller high field designs. There are however greater problems in achieving high luminosity with such a machine than with the higher field designs.

Circular e^+e^- Machines

Luminosity. The luminosities of most circular electron-positron colliders has been between 10^{31} and 10^{32} cm^{-2}s^{-1} (see Fig.1), CESR is fast approaching 10^{33} cm^{-2}s^{-1} and machines are now being constructed with even high values. Thus, at least in principle, luminosity does not seem to be a limitation (although it may be noted that the 0.2 TeV electron-positron collider LEP has a luminosity below the requirement of Eq.3).

Size and Cost. At energies below 100 MeV, using a reasonable bending field, the size and cost of a circular electron machine is approximately proportional to its energy. But at higher energies, if the bending field B is maintained, the energy lost ΔV_{turn} to synchrotron radiation rises rapidly

$$\Delta V_{\text{turn}} \propto \frac{E^4}{R\, m^4} \propto \frac{E^3\, B}{m^4} \tag{5}$$

and soon becomes excessive (R is the radius of the ring). A cost minimum is then obtained when the cost of the ring is balanced by the cost of the rf needed to replace the synchrotron energy loss. If the ring cost is proportional to its circumference, and the rf is proportional to its voltage then the size and cost of an optimized machine rises as the square of its energy. This relationship is well demonstrated by the parameters of actual machines as shown later in Fig. 7.

The highest circular e^+e^- collider is the LEP at CERN which has a circumference of 27 km, and will achieve a maximum center of mass energy of about 0.2 TeV. Using the predicted scaling, a 0.5 TeV circular collider would have to have a 170 km circumference, and would be very expensive.

e^+e^- Linear Colliders

Size and Cost. So, for energies much above that of LEP (0.2 TeV) it is probably impractical to build a circular electron collider. The only possibility then is to build two electron linacs facing one another. Interactions occur at the center, and the electrons, after they have interacted, must be discarded.

Luminosity. The luminosity $\mathcal{L}$ of a linear collider can be written:

$$\mathcal{L} = \frac{1}{4\pi E} \frac{N}{\sigma_x} \frac{P_{beam}}{\sigma_y} n_{collisions} \qquad (6)$$

where σ_x and σ_y are average beam spot sizes including any pinch effects, and we take σ_x to be much greater than σ_y. E is the beam energy, P_{beam} is the total beam power, and, in this case, $n_{collisions} = 1$. This can be expressed[3] as,

$$\mathcal{L} \approx \frac{1}{4\pi E} \frac{n_\gamma}{2r_o \alpha \, U(\Upsilon)} \frac{P_{beam}}{\sigma_y} \qquad (7)$$

where the quantum correction $U(\Upsilon)$ is given by

$$U(\Upsilon) \approx \sqrt{\frac{1}{1+\Upsilon^{2/3}}} \qquad (8)$$

with

$$\Upsilon \approx \frac{2F_2 r_e^2}{\alpha} \frac{N \, \gamma}{\sigma_z \sigma_x} \qquad (9)$$

$F_2 \approx 0.43$, r_o is the classical electromagnetic radius, α is the fine-structure constant, and σ_z is the rms bunch length. The quantum correction Υ is close to unity for all proposed machines with energy less than 2 TeV, and this term is often omitted[4]. Even in a 5 TeV design[5], an Υ of 21 gives a suppression factor of only 3.

n_γ is the number of photons emitted by one electron as it passes through the other bunch. If n_γ is significantly greater than one, then problems are incountered with backgrounds of electron pairs and mini-jets, or with unacceptable beamstrahlung energy loss. Thus n_γ can be taken as a rough criterion of these effects and constrained to a fixed value. We then find:

$$\mathcal{L} \propto \frac{1}{E} \frac{P_{beam}}{\sigma_y \, U(\Upsilon)}$$

which may be compared to the required luminosity that increases as the square of energy, giving the requirement:

$$\frac{P_{beam}}{\sigma_y \, U(\Upsilon)} \propto E^3. \qquad (10)$$

It is this requirement that makes it hard to design very high energy linear colliders. High beam power demands high efficiencies and heavy wall power consumption. A small σ_y requires tight tolerances, low beam emittances and strong final focus and a small value of $U(\Upsilon)$ is hard to obtain because of its weak dependence on Υ ($\propto \Upsilon^{-1/3}$).

Conventional RF. The gradients for structures have limits that are frequency dependent. Fig. 4 shows the gradient limits from breakdown, fatigue and dark current

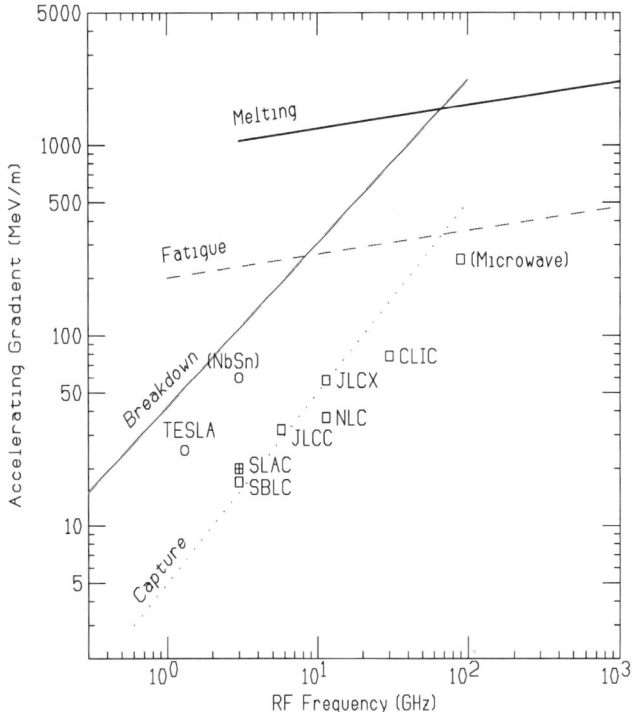

Figure 4. Gradient values and limits in linear collider electron linacs, superconducting examples are indicated as circles

capture, plotted against the operating rf frequency. Operating gradients and frequencies of several linear collider designs[6] are also indicated.

One sees that for conventional structure designs (indicated as squares in Fig. 4), the proposed gradients fall well below the limits, except for the dark current capture threshold. Above this threshold, in the absence of focusing fields, dark current electrons emitted in one cavity can be captured and accelerated down the entire linac causing loading problems. We note, however, that the superconducting TESLA design is well above this limit, and a detailed study[7] has shown that the quadrupole fields in a focusing structure effectively stop the build up of such a current.

The real limit on accelerating gradients in these designs come from a trade off between the cost of rf power against the cost of length. The use of high frequencies reduces the stored energy in the cavities, reducing the rf costs and allowing higher accelerating gradients: the optimized gradients being roughly proportional to the frequency. One might thus conclude then that higher frequencies should be preferred. There are however counterbalancing considerations from the requirements of luminosity.

Fig. 5, using parameters from the linear collider proposals [6], plots some relevant parameters against the rf frequency. One sees that as the frequencies rise,

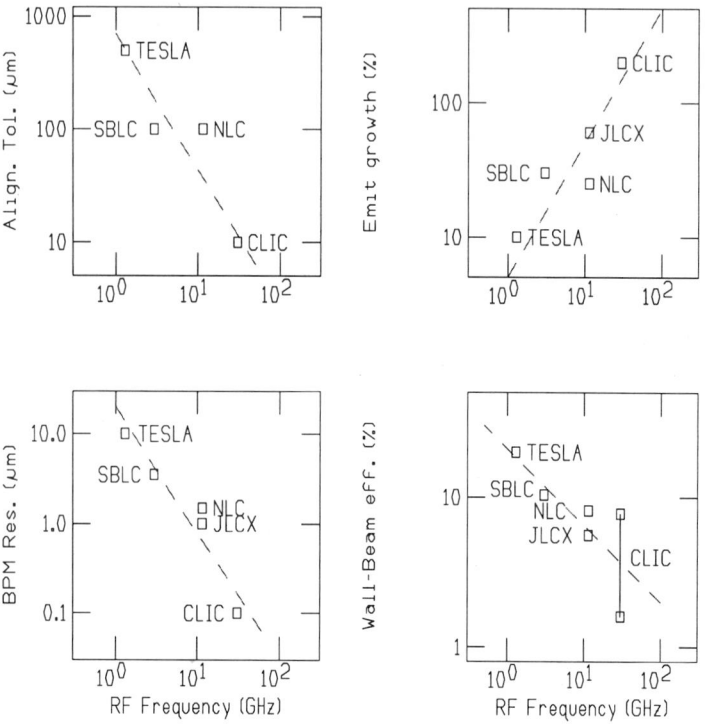

Figure 5. Dependence of some sensitive parameters as a function of linear collider rf frequency.

- the required alignment tolerances are tighter;

- the resolution of beam position monitors must also be better; and

- despite these better alignments, the calculated emittance growth during acceleration is worse; and

- the wall-power to beam-power efficiencies are also less.

Thus while length and cost considerations may favor high frequencies, yet luminosity considerations demand lower frequencies.

Superconducting RF. If, however, the rf costs can be reduced, for instance when superconducting cavities are used, then there will be no trade off between rf power cost and length and higher gradients should be expected to lower the length and cost. The removal of the constraint applied by rf power considerations is evident for the TESLA gradient plotted in Fig. 4. Its value is well above the trend of conventional rf designs. Unfortunately the gradients achievable in currently operating niobium superconducting cavities is lower than that planned in the higher frequency conventional rf colliders. Theoretically the limit is about 40 MV/m, but practically 25 MV/m is as high as seems possible. Nb$_3$Sn and high Tc materials may allow higher field gradients in the future. A possible value for Nb$_3$Sn is also indicated on Fig. 4.

In either case, the removal of the requirements for huge peak rf power allows the choice of longer wavelengths (the TESLA collaboration is proposing 23 cm at 1.3 GHz) and greatly relieves the emittance requirements and tolerances, with no loss of luminosity.

At the current 25 MeV per meter gradients, the length and cost of a superconducting machine is probably higher than for the conventional rf designs. With greater luminosity more certain, its proponents can argue that it is worth it the greater price. If higher gradients become possible, using new superconductors, then the advantages of a superconducting solution could become overwhelming.

At Higher Energies. At higher energies (as expected from Eq. 10), obtaining the required luminosity gets harder. Fig.6 shows the dependency of some example machine parameters with energy. SLC is taken as the example at 0.1 TeV, NLC parameters at 0.5 and 1 TeV, and 5 and 10 TeV examples are taken from a review paper by one of the authors[5]. One sees that:

- the assumed beam power rises approximately as E;

- the vertical spot sizes fall approximately as E^{-2};

- the vertical normalized emittances fall even faster: $E^{-2.5}$; and

- the momentum spread due to beamstrahlung has been allowed to rise almost linearly with E.

These trends are independent of the acceleration method, frequency, etc, and indicate that as the energy and required luminosity rise, so the required beam powers, efficiencies, emittances and tolerances will all get harder to achieve. The use of higher frequencies or exotic technologies that would allow the gradient to rise, will, in general, make the achievement of the required luminosity even more difficult. It may well prove impractical to construct linear electron-positron colliders, with adequate luminosity, at energies above a few TeV.

$\gamma - \gamma$ Colliders

A gamma-gamma collider[8] would use opposing electron linacs, as in a linear electron collider, but just prior to the collision point, laser beams would be Compton backscattered off the electrons to generate photon beams that would collide at the IP instead of the electrons. If suitable geometries are used, the mean photon-photon energy could be 80% or more of that of the electrons, with a luminosity about 1/10th.

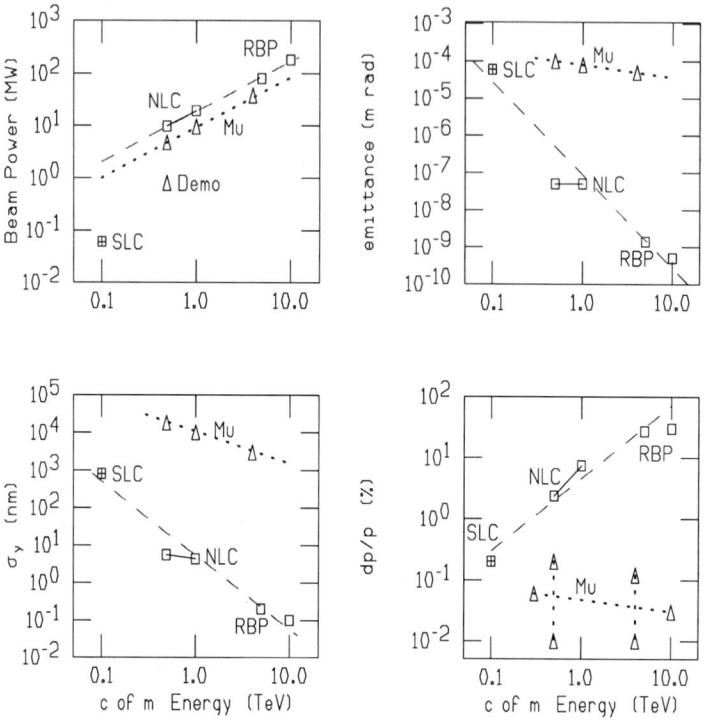

Figure 6. Dependence of some sensitive parameters on linear collider energy, with comparison of same parameters for $\mu^+\mu^-$ colliders.

If the electron beams, after they have backscattered the photons, are deflected, then backgrounds from beamstrahlung can be eliminated. The constraint on N/σ_x in Eq.6 is thus removed and one might hope that higher luminosities would now be possible by raising N and lowering σ_x. Unfortunately, to do this, one needs sources of larger number of electron bunches with smaller emittances, and one must find ways to accelerate and focus such beams without excessive emittance growth. Conventional damping rings will have difficulty doing this[9]. Exotic electron sources would be needed, and methods using lasers to generate[10] or cool[11] the electrons and positrons are under consideration.

Thus, although gamma-gamma collisions can and should be made available at any future electron-positron linear collider, to add physics capability, they may not give higher luminosity for a given beam power.

$\mu^+\mu^-$ Colliders

There are two advantages of muons, as opposed to electrons, for a lepton collider.

- The synchrotron radiation, that forces high energy electron colliders to be linear, is (see Eq. 5) inversely proportional to the fourth power of mass: It is negligible in muon colliders with energy less than 10 TeV. Thus a muon collider, up to such energy, can be circular. In practice this means in can be smaller. The linacs for a 0.5 TeV NLC would be 20 km long. The ring for a muon collider of the same energy would be only about 1.2 km circumference.

- The luminosity of a muon collider is given by the same formula (Eq. 6) as given above for an electron positron collider, but there are two significant changes: 1) The classical radius r_o is now that for the muon and is 200 times smaller; and 2) the number of collisions a bunch can make $n_{collisions}$ is no longer 1, but is now related to the average bending field in the muon collider ring, with

$$n_{collisions} \approx 150\ B_{ave}$$

With an average field of 6 Tesla, $n_{collisions} \approx 900$. Thus these two effects give muons an *in principle* luminosity advantage of more than 10^5.

As a result of these gains, the required beam power, spot sizes, emittances and energy spread are far less in $\mu^+\mu^-$ colliders than in e^+e^- machines of the same energy. The comparison is made in Fig. 6 above.

But there are problems with the use of muons:

- Muons can be best be obtained from the decay of pions, made by higher energy protons impinging on a target. A high intensity proton source is thus required and very efficient capture and decay of these pions is essential.

- Because the muons are made with very large emittance, they must be cooled and this must be done very rapidly because of their short lifetime. Conventional synchrotron, electron, or stochastic cooling is too slow. Ionization cooling is the only clear possibility, but does not cool to very low emittances.

- Because of their short lifetime, conventional synchrotron acceleration would be too slow. Recirculating accelerators or pulsed synchrotrons must be used.

- Because they decay while stored in the collider, muons radiate the ring and detector with their decay products. Shielding is essential and backgrounds will certainly be significant.

These problems and their possible solutions will be discussed in more detail in the following chapters. Parameters will be given there of a 4 TeV center of mass collider, and of a 0.5 TeV demonstration machine.

Comparison of Machines

Length. In Fig. 7, the effective physics energies (as defined by Eq. 4) of representative machines are plotted against their total tunnel lengths. We note:

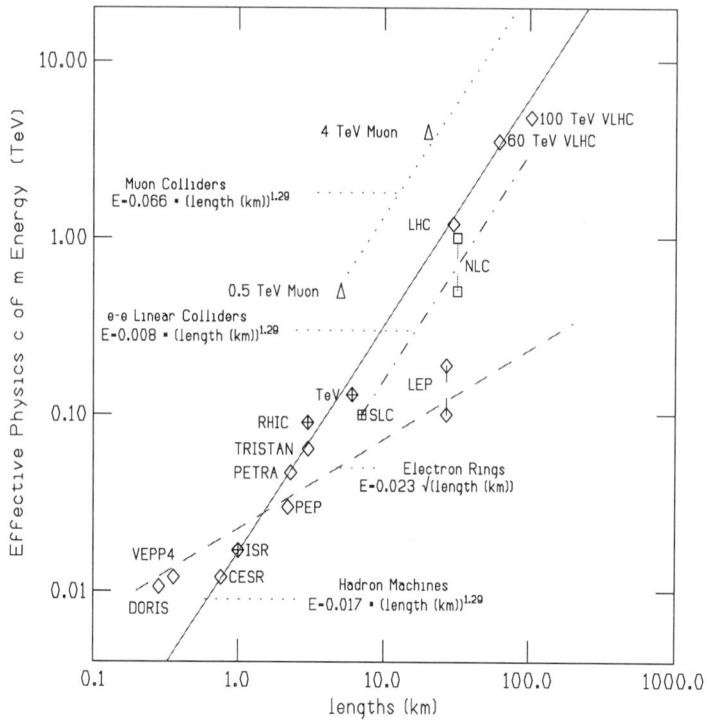

Figure 7. Effective physics energies of colliders as a function of their total length.

- Hadrons Colliders: It is seen that the energies of machines rise with their size, but that this rise is faster than linear ($E_{\text{eff}} \propto L^{1.3}$). This extra rise is a reflection of the steady increase in bending magnetic fields used as technologies and materials have become available.

- Circular Electron-Positron Colliders: The energies of these machines rise approximately as the square root of their size, as expected from the cost optimization discussed above.

- Linear Electron-Positron Colliders: The SLC is the only existing machine of this type and only one example of a proposed machine (the NLC) is plotted. The line drawn has the same slope as for the hadron machines and implies a similar rise in accelerating gradient, as technologies advance.

- Muon-Muon Colliders: Only the 4 TeV collider, discussed above, and the 0.5 TeV *demonstration machine* have been plotted. The line drawn has the same slope as for the hadron machines.

It is noted that the muon collider offers the greatest energy per unit length. This is also apparent in Fig. 8, in which the footprints of a number of proposed machines are given on the same scale. But does this mean it will give the greatest energy per unit of cost ?

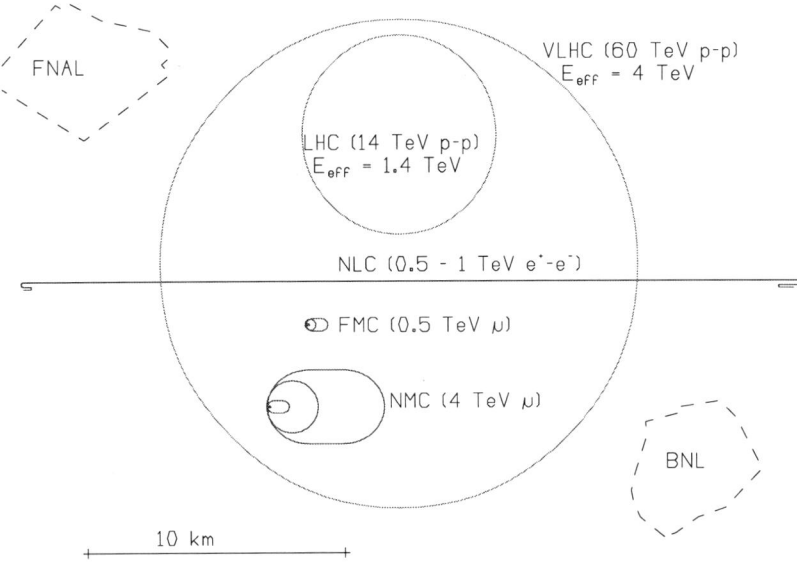

Figure 8. Approximate sizes of some possible future colliders.

Cost. Fig. 9 plots the cost of a sample of machines against their size. Before examining this plot, be warned: the numbers you will see will not be the ones you are familiar with. The published numbers for different projects use different accounting procedures and include different items in their costs. Not very exact corrections and escalation have been made to obtain estimates of the costs under fixed criteria: 1996 $'s, US accounting, no detectors or halls. The resulting numbers, as plotted, must be considered to have errors of at least ± 20%.

The costs are seen to be surprisingly well represented by a straight line. Circular electron machines, as expected, lie significantly lower. The only plotted muon collider (the 0.5 TeV demonstration machine's very preliminary cost estimate) lies above the line. But the clear indication is that length is, or at least has been, a good estimator of approximate cost. It is interesting to note that the fitted line indicates costs rising, not linearly, but as the 0.85 th power of length. This can be taken as a measure of economies of scale.

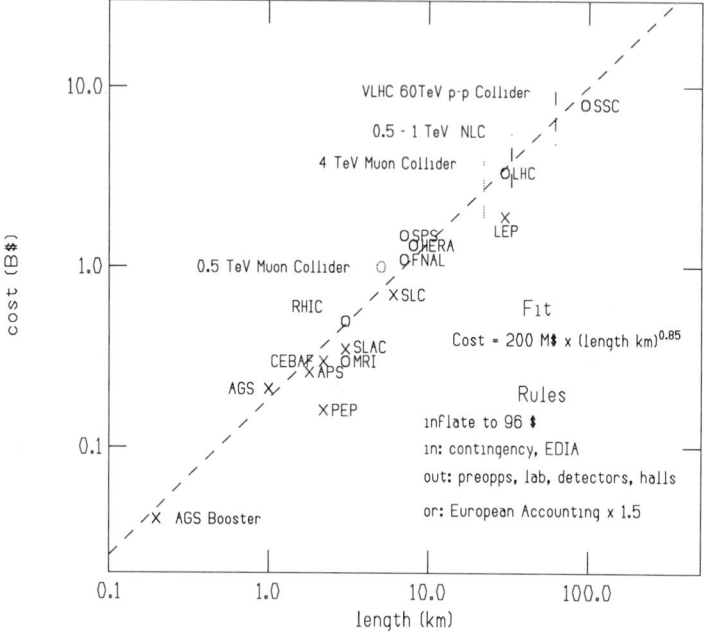

Figure 9. Costs of some machines as a function of their total lengths.

Conclusions

Our conclusions for this chapter, with the caveat that they are indeed only our opinions, are:

- The LHC is a well optimized and appropriate next step towards high *effective physics* energy.

- A Very Large Hadron Collider with energy greater than the SSC (e.g. 60 TeV c-of-m) and cost somewhat less than the SSC, may well be possible with the use of high T_c superconductors that may become available.

- A "Next Linear Collider" is the only clean way to complement the LHC with a lepton machine, and the only way to do so soon. But it appears that even a 0.5 TeV collider will be more expensive than the LHC, and it will be technically challenging: obtaining the design luminosity may not be easy.

- Extrapolating conventional rf e^+e^- linear colliders to energies above 1 or 2 TeV will be very difficult. Raising the rf frequency can reduce length and probably cost for a given energy, but obtaining luminosity increasing as the square of energy, as required, may not be feasible.

- Laser driven accelerators are becoming more realistic and can be expected to have a significantly lower cost per TeV. But the ratio of luminosity to wall power and the ability to preserve very small emittances, is likely to be significantly worse than for conventional rf driven machines. Colliders using such technologies are thus unlikely to achieve very high luminosities and are probably unsuitable for higher (above 2 TeV) energy physics research.

- A higher gradient superconducting Linac collider using Nb$_3$Sn or high T$_c$ materials, if it becomes technically possible, could be the only way to attain the required luminosities in a higher energy e^+e^-collider.

- Gamma-gamma collisions can and should be obtained at any future electron-positron linear collider. They would add physics capability to such a machine, but, despite their freedom from the beamstrahlung constraint, may not achieve higher luminosity.

- A Muon Collider, being circular, could be far smaller than a conventional electron-positron collider of the same energy. Very preliminary estimates suggest that it would also be significantly cheaper. The ratio of luminosity to wall power for such machines, above 2 TeV, appears to be better than that for electron positron machines, and extrapolation to a center of mass energy of 4 TeV or above does not seem unreasonable. If research and development can show that it is practical, then a 0.5 TeV muon collider could be a useful complement to e^+e^-colliders, and, at higher energies (e.g. 4 TeV), could be a viable alternative.

PHYSICS CONSIDERATIONS
Introduction

The physics opportunities and possibilities of the muon collider have been well documented in the Feasibility Study[12] and by additional papers[13]. For most reactions the physics capabilities of $\mu^+\mu^-$ and e^+e^- colliders with the same energy and luminosity are similar, so that the choice between them will depend mainly on the feasibility and cost of the accelerators. But for some reactions, the larger muon mass does provide some advantages:

- The suppression of synchrotron radiation induced by the opposite bunch (beamstrahlung) allows, in principle, the use of beams with very low momentum spread

- QED radiation is reduced by a factor of $[\ln(\sqrt{s}/m_\mu)/\ln(\sqrt{s}/m_e)]^2$, leading to smaller $\gamma\gamma$ backgrounds and a smaller effective beam energy spread.

- s-channel Higgs production is enhanced by a factor of $(m_\mu/m_e)^2 \approx 40000$.

- The suppression of synchrotron radiation, allowing acceleration and storage of muons in a ring, combined with the suppression of beamstrahlung, may allow the construction of $\mu^+\mu^-$ colliders at higher energy than e^+e^- machines.

The disadvantages are:

- Less polarization appears practical in a $\mu^+\mu^-$ collider than in an e^+e^- machine, and some luminosity loss is likely.

- The $\mu^+\mu^-$ machine will have considerably worse background and probably require a shielding cone, extending down to the vertex, that takes up a larger solid angle than that needed in an e^+e^- collider.

In the following sections we will give examples of physics for which there is a advantage in $\mu^+\mu^-$. These examples are taken from the discussion in section II of the $\mu^+\mu^-$ Collider Feasibility Study [12]. For a discussion of the other physics, SUSY particle identification in particular, the reader is refered to the physics sections of the Next Linear Collider Zeroth Order Design Report (ZDR)[14].

Precision Threshold Studies. The high energy resolution and suppression of Initial State Radiation (ISR) in a $\mu^+\mu^-$ collider makes it particularly well suited to threshold studies. As an example, Fig. 10 shows the threshold curves for top quark production for both $\mu^+\mu^-$ and e^+e^- machines, with and without beam smearing. (An rms energy spread of 1 % is assumed for e^+e^- and 0.1 % for $\mu^+\mu^-$). The rms mass resolution Δm_t obtained with $10\,\text{fb}^{-1}$ in a $\mu^+\mu^-$ Collider is estimated to be $\pm 0.3\,\text{GeV}$. This can be compared with 4 GeV for the Tevatron, 2 GeV for the LHC, and 0.5 GeV for NLC.

Studies of Standard Model, or SUSY Model Light, Higgs h. The feature that has attracted most theoretical interest is the possibility of s-channel studies of Higgs production. This is possible with μ's, but not with e's, due to the strong coupling of muons to the Higgs channel that is proportional to the mass of the lepton. If the Higgs sector is more complex than just a simple standard model (SM) Higgs, it will be necessary to measure the widths and quantum numbers of any newly discovered particles to ascertain the nature of those particles and the structure of the theory. In addition to the increased coupling strength of the muons, the beamstrahlung is much reduced for muons allowing much better definition of the beam energy.

The cross sections for Higgs production with a $\mu^+\mu^-$ collider are substantial. Fig. 11 shows a) the Higgs signal, b) the background, and c) the luminosity required

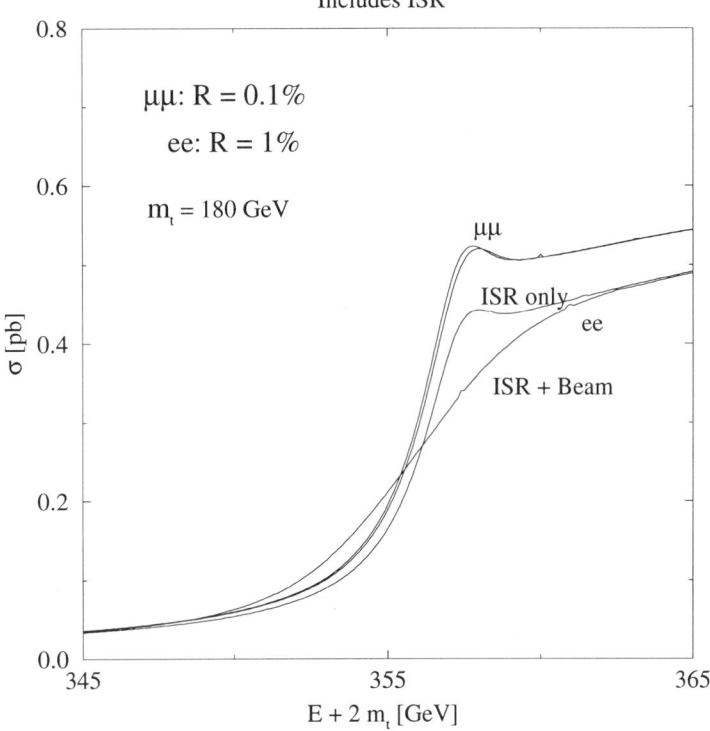

Figure 10. The threshold curves are shown for $\mu^+\mu^-$ and e^+e^- machines including ISR and with and without beam smearing. Beam smearing has only a small effect at a muon collider, whereas at an electron collider the threshold region is significantly smeared (An rms energy spread of 1 % is assumed for e^+e^- and 0.1 % for $\mu^+\mu^-$).

for a 5σ signal significance, for two different rms energy spreads of the muon beam: 0.01 % and 0.06 %. Signals are shown for three final states: $b\bar{b}$, WW$^{(*)}$ and ZZ$^{(*)}$ (reconstructable, non- 4 jet, with channel isolation efficiency $\epsilon = 0.5$). It is seen that:

- For an rms energy resolution of 0.01 %, a luminosity of only 0.1 fb^{-1} is required to yield a detectable signal for all $m_{h_{SM}}$ above the current LEP limit, except in the region of the Z peak, where 1 fb^{-1} is required.

- For an rms energy resolution of 0.06 %, the luminosity required is 20- 30 times larger, indicating that the higher resolution is desirable even at significant loss of luminosity.

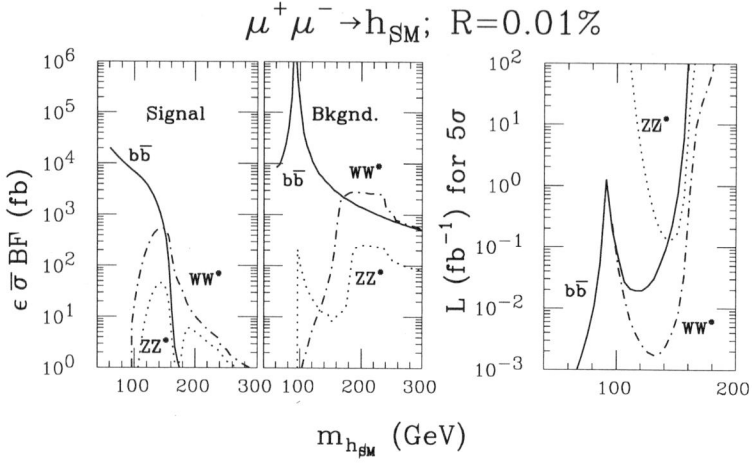

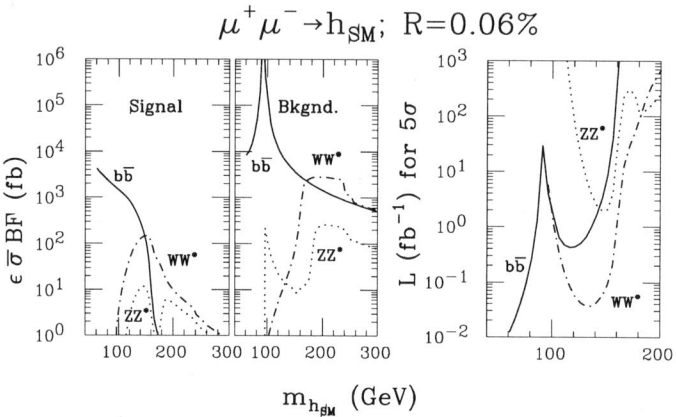

Figure 11. The (a) h_{SM} signal and (b) background cross sections, $\epsilon \bar{\sigma} BF(X)$, for $X = b\bar{b}$, and useful (reconstructable, non-4j) $WW^{(\star)}$ and $ZZ^{(\star)}$ final states (including a channel-isolation efficiency of $\epsilon = 0.5$) versus $m_{h_{SM}}$ for SM Higgs s-channel production. Also shown: (c) the corresponding luminosity required for a $S/\sqrt{B} = 5$ standard deviations signal in each of the three channels. Results for $R = 0.01\%$ and $R = 0.06\%$ are given.

Fig. 12 shows the total widths of standard model and MSSM Higgs. In the case of MSSM masses are plotted for the stop quark mass $m_{stop} = 1$ TeV, $\tan\beta = 2$ and 20. Two loop corrections have been included, but no squark mixing or SUSY decay channels.

The standard model Higgs with mass below m_t is seen to be very narrow. For 110 GeV it is ≈ 3 MeV. A Supersymmetric model Higgs would be wider, but might be only a little wider. It could be important to measure the width of a low mass

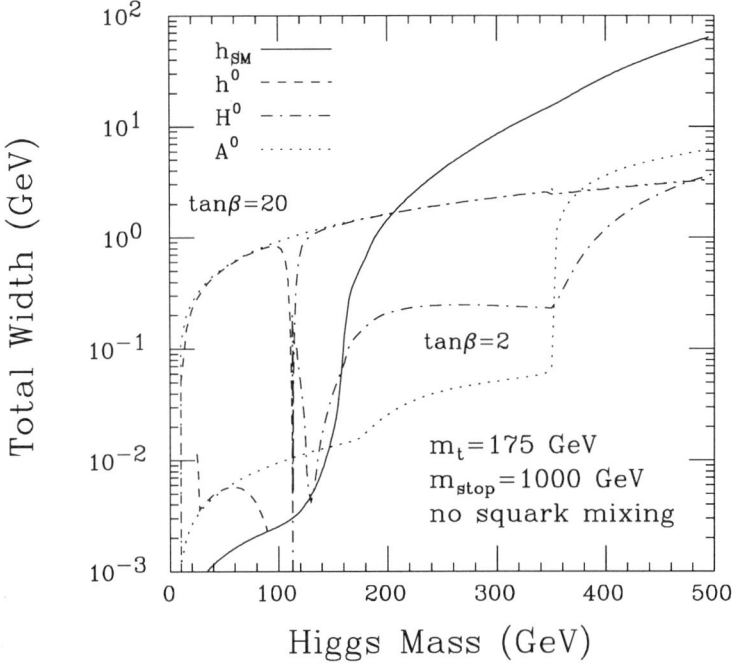

Figure 12. Total width vs mass of the SM and MSSM Higgs bosons for $m_t = 175$ GeV. In the case of the MSSM, we have plotted results for $\tan\beta = 2$ and 20, taking $m_{\tilde{t}} = 1$ TeV and including two-loop corrections.

Higgs to determine its character. It has been shown that a muon collider with an rms energy spread of 0.01 % could measure the width of a 110 GeV standard model Higgs to ±1 MeV with only 2 inverse femtobarns. Only if the Higgs mass is close to that of the Z does it become difficult to make such a determination without a large amount of data (200 inverse femtobarns). This could be a very important measurement (that could not be done in any other way) since it would distinguish clearly the nature of the boson seen. Together with branching ratio measurements (also possible with a muon collider), it could even predict the mass of the other SUSY Higgs bosons: H and A.

Studies of SUSY Model Heavy Higgs Particles: H and A. The H and A SUSY Higgs bosons are expected to be significantly heavier than the lightest h, and might have quite similar masses. If $\tan\beta$ is small (< 3) then they can easily be identified at the LHC, but may not be identified there if $\tan\beta$ is large. They could be searched for in an e^+e^- machine in $e^+e^- \to H, A$, (h,A or h, A are depressed) but only up to about $m_{H,A} \approx \sqrt{s}/\sqrt{2}$ or even less. In a muon collider, on the other hand, they could,

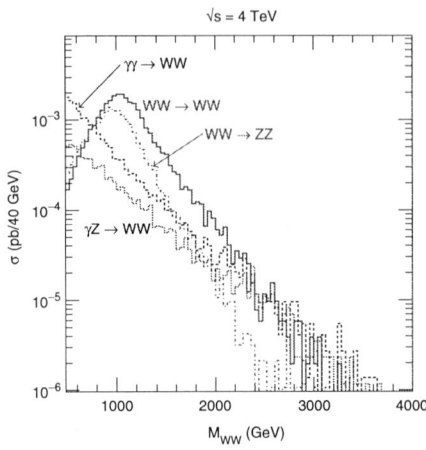
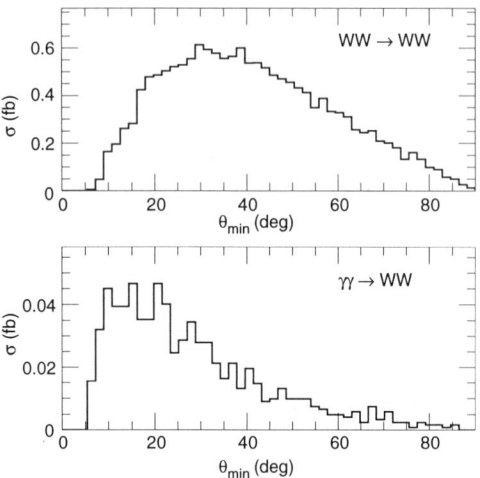

Figure 13: Signals and physics backgrounds for a 1 TeV Higgs boson at a $\mu\mu$ collider, including the effect of a 20° dead cone around the beamline.

Figure 14: $WW \to WW$ signal and $\gamma\gamma \to WW$ background vs. the minimum angle, $\theta_{\min}$, of the W.

providing $\tan\beta$ is large (> 3), be easily observed in the s-channel, up to masses equal to $\sqrt{s}$.

Studies of Non-SUSY Model Strong WW Interactions. If SUSY does not exist and we are forced to a much higher mass scale to study the symmetry breaking process then a 4 TeV muon collider is a viable choice to study WW scattering as it becomes a strong reaction.

Fig. 13 shows the mass distribution for the 1 TeV Higgs signals and physics backgrounds from PYTHIA in a toy detector, which includes segmentation of $\Delta\eta = \Delta\phi = 0.05$ and the angular coverage, $20° < \theta < 160°$, assumed in the machine background calculations. Since the nominal luminosity is $1000\,\mathrm{fb}^{-1}$, there are $\gtrsim 1000$ events per bin at the peak. The loss in signal from the 20° cone is larger for this process than for s-channel processes but is still fairly small, as can be seen in Fig. 14. The dead cone has a larger effect on $\gamma\gamma \to WW$ and thus the accepted region has a better signal to background ratio. It would be desirable to separate the WW and ZZ final states in purely hadronic modes by reconstructing the masses. Whether this is possible or not will depend on the details of the calorimeter performance and the level of the machine backgrounds. If it is not, then one can use the $\sim 12\%$ of events in which one $Z \to ee$ or $\mu\mu$ to determine the Z rate. Clearly there is a real challenge to try to measure the hadronic modes.

The background from $\gamma\gamma$ and γZ processes is smaller at a muon collider than at an electron collider but not negligible. Since the p_T of the photons is usually very small while the WW fusion process typically gives a p_T of order M_W, these backgrounds can be reduced by making a cut $p_{T,WW} > 50\,\text{GeV}$. This cut keeps most of the signal while significantly reducing the physics background.

Summary. For many reactions, SUSY particle discovery for example, an e^+e^- collider, with its higher polarization and lower background, would be preferable to a $\mu^+\mu^-$ machine of the same energy and luminosity. There are however specific reactions, s-channel Higgs production for example, where the $\mu^+\mu^-$ machine would have unique capabilities. Ideally both machines would be built and they would be complementary. Whether both machines could be built, at both moderate and multi TeV energies, and whether both could be afforded, remains to be determined.

There are several hardware questions that must be carefully studied. The first is the question of the luminosity available when the beam momentum spread is decreased. In addition there will have to be good control of the injected beam energy as there is not time to make large adjustments in the collider ring. Precision determination of the energy and energy spread will be mandatory: presumably by the study of spin precession. Finally, the question of luminosity vs. percent polarization needs additional study; unlike the electron collider, both beams can be polarized but as shown later in this report, but the luminosity decreases as the polarization increases.

MUON COLLIDER COMPONENTS
Introduction

The possibility of muon colliders was introduced by Skrinsky et al.[15] and Neuffer[16] and has been aggressively developed over the past two years in a series of meetings and workshops[17, 18, 19, 20].

A collaboration, lead by BNL, FNAL and LBNL, with contributions from 18 institutions has been studying a 4 TeV, high luminosity scenario and presented a Feasibility Study[12] to the 1996 Snowmass Workshop. The basic parameters of this machine are shown schematically in Fig. 15 and given in Tb. 2. Fig. 16 shows a possible layout of such a machine.

Tb. 2 also gives the parameters of a 0.5 TeV demonstration machine based on the AGS as an injector. It is assumed that a demonstration version based on upgrades of the FERMILAB, or CERN machines would also be possible.

The main components of the 4 TeV collider would be:

- A proton source with KAON like parameters (30 GeV, 10^{14} protons per pulse, at 15 Hz).

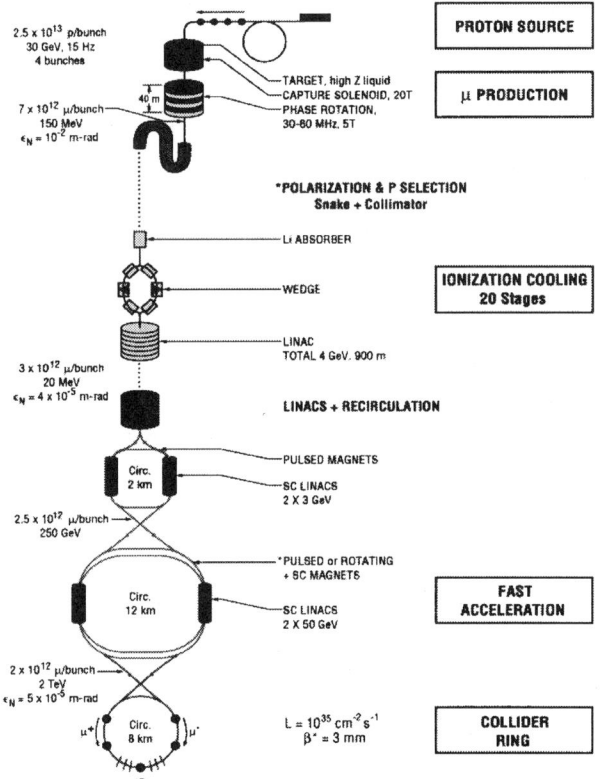

Figure 15. Overview of a 4 TeV Muon Collider

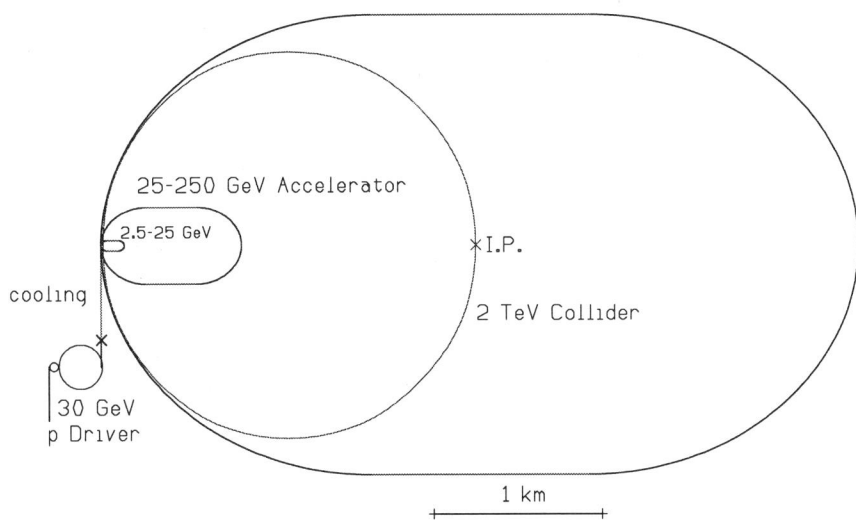

Figure 16. Layout of the collider and accelerator rings.

Table 2. Parameters of Collider Rings

c-of-m Energy	TeV	4	.5
Beam energy	TeV	2	.25
Beam γ		19,000	2,400
Repetition rate	Hz	15	2.5
Proton driver energy	GeV	30	24
Protons per pulse		10^{14}	10^{14}
Muons per bunch	10^{12}	2	4
Bunches of each sign		2	1
Beam power	MW	38	.7
Norm. rms emit. ϵ_N	π mm mrad	50	90
Bending Field	T	9	9
Circumference	Km	8	1.3
Ave. ring field B	T	6	5
Effective turns		900	800
β^* at intersection	mm	3	8
rms I.P. beam size	μm	2.8	17
Chromaticity		2000-4000	40-80
β_{max}	km	200-400	10-20
Luminosity	$cm^{-2}s^{-1}$	10^{35}	10^{33}

- A liquid metal target surrounded by a 20 T hybrid solenoid to make and capture pions.

- A 5 T solenoidal channel within a sequence of rf cavities to allow the pions to decay into muons and, at the same time, decelerate the fast ones that come first, while accelerating the lower momentum ones that come later. Muons from pions in the 100-500 MeV range emerge in a 6 m long bunch at 150 ± 30 MeV.

- A solenoidal snake and collimator to select the momentum, and thus polarization, of the muons.

- A sequence of 20 ionization cooling stages, each consisting of: a) energy loss material in a strong focusing environment for transverse cooling; b) linac reacceleration and c) lithium wedges in a dispersive environment for cooling in momentum space.

- A linac and/or recirculating linac pre-accelerator, followed by a sequence of pulsed field synchrotron accelerators using superconducting linacs for rf.

- An isochronous collider ring with locally corrected low beta (β=3 mm) insertion.

Proton Driver

The specifications of the proton drivers are given in Tb 3. In the 4 TeV example, it is a high-intensity (4 bunch, 2.5×10^{13} protons per pulse) 30 GeV proton synchrotron. The preferred cycling rate would be 15 Hz, but for a demonstration machine using the AGS[21], the repetition rate would be limited to 2.5 Hz and the energy to 24 GeV. For the lower energy machine, 2 final bunches are employed (one to make μ^-'s and the other to make μ^+'s). For the high energy collider, four are used (two μ bunches of each sign).

Table 3. Proton Driver Specifications

		4 TeV	.5 TeV Demo
Proton energy	GeV	30	24
Repetition rate	Hz	15	2.5
Protons per bunch	10^{13}	2.5	5
Bunches		4	2
Long. phase space/bunch	eV s	5	10
Final *rms* bunch length	ns	1	1

In order to reduce the cost of the muon phase rotation section, minimize the final muon longitudinal phase space and maximize the achievable polarization,, it appears that the final proton bunch length should be of the order of 1 ns. Is this practical ?

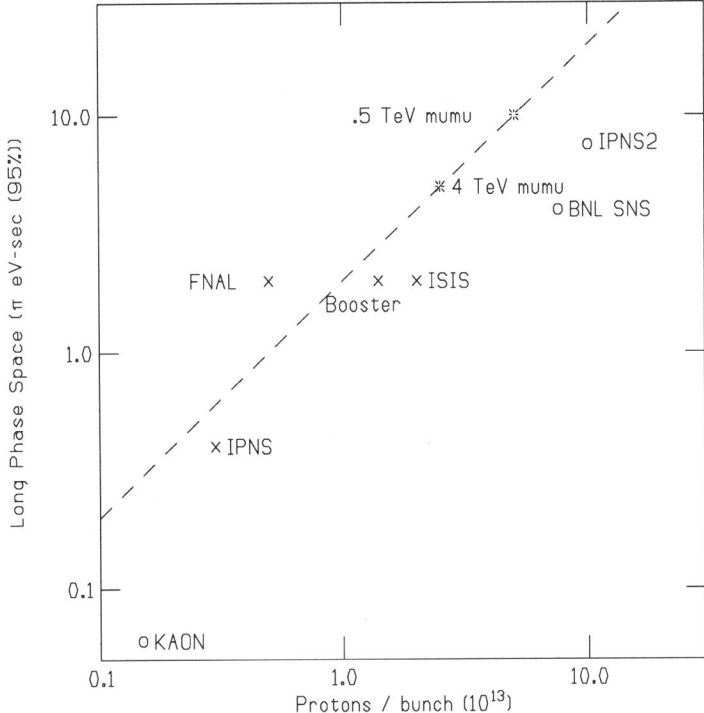

Figure 17. Longitudinal Phase Space of Bunches vs. number of Protons; x's indicate existing machines, circles proposed machines, and stars the values used here

There appears to be a relationship between the number of protons in a bunch and the longitudinal phase space of that bunch that can be maintained stability in a circular machine. Fig. 17 shows values obtained and those planned in a number of machines. The conservative assumption is that phase space densities will be similar to those already achieved: around 2 eV seconds per 10^{13} protons, as indicated by the line in Fig. 17. The required bunches of 2.5 10^{13} protons would thus be expected to have a phase space of 5 eVs (at 95%) = $6\pi\,\sigma_t\sigma_E$ eVs rms. A 1 ns rms bunch at 30 GeV with this phase space will have an rms momentum spread of 0.8 %, (2 % at 95%), and the space charge tune shift just before extraction would be ≈ 0.5. Provided the rotation can be performed rapidly enough, this should not be a problem. For the 0.5 TeV machine the bunch intensity, and thus area, would be double, leading to a final spread of 1.6 % rms (4 % at 95 %).

An attractive technique[22] for bunch compression would be to generate a large momentum spread with modest rf at a final energy close to transition. Pulsed quads would then be employed to move the operating point away from transition, resulting in rapid compression.

Earlier studies had suggested that the driver could be a 10 GeV machine with the same charge per fill, but a repetition rate of 30 Hz. This specification was almost identical to that studied[23] at ANL for a spallation neutron source. Studies at FNAL[24] have further established that such a specification is reasonable. But if 10 GeV protons are used, then approximately twice as many protons per bunch are required for the same pion production: 5×10^{13} per bunch for the 4 TeV case, 1×10^{14} per bunch for the 0.5 TeV case; the phase space of the bunches would be expected to be twice as big and the resulting % momentum spread for the 1 ns bunch 6 times as large: i.e. 12 % (at 95%) which may be hard to achieve. For the 0.5 TeV specification, this rises to 24 %: *clearly unreasonable*.

Target and Pion Capture

Pion Production. Predictions of the nuclear Monte-Carlo program ARC[25] suggest that π production is maximized by the use of heavy target materials, and that the production is peaked at a relatively low pion energy ($\approx$ 100 MeV), substantially independent of the initial proton energy. Fig.18 shows the forward π^+ production as a function of proton energy and target material; the π^- distributions are similar.

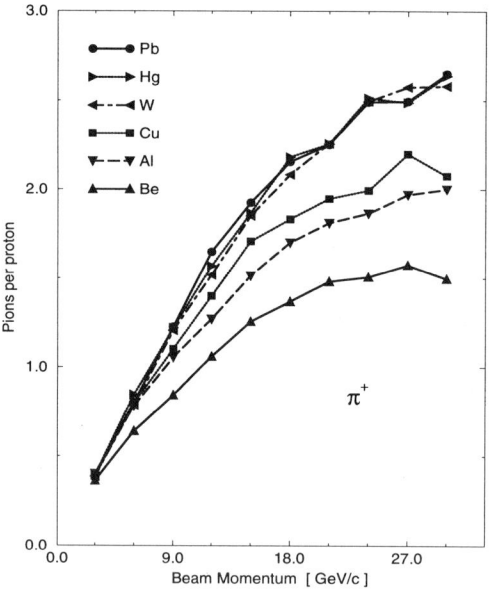

Figure 18. ARC forward π^+ production vs. proton energy and target material.

Other programs[26],[27] do not predict such a large low energy peak,(see for instance Fig. 19) and there is currently very little data to indicate which is right. An experiment (E910)[28][29], currently running at the AGS, should decide this question, and thus settle at which energy the capture should be optimized.

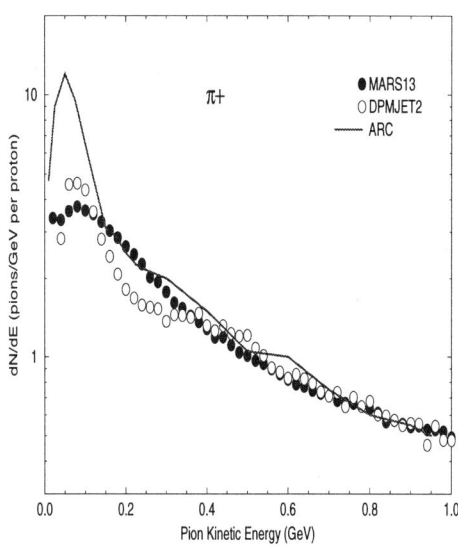

Figure 19. π^+ energy distribution for 24 GeV protons on Hg.

Target. For a low repetition rate the target could probably be made of Cu, approximately 24 cm long by 2 cm diameter. A study[30] indicates that, with a 3 mm rms beam, the single pulse instantaneous temperature rise is acceptable, but, if cooling is only supplied from the outside, the equilibrium temperature, at our required repetition rate, would be excessive. Some method must be provided to give cooling within the target volume. For instance, the target could be made of a stack of relatively thin copper disks, with water cooling between them. A graphite target could be used, but with significant loss of pion production, or a liquid metal target. Liquid lead and gallium are under consideration. In order to avoid shock damage to a container, the liquid could be in the form of a jet.

It appears that for maximum muon yield, the target (and incoming beam) should be at an angle to the axis of the solenoid and outgoing beam. The introduction of such an angle reduces the loss of pions when they reenter the target after being focused by the solenoid. A Monte Carlo simulation[31] gave a muon production increase of 60 % with at an angle 150 milliradians. The simulation assumed a copper target (interaction length 15 cm), ARC[25] pion production spectra, a fixed pion absorption cross section, no secondary pion production, a 1 cm target radius, and the capture solenoid, decay channel, phase rotation and bunch defining cuts described below. Fig. 20 shows the final muon to proton ratio as a function of the skew angle for a target whose length (45 cm) and transverse position (front end displaced - 1.5 cm from the axis) had been

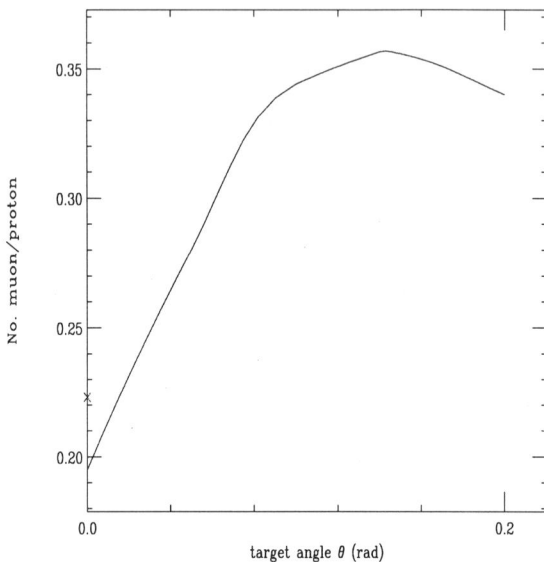

Figure 20. The muon to proton ratio as a function of the skew angle for a target whose length and transverse position has been reoptimized for the skew case. The single x indicates the production ratio at zero angle with the original optimization.

reoptimized for the skew case. The single X indicates the production ratio at zero angle with the original optimization (target length 30 cm, on axis). One notes that the reoptimized target length is 3 interaction lengths long, and thus absorbs essentially all of the initial protons.

Capture. Several capture methods were studied[32]. Pulsed horns were effective at the capture of very high energy pions. Multiple lithium lenses were more effective at lower pion energies, but neither was as effective as a high field solenoid at the 100 MeV peak of the pion spectrum. Initially, a 15 cm diameter, 28 T field was considered. Such a magnet could probably be built using superconducting outer coils and a Bitter, or other immersed sheet conductor inner coil, but such an immersed coil would probably have limited life[33]. A 15 cm diameter, 20 T solenoid could use a more conventional hollow conductor inner coil and was thus chosen despite the loss of 24 % pion capture (see Fig. 21)

A preliminary design[33] (see Fig. 22) has an inner Bitter magnet with an inside diameter of 24 cm (space is allowed for a 4 cm heavy metal shield inside the coil) and an outside diameter of 60 cm; it provides half (10T) of the total field, and would consume approximately 8 MW. The superconducting magnet has a set of three coils, all with inside diameters of 70 cm and is designed to give 10 T at the target and provide the required tapered field to match into the periodic superconducting solenoidal decay

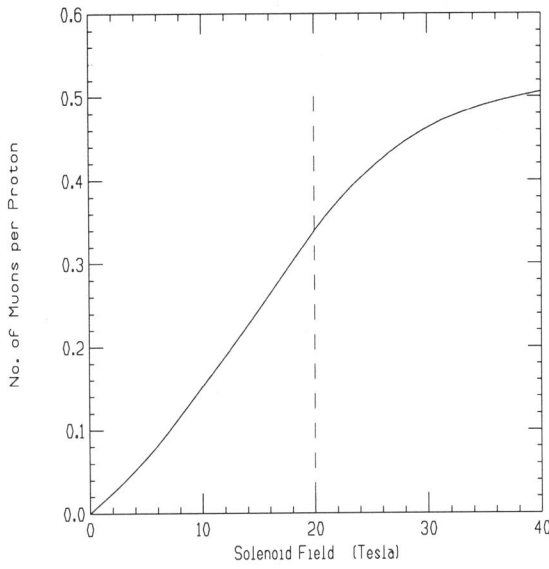

Figure 21. The muon to proton ratio as a function of capture solenoid field

channel (5 T and radius = 15 cm). A similar design has been made at LBL[34].

A new design[35] using a hollow conductor insert is now in progress. The resistive coil would give 6 T and consume 4 MW. The superconducting coils will supply 14 T.

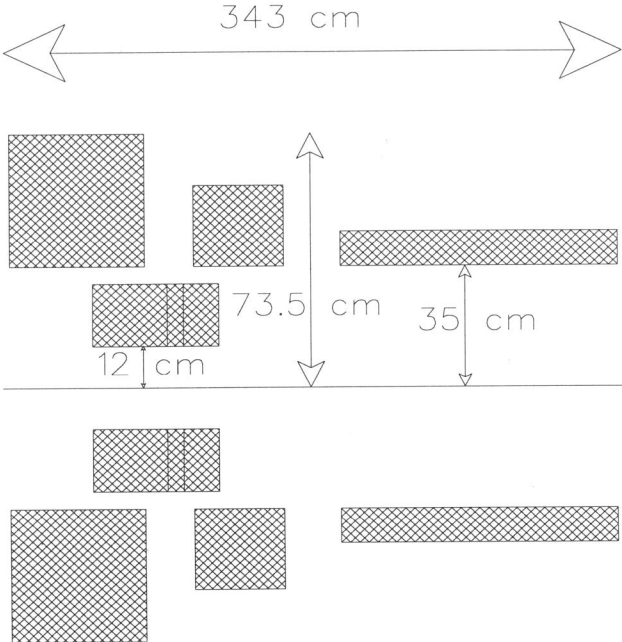

Figure 22. Schematic of a hybrid magnet solenoid system for π capture and matching.

Monte Carlo studies indicate a yield of 0.4–0.6 muons, of each sign, per initial proton, captured in the decay channel. Surprisingly, this conclusion seems relatively independent of whether the system is optimized for energies of 50 to 500 MeV (using ARC), or 200 to 2000 MeV (using MARS).

Use of Both Signs. Protons on the target produce pions of both signs, and a solenoid will capture both, but the required subsequent phase rotation rf systems will have opposite effects on each. One solution is to break the proton bunch into two, aim them on the same target one after the other, and adjust the rf phases such as to act correctly on one sign of the first bunch and on the other sign of the second. This is the solution assumed in the parameters of this paper.

A second possibility would be to separate the charges into two channels, delay the particles of one charge by introducing a chicane in one of the channels, and then recombine the two channels so that the particles of the two charges are in line, but separated longitudinally (i.e. box cared). Both charges can now be phase rotated by a single linac with appropriate phases of rf.

A third solution is to separate the pions of each charge prior to the use of rf, and feed the beams of each charge into different channels.

In either of the latter two solutions, there is a problem in separating the beams. After the target, and prior to the use of any rf or cooling, the beams have very large emittances and energy spread. Conventional charge separation using a dipole is not practical. But if a solenoidal channel is bent, then the particles trapped within that channel will drift[30],[36], in a direction perpendicular to the bend (this effect is discussed in more detail in the section on Options below). With our parameters this drift is dominated by a term (curvature drift) that is linear with the forward momentum of the particles, and has a direction that depends on the sign of the charges. If sufficient bend is employed[30], the two charges could be separated by a septum and captured into two separate channels. When these separate channels are bent back to the same forward direction, the momentum dispersion is separately removed in each new channel.

Although this idea is very attractive, it has some problems:

- If the initial beam has a radius r=0.15 m, and if the momentum range to be accepted is $F = \frac{p_{max}}{p_{min}} = 3$, then the required height of the solenoid just prior to separation is $2(1+F)r=1.2$ m. Use of a lesser height will result in particle loss. Typically, the reduction in yield for a curved solenoid compared to a straight solenoid is about 25 % (due to the loss of very low and very high momentum pions), but this must be weighed against the fact that both charge signs are captured for each proton on target.

- The system of bend, separation, and return bend will require significant length and must occur prior to the start of phase rotation (see below). Unfortunately, it appears that the cost of the phase rotation rf is strongly dependent on keeping this distance as short as possible.

Clearly, compromises will be involved, and more study of this concept is required.

Phase Rotation Linac

The pions, and the muons into which they decay, have an energy spread from about 0 - 500 MeV, with an rms/mean of $\approx$ 100%, and peak at about 100 MeV. It would be difficult to handle such a wide spread in any subsequent system. A linac is thus introduced along the decay channel, with frequencies and phases chosen to deaccelerate the fast particles and accelerate the slow ones; i.e. to phase rotate the muon bunch. Tb. 4 gives an example of parameters of such a linac. It is seen that the lowest frequency is 30 MHz, a low but not impossible frequency for a conventional structure.

Table 4. Parameters of Phase Rotation Linacs

Linac	Length m	Frequency MHz	Gradient MeV/m
1	3	60	5
2	29	30	4
3	5	60	4
4	5	37	4

A design of a reentrant 30 MHz cavity is shown in Fig.23. Its parameters are given in Tb.5. It has a diameter of approximately 2 m, only about one third of that of a

Table 5. Parameters of 30 MHz rf Cavity

Cavity Radius	cm	101
Cavity Length	cm	120
Beam Pipe Radius	cm	15
Accelerating Gap	cm	24
Q		18200
Average Acceleration Gradient	MV/m	3
Peak rf Power	MW	6.3
Average Power (15 Hz)	KW	18.2
Stored Energy	J	609

conventional pill-box cavity. To keep its cost down, it would be made of aluminum.

Multipactoring would probably be suppressed by stray fields from the 5 T focusing coils, but could also be controlled by an internal coating of titanium nitride.

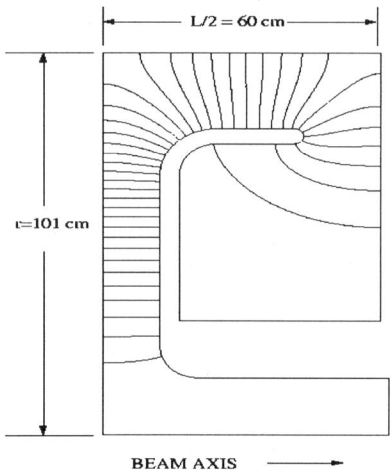

Figure 23. 30 MHz cavity for use in phase rotation and early stages of cooling.

Figs. 24 and 25 show the energy vs. c t at the end of the decay channel with and without phase rotation. Note that the c t scales are very different: the rotation both compacts the energy spread and limits the growth of the bunch length.

After this phase rotation, a bunch can be selected with mean energy 150 MeV, rms bunch length 1.7 m, and rms momentum spread 20 % (95 %, $\epsilon_L = 3.2$ eVs). The number of muons per initial proton in this selected bunch is 0.35, about half the total number of pions initially captured. As noted above, since the linacs cannot phase rotate both signs in the same bunch, we need two bunches: the phases are set to rotate the μ^+'s of one bunch and the μ^-'s of the other. Prior to cooling, the bunch is accelerated to 300 MeV, in order to reduce the momentum spread to 10 %.

Cooling

For a collider, the phase-space volume must be reduced within the μ lifetime. Cooling by synchrotron radiation, conventional stochastic cooling and conventional electron cooling are all too slow. Optical stochastic cooling[37], electron cooling in a

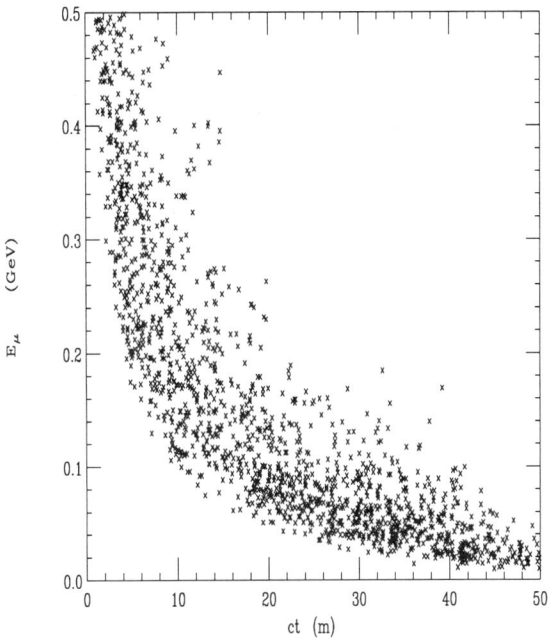

Figure 24. Energy vs. ct of Muons at End of Decay Channel without Phase Rotation.

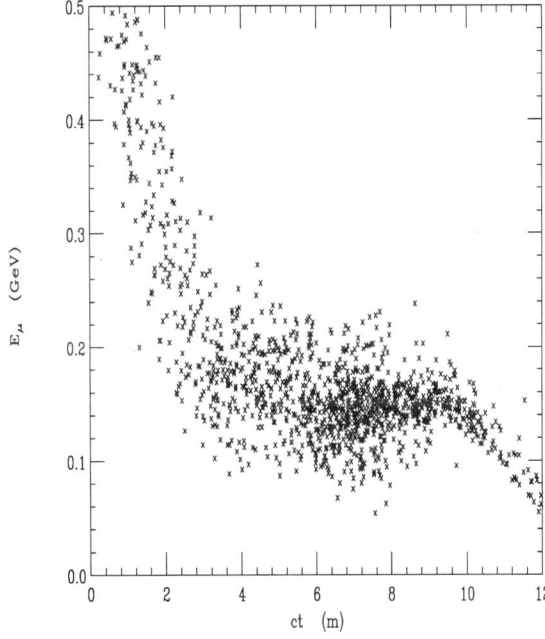

Figure 25. Energy vs. ct of Muons at End of Decay Channel with Phase Rotation.

plasma discharge[38] and cooling in a crystal lattice[39] are being studied, but appear very difficult. Ionization cooling[40] of muons seems relatively straightforward.

Ionization Cooling Theory. In ionization cooling, the beam loses both transverse and longitudinal momentum as it passes through a material medium. Subsequently, the longitudinal momentum can be restored by coherent reacceleration, leaving a net loss of transverse momentum. Ionization cooling is not practical for protons and electrons because of nuclear interactions (p's) and bremsstrahlung (e's), but is practical for μ's because of their low nuclear cross section and relatively low bremsstrahlung.

The approximate equation for transverse cooling (with energies in GeV) is:

$$\frac{d\epsilon_n}{ds} \approx -\frac{dE_\mu}{ds}\frac{\epsilon_n}{E_\mu} + \frac{\beta_\perp (0.014)^2}{2\, E_\mu m_\mu\, L_R}, \qquad (11)$$

where ϵ_n is the normalized emittance, $\beta_\perp$ is the betatron function at the absorber, dE_μ/ds is the energy loss, and L_R is the radiation length of the material. The first term in this equation is the coherent cooling term, and the second is the heating due to multiple scattering. This heating term is minimized if $\beta_\perp$ is small (strong-focusing) and L_R is large (a low-Z absorber). From Eq. 11 we find a limit to transverse cooling, which occurs when heating due to multiple scattering balances cooling due to energy loss. The limits are $\epsilon_n \approx 0.6\ 10^{-2}\ \beta_\perp$ for Li, and $\epsilon_n \approx 0.8\ 10^{-2}\ \beta_\perp$ for Be.

The equation for energy spread (longitudinal emittance) is:

$$\frac{d(\Delta E)^2}{ds} \approx -2\frac{d\left(\frac{dE_\mu}{ds}\right)}{dE_\mu} <(\Delta E_\mu)^2> + \frac{d(\Delta E_\mu)^2_{\text{straggling}}}{ds} \qquad (12)$$

where the first term is the cooling (or heating) due to energy loss, and the second term is the heating due to straggling.

Cooling requires that $\frac{d(dE_\mu/ds)}{dE_\mu} > 0$. But at energies below about 200 MeV, the energy loss function for muons, dE_μ/ds, is decreasing with energy and there is thus heating of the beam. Above 400 MeV the energy loss function increases gently, giving some cooling, but not sufficient for our application.

Energy spread can also be reduced by artificially increasing $\frac{d(dE_\mu/ds)}{dE_\mu}$ by placing a transverse variation in absorber density or thickness at a location where position is energy dependent, i.e. where there is dispersion. The use of such wedges can reduce energy spread, but it simultaneously increases transverse emittance in the direction of the dispersion. Six dimensional phase space is not reduced, but it does allow the exchange of emittance between the longitudinal and transverse directions.

In the long-path-length Gaussian-distribution limit, the heating term (energy straggling) is given by[41]

$$\frac{d(\Delta E_\mu)^2_{\text{straggling}}}{ds} = 4\pi\,(r_e m_e c^2)^2\, N_o\, \frac{Z}{A}\, \rho\gamma^2 \left(1 - \frac{\beta^2}{2}\right), \qquad (13)$$

where N_o is Avogadro's number and ρ is the density. Since the energy straggling increases as γ^2, and the cooling system size scales as γ, cooling at low energies is desired.

Low $\beta_\perp$ Lattices for Cooling

We have seen from the above that for a low equilibrium emittance we require energy loss in a strong focusing (low $\beta_\perp$) region. Three sources of strong focusing have been studied:

Solenoid. The simplest solution would appear to be the use of a long high field solenoid in which both acceleration and energy loss material could be contained. There is, however, a problem: when particles enter a solenoid other than on the axis, they are given angular momentum by the radial field components that they must pass. This initial angular momentum is proportional to the solenoid field strength, and to the particles' radius. In the absence of material, this extra angular momentum is maintained proportional to the tracks' radius as they pass along the solenoid until they are exactly corrected by the radial fields at the exit. But if material is introduced, all transverse momenta are "cooled", including the extra angular momentum given by these radial fields. When the cooled particles now leave the solenoid, then the end fields overcorrect them, leaving the particles with a finite added angular momentum. In practice, this angular momentum is equivalent to a significant heating term that limits the maximum emittance reduction to a quite small factor. The problem can only be averted if the direction of the solenoid field is periodically reversed.

Alternating Solenoid (FOFO) Lattice. An interesting case of such periodic solenoid field reversals is a lattice with rapid reversal that, for example, might approximate sinusoidal variations. We describe such a lattice as FOFO (focus focus) in analogy with quadrupole lattices that are FODO (focus defocus). Not only do such lattices avoid the angular momentum problems of a long solenoid, but they can, if the phase advance per cell approaches π, provide $\beta_\perp$'s at the zero field points, that are less than the same field would provide in the long solenoid case.

But as noted above, for cooling to be effective, the ratio of emittance to $\beta_\perp$ must remain above a given value. This implies that the angular amplitude of the particles has to be relatively large (typically greater than 0.1 radians rms). When tracking of such distributions was performed on realistic lattices three apparent problems were observed:

1. Particles entering with large amplitude (radius or angle) were found[42] to be lost or reflected by the fringe fields of the lenses. The basic problem is that there are strong non-linear effects that focus the large angle particles more strongly than those at small angles (this is known as a second order tune shift). The stronger focus causes an increase in the phase advance per cell resulting in resonant behavior, emittance growth and particle loss.

2. A bunch, even when monoenergetic, passing along such a lattice would be seen[43] to rapidly grow in length because the larger amplitude particles, traveling longer orbits, would fall behind the small amplitude ones.

3. With material present, the energy spread of a bunch grew because the high amplitude particles were passing through more material than the low amplitude ones.

Surprisingly however, none of these turns out to be a real problem. If the particles are matched, as they must be, into rf buckets, then all particles at the centers of these buckets must be traveling with the same average forward velocity. If this were not so then they would be arriving at the next rf cavity with different phases and would not be at the center of the bucket. It follows that large amplitude particles (whose trajectories are longer) must have higher momenta than those with lower amplitude. The generation of this correlation is part of the matching requirement, and would be naturally generated if an adiabatic application of FOFO strength were introduced. It could also be generated by a suitable gradation of the average radial absorber density.

Since higher amplitude particles will thus have higher momenta, they will, as a result, be less strongly focused: an effect of the opposite sign to the second order tune shift natural to the lattice. Can the effects cancel? In practice they are found to cancel almost exactly at a specific momentum: close to 100 MeV/c for a continuous sinusoidal FOFO lattice (the exact momentum will depend on the lattice).

A second, but only partial, cancelation also occurs: the higher amplitude, and now higher momentum, particles lose less energy in the absorber because of the natural energy dependence of the energy loss. This difference of energy loss, at 100 MeV/c, actually overcorrects the difference in energy loss from the difference in trajectories in the material. But this too is no problem. The natural bucket center for large amplitude particles will be displaced not only up in energy, but also over in phase, so as to be in a different accelerating field, and thus maintain their energy. Again, this would occur naturally if the lattice is introduced adiabatically and can also be generated by a combination of radially graded absorbers and drifts. Particles of differing momentum or phase will, as in normal synchrotron oscillation, gyrate about their bucket centers, but now each amplitude has a different center.

Using particles so matched, a simulation using fully Maxwellian sinusoidal field has been shown to give continuous transverse cooling without significant particle loss (see Fig. 26). In this simulation, the axial field has been gradually increased, and its period decreased, so as to maintain a constant rms angular spread as the emittance falls. The peak rf accelerating fields were 10 MeV/m, their frequency 750 MHz, the absorbing material was lithium, placed at the zero magnetic field positions, with lengths such that they occupied 5 % of the length. The mean momentum was 110 MeV/c, and rms width 2 %. 500 particles were tracked; none were lost.

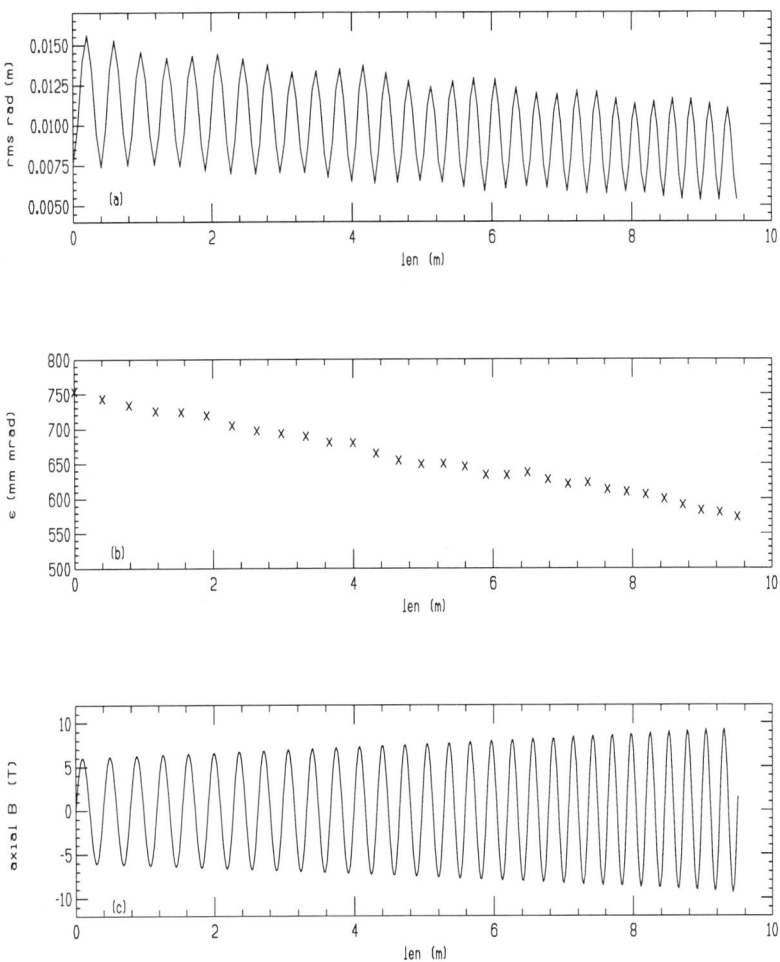

Figure 26. Cooling in a FOFO Lattice: (a) rms radius; (b) normalized emittance; and (c) axial magnetic fields; all plotted vs axial length.

Lithium Rods. The third method of providing strong focusing and energy loss is to pass the particles along a current carrying lithium rod (a long lithium lens). The rod serves simultaneously to maintain the low $\beta_\perp$, and attenuate the beam momenta. Similar lithium rods, with surface fields of 10 T, were developed at Novosibirsk[44] and have been used as focusing elements at FNAL[45] and CERN[46]. At the repetition rates required here, cooling of a solid rod will not be possible, and circulating liquid columns will have to be used. A small lens using such liquid cooling has also been tested at Novosibirsk. It is also hoped[47] that because of the higher compressibility of the liquid, surface field up to 20 T may be possible.

Lithium lenses will permit smaller $\beta_\perp$ and therefore cooling to lower emittances than in a practicable FOFO lattice, and such rods are thus preferred for the final cooling stages. But they are pulsed devices and consequently they are likely to have significant life time problems, and are thus not preferred for the earlier stages where they are not absolutely needed.

Such rods do not avoid the second order tune shift complications discussed above for the FOFO lattices. The rods must be alternated with acceleration sections and thus the particles must periodically be focused into and out of the rods. All three of the nonlinear effects enumerated above will be encountered. It is reasonable to believe that they can be controlled by the same mechanisms, but a full simulation of this has not yet been done.

Emittance Exchange Wedges. Emittance exchange in wedges to reduce the longitudinal emittance has been modeled with Monte Carlo calculations and works as theoretically predicted. But the lattices needed to generate the required dispersions and focus the particles onto the wedges have yet to be designed. The nonlinear complications discussed above will again have to be studied and corrected.

Emittance exchange in a bent current carrying rod has also been studied, both for a rod of uniform density[48] (in which the longer path length on the outside of the helix plays the role of a wedge; and where the average rod density is made greater on the outside of the bends by the use of wedges of a more dense material[49].

Reverse Emittance Exchange. At the end of a sequence of a cooling elements, the transverse emittance may not be as low as required, while the longitudinal emittance, has been cooled to a value less than is required. The additional reduction of transverse emittance can then be obtained by a reverse exchange of transverse and longitudinal phase-spaces. This can be done in one of several ways:

1. by the use of wedged absorbers in dispersive regions between solenoid elements.

2. by the use of septa that subdivide the transverse beam size, acceleration that shifts the energies of the parts, and bending to recombine the parts[49].

3. by the use of lithium lenses at very low energy: at very low energies the $\beta_\perp$'s, and thus equilibrium emittances, can be made arbitrarily low; but the energy spread is blown up by the steep rise in dE/dx. If this blow up of dE/dx is left uncorrected, then the effect can be close to an emittance exchange.

Model Cooling System

We require a reduction of the normalized transverse emittance by almost three orders of magnitude (from 1×10^{-2} to 5×10^{-5} m-rad), and a reduction of the longitudinal emittance by one order of magnitude.

A *model example* has been generated that uses no recirculating loops, and it is assumed for simplicity that the beams of each charge are cooled in separate channels (it may be possible to design a system with both charges in the same channel). The cooling is obtained in a series of cooling stages. In the early stages, they each have two components:

1. FOFO lattice consisting of spaced axial solenoids with alternating field directions and lithium hydride absorbers placed at the centers of the spaces between them where the $\beta_\perp$'s are minimum. RF cavities are introduced between the absorbers along the entire length of the lattice.

2. A lattice consisting of more widely separated alternating solenoids, and bending magnets between them to generate dispersion. At the location of maximum dispersion, wedges of lithium hydride are introduced to interchange longitudinal and transverse emittance.

In the last stages, reverse emittance exchange is achieved using current carrying lithium rods. The energy is allowed to fall to 15 MeV, thus increasing the focussing strength and lowering $\beta_\perp$.

The design is based on analytic calculations. The phase advance in each cell of the FOFO lattice is made as close to π as possible in order to minimize the $\beta_\perp$'s at the location of the absorber. The following effects are included: space charge transverse defocusing and longitudinal space charge forces; a 3σ fluctuation of momentum and 3σ fluctuations in amplitude.

The emittances, transverse and longitudinal, as a function of stage number, are shown in Fig.27, together with the beam energy. In the first 15 stages, relatively strong wedges are used to rapidly reduce the longitudinal emittance, while the transverse emittance is reduced relatively slowly. The objective is to reduce the bunch length, thus allowing the use of higher frequency and higher gradient rf in the reacceleration linacs. In the next 10 stages, the emittances are reduced close to their asymptotic limits. In the final three stages, lithium rods are used to produce an effective emittance exchange, as described above.

Individual components of the lattices have been defined, but a complete lattice has not yet been specified, and no complete Monte Carlo study of its performance has yet been performed. Wake fields, resistive wall effects, second order rf effects and some higher order focus effects are not yet included in this design of the system.

The total length of the system is 750 m, and the total acceleration used is 4.7 GeV. The fraction of muons that have not decayed and are available for acceleration is calculated to be 55%.

It would be desirable, though not necessarily practical, to economize on linac sections by forming groups of stages into recirculating loops.

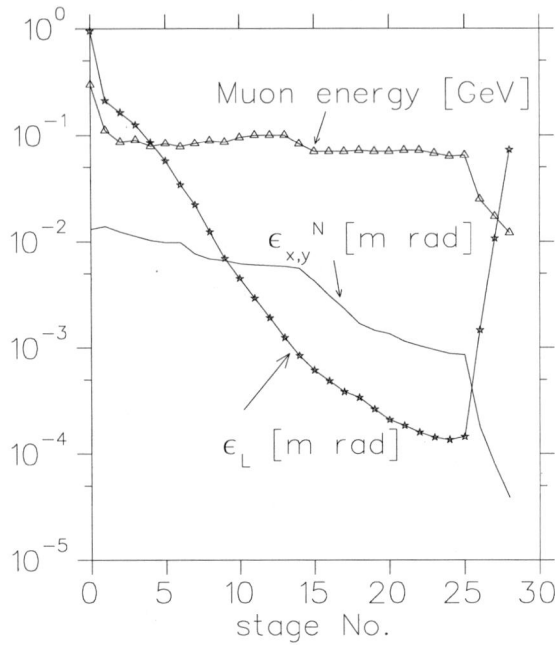

Figure 27. $\epsilon_\perp$, $\frac{\epsilon_L c}{\langle E_\mu \rangle}$ and E_μ [GeV] vs. stage number in the cooling sequence.

Acceleration

Following cooling and initial bunch compression the beams must be rapidly accelerated to full energy (2 TeV, or 250 GeV). A sequence of linacs would work, but would be expensive. Conventional synchrotrons cannot be used because the muons would decay before reaching the required energy. The conservative solution is to use a sequence of recirculating accelerators (similar to that used at CEBAF). A more economical solution would be to use fast rise time pulsed magnets in synchrotrons, or synchrotrons with rapidly rotating permanent magnets interspersed with high field fixed magnets.

Recirculating Acceleration. Tb. 6 gives an example of a possible sequence of recirculating accelerators. After initial linacs, there are two conventional rf recirculating accelerators taking the muons up to 75 GeV, then two superconducting recirculators going up to 2000 GeV.

Criteria that must be considered in picking the parameters of such accelerators are:

Table 6. Parameters of Recirculating Accelerators

		Linac	#1	#2	#3	#4
initial energy	GeV	0.20	1	8	75	250
final energy	GeV	1	8	75	250	2000
nloop			1	12	18	18	18
freq.	MHz	100	100	400	1300	2000
linac V	GV	0.80	0.58	3.72	9.72	97.20
grad		5	5	10	15	20
dp/p initial	%	12	2.70	1.50	1	1
dp/p final	%	2.70	1.50	1	1	0.20
σ_z initial	mm	341	333	82.52	14.52	4.79
σ_z final	mm	303	75.02	13.20	4.36	3.00
η	%	1.04	0.95	1.74	3.64	4.01
N_μ	10^{12}	2.59	2.35	2.17	2.09	2
τ_{fill}	μs	87.17	87.17	10.90	s.c.	s.c.
beam t	μs	0.58	6.55	49.25	103	805
decay survival		0.94	0.91	0.92	0.97	0.95
linac len	km	0.16	0.12	0.37	0.65	4.86
arc len	km	0.01	0.05	0.45	1.07	8.55
tot circ	km	0.17	0.16	0.82	1.72	13.41
phase slip	deg	0	38.37	7.69	0.50	0.51

- The wavelengths of rf should be chosen to limit the loading, η, (it is restricted to below 4 % in this example) to avoid excessive longitudinal wakefields and the resultant emittance growth.

- The wavelength should also be sufficiently large compared to the bunch length to avoid excessive second order effects (in this example: 10 times).

- For power efficiency, the cavity fill time should be long compared to the acceleration time. When conventional cavities cannot satisfy this condition, superconducting cavities are specified.

- In order to minimize muon decay during acceleration (in this example 73% of the muons are accelerated without decay), the number of recirculations at each stage should be kept low, and the rf acceleration voltage correspondingly high. For minimum cost, the number of recirculations appears to be of the order of 18. In order to avoid a large number of separate magnets, multiple aperture magnets can be designed (see Fig.28).

Note that the linacs see two bunches of opposite signs, passing through in opposite directions. In the final accelerator in the 2 TeV case, each bunch passes through the linac 18 times. The total loading is then $4 \times 18 \times \eta = 288\%$. With this loading, assuming 60% klystron efficiencies and reasonable cryogenic loads, one could probably achieve 35% wall to beam power efficiency, giving a wall power consumption for the rf in this ring of 108 MW.

A recent study[50] tracked particles through a similar sequence of recirculating accelerators and found a dilution of longitudinal phase space of the order of 15% and negligible particle loss.

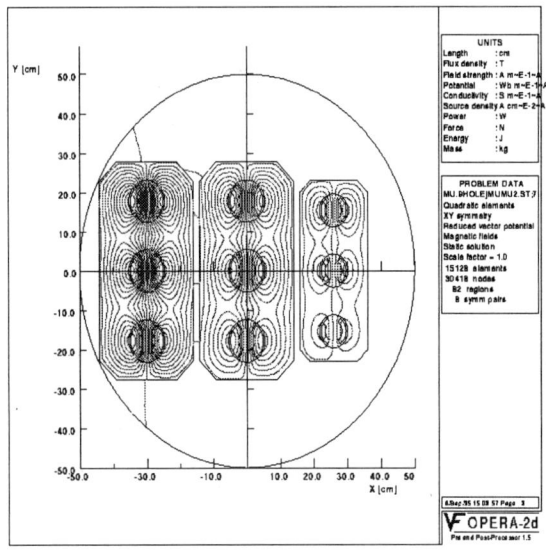

Figure 28. A cross section of a 9 aperture sc magnet.

Pulsed Magnet Acceleration. An alternative to recirculating accelerators for stages #2 and #3 would be to use pulsed magnet synchrotrons with rf systems consisting of significant lengths of superconducting linac.

The cross section of a pulsed magnet for this purpose is shown in Fig. 29. If desired, the number of recirculations could be higher in this case, and the needed rf voltage correspondingly lower, but the loss of particles from decay would be somewhat more. The cost for a pulsed magnet system appears to be significantly less than that of a multi-hole recirculating magnet system, and the power consumption is moderate for energies up to 250 GeV. Unfortunately, the power consumption is impractical at energies above 500 GeV.

Pulsed and Superconducting Hybrid. For the final acceleration to 2 TeV in the high energy machine, the power consumed by a ring using only pulsed magnets would be excessive, but a hybrid ring with alternating pulsed warm magnets and fixed superconducting magnets[51][52] should be a good alternative.

Tb. 7 gives an example of a possible sequence of such accelerators. Fig. 16 used a layout of this sequence. The first two rings use pulsed cosine theta magnets with peak fields of 3 T and 4 T. Then follow two hybrid magnet rings with 8 T fixed magnets alternating with ±2 T iron yoke pulsed magnets. The latter two rings share the same tunnel, and might share the same linac too. The survival from decay after all four rings is 67 %. Phase space dilution should be similar to that determined for the recirculating accelerator design above.

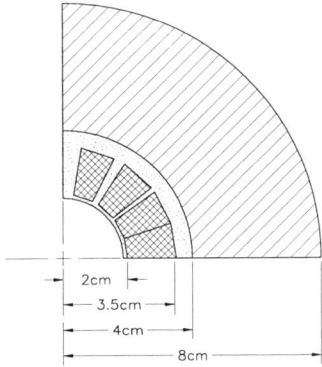

Figure 29. Cross section of pulsed magnet for use in the acceleration to 250 GeV.

Collider Storage Ring

After acceleration, the μ^+ and μ^- bunches are injected into a storage ring that is separate from the accelerator. The highest possible average bending field is desirable, to maximize the number of revolutions before decay, and thus maximize the luminosity. Collisions would occur in one, or perhaps two, very low-β^* interaction areas. Parameters of the ring were given earlier in Tb.2.

Table 7. Parameters of Pulsed Accelerators

	Ring	1	2	3	4
E_{init}	(GeV)	2.5	25	250	1350
E_{final}	(GeV)	25	250	1350	2000
fract pulsed	%	100	100	73	44
B_{pulsed}	(T)	3	4	±2	±2
Acc/turn	(GeV)	1	7	40	40
Acc Grad	(MV/m)	10	12	20	20
RF Freq	(MHz)	100	400	1300	1300
circumference	(km)	0.4	2.5	12.8	12.8
turns		22	32	27	16
acc. time	(μs)	26	263	1174	691
ramp freq	(kHz)	12.5	1.3	0.3	0.5
loss	(%)	13.4	13.2	9.0	2.2

Bending Magnet Design. The magnet design is complicated by the fact that the μ's decay within the rings ($\mu^- \to e^- \overline{\nu_e} \nu_\mu$), producing electrons whose mean energy is approximately 0.35 that of the muons. These electrons travel toward the inside of the ring dipoles, radiating a fraction of their energy as synchrotron radiation towards the outside of the ring, and depositing the rest on the inside. The total beam power, in the 4 TeV machine, is 38 MW. The total power deposited in the ring is 13 MW, yet the maximum power that can reasonably be taken from the magnet coils at 4 K is only of the order of 40 KW. Shielding is required.

The beam is surrounded by a thick warm shield, located inside a large aperture magnet. Fig.30 shows the attenuation of the heating produced as a function of the thickness of a warm tungsten liner[53]. If conventional superconductor is used, then the thicknesses required in the two cases would be as given in Tb.8. If high Tc superconductors could be used, then these thicknesses could probably be halved.

Table 8. Thickness of Shielding for Cos Theta Collider Magnets.

		2TeV	0.5 TeV Demo
Unshielded Power	MW	13	.26
Liner inside rad	cm	2	2
Liner thickness	cm	6	2
Coil inside rad	cm	9	5
Attenuation		400	12
Power leakage	KW	32	20
Wall power for $4 K$	MW	26	16

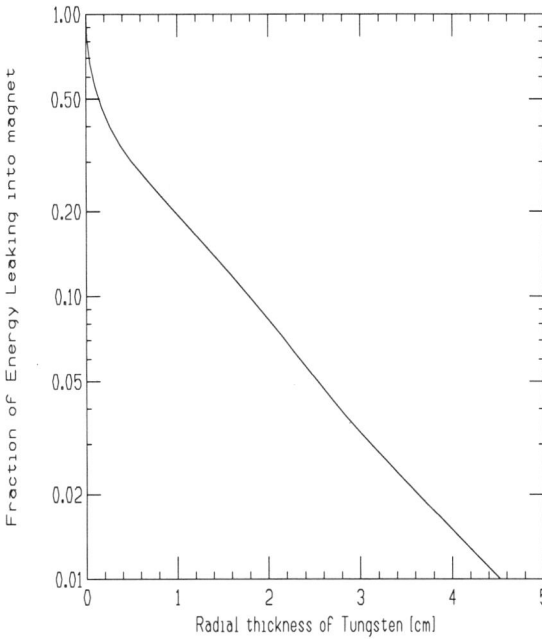

Figure 30. Energy attenuation vs. the thickness of a tungsten liner.

The magnet could be a conventional cosine-theta magnet (see Fig.31), or, in order to reduce the compressive forces on the coil midplane, a rectangular block design.

The power deposited could be further reduced if the beams are kicked out of the ring prior to their their complete decay. Since the luminosity goes as the square of the number of muons, a significant power reduction can be obtained for a small luminosity loss.

Quadrupoles. The quadrupoles could have warm iron poles placed as close to the beam as practical. The coils could be either superconducting or warm, as dictated by cost considerations. If an elliptical vacuum chamber were used, and the poles were at 1 cm radius, then gradients of 150 T/m should be possible.

Lattice Design.

1. **Arcs:** In a conventional 2 TeV FODO lattice the tune would be of the order of 200 and the momentum compaction α around 2×10^{-3}. In this case, in order to maintain a bunch with rms length 3 mm, 45 GeV of S-band rf would be required. This would be excessive. It is thus proposed to use an approximately isochronous lattice of the dispersion wave type[54]. Ideally one would like an α of the order of 10^{-7}. In this case the machine would behave more like a linear beam transport and rf would be needed only to correct energy spread introduced by wake effects.

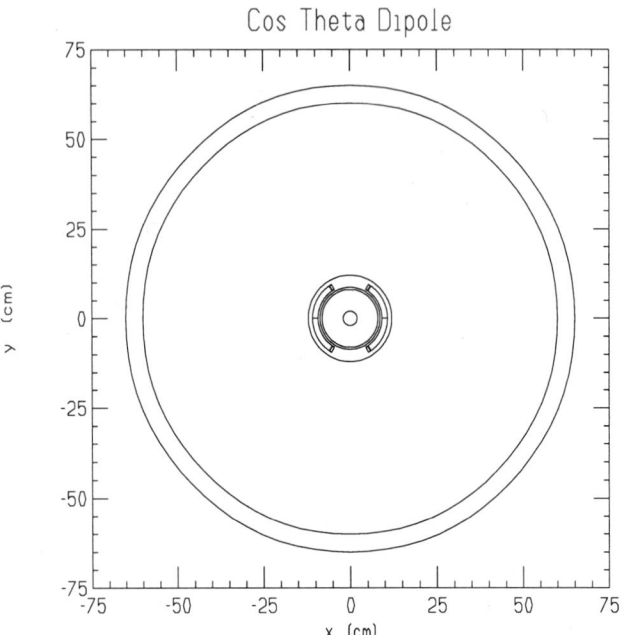

Figure 31. Cos Theta Arc Bending Magnet

It appears easy to set the zero'th order slip factor η_0 to zero, but if nothing is done, there is a relatively large first order slip factor η_1 yielding a minimum α of the order of 10^{-5}. The use of sextupoles appears able to correct this η_1 yielding a minimum α of the order of 10^{-6}. With octupoles it may be possible to correct η_2, but this remains to be seen. But even with an α of the order of 10^{-6} very little rf is needed.

It had been feared that amplitude dependent anisochronisity generated in the insertion would cause bunch growth in an otherwise purely isochronous design. It has, however, been pointed out[55] that if chromaticity is corrected in the ring, then amplitude dependent anisochronisity is automatically removed.

2. **Low β Insertion:** In order to obtain the desired luminosity we require a very low beta at the intersection point: $\beta^* = 3$ mm for 4 TeV, $\beta^* = 8$ mm for the .5 TeV design. An initial final focusing quadruplet design used 6.4 T maximum fields at 4σ. This would allow a radiation shield of the order of 5 cm, while keeping the peak fields at the conductors less than 10 T, which should be possible using Nb_3Sn conductor. The maximum beta's in both x and y were of the order of 400 km in the 4 TeV case, and 14 km in the 0.5 TeV machine. The chromaticities ($1/4\pi \int \beta dk$) are approximately 6000 for the 4 TeV case, and 600 for the .5 TeV machine. A later design[56] has lowered these chromaticities somewhat, but in either case the

chromaticities are too large to correct within the rest of a conventional ring and therefore require local correction[57][58].

It is clear that there is a great advantage in using very powerful final focus quadrupoles. The use of niobium tin or even more exotic materials should be pursued.

3. **Model Designs:** Initially, two lattices were generated[59][60],[61], one of which[61], with the application of octupole and decapole correctors, had an adequate calculated dynamic aperture. More recently, a new lattice and IR section has been generated[56] with much more desirable properties than those in the previously reported versions. Stronger final focusing quadrupoles were employed to reduce the maximum β's and chromaticity, the dispersion was increased in the chromatic correction regions, and the sextupole strengths reduced. It was also discovered that, by adding dipoles near the intersection point, the background in the detector could be reduced.[56]

Instabilities. Studies[62] of the resistive wall impedance instabilities indicate that the required muon bunches (eg. for 2 TeV: $\sigma_z = 3\ mm$, $N_\mu = 2 \times 10^{12}$) would be unstable in a conventional ring. In any case, the rf requirements to maintain such bunches would be excessive.

If one can obtain momentum-compaction factor $\alpha \leq 10^{-7}$, then the synchrotron oscillation period is longer than the effective storage time, and the beam dynamics in the collider behave like that in a linear beam transport[63][64]. In this case, beam breakup instabilities are the most important collective effects. Even with an aluminum beam pipe of radius $b = 2.5$ cm, the resistive wall effect will cause the tail amplitude of the bunch to double in about 500 turns. For a broad-band impedance of $Q = 1$ and $Z_\parallel/n = 1$ Ohm, the doubling time in the same beam pipe is only about 130 turns; which is clearly unacceptable. But both these instabilities can easily be stabilized using BNS[65] damping. For instance, to stabilize the resistive wall instability, the required tune spread, calculated[63] using the two particle model approximation, is (for Al pipe)

$$\frac{\Delta\nu_\beta}{\nu_\beta} = \begin{cases} 1.58\ 10^{-4} & b = 1.0\,\mathrm{cm} \\ 1.07\ 10^{-5} & b = 2.5\,\mathrm{cm} \\ 1.26\ 10^{-6} & b = 5.0\,\mathrm{cm} \end{cases} \quad (14)$$

This application of the BNS damping to a quasi-isochronous ring, where there are other head-tail instabilities due to the chromaticities ξ and η_1, needs more careful study.

If it is not possible to obtain an α less than 10^{-7}, then rf must be introduced and synchrotron oscillations will occur. The above instabilities are then somewhat stabilized because of the interchanging of head and tail, but the impedance of the rf now adds to the problem and simple BNS damping is no longer possible.

If, for example, a momentum-compaction factor $|\alpha| \approx 1.5 \times 10^{-5}$ is obtained, then rf of ~ 1.5 GV is needed which gives a synchrotron oscillation period of 150 turns. Three different impedance models: resonator, resistive wall, and a SLAC-like or a CEBAF-like rf accelerating structure have been used in the estimation for three sets of design parameters. The impedance of the ring is dominated by the rf cavities, and the microwave instability is well beyond threshold. Two approaches are being considered to control these instabilities: 1) BNS damping applied by rf quadrupoles as suggested by Chao[66]; and 2) applying an oscillating perturbation on the chromaticity[67].

When the ring is nearly isochronous, a longitudinal head-tail (LHT) instability may occur because the nonlinear slip factor η_1 becomes more important than the first order η_0. The growth time for the rf impedance when $\eta \simeq 10^{-5}$ is about $0.125\,b\,\eta_0/\eta_1$ s, where b is the pipe radius in cm. This would be longer than the storage time of ~ 41 ms if $\eta_1 \sim \eta_0$. However, if $\eta_1 \sim \eta_0/\delta$, with $\delta \sim 10^{-3}$, then the growth time is about $0.125b$ ms, which is much shorter than the storage time. More study is needed.

BACKGROUND AND DETECTOR
Design of the Intersection Region

The design of the Intersection Region[68] is driven by the desire to reduce the background from muon decays in the detector as much as possible. For this study a 130 m final focus section (Fig. 32) which included four final quadrupoles, three toroids, a 2 T solenoidal field for the detector and the connecting beam pipe and shielding was modeled in GEANT with all the appropriate magnetic fields and shielding materials. The parameters used were taken from [59][60]. Trajectories of particles with and without decay are shown later in Figs.33 and 34. Studies of the effects of high energy electrons hitting specific edges and surfaces were carried out and the shielding adjusted or augmented to mitigate the apparent effects of particular background problems. Effects due to electrons, photons, neutrons and charged hadrons and muons were considered in turn to try to optimized the design. While the current design is not fully optimized, it is a marked improvement over a much simpler design which had been used in the past. More importantly, it helped develop the tools and strategy to do such an optimization as the lattice is further developed. A second study[69] using a somewhat different final focus design and selecting shielding parameters has given results that are of the same order of magnitude as those that will be discussed in detail here.

The final focus may be thought to be composed of 3 separate regions. The longest of these, from 130 m to approximately 6.5 m contains the quadrupole magnets which

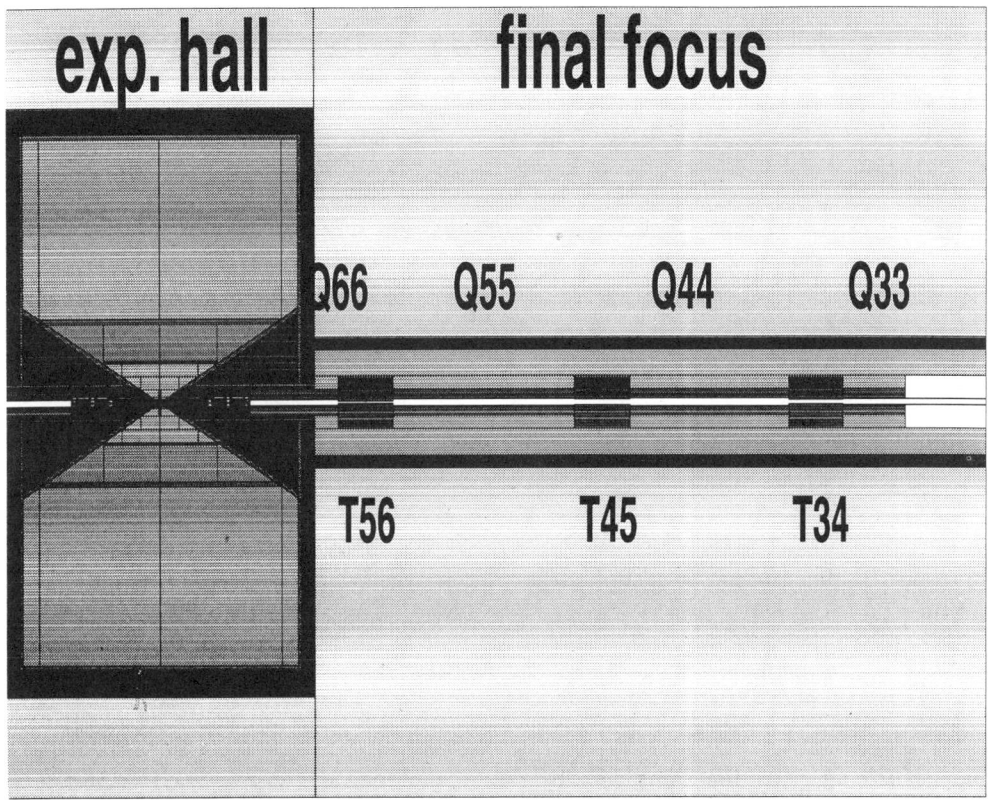

Figure 32. Region around the Intersection Region modeled in GEANT. The black regions represent tungsten shielding. The final quadrupoles (Q) and toroids (T) on one side of the detector enclosure are shown. The shaded areas around the intersection point represent the various detector volumes used in calculating particle fluences.

bring the beam to the final focus in the intersection region. The space available between the four quadrupoles was used to install toroids. They fulfill a double role: first they are used as scrapers for the electromagnetic debris; secondly, they serve as magnetic deflectors for the Bethe–Heitler(BH) muons generated upstream. The effect of the toroids on the BH muons will be discussed later. In order to optimize the inner aperture of the toroids, the σ_x and σ_y envelope of the muon bunch at every exit of the quadrupoles has been estimated. The inner aperture of each toroid was chosen to match the 4 σ ellipse of the muon bunch at that point. The second region, from 6.5 m to 1.1 m contains tungsten plus additional shielding boxes to help contain neutrons produced by photons in the electromagnetic showers. A shielding box consists of a block of Cu surrounded by polyboron. The shielding here is designed with inverted cones to reduce the probability of electrons hitting the edges of collimators or glancing off shielding surfaces (Fig. 35). The beam aperture at the entrance to this section is reduced to 2.5

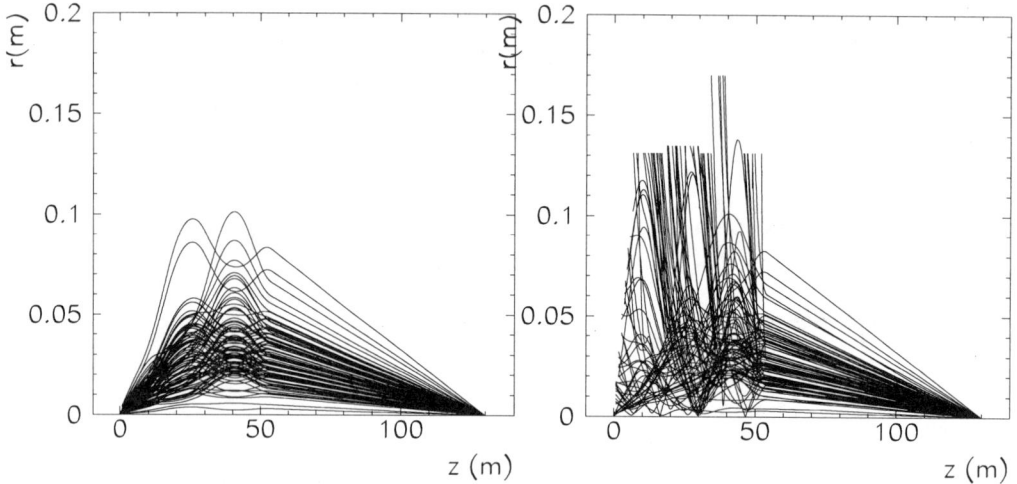

Figure 33: Trajectories in the Final Focus Region with Muon decays turned off

Figure 34: Trajectories in the Final Focus Region with Muon Decays allowed. The decay electrons are tracked until they reach either a magnet or shielding.

cm and by the exit of the section to 4.5 mm. This profile follows the beam envelope as the particles approach the intersection region. The intersection region itself (Fig. 36) is designed as an inverse cone to prevent electrons which reach this region from hitting any shielding as this region is directly viewed by the detector. A 20° tungsten cone around the intersection region is required for the reduction of the electromagnetic component of the background. The cone is lined, except very near the intersection region with polyboron to reduce the slow neutron flux. In the shielding calculations it is also assumed that there is a polyboron layer before the calorimeter and surrounding the muon system. In earlier designs this cone was only 9°. Whether or not the full 20° is required is still under study and work is ongoing to evaluate the physics impact of this choice of the shielding cone angle. It is likely that, after optimization is completed, the cone angle will be reduced.

Muon Decay Backgrounds.

Results using GEANT Simulation. The backgrounds in the detector are defined as the fluence of particles (number of particles per cm^2 per beam crossing) across surfaces which are representative of the various kinds of detectors which might be considered. For this study the calorimeter was assumed to be a composition of copper and liquid argon in equal parts by volume which represents a good resolution calorimeter with approximately 20% sampling fraction. The other volumes of the detector were vac-

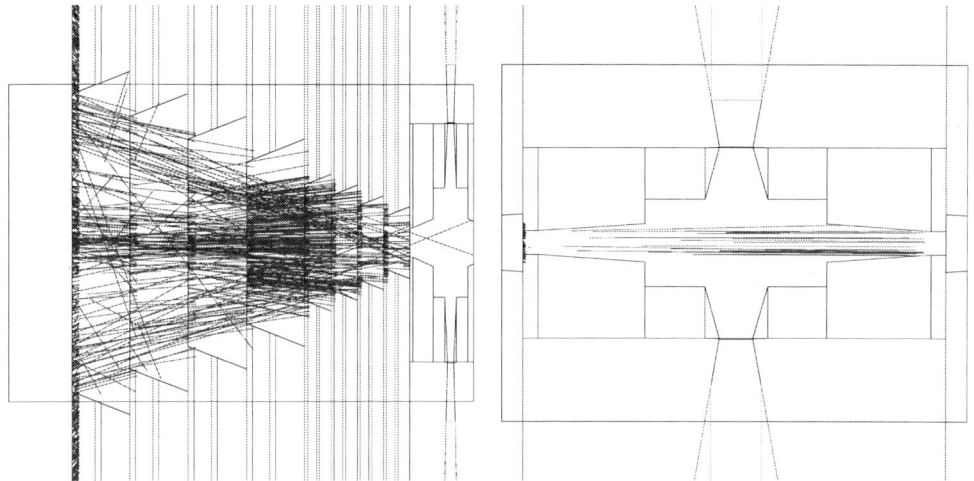

Figure 35: Expanded View of Region (2) near the Intersection point. The lines represent electrons from a random sample of muon decays.

Figure 36: Detailed View of Region (1), the Intersection Region. The lines represent electrons from a random sample of muon decays.

uum. The calorimeter starts from a radius of 150 cm and is 150 cm deep. The tracker volume is defined from 50 to 150 cm . An array of horizontal and vertical planes were placed in the detector volumes. These planes were used for flux calculations; their positions are evident in the tables of rates below.

Table 9. Longitudinal Particle Fluences from Muon Decays and Interactions from the GEANT Calculation. Fluence = particles/cm^2/crossing for two bunches of 2 $10^{12}\mu$'s each.

Detector	Radius(cm)	γ's	neutrons	$e^{\pm}$	$\pi^{\pm}$	protons	$\mu^{\pm}$
Vertex	5-10	15800	2200	138	28.8	1.6	3.0
	10-15	6200	2400		7.4	0.1	1.0
	15-20	3200	2000		9.2	8.0	4.6
Tracker	20-50	900	1740		1.6	7.8	0.6
	50-100	240	1040		0.2	4.4	0.12
	100-150	260	660		0.006	0.8	0.02
Calorimeter	160-310						0.004
Muon	310-10000						0.0004

Table 10. Radial Particle Fluences from Muon Decays and Interactions from the GEANT Calculation. Fluence = particles/cm^2/crossing for two bunches of $2 \cdot 10^{12}$ μ's each.

Detector	Radius(cm)	γ's	neutrons	$e^\pm$	$\pi^\pm$	protons	$\mu^\pm$
Vertex	5	34000	3200	168.0	19	3.4	.7
	10	9600	3400	19	9	2.8	0.86
	15	4400	3400	4.2	4.2	2.2	0.66
	20	2500	3400		2.6	3.8	0.40
Tracker	50	880	3000		0.44	8.4	0.064
	100	320	720		0.08	1.6	0.016

Table 11. Mean kinetic energies and momenta of particles as calculated by GEANT.

Detector	Radius	μ GeV	γ MeV	p MeV	$\pi^\pm$ MeV	n MeV
Vertex	10-20	24	1	30	240	10
Tracker	50-100	66	"	"	"	"
	100-150	31	"	"	"	"
Calorimeter	160-310	19	"	"	"	"

Detector Specifications and Design

The physics requirements of a muon collider detector are similar to those of an electron collider. The main difference has to do with the machine related backgrounds and the added shielding that is needed near the beam pipe.

At this time little detailed work has been done on the design of a complete detector. Most of the discussion has centered around the types of detector elements which might function well in this environment. The background levels detailed in the previous section are much higher than the comparable levels calculated for the SSC detectors and appear to be somewhat in excess of the levels expected at the LHC. Clearly segmentation is the key to successfully dealing with this environment. One major advantage of this muon collider over high energy hadron colliders is the long time between beam crossings; the LHC will have crossings every 25 ns compared to the 10 μs expected for the 4 TeV μ-collider. Much of the detector discussion has focused on ways to exploit this time between crossings to increase the segmentation while holding the number of readout elements to manageable levels.

The real impact of the backgrounds will be felt in the inner tracking and vertex systems. One attractive possibility for a tracking system is a Time Projection Chamber (TPC)[70]. This is an example of a low density, high precision device which takes advantage of the long time between crossings to provide low background and high segmentation with credible readout capability.

Silicon, if it can withstand damage from the neutron fluxes, appears to be an adequate option for vertex detection. Again, because of the time between beam crossings, an attractive option is the Silicon Drift Detector[71]. Short drift TPC's with microstrip[72] readout could also be considered for vertex detection.

An interesting question which has yet to be addressed is whether or not it is possible to tag high energy muons which penetrate the tungsten shielding which, in the present design, extends to 20° from the beam axis. For example, in the case of $\mu\mu \to \nu\nu W^+W^-$ the primary physics background is due to $\mu\mu \to \mu\mu W^+W^-$. To reduce the background, in addition to a high p_T cut on the WW pair, in might be advantageous to tag forward going muons. These μ's would penetrate the shielding.

Strawman Detector

Table 12. Detector Performance Requirements.

Detector Component	Minimum Resolution/Characteristics
Magnetic Field	Solenoid; B$\geq$2 T
Vertex Detector	b-tagging, small pixels
Tracking	$\Delta p/p^2 \sim 1\times 10^{-3} (\text{GeV})^{-1}$ at large p
	High granularity
EM Calorimeter	$\Delta E/E \sim 10\%/\sqrt{E} \oplus 0.7\%$
	Granularity: longitudinal and transverse
	Active depth: 24 X_0
Hadron Calorimeter	$\Delta E/E \sim 50\%/\sqrt{E} \oplus 2\%$
	Granularity: longitudinal and transverse
	Total depth (EM + HAD)$\sim 7\lambda$
Muon Spectrometer	$\Delta p/p \sim 20\%$ 1 TeV

The detector performance criteria that are used for the design of the detector are summarized in Tb. 12. The object of this present exercise is to see if a detector can be built using state-of-the-art (or not far beyond) technology to satisfy the physics needs of the muon collider.

A layout of the strawman detector is shown in Fig. 37. A large cone (20°) that is probably not instrumented and is used to shield the detector from the machine induced background.

The main features of the detector are: The element nearest to the intersection region is the vertex detector located at as small a radius as possible. A number of technologies including Silicon Drift Detectors(SDD), Silicon Pixels[73], and CCD detectors have been considered, as well as short drift TPC's. SDD seem especially attractive because of the reduced number of readout channels and potentially easier construction. A micro TPC would have lower occupancy and greater radiation resistance. Tracking technologies considered were cathode pad chambers, silicon strips and TPCs. The use of a TPC is interesting as the amount of material is minimized and thus the detector does not suffer as much from low energy photon and neutron backgrounds.

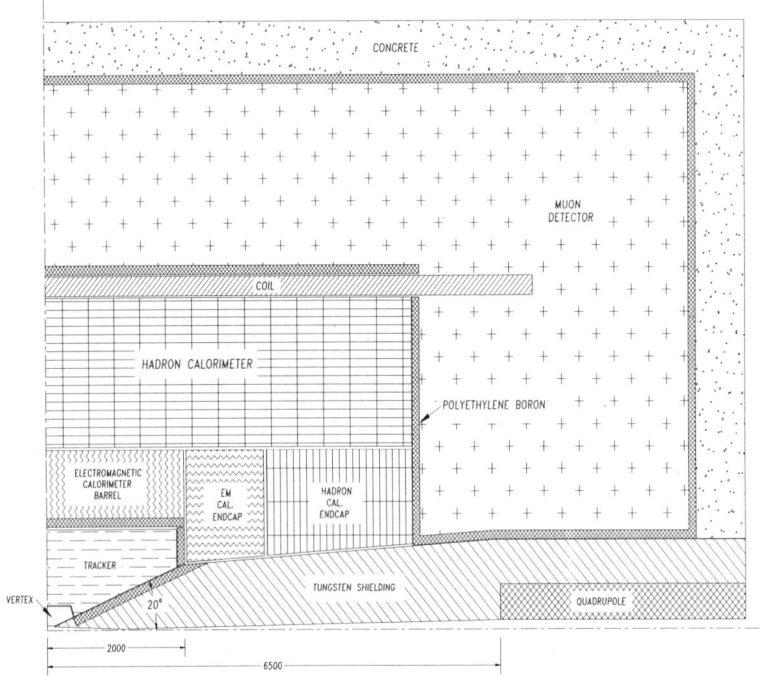

Figure 37. Strawman Detector

For the calorimeter system there are many options. A liquid argon accordion calorimeter for the EM part[73],[74] and a scintillator hadronic calorimeter appear appropriate. This combination gives a hadronic resolution that is of order $50\%/\sqrt{E}$ which may be good enough. The high granularity of the EM section allows good electron identification which will be of help in tagging b-jets. In addition the longitudinal and transverse granularity allow for corrections on an event by event basis to compensate for the fact that the calorimeter is non-compensating as well as to allow the identification of catastrophic muon bremsstrahlung.

There is a single solenoid magnet, with a field of 2 T in the tracking region. The magnet is located behind the hadron calorimeter.

The muon system is a stand-alone system. The chambers in the muon system are Cathode Strip Chambers(CSC) that can be used for both a two dimensional readout as well as a trigger. These chambers have good timing resolution and relatively short drift time which minimizes neutron background problems. We now consider these components in detail:

Silicon Drift Vertex Detector

The best silicon detectors are capable of achieving a position resolution of 4 μm with a detector 300 μm thick. However, these results were obtained only with normally incident charged particles. For other angles of incidence, the resolution can be degraded by the fluctuations in the density of the ionization (Landau) in the silicon. The presence of a magnetic field modifies the trajectories of electrons in a silicon drift detector and normal incidence is no longer the ideal one. There is, however, a proper incidence angle which does not give any degradation of the position resolution.

The granularity of the proposed detector seems be reasonable for the rates of background particles. Based on rates in Tbs. 9,10,11, a layer located 10 cm from the beam would have 22 hits per cm^2 from the interaction of neutral particles and 32 from crossings of charged particles. For a pixel size of 316 × 316 μm^2 the number of pixel per cm^2 is 1000. In this case the occupancy of background hits is less than 6%.

The damage due to the radiation dose may be a serious problem. Only about 1/3 of the neutrons have energies above 100 KeV and contribute to this damage, but the integrated flux per year for this example would still be 3 10^{14} n's cm^{-2}year^{-1}. If detectors are produced from an n–type silicon with a bulk doping level of $1.5 \times 10^{12}/cm^3$ the detectors would have to be replaced after a year of operation. The use of p–type material seems to be more appropriate for this application. P–type silicon drift detectors are being developed in LBNL. These detectors are supposed to be much more radiation resistant. Some R&D may be required.

Time Projection Chamber (TPC)

An interesting candidate for tracking at a muon collider is a Time Projection Chamber (TPC). This device has good track reconstruction capabilities in a low density environment, good 3-dimensional imaging and provides excellent momentum resolution and track pointing to the vertex region. It is perhaps particularly well suited to this environment as the long time between bunch crossings ($\sim$ 10 μs) permits drifts of $\sim$1 m and the average density of the device is low compared to more conventional trackers which helps to reduce the measured background rates in the device. In the present detector considerations the TPC would occupy the region between the conic tungsten absorber and electromagnetic calorimeter in the region from 35 cm to 120 cm, divided into two parts, each 1 m long.

To reduce background gamma and neutron interactions in the detector volume, a low density gas mixture should be chosen as the detection medium of the TPC. Another important parameter is the electron drift velocity. Since the time between beam crossings is fixed (10 μs in the present design) the drift velocity should be high enough to collect all the ionization deposited in the drift region. Finally the detection medium should not contain low atomic number gases to help reduce the transfer energy to the recoil nucleus and in this way to reduce its range in the gas. The gas mixture 90% He + 10% CF_4 satisfies all these requirements and could be an excellent candidate. It does not contain hydrogen which would cause a deleterious effect from the neutrons, has a density 1.2 mg/cm^3 and a drift velocity of 9.4 cm/μs. The single electron longitudinal diffusion for this gas is

$$\sigma_l = 0.15 \, \text{mm}/\sqrt{\text{cm}}. \tag{15}$$

The transverse diffusion, which is strongly suppressed by the 2 T magnetic field is given by,

$$\sigma_t = \frac{\sigma_t(B=0)}{\sqrt{1+(\omega\tau)^2}} = 0.03 \, \text{mm}/\sqrt{\text{cm}} \tag{16}$$

Each time slice will contain about 25 ionization electrons, and the expected precision in r- ϕ and z coordinates is,

$$\sigma_\phi = \sqrt{\frac{Z\sigma_t^2}{25} + (50)^2} \approx 100 \, (\mu\text{m}) \qquad \sigma_Z = \sqrt{\frac{Z\sigma_l^2}{25} + (150)^2} \approx 300 \, (\mu\text{m}) \tag{17}$$

where Z, the drift length is 1 m. The precision of r-coordinate is defined by the anode wire pitch - 3 mm.

Occupancy from Photons. Low energy photons, neutrons and charged particles produce the main backgrounds in the tracker. Photons in the MeV region interact with matter mainly by Compton scattering. For a 1 MeV photon the probability of producing a Compton electron in 1 cm of gas is $\xi_\gamma = 4.5 \times 10^{-5}$. For an average photon fluence $h_\gamma = 200$ cm^{-2} about $N_\gamma = 8 \times 10^4$ electron tracks are created in the chamber volume. Because the transverse momentum of Compton electrons is rather small the electrons are strongly curled by the magnetic field and move along the magnetic field lines. Most of the electrons have a radius less than one millimeter and their projection on the readout plane covers not more than one readout pitch, 0.3 × 0.4 cm^2. The average length of the Compton electron tracks in the TPC is 0.5 meter and therefore, the volume occupied by electron tracks is $v_{comp.e} = 4.8 \times 10^5$ cm^3. Since the total chamber volume is 10^7 cm^3, the average occupancy due to background photon interactions is equal to,

$$< occupancy >_\gamma = \frac{V_{comp.e}}{V_{total}} = 4.4 \times 10^{-2} \quad (18)$$

and could be further reduced by subdividing the chamber in length, thus shortening the drift distances. Indeed this may be required to reduce an excessive space charge distortion from the accumulation of ions.

These Compton tracks can easily be identified and removed. Because almost all points of a Compton track lie along the z-axis most of them will be projected into one cell and therefore the number of points in this cell will be very different from hit cells from non-background tracks. To remove low momentum electron tracks, all cells containing more than some threshold number of points should be excluded. Applying this procedure a few percent of volume is lost but the quality of the high momentum tracks is not substantially changed. This is illustrated in Fig. 38 where one sector of the TPC is shown after the application of different value threshold cuts.

Occupancy from Neutrons. For neutrons in the MeV region the primary interaction with matter is elastic collisions. In this case the energy transfer to the nucleus has a flat distribution and the maximum transfer energy is given by $4E_n A/(A+1)^2$ or $4E_n/A$ when $A \gg 1$.

The calculated mean energy of background neutrons is $E_n = 27$ MeV. In this case, for hydrogen, their mean range in the gas is several meters, but for the gas chosen, the mean length of the recoil nucleus tracks will only be a few millimeters.

The calculated neutron fluence is $< n >= 2 \times 10^3$ cm^{-2}. The track of the recoil nucleus occupies, typically, not more than one volume cell of the TPC, $v_n = 0.3 \times 0.4 \times 1.0$ cm^3. The probability of a background neutron interacting in 1 cm of the gas is $\xi_n = 2 \times 10^{-5}$, the number of recoil tracks $N_n = < n > \cdot \xi_n \cdot V_{total} = 4 \times 10^5$ and therefore the neutron occupancy is,

$$< occupancy >_n = \frac{N_n \cdot v_n}{V_{total}} = 0.48 \times 10^{-2} \quad (19)$$

It is easy to clean out these recoil tracks owing to their large ionization density per cell. Only a simple cut to remove all volume cells which contain a charge in excess of some preset threshold is required. This cut will only eliminate about 1% of the TPC volume.

Micro TPC for vertex detection

TPCs with very short drift distances (3 cm) and microstrip readout (0.2 x 2 mm pads) might be an interesting alternative to silicon drift chambers for vertex detection. The resolution of such chambers would be somewhat worse (of the order of 30 μm,

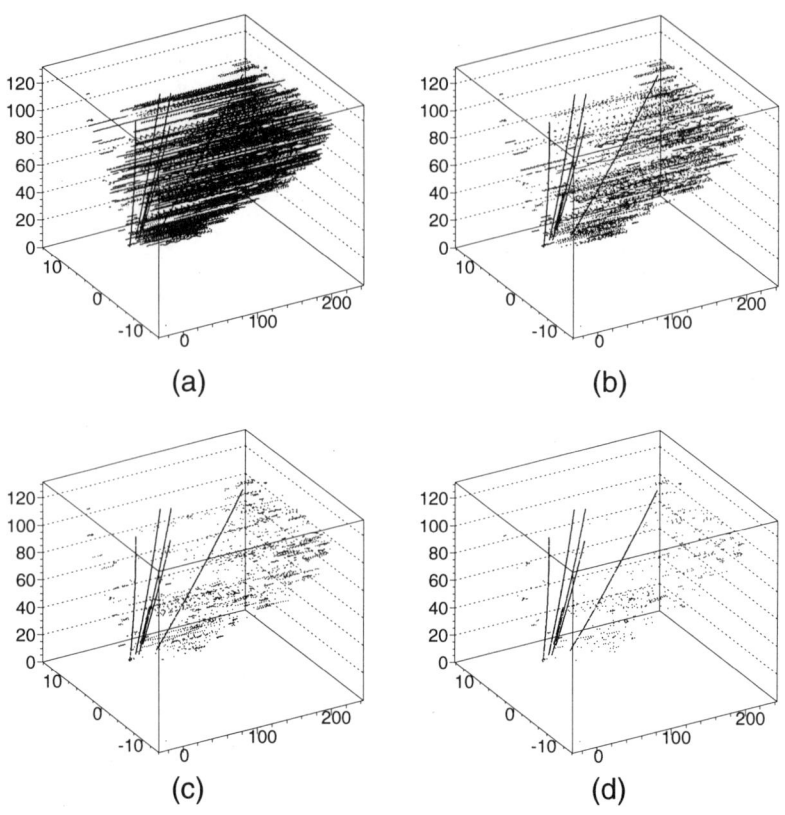

Figure 38. Charge distribution from a signal event and background Compton electrons with threshold cuts; (a) has no cut applied while (d) has the most stringent cut.

compared with 4 μm), but the greater number of points would compensate for this to some extent. Such chambers would have less occupancy and much greater radiation resistance than silicon devices.

Electromagnetic Calorimeter

An accordion liquid argon calorimeter is being developed for the ATLAS collaboration[73] A similar calorimeter designed for the GEM Collaboration at the SSC.

From the GEANT background calculations, the total energy deposited from the electromagnetic debris is ~ 13 TeV but relatively uniformly distributed. If one divides the calorimeter into $\sim 2 \times 10^5$ cells, the mean energy would be about 65 MeV/cell. Certainly, energetic electromagnetic showers from γ's or electrons or the core of jets will stand out above this uniform noise. Since the readout is every 10 μs, multiplexing is possible to reduce costs compared to the LHC where collisions occur every 25 ns.

Hadron Calorimeter

A good choice for the hadron calorimeter is a scintillator tile device being designed for ATLAS[73]. It uses a novel approach where the tiles are arranged perpendicular to the beam direction to allow easy coupling to wave-length shifting fibers[75]. With a tile calorimeter of the type discussed here it should be possible to achieve a resolution of $\Delta E/E \sim 50\%/\sqrt{E}$, satisfying the requirements in Tb. 12.

From the GEANT background calculations, the total energy deposited in the calorimeter from electromagnetic and hadronic showers and muons is about 200 TeV. Again, this is rather uniform with and if subdivided into 10^5 towers would introduce 2 GeV pedestals with 300 MeV fluctuations: also acceptable. But the muons, arising from Bethe-Heitler pair production in EM showers or from a halo in the machine, though modest in number, have high average energies. They would not be a problem in the tracking detectors, but in the calorimeters, they would occasionally induce deeply inelastic interactions, depositing clumps of energy deep in the absorbers. If a calorimeter is not able to recognize the direction of such interactions (they will be pointing along the beam axis) then they would produce unacceptable fluctuations in hadron energy determination. Segmenting the calorimetry in depth should allow these interactions to be subtracted. We are studying various solutions, including the use of fast time digitizing[76] to provide such segmentation, but ultimately there will have to be some hardware tests to verify the MC study.

Muon Spectrometer

Triggering is probably the most difficult aspect of muon spectrometers in large, 4π detectors in both lepton and hadron colliders. In addition, a muon system should be able to cope with the larger than usual muon backgrounds that would be encountered in a muon collider. Segmentation is, again, the key to handling these high background rates. Cathode Strip Chambers (CSC) are an example of a detector that could be used in the muon system of a muon collider experiment. This detector performs all functions necessary for a muon system:

- Precision coordinate(50 to 70 μm)

- Transverse coordinate(of order mm or coarser as needed)

- Timing (to a few ns)

- Trigger primitives

In addition, the cathodes can be lithographically segmented almost arbitrarily resulting in pixel detectors the size of which is limited only by the density and signal routing of the readout electronics.

Halo Background

There could be a very serious background from the presence of even a very small halo of near full energy muons in the circulating beam. The beam will need careful preparation before injection into the collider, and a collimation system will have to be designed to be located on the opposite side of the ring from the detector.

Pair Production

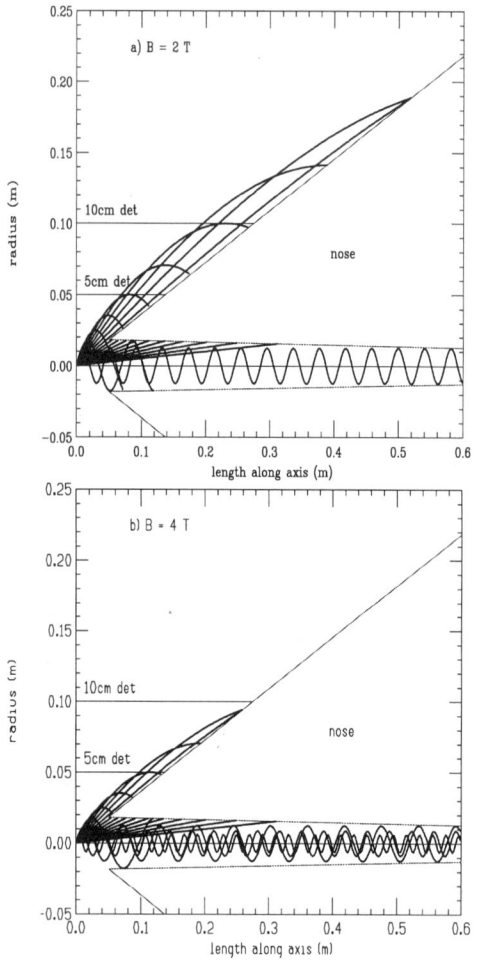

Figure 39. Radius vs. length of electron pair tracks for initial momenta from 3.8 to 3000 MeV in geometric steps of $\sqrt{2}$; (a) for a solenoid field of 2 T, (b) for 4 T.

There is also a background from incoherent (i.e. $\mu^+\mu^- \to e^+e^-$) pair production in the 4 TeV Collider case[77]. The cross section is estimated to be 10 mb[78], which would give rise to a background of $\approx 3 \times 10^4$ electron pairs per bunch crossing. The electrons at production, do not have significant transverse momentum, but the fields of the oncoming 3 μm bunch can deflect them towards the detector. A simple program was written to track electrons from close to the axis (the worst case) as they are deflected away from the bunch center. Once clear of the opposing bunch, the tracks spiral under the influence of the experimental solenoid field. Fig. 39 (a) shows the radii vs, length of these electron tracks for initial momenta from 3.8 to 3000 MeV in geometric steps of $\sqrt{2}$ and a solenoid field of 2 T, see Fig.39a (Fig. 39b it is 4 T). In the 2 T case, tracks with initial energy below 30 MeV do not make it out to a detector at 10 cm, while those above 100 MeV have too small an initial angle and remain within the shield. Approximately 10 %(3000) of these are in this energy range and pass through a detector at 10 cm. The track fluence at the ends of the detector are less than 10 tracks per cm^2 which should not present a serious problem. At 5 cm, there are 4500 tracks giving a fluence of 30 per cm^2, which is also probably acceptable. If the detector solenoid field is raised to 4 T then no electrons reach 10 cm and the flux at 5 cm is reduced by a factor of 2.

There remains some question about the coherent pair production generated by virtual photons interacting with the coherent electromagnetic fields of the entire oncoming bunch. A simple Weizsäcker-Williams calculation[77] yields a background that would consume the entire beam at a rate comparable with its decay. However, I. Ginzburg[78] and others have argued that the integration must be cut off due to the finite size of the final electrons. If this is true, then the background becomes negligible. A more detailed study of this problem is now underway[79][80].

If the coherent pair production problem is confirmed, then there are two possible solutions:

1) one could design a two ring, four beam machine (a μ^+ and a μ^- bunch coming from each side of the collision region, at the same time). In this case the coherent electromagnetic fields at the intersection are canceled and the pair production becomes negligible.

2) plasma could be introduced at the intersection point to cancel the beam electromagnetic fields[81].

Detector and Background Conclusions.

Two independent background calculations have been used for a preliminary study of the expected background level at a 4 TeV muon collider. The optimization of the intersection region is still at its infancy, but the results of both studies show that the level of background while still large, can be managed with proper design of the intersection region and choice of detector technologies. This is in large part due to the fact that the background is composed of many very soft particles which behave like a pedestal shift in the calorimeter. The tracking and vertexing systems will have to be highly segmented to handle this flux of background particles.

A large amount of work is still needed in order to optimize the intersection region and the final focus. In particular a better understanding of the trade off between the different backgrounds is required. The strawman detector is meant only to show that the muon collider detector has unique problems and advantages. An optimized detector needs to be developed taking these problems into consideration.

Some preliminary calculations for machine related backgrounds for a lower energy collider (250 GeV x 250 GeV) have also been carried out. It appears that the backgrounds in this case are comparable to those at the 4 TeV machine. Since little attention has yet been paid to the details of the final focus for this lower energy machine it is possible that reductions in the machine related backgrounds will be achievable in the future.

OPTIONS
Introduction

Up to this point, this report has concentrated on the design of a muon collider with

1) beam energies of 2 + 2 TeV
2) operating at its maximum energy
3) with a fixed rms energy spread of 0.12
4) with no attention to maximizing polarization

In this section we discuss modifications to enhance the muon polarization's, operating parameters with very small momentum spreads, operations at energies other than the maximum for which a machine is designed, and designs of machines for different maximum energies.

Polarization

Polarized Muon Production. The specifications and components in the baseline design have not been optimized for polarization. Nevertheless, simple manipulations of parameters and the addition of momentum selection after phase rotation does generate significant polarization with relatively modest loss of luminosity. The only other significant changes required to give polarization at the interaction point are rotators in the transfer lines, and a chicane snake in the collider opposite the IP.

In the center of mass of a decaying pion, the outgoing muon is fully polarized (-1 for μ^+ and +1 for μ^-). In the lab system the polarization depends[82] on the decay angle θ_d and initial pion energy. Figure 40 shows this polarization as a function of the cosine of the center of mass decay angle, for a number of pion energies. It is seen

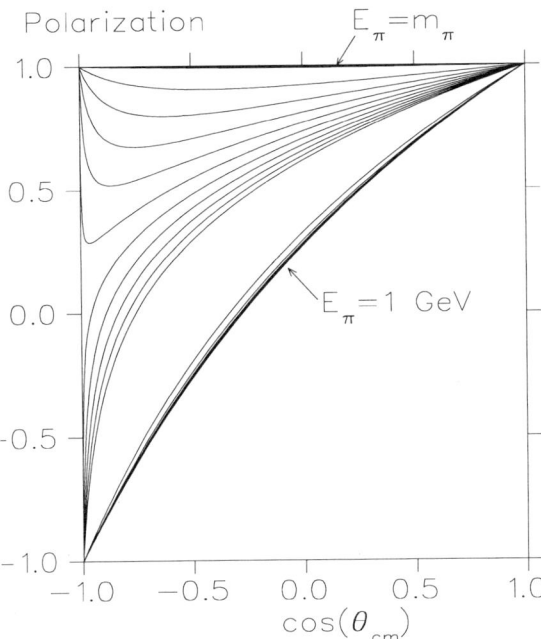

Figure 40. Polarization in the lab. frame vs. the cosine of the center of mass decay angle, for a number of pion energies.

that for pion kinetic energy larger than the pion mass, the dependence on pion energy becomes negligible and the polarization is given approximately by:

$$P_{\mu^-} \approx \cos\theta_d + 0.28(1 - \cos^2\theta_d) \qquad (20)$$

The average value of this is about 0.19. A Monte Carlo calculation[83] of the capture, decay and phase rotation discussed above gave muon polarization of approximately 0.22. The slight difference of this value from the average comes from an accidental bias towards forward decay muons.

If higher polarization is required, some deliberate selection of muons from forward pion decays ($\cos\theta_d \to 1$) is required. This could be done by selecting pions within a narrow energy range and then selecting only those muons with energy close to that of the selected pions. But such a procedure would collect a very small fraction of all possible muons and would yield a very small luminosity. Instead we wish, as in the unpolarized case, to capture pions over a wide energy range, allow them to decay, and to use rf to phase rotate the resulting distribution.

Consider the distributions in velocity vs. ct at the end of a decay channel. If the source bunch of protons is very short and if the pions were generated in the forward direction, then the pions, if they did not decay, would all be found on a single curved line. Muons from forward decays would have gained velocity and would lie above that line. Muons from backward decays would have lost velocity and would fall below the line. A real distribution will be diluted by the length of the proton bunch, and by differences in forward velocity due to the finite angles of particles propagating in the solenoid fields. In order to reduce the latter, it is found desirable to lower the solenoid field in the decay channel from 5 to 3 Tesla. When this is done, and in the absence of phase rotation, one obtains the distribution shown in Fig. 41, where the polarization P$>\frac{1}{3}$, $-\frac{1}{3} < P < \frac{1}{3}$, and P$< -\frac{1}{3}$ is marked by the symbols '+', '.' and '-' respectively. One sees that the +'s are high, and the -'s are low, all along the distribution.

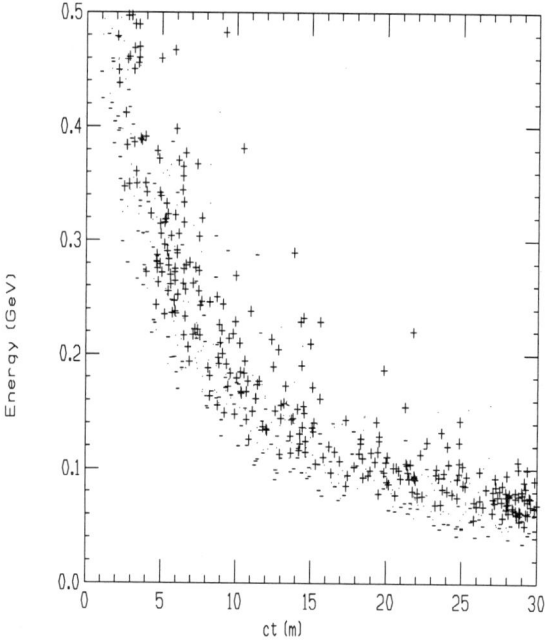

Figure 41. Energy vs. ct of μ's at end of decay channel without phase rotation; muons with polarization P$>\frac{1}{3}$, $-\frac{1}{3} < P < \frac{1}{3}$, and P$< -\frac{1}{3}$ are marked by the symbols '+', '.' and '-' respectively.

It is found that phase rotation does not remove this correlation: see Fig. 42. Now, after time cuts to eliminate decays from high and low energy pions, there is a simple correlation of polarization with the energy of the muons. If a selection is made on the minimum energy of the muons, then net polarization is obtained. The higher the cut on energy, the greater the polarization, but the less the fraction F_{loss} of muons that are selected. The cut in time can probably be obtained from the phasing of the rf used to capture the bunch. Alternatively, it could be provided by a second energy cut applied after a 90 degree longitudinal phase rotation.

In order to provide the required cut on energy, one needs to generate dispersion that is significantly larger than the beam size. Collimation from one side can then select the higher energy muons. After collimation, the remaining dispersion should

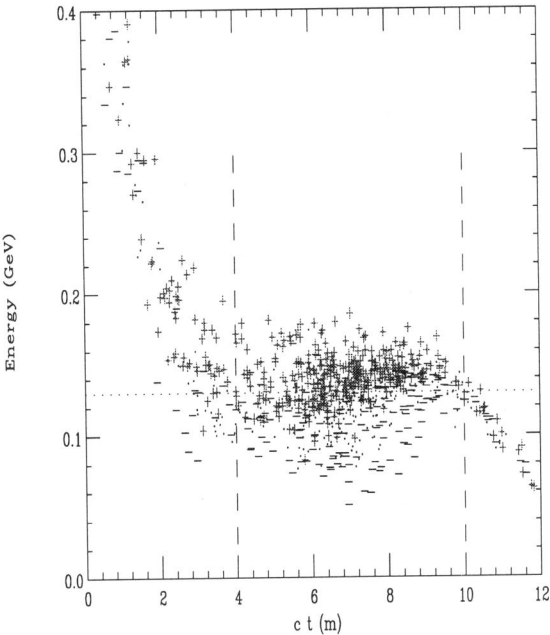

Figure 42. Energy vs. ct of μ's at end of decay channel with phase rotation; muons with polarization P$> \frac{1}{3}$, $\frac{-1}{3} < P < \frac{1}{3}$, and P$< \frac{-1}{3}$ are marked by the symbols '+', '.' and '-' respectively.

be removed. The generation of sufficient dispersion, in the presence of the very large emittance, is non-trivial. The only practical method appears to be the use of a bent solenoid (as discussed above, in the section of Muon Collider Components). First the solenoid is bent one way to generate the dispersion; the collimator is introduced; then the solenoid is bend the other way to remove the dispersion. The complete system thus looks like an "S" or "snake". Particles with momentum p_μ in a magnetic field B have a bending radius of R_B, given by:

$$R_B = \frac{(ep_\mu/mc)}{c\,B}. \tag{21}$$

If the particles are trapped in a solenoid with this field, and the solenoid is bent with a radius R_{bend}, where $R_{bend} \gg R_B$, then those particles, besides their normal helical motion in the solenoid, will drift in a direction (z) perpendicular to the bend, with a drift angle ($\theta_{drift} = dz/ds$) given by:

$$\theta_{drift} \approx \frac{R_B}{R_{bend}} \tag{22}$$

The integrated displacement in z, ie. the dispersion D, is then:

$$D = \theta_{drift}\, s \approx \phi\, R_B, \tag{23}$$

where ϕ is the total angle of solenoid bend.

As an example, we have traced typical particles with momenta of 150 and 300 MeV/c through a snake with $B = 1\,T$, $R_{bend} = 6\,m$, with a first band with $\phi = \pi$ followed by a reverse bend $\phi = -\pi$. Fig. 43 shows the trajectories of muons as viewed from the z direction. No significant dispersion is seen. The two momenta are seen to be dispersed during the right hand turn and recombined by the left hand turn.

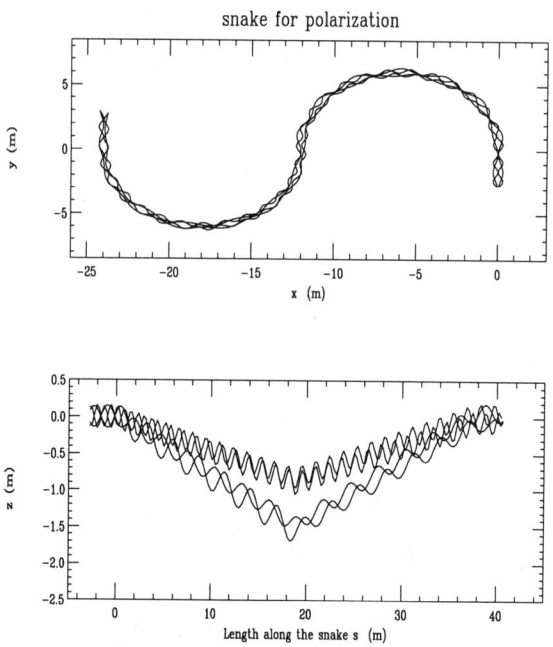

Figure 43. Dispersion Snake: trajectories as seen from the z direction (top); vertical (z) particle positions vs. length (s) along the snake (bottom).

Fig. 44 and Tb. 13 give the results of a Monte Carlo study[83] in which dispersion is introduced, and progressive cuts are applied, to the muons at the end of the phase rotation. In this calculation, in order to calculate more rapidly, the trajectories were not actually traced through a snake. Instead, the particles were propagated through 20 m of straight solenoid, followed by the application of dispersion equal to 6 times the momentum in GeV/c. A snake that would give such dispersion could have the parameters: solenoid field of 3 T (giving $R_B = 0.25\,m$ at the average momentum of 230 MeV/c), diameter of snake bends greater than 5 m and bend angles of 320°, (which would require some variations in bend curvature to avoid the solenoid crossing itself),

Tb. 13 gives results for two fields in the decay channel solenoids: 5 T, the field in the point design; and 3 Tesla, chosen to increase the polarization. It is seen that for weak cuts and small polarization, it is better to avoid the loss of muons from the lower, 3 T, field, but with stronger cuts, the lower field gives greater polarization. In Fig. 44, and subsequent plots, only data from the preferred fields are shown beyond the cross over.

Table 13. Production Polarization vs. Position. B, decay channel field; cut, the position of the colimator; P_{init} initial polarization; P_{final} polarization after dilution in the cooling section; P_{vec} effective vector polarization (see Eq.26); $R_{v/s}$ vector to scalar ratio (see Eq.27); L_{vec} luminosity enhancement for vector state; E_{ave} average final muon energy; ΔE rms final energy spread

B	cut	F_{loss}	P_{init}	P_{final}	P_{vec}	$R_{v/s}$	L_{vec}	E_{ave}	ΔE
T	m							MeV	MeV
5	0.00	1.000	0.23	0.18	0.36	1.45	1.18	130	23
5	1.00	0.960	0.27	0.21	0.41	1.54	1.21	144	23
5	1.12	0.890	0.30	0.24	0.46	1.64	1.24	147	20
5	1.24	0.759	0.36	0.29	0.53	1.80	1.29	151	18
5	1.30	0.614	0.41	0.33	0.60	1.99	1.33	157	17
5	1.40	0.360	0.48	0.39	0.67	2.26	1.39	166	15
5	1.50	0.163	0.56	0.45	0.75	2.64	1.75	177	15
3	0.00	0.801	0.22	0.18	0.34	1.43	1.18	130	22
3	1.06	0.735	0.29	0.23	0.44	1.61	1.23	133	22
3	1.16	0.673	0.35	0.28	0.52	1.77	1.28	137	19
3	1.26	0.568	0.41	0.33	0.59	1.98	1.33	141	17
3	1.32	0.417	0.50	0.40	0.69	2.32	1.40	147	15
3	1.40	0.264	0.59	0.47	0.77	2.78	1.47	151	13
3	1.48	0.126	0.70	0.56	0.86	3.58	1.56	159	13
3	1.56	0.055	0.77	0.62	0.90	4.25	1.62	168	12

It is seen from Tb. 13 that the energy cut not only increases the polarization, but also decreases the energy spread ΔE of the remaining muons. In Fig. 45 the fractional energy spread $\Delta E/E$ is plotted against the loss factor F_{loss}. The energy spread is reduced almost a factor of two for reasonable collimator positions. This reduction in energy spread would eliminate the need for the first stage of emittance cooling.

A Monte Carlo study has also been done on the effect of variations of the proton bunch length σ_t. Fig. 46a shows the polarization before cooling as a function of σ_t for three values of the loss factor F_{loss}. It is seen that serious loss of polarization can occurs when the rms width is more than 1 nsec. Fig. 46b shows the muon rms energy spread after the polarization cut. Again it is shown as a function of σ_t for three values of the loss factor F_{loss}. With no cut, the rise in energy spread would be serious ($\Delta E > 20\,\text{MeV}$ is difficult to cool) for an rms width more than 1 ns. But with polarization cuts, the energy spread is so reduced that a larger proton bunch length would not be a problem.

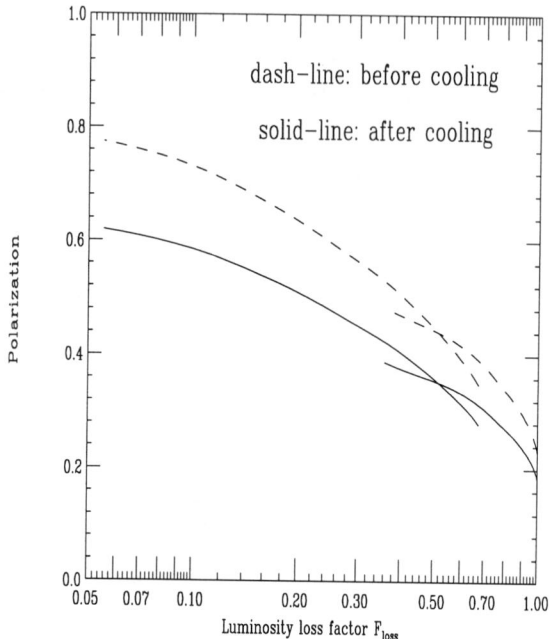

Figure 44. Polarization vs. F_{loss} of muons accepted; the dashes show polarization as selected, the line gives polarization after cooling.

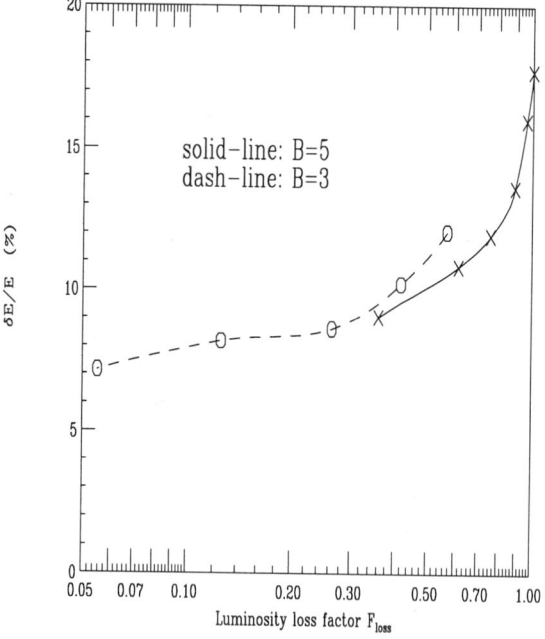

Figure 45. the fractional energy spread $\Delta E/E$ is plotted against the loss factor F_{loss}.

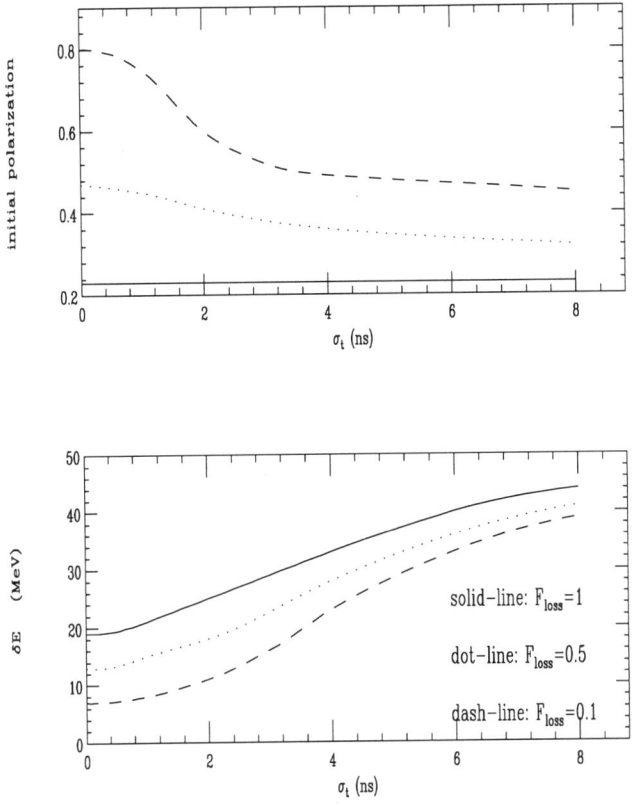

Figure 46. Polarization vs σ_t, the proton bunch length (upper plot). Muon *rms* energy spread vs. σ_t (lower plot). Both plots for three values of the loss factor F_{loss}.

Polarization Preservation

A recent paper[84] has discussed the preservation of muon polarization in some detail. During the ionization cooling process the muons lose energy in material and have a spin flip probability $\mathcal{P}$, where

$$\mathcal{P} \approx \int \frac{m_e}{m_\mu} \beta_v^2 \frac{\Delta E}{E} \qquad (24)$$

where β_v is the muon velocity divided by c, and $\Delta E/E$ is the fractional loss of energy due to ionization. In our case the integrated energy loss is approximately 3 GeV and the typical energy is 150 MeV, so the integrated spin flip probability is close to 10%. The change in polarization dP/P is twice the spin flip probability, so the reduction in polarization is approximately 20 %. This dilution is included in the "P_{final}" column in Tb. 13 and is plotted as the line in Fig. 44.

During circulation in any ring, the muon spins, if initially longitudinal, will precess by $(g-2)/2\gamma$ turns per revolution in the ring; where $(g-2)/2$ is 1.166×10^{-3}. An energy spread $d\gamma/\gamma$ will introduce variations in these precessions and cause dilution of the polarization. But if the particles remain in the ring for an exact integer number of synchrotron oscillations, then their individual average γ's will be the same and no dilution will occur. It appears reasonable to use this "synchrotron spin matching"[84] to avoid dilution during acceleration.

In the collider, however, the synchrotron frequency will be too slow to use "synchrotron spin matching", so one of two methods must be used.

- Bending can be performed with the spin orientation in the vertical direction, and the spin rotated into the longitudinal direction only for the interaction region. The design of such spin rotators appears relatively straightforward. The example given in the above reference would only add 120 m of additional arc length, but no design has yet been incorporated into the lattice.

- The alternative is to install a Siberian Snake[85] at a location exactly opposite to the intersection point. Such a snake reverses the sign of the horizontal polarization and generates a cancelation of the precession in the two halves of the ring.

Provision must also be made to allow changes in the relative spins of the two opposing bunches. This could be done, prior to acceleration, by switching one of the two beams into one or the other of two alternative injection lines.

Benefits of Polarization of Both Beams

We consider two examples of the general advantage of having polarization in both beams. Individual physics experiments would have to be considered to determine how important such advantages are.

Consider the polarization of a vector spin state generated by the annihilation of the two muons.

$$P_{vec} = \frac{F^{++} - F^{--}}{F^{++} + F^{--}} \qquad (25)$$

When only one beam has polarization P_1, then $P_{vec} = P_1$. But if both beams have polarization P in the same direction (ie. with opposite helicities), then

$$P_{vec} = \frac{(P+1)^2 - (P-1)^2}{(P+1)^2 + (P-1)^2} \qquad (26)$$

In Fig. 47 both the polarization of each beam P, and the resulting polarization of a vector state P_{vector} are plotted against the loss factor F_{loss}.

A second advantage is that the ratio $R_{v/s}$ of vector to scalar luminosity can be manipulated to enhance either the vector or the scalar state. If the polarization directions have been chosen to enhance the ratio of vector to scalar states, then:

$$R_{v/s} = \frac{1+P}{1-P}. \qquad (27)$$

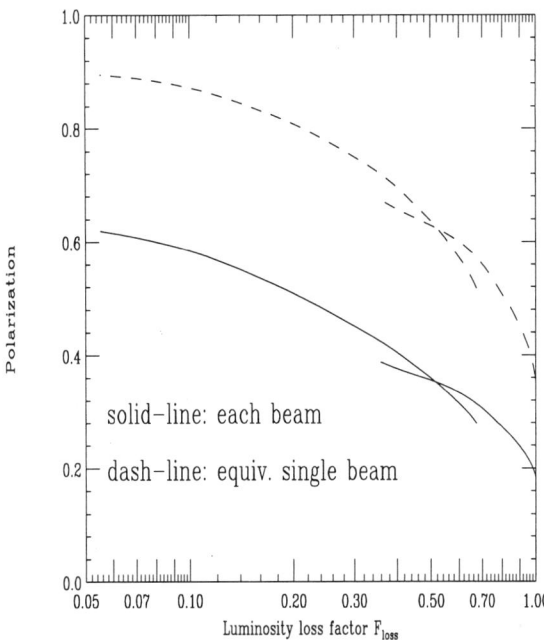

Figure 47. Polarization of each beam P, and the resulting polarization of a vector state P_{vec} vs. the loss factor F_{loss}.

Tb. 13 and Fig. 48 show this ratio as a function of the loss factor F_{loss}.

Tb. 13 also shows that the fraction of total luminosity in a given state can be enhanced. If polarizations are chosen to enhance the vector state, then the fraction of vector luminosity is increased from $1/2$ to $(1+P)/2$, giving a gain factor of

$$\mathcal{L}_{vec} = 1 + P \tag{28}$$

Luminosity loss

If nothing else is done, then the luminosity, which is proportional to n_μ, will drop as F_{loss}^2; where F_{loss} is the fraction muons lost by the muon momentum cut. At the same time, however, the space charge, wakefield, and loading during the cooling and acceleration will all be reduced; as will the beam beam tune shift in the collider. Clearly, the machine parameters should be reoptimized and some part of the lost luminosity recovered.

One way to recover the luminosity would be to increase the proton bunch intensity by the factor F_{loss}. If this were done, then the original number of muons per bunch would be generated; all the wake field, loading and space charge effects would be the same; and the luminosity per bunch crossing would be the same. If we assume that the total proton current is determined by the driver, then such an increase in proton

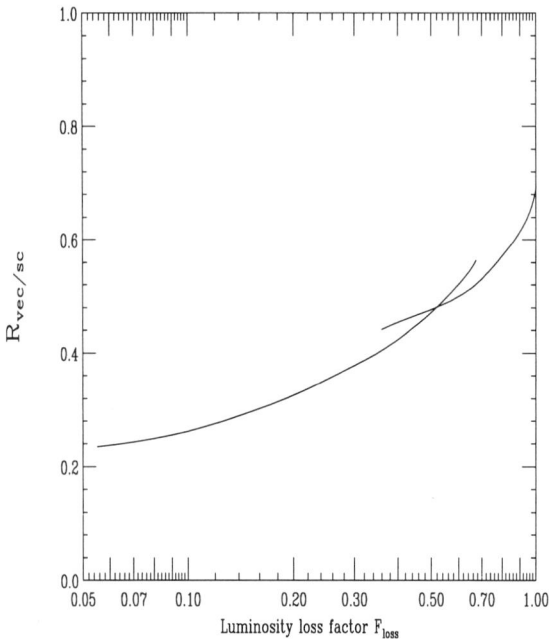

Figure 48. Ratio of vector to scalar states, $R_{v/s}$ vs. the loss factor F_{loss}.

intensity per bunch will necessitate a reduction in the number of bunches or repetition rate, by the same factor F_{loss}. The luminosity will then fall by F_{loss} and not by F_{loss}^2 as before.

For instance, in the unpolarized case of the 4 TeV collider, there were two bunches of each sign. If the momentum cut is chosen to give a value of $F_{loss} = 1/2$, and the proton beam is distributed into 2 instead of 4 initial bunches, then the final number of muons per bunch, the loading, beam beam tune shift, etc. would all be the same as in the unpolarized case. The luminosity would be down by a factor of only two, for a polarization of 34 % in both beams.

For higher polarization at good luminosity it would be desirable to have a proton source with the option of a lower repetition rate, but even larger numbers of protons per spill. For example 4 x 10^{14} protons per pulse at 4 Hz. It should then be possible to extend this method to an operation with $F_{loss} = 1/8$, and polarization of both beams of 57 %.

One also notes that the luminosity could be maintained at the full unpolarized value if the proton source intensity could be increased. Such an increase in proton source intensity in the unpolarized case would be impractical because of the resultant excessive high energy muon beam power, but this restriction does not apply if the increase is used to offset losses in generating polarization. If, for instance, the driver repetition rate were increased from 15 to 30 Hz, the fractions F_{loss} set at 0.5, and the number of bunches reduced to one, then the full luminosity of 10^{35} $(cm^{-2}s^{-1})$ would be maintained with polarization of both beams of 34%.

Luminosity

The bunch populations decay exponentially, yielding an integrated luminosity equal to its initial value multiplied by an *effective* number of turns $n_{\text{eff}} \approx 150\ B$, where B is the mean bending field in T.

The luminosity is given by:

$$\mathcal{L} = \frac{n_\mu^2\ n_b f_{\text{rep}}\ n_{\text{eff}} \gamma}{4\pi\ \beta_\perp^*\ \epsilon_n} H(A, D) \tag{29}$$

where n_μ is the number of muons per bunch, n_b is the number of bunches, γ is the normalize energy, $\beta_\perp^*$ is the beta function at the IP and ϵ_n is the transverse normalize emittances (assumed vertical and horizontal to be equal), and the enhancement factor $H(A, D)$ is

$$H(A, D) \approx 1 + D^{1/4} \left[\frac{D^3}{1+D^3}\right] \left\{\ln\left(\sqrt{D}+1\right) + 2\ln\left(\frac{0.8}{A}\right)\right\}, \tag{30}$$

$$A = \sigma_z/\beta^*, \tag{31}$$

and

$$D = \frac{\sigma_z n_\mu}{\gamma \sigma_t^2} r_e \left(\frac{m_e}{m_\mu}\right) \tag{32}$$

In the cases we are considering[86]: A = 1, D ≈ .5 and H(A,D) ≈ 1.

Luminosity vs. Energy, for a Given Ring

For a fixed collider lattice, operating at energies lower than the design value, the luminosity will fall as γ^3. One power comes from the γ in Eq. 29; a second comes from n_{eff}, the effective number of turns, that is proportional to γ; the third factor comes from β^*, which must be increased proportional to γ in order to keep the beam size constant within the focusing magnets. The bunch length σ_z must also be increased proportional to γ so that the required longitudinal phase space is not decreased; so A = σ_z/β^* remains constant.

Scaling for Collider Rings for Different Energies

As noted above, the luminosity in a given ring will fall as the third power of the energy at which it is operated. Such a drop is more rapid than the gain in typical cross sections, and, as we shall see, it is more rapid than the drop in luminosity obtained with rings designed for the lower energies. It would thus be reasonable, having invested in a muon source and accelerator, to build a sequence of collider rings at spacings of factors of 2-3 in maximum energy. We will now derive scaling rules for such collider rings.

The luminosity

$$\mathcal{L} = \frac{n_\mu^2 \, n_{\text{eff}} \, n_b \, f_{\text{rep}} \, \gamma}{4 \pi \, \epsilon_n \, \beta_\perp^*} \propto \frac{n_\mu \, I_\mu \, \gamma}{\epsilon_n \, \beta^*} \tag{33}$$

where $I_\mu = n_\mu n_b f_{\text{rep}}$, is the muon flux and which, since $\Delta\nu_{bb}$, the beam beam tune shift is given by:

$$\Delta\nu_{bb} \propto \frac{n_\mu}{\epsilon_n}, \tag{34}$$

gives:

$$\mathcal{L} \propto \frac{I_\mu \, \Delta\nu_{bb} \, \gamma}{\beta_\perp^*} \tag{35}$$

If a final focus multiplet is scaled keeping the relative component lengths and the pole tip fields constant, then one obtains:

$$\ell^* \propto \sqrt{a_{max} \, \gamma} \tag{36}$$

$$\theta^* \propto \sqrt{\frac{a_{max}}{\gamma}} \propto \sqrt{\frac{\epsilon_n}{\beta_\perp^* \, \gamma}} \tag{37}$$

$$\beta_\perp^* \propto \frac{\epsilon_n}{a_{max}} \tag{38}$$

where θ^* is the rms angle of muons diverging from the focus, ℓ^* is the free space from the target to the first quadrupole (proportional to all quadrupole lengths in the multiplet), and a_{max} is the maximum aperture of any quadrupole (proportional to all apertures in the multiplet).

The normalized emittance ϵ_n is constrained by the ionization cooling, but since one can exchange transverse and longitudinal emittance, it is, in principle, the six dimensional emittance ϵ_6 that is constrained. Extending the lepton emittance conventions, we define:

$$\epsilon_6 = (\epsilon_n)^2 \, \frac{dp}{p} \sigma_z \gamma \beta_v. \tag{39}$$

With this definition, the six dimensional phase space $\Phi_6 = \pi^3 \, m_\mu^3 \, \epsilon_6$. σ_z cannot be large compared with the focus parameter β^*, so, taking them to be proportional to one another, and taking the $\beta_v = 1$, then:

$$\epsilon_6 \propto (\epsilon_n)^2 \, \frac{dp}{p} \beta^* \gamma \tag{40}$$

and from the above:

$$(\epsilon_n)^3 \propto \frac{\epsilon_6 \, a_{max}}{\gamma \, \frac{dp}{p}} \tag{41}$$

$$(\beta_\perp^*)^3 \propto \frac{\epsilon_6}{\gamma \, \frac{dp}{p} \, a_{max}^2} \tag{42}$$

Six Dimensional Emittance dependence on n_μ and ϵ_n

The six dimensional emittance ϵ_6 obtained from the cooling will, because of more detailed constraints, depend to some extent on the number of muons n_μ, and on the final transverse emittance ϵ_n.

The approximate dependence on the number of muons is relatively transparent. As the number of muons per bunch rises, the longitudinal space charge forces increase and it becomes impossible, without changing the rf gradients, to maintain the same bunch lengths. As a result the bunch lengths must be increased by the square root of the number of muons.

A study, with the analytic formulae used for the model cooling system discussed before, was used again to derive cooling sequences with different final parameters. First, sequences were calculated with numbers of initial muons per bunch of 1, 2, 3.75, 7.5, and 15 x 10^{12} (corresponding to muons in the collider of 0.1, 0.2, 1, 2, and 4 x 10^{12}). The final transverse emittance at the end of the cooling was required to be 4×10^{-5} m, (corresponding to an emittance in the collider of 5×10^{-5} m). The six dimensional emittances obtained are plotted in Fig. 49a. It is seen that for $n_\mu > 10^{12}$ the six dimensional emittances are indeed approximately proportional to the root of the number of muons (the line shows this dependence).

The study also obtained cooling sequences giving six dimensional emittances for a range of final transverse emittances. The dependence here is more complicated. If emittance exchange between longitudinal and transverse emittances could be achieved without material then the six dimensional emittance should be independent of the final transverse emittance chosen. But the exchange does require material and Coulomb scattering in this material increases the six dimensional emittances; and it does so to a greater extent if the transverse emittance is small. In Fig. 49b, we show the six dimensional emittances obtained for 5 representative transverse emittances. Over the range of interest the dependence of ϵ_6 is approximately the inverse root of ϵ_n (the line shows this dependence).

For the purposes of this study, we may thus assume that:

$$\epsilon_6 \propto \sqrt{\frac{n_\mu}{\epsilon_n}} \qquad (43)$$

Energy Scaling, allowing the emittances to vary

If n_μ is limited by the beam beam tune shift:

$$n_\mu \propto \epsilon_n \Delta \nu_{bb} \qquad (44)$$

substituting this in Eq. 43:

$$\epsilon_6 \propto \sqrt{\Delta \nu_{bb}} \qquad (45)$$

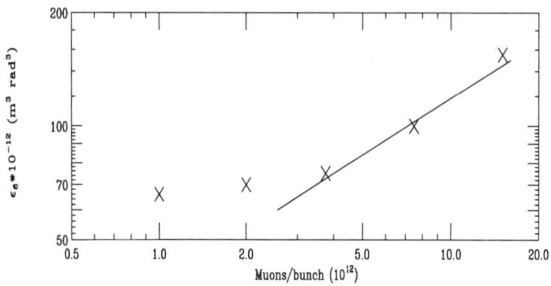

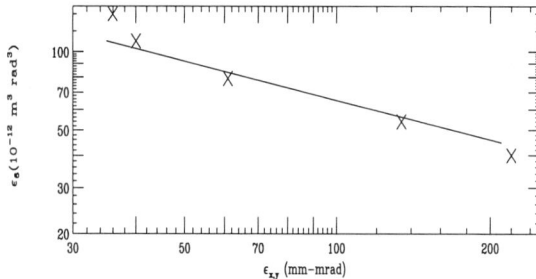

Figure 49. Six dimensional emittance ϵ_6 vs. a) muon intensity n_μ entering the cooling, and b) the transverse emittance ϵ_n at the end of the cooling

giving:

$$\epsilon_n \propto \Delta\nu_{bb}^{1/6} \left(\frac{a_{max}}{\gamma \, dp}\right)^{1/3} \tag{46}$$

$$\beta_\perp^* \propto \frac{\epsilon_n}{a_{max}} \tag{47}$$

$$n_\mu \propto (\Delta\nu_{bb})^{1\frac{1}{6}} \left(\frac{a_{max}}{\gamma \, dp}\right)^{1/3} \tag{48}$$

so:

$$\mathcal{L}(\Delta\nu) \propto I_\mu \, \gamma^{4/3} \, \Delta\nu_{bb}^{5/6} \, a_{max}^{2/3} \, dp^{1/3} \tag{49}$$

One notes however that as γ or dp fall the required number of muons n_μ rises, and will at some point become unreasonable. If we impose a maximum number of muons n_{max}, then, when this bound is reached,

$$\epsilon_n \propto n_{max}^{1/7} \left(\frac{a_{max}}{\gamma \, dp}\right)^{2/7} \tag{50}$$

$$\beta_\perp^* \propto \frac{\epsilon_n}{a_{max}} \tag{51}$$

and:

$$\mathcal{L}(n) \propto I_\mu \, n_{max}^{12/7} \, \gamma^{11/7} \, a_{max}^{3/7} \, dp^{4/7} \tag{52}$$

Using the above relationships. and assuming a constant value of a_{max} we obtain the scaled parameters for a sequence of colliding rings given in Tb. 14. Fig. 50 shows the luminosities that would be available at all energies, including those requiring the use of rings at energies less than their maximum. The lines and dashed lines indicate the luminosities with a bound on n_μ of 4×10^{12}. The line gives luminosities for the nominal rms dp/p of 0.12%, while the dashed line is for a dp/p of 0.01%.

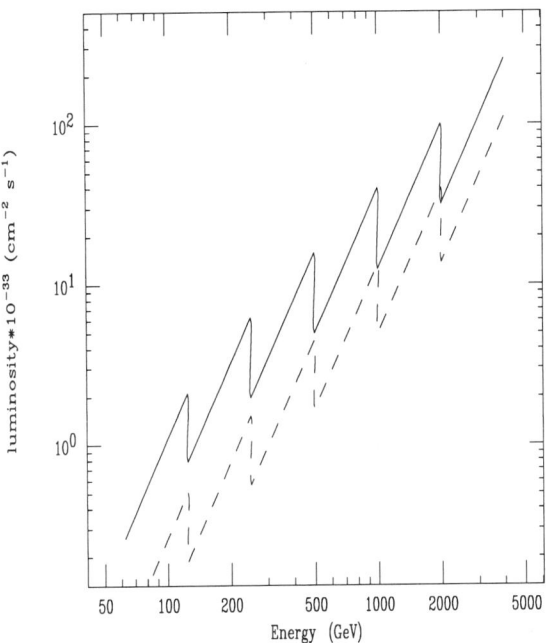

Figure 50. Luminosity vs. energy assuming rings spaced by factors of two in energy; the line is for $\Delta E/E = 0.12\%$, the dashed line is for $\Delta E/E = 0.01\%$.

RESEARCH AND DEVELOPMENT PLAN

In this section we discuss a Research and Development plan aimed at the operation of a 0.5 TeV demonstration machine by the year 2010, and of the 4 TeV machine by year 2020. It assumes 5 years of theoretical study, component modeling and critical subsystem demonstration; followed by 4 years of component development and demonstration machine design. Construction of the demonstration machine would follow and take about 4 years. The high energy machine would come a decade later.

Theoretical Studies

Much progress has been made during the last year. New problems continue to be uncovered, but new solutions have been found. Much work remains to be done: The

Table 14. Scaling of Parameters with Energy and Momentum spread.

E GeV	Luminosity $cm^{-2}s^{-1}$	emittance π m rad	n_μ 10^{12}	$\delta\nu_{bb}$	$\beta^*_\perp$ mm	len* m	β_{max} km	chrom	$\Delta E/E$ %
4000	2.5E+35	4.0E-05	1.6	0.040	2.4	9.2	882	12829	0.12
2000	1.0E+35	5.0E-05	2.0	0.040	3.0	6.5	350	3600	0.12
1000	4.0E+34	6.3E-05	2.5	0.040	3.8	4.6	139	1010	0.12
500	1.6E+34	7.9E-05	3.2	0.040	4.8	3.3	55	283	0.12
250	6.3E+33	1.0E-04	4.0	0.040	6.0	2.3	22	80	0.12
125	2.1E+33	1.2E-04	4.0	0.033	7.3	1.6	9	23	0.12
4000	1.1E+35	9.1E-05	3.6	0.040	5.5	9.2	385	5604	0.01
2000	4.0E+34	1.1E-04	4.0	0.036	6.7	6.5	156	1603	0.01
1000	1.3E+34	1.4E-04	4.0	0.029	8.2	4.6	64	465	0.01
500	4.5E+33	1.7E-04	4.0	0.024	10.0	3.3	26	135	0.01
250	1.5E+33	2.0E-04	4.0	0.020	12.2	2.3	11	39	0.01
125	5.1E+32	2.5E-04	4.0	0.016	14.9	1.6	4	11	0.01

first object will be to define a single self consistent set of parameters for the 4 TeV collider. Items needing study include:

1. Define parameters for the proton source, target, capture and phase rotation systems.

2. Incorporate operating parameters for the optional operation with polarized, or very low energy spread, beams.

3. Define and simulate a complete cooling scenario.

4. Define a preferred acceleration scenario and perform complete simulations. Study the required shielding of the superconducting cavities from muon decay electrons.

5. Design a halo scraping system for the collider ring.

6. Continue work on the collider lattice, including a study of the effect of lattice errors, and an investigation of the use of higher order multipole correctors. Continue the study of the stability of the proposed beams and design an rf system for BNS damping.

7. Continue optimization of the shielding of the detector.

8. Design a "strawman" detector with all components capable of withstanding the backgrounds, and simulate some representative physics observations.

9. Study safety and radiation exposures both on and off site, including the hazards from neutrino fluxes.

It is estimated (see Tb. 15) that the current effort is about 22 full time equivalents, but only a few of these are funded specifically for such work. Not only should the effort be legitimized, but, if we are to determine if such machines are practical, it needs to be expanded. The machine is complex and unconventional. Many separate systems need study. Some have hardly been looked at yet.

Table 15. Required Base Manpower

	Now	Required
ANL	1	2
BNL	8	16
FNAL	7	16
LBNL	4	8
BINP	1	3
Other US	1	3
Total FTE's	22	48

Component Development and Demonstrations

Theoretical studies alone will not be sufficient to determine the practicality of a muon collider. Experimental studies are essential. Some such studies can be undertaken without new funding, but the major efforts will require specific support. We attempt below to estimate what will be required.

Proton Driver Experimental R & D. Beam dynamic experiments at the BNL AGS will be needed, but should not be expensive. A modification of the AGS to avoid transition in that machine, and study the resulting improvements in phase space density would be very desirable, but the cost should probably be justified as an AGS improvement, rather than as a muon collider experiment, and it has not been included in this estimate.

Target, Capture and Decay Channel Experimental R & D. An experiment[28, 29] has taken data and is currently being analyzed, to determine pion production at its low energy maximum. This data, together with assumptions on pion reabsorbtion should allow more realistic Monte-Carlo calculations of total pion yield and capture in a solenoid. Nevertheless, there are several reasons why a demonstration of such capture is desirable:

- Thermal cooling requirements dictate that the target be liquid: liquid lead and gallium are under consideration. In order to avoid shock damage to a container, the liquid may need to be in the form of a jet. Since the magnetic field over the target will effect both the heat distribution in, and forces on, such a jet, an experiment is required.

- The simulation must make assumptions on the cross sections for secondary pion production by products of the primary interaction. This information is needed at low final energies and large angles where data is inadequate. A conventional experiment to determine all such cross sections would be expensive.

- We need to know the total radiation directed towards the capture and focusing solenoids. Shielding will have to be provided to protect the insulation of the inner resistive solenoid, and limit heating of the outer superconducting magnets. Only direct measurement of such radiation can provide a reliable determination.

- In the current design of phase rotation, the first rf cavity is placed 3 m from the target. If unshielded, the radiation level at this point will be very high. We have little data on the performance of a cavity under such conditions and thus have difficulty calculating the shielding requirements.

Ionization Cooling Experimental R & D. Although the principals of ionization cooling are relatively simple, there are practical problems in designing lattices that can transport, and focus the large emittances without exciting betatron resonances that blow up the emittance and attenuate the beam. There will also be problems with space charge and wake field effects.

After a design has been defined and simulated, demonstrations will be required. They will require significant rf acceleration ($\approx$ 100 MeV) and several meters of high field solenoids interspersed with bending magnets and, for a final stage demonstration, current carrying lithium rods. Such an experiment has not been designed yet. It has been suggested that this experiment might be carried out at FNAL.

An R & D program would also be required to develop the current carrying rods. This could be undertaken in a collaboration between BINP, Novosibirsk, and FNAL.

Magnet Design and Acceleration Experimental R & D. R & D programs are required both for the high field pulsed cosine theta magnets and for the lower field pulsed field magnets. The R & D on the former is somewhat more urgent since they are less conventional.

Some R & D work is also needed to determine the performance of the required superconducting cavities when excited for the relatively short pulse durations required. Studies of their sensitivity to muon decay electrons may also be needed.

Collider Ring Experimental R & D. The insertion quadrupoles need urgent R & D because the lattice design work depends on the gradients that are achieved. Nb$_3$Sn, or other higher field conductor will be prefered. Since the magnets operate at a constant field, metallic insulation may be acceptable, which would obviate the need for impregnation and thus provide better cooling. High T_c materials should be considered.

The dipole magnets, if of cosine theta design, may develop excessive mid plane compression in their coils. Block conductor arrangements may need to be developed. The use of Nb$_3$Sn will again be prefered for its high field capability.

Detector Experimental R & D. Detector R & D is required to develop the required detectors and confirm that they can both withstand the expected radiation and separate the tracks of interest from the background.

CONCLUSION

- The initial motive for the study of $\mu^+\mu^-$ colliders was:
  - The lack of beamstrahlung constraints allowing the circulation of muons, and the suppression of beamstrahlung which could, in principle, give a luminosity an advantage relative to e^+e^- at high energies.
  - The realization that, despite the problems in using muons whose lifetime is short and production diffuse, it is possible to sketch a design for $\mu^+\mu^-$ collider with parameters:
    * Energy = 4 TeV
    * Luminosity = 10^{35} $(cm^{-2}sec^{-1})$
    * More moderate power requirements and tolerances than those in an e^+e^- collider with the same specification.

- A $\mu^+\mu^-$ collider would have some unique Physics Advantages:
  - Because of the lack of beamstrahlung, a $\mu^+\mu^-$ collider could have very narrow energy spread: dE/E = 0.1 - .01 %
  - Observed reactions would have low Radiative Corrections;
  - Cross Section of $\mu^+\mu^-$ to S-channel production of any Higgs Boson (h, H, A) would be approximately 40,000 times higher than for e^+e^-. This together with the above items, would allow precision measurements of masses and widths not possible with e^+e^-.

- We note that:

- Although a $\mu^+\mu^-$ collider is radically different from existing machines, yet it requires no "exotic" technology; rather, its components would be modest extensions of existing technology, though used in an unusual manner.
- A $\mu^+\mu^-$ collider would be a multipurpose facility: besides $\mu^+\mu^-$ collisions, μ-p and μ-Ion collisions could be possible. Its proton driver could be a substantial source of spallation neutrons, and intense beams of pions, kaons, neutrinos and muons would be available.
- A $\mu^+\mu^-$ collider would be an order of magnitude smaller in overall size, and about a factor of 6 less in total tunnel length, than current e^+e^- collider designs. Because of its small size, it would fit on one of several existing lab sites.
- Consistent with its smaller size, it is estimated that a $\mu^+\mu^-$ collider would be significantly cheaper to construct. It might thus become affordable in a fiscal enviroment which may not allow larger "mega- science" projects.

- But we recognize disadvantages compared to an e^+e^- machine:
    - A $\mu^+\mu^-$ collider would have more background than e^+e^-.
    - The muons would have less polarization than electons, although, in partial compensation, both μ^+ and μ^- would be equally polarized.
    - A gamma-gamma capability would not be possible.
    - Although much progress has been made, the concept of a $\mu^+\mu^-$ collider is immature and there could yet be a fatal flaw or some problem could make it impossibly expensive.

- The $\mu^+\mu^-$ collider needs much R & D.
    - Both theoretical, the highest proiority items being:
        * Design and simulate a complete lattice for cooling.
        * continue to study instabilities in the collider ring.
        * design collider injection and beam halo scraping.
    - and experimental: the highest priority items being
        * A demonstrate of muon cooling cooling is essential to show that hardware can operate and be stable.
        * Pion capture and rf phase rotation must be demonstrated.
        * Many components need modelling, in particular: lithium lenses for cooling, pulsed magnets for acceleration, high field quadrupoles for the final focus, large aperture diploles for the collider ring, and muon collimators.

– We estimate that about five years of R & D is needed.

- If this R & D is successful then we believe a).5 TeV demonstration collider, with significant physics potential, could be built by 2010; and a 4 TeV collider might be possible a decade later.

Acknowledgment

We acknowledge important contributions from many colleagues, especially those that contributed to the feasibitity study submitted to the Snowmass Workshop 96 Proceedings[12] from which much of the material and some text, for this report has been taken: C. Ankenbrandt (FermiLab) , A. Baltz (BNL) , V. Barger (Univ. of Wisconsin) , O. Benary (Tel-Aviv Univ.) , M. S. Berger (Indiana Univ.) , A. Bogacz (UC, Los Angeles) , W-H Cheng (LBNL) , D. Cline (UC, Los Angeles) , E. Courant (BNL) , D. Ehst (ANL) , T. Diehl (Univ. of Illinois, Urbana) , R. C. Fernow (BNL) , M. Furman (LBNL) , J. C. Gallardo (BNL) , A. Garren (LBNL) , S. Geer (FermiLab) , I. Ginzburg (Inst. of Math., Novosibirsk) , H. Gordon (BNL) , M. Green (LBNL) , J. Griffin (FermiLab) , J. F. Gunion (UC, Davis) , T. Han (UC, Davis) , C. Johnstone (FermiLab) , D. Kahana (BNL) , S. Kahn (BNL) , H. G. Kirk (BNL) , P. Lebrun (FermiLab) , D. Lissauer (BNL) , A. Luccio (BNL) , H. Ma (BNL) , A. McInturff (LBNL) , F. Mills (FermiLab) , N. Mokhov (FermiLab) , A. Moretti (FermiLab) , G. Morgan (BNL) , M. Murtagh (BNL) , D. Neuffer (FermiLab) , K-Y. Ng (FermiLab) , R. J. Noble (FermiLab) , J. Norem (ANL) , B. Norum (Univ. Virginia) , I. Novitski (FermiLab), K. Oide (KEK), F. Paige (BNL) , J. Peterson (LBNL) , V. Polychronakos (BNL) , M. Popovic (FermiLab) , S. Protopopescu (BNL) , Z. Qian (FermiLab) , P. Rehak (BNL) , R. Roser (Univ. of Illinois, Urbana) , T. Roser (BNL) , R. Rossmanith (DESY) , Q-S Shu, (CEBAF) , A. Skrinsky (BINP) , I. Stumer (BNL) , S. Simrock (CEBAF) , D. Summers (Univ. of Mississippi) , H. Takahashi (BNL) , H. Takai (BNL) , V. Tchernatine (BNL) , Y. Torun (SUNY, Stony Brook), D. Trbojevic (BNL) , W. C. Turner (LBNL), A. Van Ginneken (FermiLab) , E. Willen (BNL) , W. Willis (Columbia Univ.) , D. Winn (Fairfield Univ.) , J. S. Wurtele (UC, Berkeley) , Y. Zhao (BNL). In particular we acknowledge the contributions of the Editors of each one of the chapters of the $\mu^+\mu^-$ Collider: A Feasibility Study: V. Barger, J. Norem, R. Noble, H. Kirk, R. Fernow, D. Neuffer, J. Wurtele, D. Lissauer, M. Murtagh, S. Geer, N. Mokhov and D. Cline. This research was supported by the U.S. Department of Energy under Contract No. DE-ACO2-76-CH00016 and DE-AC03-76SF00515.

REFERENCES

1. A. W. Chao, R. B. Palmer, L. Evans, J, Gareyte, R. H. Siemann, *Hadron Colliders (SSC/LHC)*, Proc.1990 Summer Study on High Energy Physics, Snowmass, (1990) p 667.
2. S. Holmes for the RLHC Group, *Summary Report*, presentation at the Snowmass Workshop 96, to be published.
3. K. Yokoya and P. Chen, *Beam-Beam Phenomena in Linear Colliders* in Frontiers of Particle Beams: Luminosity Limitations, Ed. M. Dienes, et al., Lecture Notes in Physics **400**, Springer-Verlag, 1990.
4. See for example, H. Murayama and M. Peskin, *Physics Opportunities of e^+e^- Linear Colliders*, SLAC-PUB-7149/LBNL-38808/UCB-PTH-96/18, June 1996; to appera in Annual Review of Nuclear and Particle Physics.
5. R. B. Palmer,*Prospects for High Energy e^+e^- Linear Colliders*, Annu. Rev. Nucl. Part. Sci. (1990) 40, p 529-92.
6. *International Linear Collider Technical Review Committee Report*, SLAC-R-95-471, (1995)
7. N. Akasaka, *Dark current simulation in high gradient accelerating structure* EPAC96 Proceedings, pp. 483 Sitges, Barcelona, Spain, June 1996), Institute of Physics Publishing
8. V. Telnov, Nucl. Instr. and Meth. A294, (1990) 72; *A Second Interaction Region for Gamma-Gamma, Gamma-Electron and Electron-Electron Collisions for NLC*, Ed. K-J Kim, LBNL-38985, LLNL-UCRL-ID 124182, SLAC-PUB-95-7192.
9. R. B. Palmer,*Accelerator parameters for $\gamma - \gamma$ colliders*; Nucl. Inst. and Meth., A355 (1995) 150-153.
10. P. Chen and R. Palmer, *Coherent Pair Creation as a Posittron Source for Linear Colliders*, AIP Press, ed. J. Wurtele, Conference Proceedings 279, 1993.
11. V. Telnov, *Laser Cooling of Electron Beams for linear colliders*; NSF-ITP-96-142 and SLAC-PUB 7337
12. $\mu^+\mu^-$ *collider, A Feasibility Study*, BNL-52503, FermiLab-Conf-96/092, LBNL-38946, submitted to the Proceedings of the Snowmass96 Workshop.
13. V. Barger, et al. and J. Gunion et al., Snowmass Workshop 96 Proceedings, unpublished. V. Barger, *New Physics Potential of Muon-Muon Collider*, Proceedings of the 9th Advanced ICFA Beam Dynamics Workshop, Ed. J. C. Gallardo, AIP Press, Conference Proceedings 372 (1996).
14. Zeroth-order Design Report for the Next Linear Collider, LBNL-PUB-5424, SLAC Report 474 and UCRL-ID-124161
15. E. A. Perevedentsev and A. N. Skrinsky, Proc. 12th Int. Conf. on High Energy Accelerators, F. T. Cole and R. Donaldson, Eds., (1983) 485; A. N. Skrinsky and V.V. Parkhomchuk, Sov. J. of Nucl. Physics **12**, (1981) 3; *Early Concepts for $\mu^+\mu^-$ Colliders and High Energy μ Storage Rings, Physics Potential & Development of $\mu^+\mu^-$ Colliders. 2nd Workshop*, Sausalito, CA, Ed. D. Cline, AIP Press, Woodbury, New York, (1995).
16. D. Neuffer, IEEE Trans. **NS-28**, (1981) 2034.
17. *Proceedings of the Mini-Workshop on $\mu^+\mu^-$ Colliders: Particle Physics and Design*, Napa CA, Nucl Inst. and Meth., **A350** (1994) ; Proceedings of the Muon Collider Workshop, February 22, 1993, Los Alamos National Laboratory Report LA- UR-93-866 (1993) and *Physics Potential & Development of $\mu^+\mu^-$ Colliders 2nd Workshop*, Sausalito, CA, Ed. D. Cline, AIP Press, Woodbury, New York, (1995).
18. *Transparencies at the 2 + 2 TeV $\mu^+\mu^-$ Collider Collaboration Meeting*, Feb 6-8, 1995, BNL, compiled by Juan C. Gallardo; transparencies at the *2 + 2 TeV $\mu^+\mu^-$ Collider Collaboration*

Meeting, July 11-13, 1995, FERMILAB, compiled by Robert Noble; Proceedings of the 9th Advanced ICFA Beam Dynamics Workshop, Ed. J. C. Gallardo, AIP Press, Conference Proceedings 372 (1996).

19. D. V. Neuffer and R. B. Palmer, Proc. European Particle Acc. Conf., London (1994); M. Tigner, in Advanced Accelerator Concepts, Port Jefferson, NY 1992, AIP Conf. Proc. **279**, 1 (1993).

20. R. B. Palmer et al., *Monte Carlo Simulations of Muon Production, Physics Potential & Development of $\mu^+\mu^-$ Colliders 2nd Workshop*, Sausalito, CA, Ed. D. Cline, AIP Press, Woodbury, New York, pp. 108 (1995); R. B. Palmer, et al., *Muon Collider Design*, in Proceedings of the Symposium on Physics Potential & Development of $\mu^+\mu^-$ Colliders, Nucl. Phys B (Proc. Suppl.) 51A (1996)

21. T. Roser, *AGS Performance and Upgrades: A Possible Proton Driver for a Muon Collider*, Proceedings of the 9th Advanced ICFA Beam Dynamics Workshop, Ed. J. C. Gallardo, AIP Press, Conference Proceedings 372 (1996) .

22. T. Roser and J. Norem, private communication and Chapter 3 in reference[12]

23. Y. Cho, et al., *A 10-GeV, 5-MeV Proton Source for a Pulsed Spallation Source*, Proc. of the 13th Meeting of the Int'l Collaboration on Advanced Neutron Sources, PSI Villigen, Oct. 11-14 (1995); Y. Cho, et al., *A 10-GeV, 5-MeV Proton Source for a Muon-Muon Collider*, Proceedings of the 9th Advanced ICFA Beam Dynamics Workshop, Ed. J. C. Gallardo, AIP Press, Conference Proceedings 372 (1996).

24. F. Mills, et al., presentation at the 9th Advanced ICFA Beam Dynamics Workshop, unpublished; see also second reference in [18].

25. D. Kahana, et al., Proceedings of Heavy Ion Physics at the AGS-HIPAGS '93, Ed. G. S. Stephans, S. G. Steadman and W. E. Kehoe (1993); D. Kahana and Y. Torun, *Analysis of Pion Production Data from E-802 at 14.6 GeV/c using ARC*, BNL Report # 61983 (1995).

26. N. V. Mokhov, *The MARS Code System User's Guide*, version 13(95), Fermilab-FN-628 (1995).

27. J. Ranft, DPMJET Code System (1995).

28. See, http://www.nevis1.nevis.columbia.edu/heavyion/e910

29. H. Kirk, presentation at the Snowmass96 Workshop, unpublished.

30. N. Mokhov, R. Noble and A. Van Ginneken, *Target and Collection Optimization for Muon Colliders*, Proceedings of the 9th Advanced ICFA Beam Dynamics Workshop, Ed. J. C. Gallardo, AIP Press, Conference Proceedings 372 (1996).

31. R. B. Palmer, et al., *Monte Carlo Simulations of Muon Production*, Proceedings of the Physics Potential & Development of $\mu^+\mu^-$ Colliders Workshop, ed. D. Cline, AIP Press Conference Proceedings 352 (1994).

32. See reference [31]

33. R. Weggel, presentation at the Snowmass96 Workshop, unpublished; Physics Today, pp. 21-22, Dec. (1994).

34. M. Green, *Superconducting Magnets for a Muon Collider*, Nucl. Phys. B (Proc. Suppl.) 51A (1996)

35. R. Weggel, private communication

36. F. Chen, *Introduction to Plasma Physics*, Plenum, New York, pp. 23-26 (9174); T. Tajima, *Computational Plasma Physics: With Applications to Fusion and Astrophysics*, Addison-Wesley Publishing Co., New York, pp. 281-282 (1989). 37. A. A. Mikhailichenko and M. S. Zolotorev, Phys. Rev. Lett. **71**, (1993) 4146; M. S. Zolotorev and A. A. Zholents, SLAC-PUB-6476 (1994).

38. A. Hershcovitch, Brookhaven National Report AGS/AD/Tech. Note No. 413 (1995).

39. Z. Huang, P. Chen and R. Ruth, SLAC-PUB-6745, *Proc. Workshop on Advanced Accelerator*

Concepts, Lake Geneva, WI , June (1994); P. Sandler, A. Bogacz and D. Cline, *Muon Cooling and Acceleration Experiment Using Muon Sources at Triumf, Physics Potential & Development of $\mu^+\mu^-$ Colliders 2nd Workshop*, Sausalito, CA, Ed. D. Cline, AIP Press, Woodbury, New York, pp. 146 (1995).

40. Initial speculations on ionization cooling have been variously attributed to G. O'Neill and/or G. Budker see D. Neuffer, Particle Accelerators, **14**, (1983) 75; D. Neuffer, Proc. 12th Int. Conf. on High Energy Accelerators, F. T. Cole and R. Donaldson, Eds., 481 (1983); D. Neuffer, in Advanced Accelerator Concepts, AIP Conf. Proc. 156, 201 (1987); see also [15].

41. U. Fano, Ann. Rev. Nucl. Sci. 13, 1 (1963).

42. D. Neuffer and A. van Ginneken, private communication

43. R. Fernow, private communication

44. G. Silvestrov, Proceedings of the Muon Collider Workshop, February 22, 1993, Los Alamos National Laboratory Report LA-UR-93-866 (1993); B. Bayanov, J. Petrov, G. Silvestrov, J. MacLachlan, and G. Nicholls, Nucl. Inst. and Meth. **190**, (1981) 9.

45. M. D. Church and J. P. Marriner, Annu. Rev. Nucl. Sci. 43 (1993) 253.

46. Colin D. Johnson, Hyperfine Interactions, **44** (1988) 21.

47. G. Silvestrov, *Lithium Lenses for Muon Colliders*, Proceedings of the 9th Advanced ICFA Beam Dynamics Workshop, Ed. J. C. Gallardo, AIP Press, Conference Proceedings 372 (1996).

48. F. Mills, presentation at the Ionization Cooling Workshop, BNL August 1996, unpublished and private communication.

49. A. Skrinsky, presentation at the Ionization Cooling Workshop, BNL August 1996, unpublished and private communication.

50. D. Neuffer, *Acceleration to Collisions for the $\mu^+\mu^-$ Collider*, Proceedings of the 9th Advanced ICFA Beam Dynamics Workshop, Ed. J. C. Gallardo, AIP Press, Conference Proceedings 372 (1996).

51. D. Summers, presentation at the 9th Advanced ICFA Beam Dynamics Workshop, unpublished.

52. D. Summers, *Hybrid Rings of Fixed $8\,T$ Superconducting Magnets and Iron Magnets Rapidly Cycling between $-2\,T$ and $+2\,T$ for a Muon Collider* submitted to the Proceedings of the Snowmass Workshop 96, unpublished.

53. I. Stumer, presentation at the BNL-LBL-FNAL Collaboration Meeting, Feb 1996, BNL, unpublished.

54. S.Y. Lee, K.-Y. Ng and D. Trbojevic, FNAL Report FN595 (1992); Phys. Rev. **E48**, (1993) 3040; D. Trbojevic, et al., *Design of the Muon Collider Isochronous Storage Ring Lattice, Micro-Bunches Workshop*, AIP Press, Conference Proceedings 367 (1996).

55. K. Oide, private communication.

56. C. Johnstone and A. Garren, Proceedings of the Snowmass Workshop 96; C. Johnstone and N. Mokhov, ibid.

57. K. L. Brown and J. Spencer, SLAC-PUB-2678 (1981) presented at the Particle Accelerator Conf., Washington, (1981) and K.L. Brown, SLAC-PUB-4811 (1988), Proc. Capri Workshop, June 1988 and J.J. Murray, K. L. Brown and T.H. Fieguth, Particle Accelerator Conf., Washington, 1987; Bruce Dunham and Olivier Napoly, *FFADA, Final Focus. Automatic Design and Analysis*, CERN Report CLIC Note 222, (1994); Olivier Napoly, it CLIC Final Focus System: Upgraded Version with Increased Bandwidth and Error Analysis, CERN Report CLIC Note 227, (1994).

58. K. Oide, SLAC-PUB-4953 (1989); J. Irwin, SLAC-PUB-6197 and LBL-33276, Particle Accelerator Conf.,Washington, DC, May (1993); R. Brinkmann, *Optimization of a Final Focus System for Large Momentum Bandwidth*, DESY-M-90/14 (1990).

59. J. C. Gallardo and R. B. Palmer, *Final Focus System for a Muon Collider: A Test Model*, in Physics Potential & Development of $\mu^+\mu^-$ Colliders, Nucl. Phys. B (Proc. Suppl.) 51A (1996), Ed. D. Cline.

60. A. Garren, et al., *Design of the Muon Collider Lattice: Present Status*, in Physics Potential & Development of $\mu^+\mu^-$ Colliders, Nucl. Phys. B (Proc. Suppl.) 51A (1996), Ed. D. Cline.

61. K. Oide, private communication

62. M. Syphers, private communication.

63. K.Y. Ng, *Beam Stability Issues in a Quasi-Isochronous Muon Collider*, Proceedings of the 9th Advanced ICFA Beam Dynamics Workshop, Ed. J. C. Gallardo, AIP Press, Conference Proceedings 372 (1996).

64. W.-H. Cheng, A.M. Sessler, and J.S. Wurtele, *Studies of Collective Instabilities, in Muon Collider Rings*, Proceedings of the 9th Advanced ICFA Beam Dynamics Workshop, Ed. J. C. Gallardo, AIP Press, Conference Proceedings 372 (1996).

65. V. Balakin, A. Novokhatski and V. Smirnov, Proc. 12*th* Int. Conf. on High Energy Accel., Batavia, IL, 1983, ed. F.T. Cole, Batavia: Fermi Natl. Accel. Lab. (1983), p. 119.

66. A. Chao, *Physics of Collective Beam Instabilities in High Energy Accelerators*, John Wiley & Sons, Inc, New York (1993).

67. W.-H. Cheng, private communication; see also Chapter 8 of reference [12].

68. I. Stumer, presentation at the BNL-LBNL-FNAL Collaboration Meeting, Feb. 1996, BNL unpublished; see also reference [18]. Presentation at the Snowmass96 Workshop, unpublished. Chapter 9 in reference [12]

69. N. Mokhovov and S. Striganov, *Simulation of Background in Detectors and Energy Deposition in Superconducting Magnets at $\mu^+\mu^-$ Colliders*, Proceedings of the 9th Advanced ICFA Beam Dynamics Workshop, Ed. Juan C. Gallardo, AIP Press, Conference Proceedings 372 (1996); N. Mokhovov, *Comparison of backgrounds in detectors for LHC, NLC, and $\mu^+\mu^-$ Colliders*, Nucl. Phys. B (Proc. Suppl.) 51A (1996).

70. *The Time Projection Chamber: A New 4π Detector for Charged Particles*, D.R. Nygren (SLAC). PEP-0144, (Received Dec 1976). 21pp. In Berkeley 1974, Proceedings, Pep Summer S tudy, Berkeley 1975, 58-78.

71. E. Gatti and P. Rehak, Nucl. Instr. and Meth. **225**, 608 (1984).

72. BaBa Notes, 39,122,171 in the WEB site http: www.slac.stanford.edu/BFROOT/doc/www/vertex.html

73. *ATLAS Technical Proposal for a General-Purpose pp Experiment at the Large Hadron Collider at CERN*, CERN/ LHCC/94-43, LHCC/P2 (15 December 1994).

74. *GEM Technical Design Report Submitted by Gammas, Electrons, and Muons Collaboration to the Superconducting Super Collider Laboratory*, GEM-TN-93-262; SSCL-SR-1219 (July 31, 1993).

75. O. Guidemeister, F. Nessi-Tadaldi and M. Nessi, Proc. 2nd Int. Conf. on Calorimetry in HEP, Capri, 1991.

76. J. Sandweiss, private communication

77. P. Chen, presentation at the 9th Advanced ICFA Beam Dynamics Workshop and Nuc. Phys. B (Proc. Suppl.) 51A (1996)

78. I. J. Ginzburg, *The e^+e^- pair production at $\mu^+\mu^-$ collider*,Nucl. Phys. B (Proc. Suppl.) 51A (1996)

79. P. Chen, *Beam-Beam Interaction in Muon Colliders*, SLAC-PUB-7161(April, 1996).

80. P. Chen and N. Kroll in preparation.

81. G. V. Stupakov and P. Chen, *Plasma Suppression of Beam-Beam Interaction in Circular Colliders*, SLAC Report: SLAC-PUB-95-7084 (1995). S. Skrinsky private communication; Juan C. Gallardo and S. Skrinsky in preparation.
82. K. Assamagan, et al., Phys Lett. **B**335, 231 (1994); E. P. Wigner, Ann. Math. **40**, 194 (1939) and Rev. Mod. Phys., **29**, 255 (1957).
83. R. B. Palmer et al., *Monte Carlo Simulations of Muon Production*, AIP Conference Proceedings 352 (1996), Ed. D. Cline.
84. B. Norum and R. Rossmanith, *Polarized Beams in a Muon Collider*, Nucl. Phys. B (Proc. Suppl.) 51A (1996), Ed. D. Cline.
85. Ya. S. Derbenev and A. M. Kondratenko, JETP **35**, 230 (1972); Par. Accel., **8**, 115 (1978).
86. P. Chen and K. Yokoya, Phys. Rev. **D**38 987 (1988); P. Chen., SLAC-PUB-4823 (1987); Proc. Part. Accel. School, Batavia, IL, 1987; AIP Conf. Proc. 184: 633 (1987).

TOP AND ELECTROWEAK PHYSICS AT HADRON COLLIDERS

M. Strovink

Physics Department and Lawrence Berkeley National Laboratory
University of California
Berkeley, CA 94720

INTRODUCTION

Because the top quark has been observed[1],[2] only recently, and progress in Fermilab Tevatron electroweak measurements has been comparably rapid, these lectures consist primarily of summaries of current experimental work. This is in contrast to the more pedagogical nature of much of the other material in these Proceedings. The initial lecture is devoted to the production and decay characteristics of the top quark. Analysis of the top quark mass is my particular interest, so I allocate two lectures to it, one to the dilepton channel, the other to the more widely discussed decay channels in which at least one W from top decays nonleptonically. The final lecture is devoted to measurement of the W mass.

Had electroweak physics occupied a larger fraction of these lectures, I would also have discussed the increasingly stringent limits set on anomalous triboson gauge couplings by the DØ and CDF experiments, as well as on the production of additional gauge bosons. However, in the interval between preparation of the lectures (Jun 96) and of this contribution to the Proceedings (Sep 96), further substantial progress has been achieved, particularly in combining results from different approaches to measuring $W\gamma$ production[3] and from an innovative assault on $Z\gamma$ production[4]. I would not have been able to report these very recent advances.

Realizing that reviews of current work are perishable, I devoted much of the time that I could allocate to the Advanced Study Institute to the task of making these lectures accessible electronically. I attempted to place enough textual material on each transparency that the basics could be understood without further explanation, and I put all transparencies on the Web[5] soon after the Institute concluded. After making this investment, I found that the only practical approach to contributing to these Proceedings was to recycle these same transparencies as figures. This accounts for the odd appearance of this contribution's pages, which consist typically of one figure containing two transparencies, with a brief caption underneath adding references where necessary. The small number in the upper right-hand corner of each transparency gives the sequence in which it was shown. In the figures that follow, a and b in the caption denotes top and bottom, respectively.

In most of these areas the DØ and CDF experiments compete vigorously. However, lecturers best serve students not *e.g.* by allocating equal space to two equally precise results, as would a conference rapporteur, but rather by explaining one result more fully. As a DØ collaborator, I have better access to and familiarity with DØ material than that from CDF. Nevertheless, I felt it appropriate in several instances to emphasize the CDF measurement, and, for some analyses, to show every relevant CDF plot made available to me.

LECTURE I. TOP PRODUCTION AND DECAY

After describing briefly how $t\bar{t}$ pairs are produced and decay, I tabulate the event selection criteria enforced by D0 and CDF. Next I tour the principal decay channels used for top cross section and branching ratio studies, including one or two example plots for each. Then I graph and tabulate the measured top production cross section, full and by channel, for DØ and CDF. I conclude with a very preliminary discussion of possible supersymmetric signatures in top final states.

The experimental results in Lecture I are drawn mainly from talks by M. Narain[6] (Mar 96) for DØ and M. Kruse[7] and S. Leone[8] (May 96) for CDF. The supersymmetric discussion is stimulated by a recent suggestion from Kane and Mrenna[9].

LECTURE II. TOP MASS IN THE DILEPTON CHANNEL

After reviewing the problem of measuring the (underconstrained) top quark mass in the dilepton channel, and presenting a general approach to that problem, I describe in detail the two analyses, using approximate methods, which have been made public in preliminary form by D0. I compare briefly the results and errors from these D0 analyses to those of a more basic method which was applied recently by CDF to their (larger) dilepton candidate sample.

Workers in this area were influenced by the early publications of Dalitz and Goldstein[16] and Kondo[17]. The general approach which I present in this lecture was described[18] in 1993, along with preliminary applications of it and of one approximate method to the first top-to-dilepton candidate with low background probability. The current DØ top mass analysis in the dilepton channel is described *e.g.* in Ref. [19] and citations therein. The corresponding CDF analysis was summarized in a talk by Rolli [20] (May 96).

I. Top production and decay

Top production
Classification by W decay
Event selection
Dilepton channel
Lepton+jets/μ tag channel
Lepton+jets/SVX tag channel
Lepton+jets/topological channel
All jets channel
Other channels
Top cross sections by channel
Supersymmetric signatures

Top production in proton-antiproton collisions

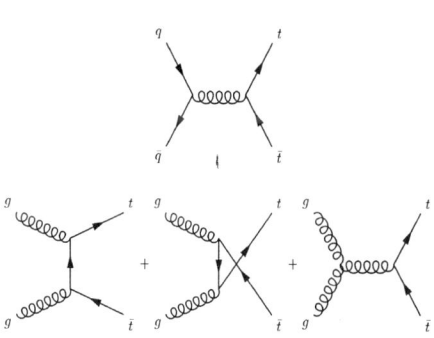

Top pairs are produced in the s-channel by quark-antiquark or gluon-gluon fusion. To leading order (diagrammed above) at the Tevatron, the quark-antiquark channel dominates by ~1 order of magnitude.

Higher-order processes are especially important for the gluon-gluon process because of initial-state emission of soft gluons. It is necessary to resum to all orders the dominant contributions from these diagrams. Variations in the resummation procedure as well as the usual QCD renormalization scale cause ~ ±20% uncertainties in the calculated cross section.

Single top is produced in the t-channel at similar rates. The backgrounds are much higher than for top pairs, so single top plays little role in present analysis.

Figure 1. (a) Outline of Lecture I. (b) Leading-order Feynman diagrams for top pair production. Higher-order calculations of $\sigma(p\bar{p} \to t\bar{t})$ have been made by Laenen et al.,[10] Berger and Contopanagos,[11] and Catani et al.[12].

Classification of top pair events by W decay channel

Assume that all top quarks decay via W+b.

W branching ratio is 1/9 per lepton and 3/9 per (colored) quark generation, leading to the chart below.

Most information on top cross section and mass comes from "lepton+jets" and "dilepton" channels, where "lepton" refers to electron and/or muon.

A significant top signal is also observed in the "all jets" channel (by CDF).

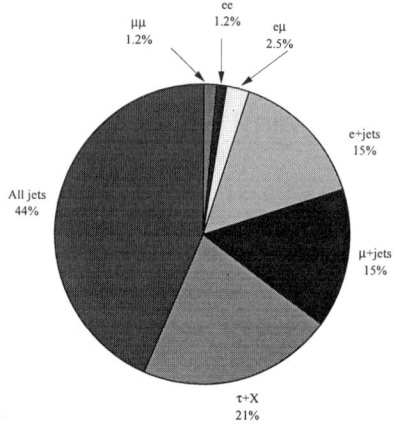

Top event selection

Both experiments use similar cuts to identify jets and W's (isolated e or μ and missing E_T for the latter).

In D0, μ's have lower background and coarser momentum resolution than e's, so are cut differently.

Additional background rejection is obtained in both experiments by requiring another e or μ, or, in D0, by applying topological cuts.

D0 cut	$e\mu$	ee	$\mu\mu$	e/μ +jets/topol	e/μ +jets/μ
minimum...					
missing E_T (GeV)	20	25*	*	25/20	20*
no. of jets	2	2	2	4	3
E_T of jet (GeV)	20	20	20	15	20
max $|\eta|$ of jet	2.5	2.5	2.5	2.0	2.0
no. of leptons	2	2	2	1	2
no. of isolated leptons	2	2	2	1	1
E_T of lepton (GeV)	15	20	15	20	20
max $|\eta|$ of lepton	2.5/1.7	2.5	1.0	2.0/1.7	2.0/1.7
A of jets+W	--	--	--	0.065	0.04
H_T (GeV)...	120	120	100	180	110
...of:	jets+e	jets+e_1	jets	jets+W	jets+W
E_T of W	--	--	--	60	--
E_T of tag μ (GeV)	--	--	--	--	4
max ΔR (jet, tag μ)	0	0	0	0	0.5

*extra cuts to reject Z's, cosmics, and mismeasured μ's

Figure 2. (a) Top pair classification by W decay channel. (b) DØ top event selection criteria.

Top event selection (cont'd)

CDF relies on topological cuts in the all jets channel, but uses them only as a cross-check for l+jets analysis.

The main method used by CDF to tag b's is the requirement of a secondary vertex in their silicon (SVX) tracker. For that purpose D0 relies on a soft muon tag.

CDF requires only 1 isolated lepton in the dilepton channels, only 3 jets in l+jets analysis, and only 5 jets in the all jets channel -- in each case one less than the nominal quota.

CDF cut minimum...	ll	l+jets	all jets
missing E_T (GeV)	25*	20	--
no. of jets	2	3	5
E_T of jet (GeV)	10 (raw)	15	15
max $|\eta|$ of jet	2.0	2.0	2.0
$|\Delta R|$ between jets	--	--	0.5
no. of leptons	2	1 (2 if SLT)	0
no. of isolated leptons	1	1	0
E_T of lepton (GeV)	20	20	--
max $|\eta|$ of lepton	~1.05	~1.05	--
$|m(ll)-m(Z)|$ (GeV)	15 (ee,$\mu\mu$)	--	--
b tag	--	SVX or SLT	SVX
$A + 0.0025 \times H_T^{\text{jets3+4+...}}$	--	--	0.54
H_T^{jets} (GeV)	--	--	300
$H_T^{\text{jets}}/\sqrt{s}$	--	--	0.75

* 50 if $\Delta\phi$ (missing E_T, nearest l or jet) < 20°

Dilepton channel

After CDF requires ≥2 jets, 10 dilepton candidates with significant missing E_T survive. One is a $\mu\mu\gamma$ with $m(\mu\mu\gamma) \sim 90$ GeV/c^2. 7 of the remaining 9 are $e\mu$'s; all but 2 $e\mu$'s have two isolated leptons.

19 events satisfying the same cuts but having only <2 jets are distributed closer to the E_T cut boundary.

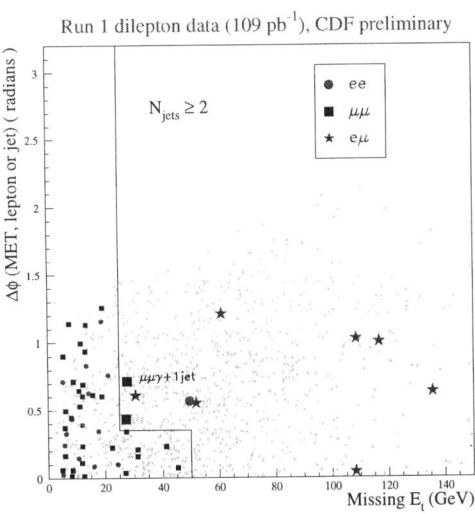

Figure 3. (a) CDF top event selection criteria. (b) Azimuth between missing E_T and the nearest lepton or jet, vs. missing E_T for CDF dilepton events. The cut boundary is irregular because missing E_T is measured more reliably when it is not collinear with another object.

Dilepton channel (cont'd)

The main backgrounds are $Z \to \tau\tau$, Drell-Yan dileptons, W pairs, and fake leptons. These have smaller $\langle H_T \rangle$ than top. CDF calculates their sum to be 2 ± 0.4.

Of the 10 CDF candidates plotted, 3 satisfy the same $E_T(\text{jet } 2)$, $H_T(e+\text{jets})$, and lepton isolation cuts survived by D0's 5 candidates. (3 of the 5 D0 candidates pass CDF cuts.)

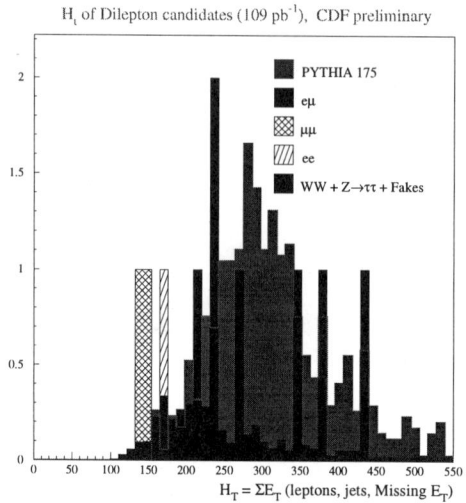

Lepton+jets channel enhanced by muon tag from b decay

~40% of top pairs yield an extra soft muon from the decay of one b or its daughter c quark, which is detected with ~50% efficiency.

The background to tag muons is particularly low in D0, with its short flight path and a thick muon filter. Only ~0.5% of generic QCD jets have a soft muon tag.

The D0 data show not only an excess over background in the signal region (at least 3 jets), but also consistency with expectation for lower jet multiplicities.

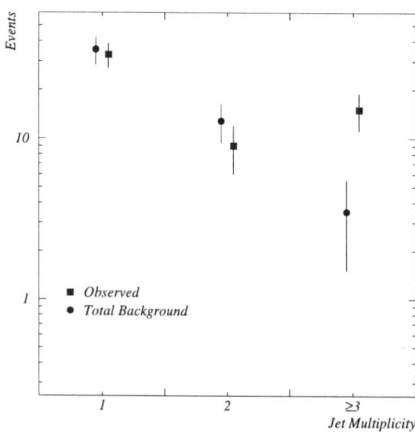

Figure 4. (a) Scalar summed E_T of leptons, jets, and missing E_T for CDF dilepton events. (b) Jet multiplicity for DØ soft muon b tagged lepton+jets events.

Lepton+jets channel enhanced by displaced vertex tag from b decay

A b hadron with p_T=10 GeV/c decays at $<\sqrt{x^2+y^2}>$ ~1 mm from the beam axis. A CDF SVX tag requires a 2 (>2) track vertex with a >3σ (>2σ) xy impact parameter.

Displayed is the $c\tau$ distribution of tagged jets. The data and top are slightly broader than the background.

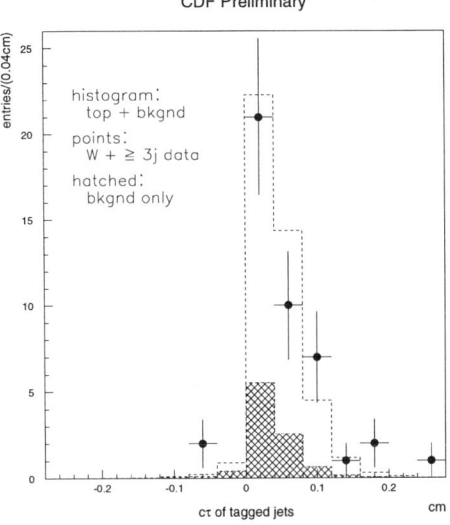

Lepton+jets channel enhanced by displaced vertex tag from b decay (cont'd)

After all cuts including the SVX tag are imposed, CDF sees a clear excess over background for ≥3 jets. This excess constrains the size of the top signal plotted.

In the 2 jet channel, 45 SVX tags are seen (6 double tagged), while 29 background + 6 top = 35 are expected. The ~1.5σ significance of the 2 jet excess increases when the double tags are taken into account.

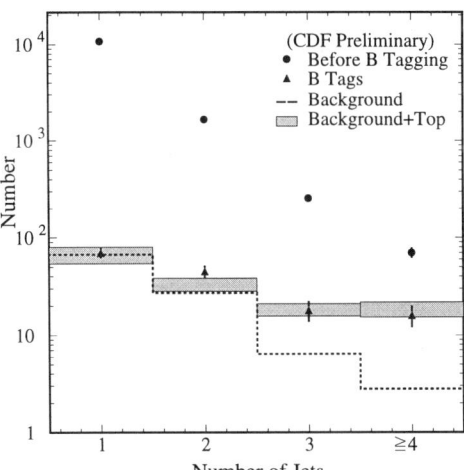

Figure 5. (a) Tagged jet $c\tau$ for CDF SVX b tagged lepton+jets events. (b) Jet multiplicity for CDF lepton+jets events before and after SVX b tag requirement.

Lepton+jets channel enhanced by topological cuts

After subtracting QCD multijet background, DØ fits the remaining data, mainly W+jets background, to a form exponential in the inclusive jet multiplicity ("Berends scaling"), using the first three points. The fit allows for a small top contribution.

Extrapolating to the signal region (point "4", with at least 4 jets) yields a "base sample" of W+jets before final cuts.

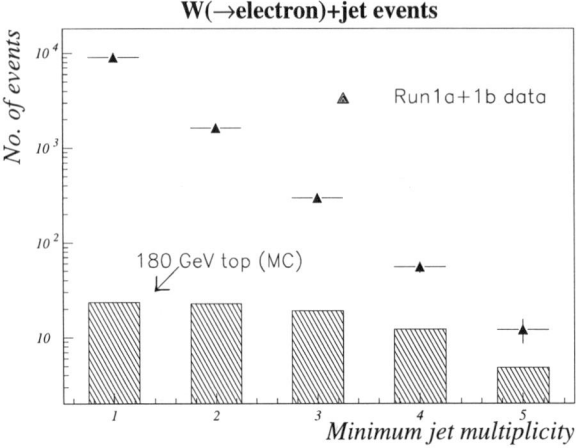

Lepton+jets channel enhanced by topological cuts (cont'd)

Shown is the distribution of DØ data, top, W+jets background, and QCD multijet background in aplanarity A vs. H_T. A is 3/2 the smallest eigenvalue of the normalized momentum tensor and H_T is the scalar transverse energy including the jets and the W.

The efficiency of the final cut $A > 0.065$ and $H_T > 180$ GeV for "base sample" top, W+jets, and QCD events is calculated by Monte Carlo simulation.

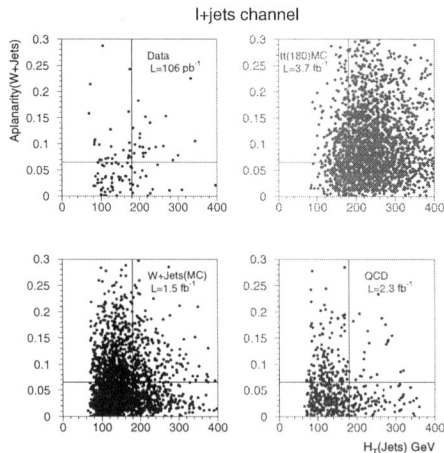

Figure 6. (a) Minimum jet multiplicity for DØ lepton+jets events before topological cuts are applied. (b) Aplanarity vs. H_T for DØ lepton+jets data, 180 GeV/c^2 top MC, W+jets MC, and non W ("QCD") background data.

CDF all jets channel enhanced by SVX tag

Background is estimated by measuring the SVX tag rate as a function of the kinematic parameters $\{K\}$ of generic QCD jets, then applying it to jets after heavy topological cuts. One must be convinced that these cuts have no effect on the tag rate for jets of fixed $\{K\}$.

230 SVX tags are observed *vs.* 160.5±10.4 expected (3.6σ significance), mainly in the 5 or 6 jet channels (below).

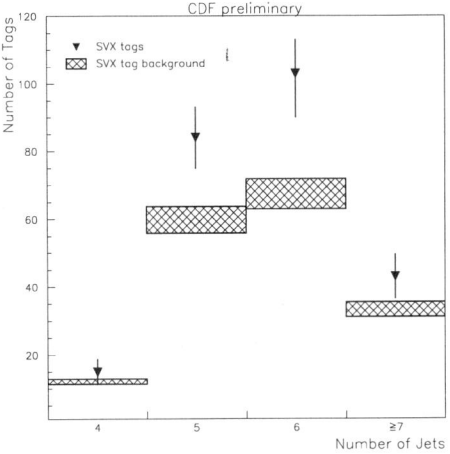

CDF all jets channel enhanced by SVX tag (cont'd)

Kinematic distributions for signal, background, and data are similar after cuts, *e.g.* for jet E_T (below).

For top 175, CDF's all jets cross section is $10.7^{+7.6}_{-4.0}$ pb, more than twice the ~5 pb calculated by Laenen *et al.*

Unfluctuated top production at the ~5 pb level would not allow a significant top signal to be observed in these data.

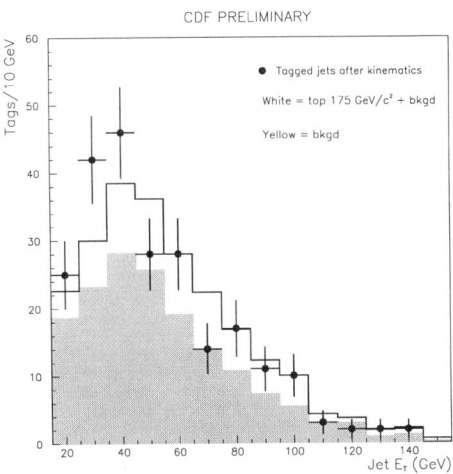

Figure 7. (a) Jet multiplicity for CDF all jets events after SVX and topological cuts are applied. (b) Jet E_T for CDF all jets data and calculated background.

Other channels

CDF analyzes the final state consisting of one isolated e or μ and one τ. The latter decays to 1 or 3 charged hadrons with 64% branching ratio. Its signature is an isolated track with $p_T > 15$ GeV/c matched to a calorimeter signal that is inconsistent with e's or μ's. After requiring two jets, significant missing E_T, and $H_T > 180$ GeV (for all objects including missing E_T), four candidates survive vs. an expected background of 2 ± 0.4. Three of these are b tagged (one twice), but two have missing E_T collinear with the principal jet. <1 top event is expected.

CDF measures $B(t \to Wb)/B(t \to Wq) > 0.61$ (95% conf), yielding no strong constraint on V_{tb} for a unitary 3×3 CKM matrix and no constraint for the 4×4 case.

D0 analyzes the "$e\nu$" final state with one isolated e and large (>50 GeV) missing E_T, along with two jets. Contributions are expected from the ee and $e\mu$ channels when one lepton escapes detection, from the $e\tau$ channel, and from the lepton+jets channel when two jets are undetected. After requiring that the "W" transverse mass exceed 115 GeV/c^2 (to reject W+jets background), two candidates survive vs. an expected background of 1.4±0.4. ~1.2 top events are expected.

D0 also analyzes the all jets channel, using topological selection and a soft μ tag. 15 candidates survive vs. an expected background of 11±2.3. ~5 events are expected; intense work on this analysis continues.

Top cross section measured by D0 (1/96)

D0's combined-channel top cross section is dominated by topologically selected lepton+jets data; the muon tagged and dilepton data are consistent within error.

It is expressed as a function of top mass because of correlation induced by the topological cuts. D0's central cross section coincides with theory for top mass ~175 GeV, but its uncertainty allows a wide mass range.

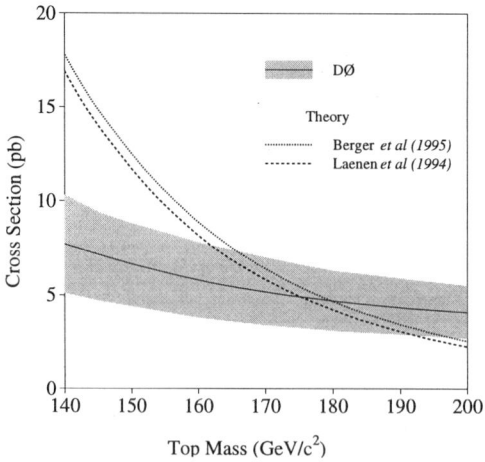

Figure 8. (a) Summary of other top analyses: (CDF) $\tau + e$ or $\mu + 2$ jets; (CDF) $\mathcal{B}(t \to Wb)/\mathcal{B}(t \to Wq)$; (DØ) e + large missing $E_T + 2$ jets; (DØ) all jets with soft muon b tag. (b) DØ $\sigma(p\bar{p} \to t\bar{t})$ vs. top mass.

Cross sections for D0 and CDF counting experiments

In the leptonic channels, D0 (CDF) expect 24 (33) top 175 events with a *S/N* of 1.8 (1) based on Laenen *et al.*'s σ_{top}.

D0 channel	pb^{-1}	Background	Expected top 180	Data	σ (top 180) (pb)
eμ	90.5	0.36±0.09	1.69±0.27	3	
ee	105.9	0.66±0.17	0.92±0.11	1	
μμ	86.7	0.55±0.28	0.53±0.11	1	
ll subtotal		1.57	3.14		4.6±3.1
e+jets/topol	105.9	3.81±1.41	6.46±1.38	10	
μ+jets/topol	95.7	5.42±2.05	6.40±1.51	11	
topol subtotal		9.23	12.86		3.9±1.9
e+jets/μ	90.5	1.45±0.42	2.43±0.42	5	
μ+jets/μ	95.7	1.13±0.23	2.78±0.92	6	
/μ subtotal		2.58	5.21		6.8±3.2
Total	98.0	13.4±3.0	21.2±3.8*	37	4.7±1.6

*~24 for top 175

CDF channel	pb^{-1}	Background	Expected top 175	Data	σ (top 175) (pb)
eμ	109	0.76±0.21		7	
ee+μμ	109	1.23±0.36		3	
ll subtotal			4.2		$9.3^{+4.4}_{-3.4}$
l+jets/SVX	110	7.96±1.37	19	34	$6.8^{+2.3}_{-1.8}$
l+jets/SLT	110	24.3±3.5	10	40	$8.0^{+4.4}_{-3.6}$
ll+*l* subtotal	110	34.3	33	84	$7.5^{+1.9}_{-1.6}$
all jets/SVX	109.4	137.1±11.3	25	192	$10.7^{+7.6}_{-4.0}$

Top cross sections by channel

The D0 and CDF total (leptonic) cross sections are mutually consistent, as are their component cross sections.

Therefore there is no strong motivation for analyzing the origin of possible variations.

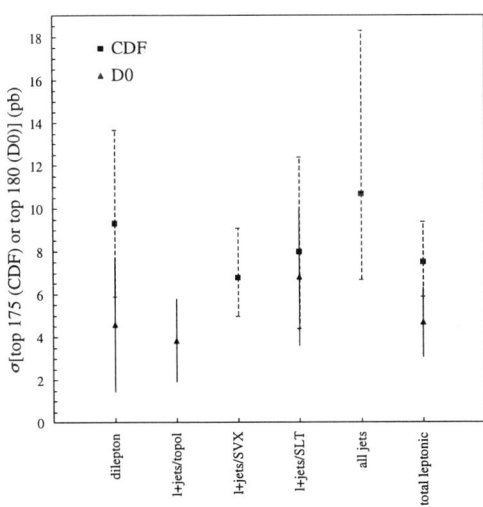

Figure 9. DØ and CDF top cross sections by channel, (a) tabulated and (b) plotted.

Top cross sections by channel (cont'd)

Caveats aside, one may sort the channel cross sections by the no. of W's (isolated leptons), no. of b's (SVX or soft lepton tags), or their sum per event.

The CDF dilepton point would plummet if one were to superimpose the D0 dilepton cuts. Then one might imagine seeing a positive correlation of apparent top cross section with the number of b's.

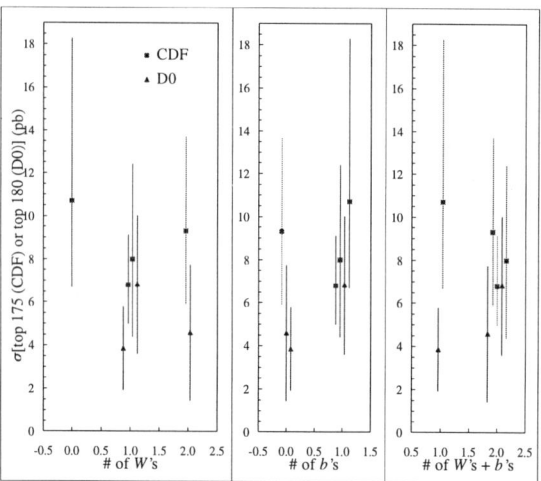

Supersymmetric signatures in top final states?

D0 and OPAL already have excluded a significant region of the N_1 (lightest neutralino) vs. stop mass plane.

The D0 signature is two acollinear jets with missing E_T from pair production of stop. Each stop is assumed always to decay into cN_1, and N_1 is assumed to be the LSP.

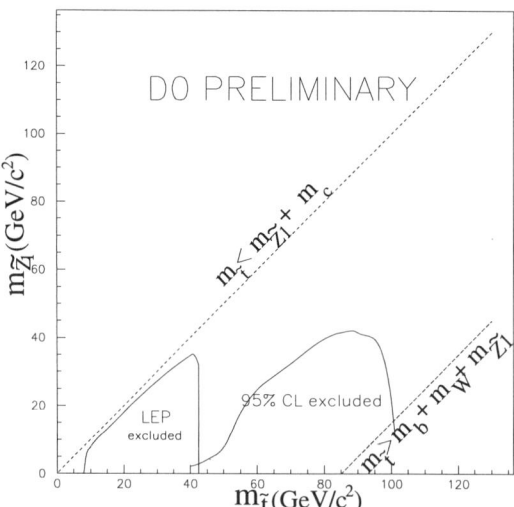

Figure 10. (a) DØ and CDF top cross sections vs. no. of isolated leptons ("W's"), b tags ("b's"), and W's + b's. (b) Regions in the lightest neutralino vs. stop mass plane, which are excluded by OPAL[13] and by DØ,[14] the latter assuming that stop always decays into a c quark plus the lightest supersymmetric particle.

Supersymmetric signatures in top final states (cont'd)

Kane and Mrenna argue that top often could decay to *stop*, and often could arise from decay of gluinos, squarks, and gauginos, without violating present bounds.

The main effect of the decay

$$\text{top} \to \text{stop} + N,$$

where N is a neutralino, would be to remove detectable top final states from the usual candidate samples, except perhaps for states with unusually large missing E_T.

To compensate, one needs extra sources of top. In order for top pairs often to be produced by

$$\text{gluino} \to \text{top} + \text{stop},$$

stop would need to be the lightest *squark*. Gauginos and other *squarks* could contribute directly and through gluino cascades to this final state.

With R-parity assumed conserved, a general feature of top pair production through supersymmetric decay is the appearance of two invisible "lightest supersymmetric particles", or LSP's. (Two additional LSP's are produced for each top that decays through *stop*, but this tends to make the event harder to detect.) These LSP's would smear out missing E_T and thus the reconstructed W transverse mass and the p_T of the top pair.

Supersymmetric signatures in top final states (cont'd)

Unfortunately, the CDF $W+\geq 3$ jet SVX tagged data appear not to be smeared out in W transverse mass. If anything, the points are more sharply peaked than the expectation from top plus background.

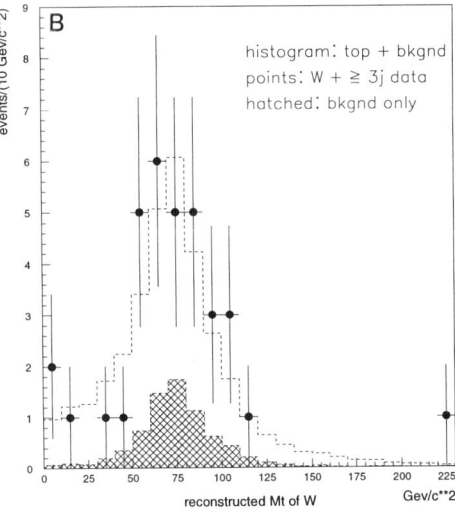

Figure 11. (a) Possible top decay and production mechanisms[9] involving supersymmetric particles. (For a discussion of supersymmetric processes which could contribute to present top-to-dilepton samples, see Barnett and Hall.[15]) (b) Reconstructed transverse mass of (e or μ plus missing E_T) in CDF lepton+jets events.

Supersymmetric signatures in top final states (cont'd)

Neither does the invariant mass m_{tt} of the top pair, nor the average transverse momentum $p_T(t)$ of the top quarks, show a significant deviation from expectation.

The DØ data below are those used for their 3/96 l+jets top mass analysis. The histograms are the data, and the points are the calculated sum of top and background.

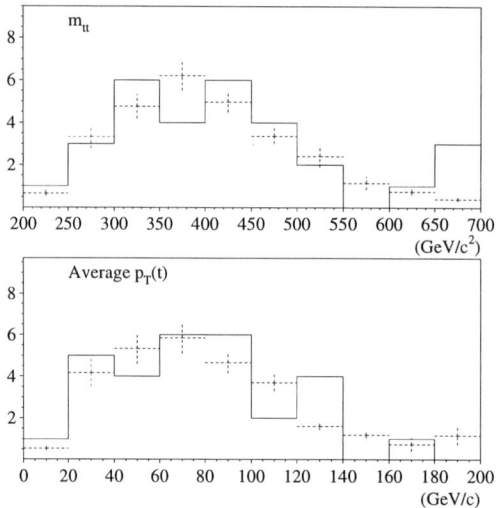

Supersymmetric signatures in top final states (cont'd)

Finally, p_T of the top pair is plotted by CDF. To the eye, the data appear to be more broadly distributed than expected.

However, due to an artifact of PAW, the bins containing no data (40-45, 45-50, 55-60 etc.) have no plotted point.

In the mean (data 24±~4.1, MC 21.2) and standard deviation (data 13±~3.2, MC 14.6), the distributions are consistent.

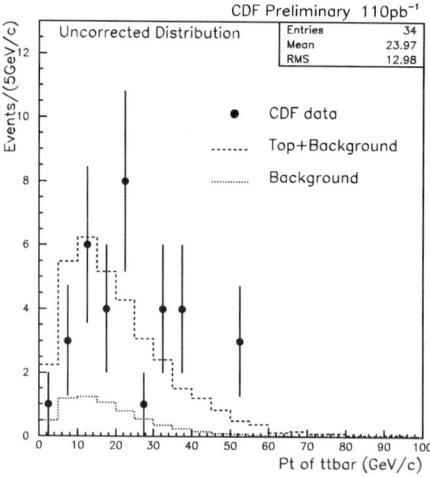

Figure 12. (a) Top quark pair invariant mass and average top quark p_T for DØ lepton+jets events compared to the calculated sum of top and background. (b) Transverse momentum of the pair of top quarks in CDF lepton+jets events.

Conclusions -- top production and decay

In the discovery channels, present observations are clearly inconsistent with background. They conform broadly to expectations from top.

Top analyses in the more difficult channels are progressing.

Measured top cross sections are in adequate mutual agreement across channels and across experiments.

Where they differ from calculated values, measured top cross sections tend to be larger. Where experiments differ, CDF finds larger cross sections than D0. These differences are generally a factor of two or less, and at present they are not significant.

The distributions shown reveal no evidence for supersymmetry in top production or decay.

Figure 13. Conclusions for Lecture I.

II. Top mass in the dilepton channel

Motivation
Reconstructing the top mass
Weighting the solutions
The following is for D0:
Approximations to the general weight
Top mass likelihoods for events
Mass resolution functions
Likelihood fitting
Ensemble tests
Comparison to CDF dilepton top mass analysis:
Event selection
CDF method and result
Systematic errors

Relevance of the dilepton channel to top mass measurement

Dilepton events have lower statistics and fewer constraints, leading to a statistical error 2-3× larger than for lepton + 4 jets events. Why bother with thorough top mass analysis in this channel?

Part of the reason is historical. Event 58796-417, collected in late 1992 and still one of its three $e\mu$ events, was the first D0 top candidate to have a low background probability. And in mid 1995, with large errors, D0 reported central top masses of 199 (145) GeV/c^2 for lepton+jets (dilepton) events.

Other reasons are of current interest.

Dilepton events have low background.

The dilepton top mass can be less sensitive to the jet energy scale than is the lepton+jets top mass.

If physics beyond the minimal Standard Model is present in top production or decay, or in non top events which partly mimic top, the apparent top mass for dileptons might be different than for lepton+jets events. Two isolated leptons rather than one are present. No b tags are required, while most lepton+jets analyses demand at least one.

Figure 14. (a) Outline of Lecture II. (b) Motivation for top mass analysis in the dilepton channel.

Reconstructing the top mass in the dilepton final state

Given m_t and neglecting jet masses, the process

$$p + \bar{p} \to t + \bar{t} + X$$
$$t \to W^+ + b_{\text{jet1}}$$
$$\bar{t} \to W^- + \bar{b}_{\text{jet2}}$$
$$W^+ \to e^+ + \nu_e$$
$$W^- \to \mu^- + \bar{\nu}_\mu$$

is 0C with 14 observables, e.g.

$$\{o_1 \ldots o_{14}\} \equiv \{\mathbf{p}(e, \mu, j1, j2), \mathbf{p}_\perp(t\bar{t})\}.$$

The 0C fit solves a quartic equation equivalent to the geometrical construction of Dalitz, Goldstein and Kondo, usually yielding 2 or 4 solutions over a wide range of m_t. One narrows this range by weighting the solutions.

Example of solution to quartic equation

For illustration I use figures prepared in 1993 for D0's event 58796-417. At that time, D0's jet and μ reconstruction put the most probable top mass for that event at ~145 rather than ~160 GeV, and yielded 3 rather than 2 jets.

The solutions to the quartic equation are the intersections of two ellipses in a space measured by differences of x vs. y neutrino p_T's. Below "nu" refers to $\nu_e + \nu_\mu$.

Shown below are the ellipses for m_t =146 GeV. The variable stepped is cos(θ$_\nu$); "nu" is the sum of ν_e and ν_μ. The density of solutions varies greatly.

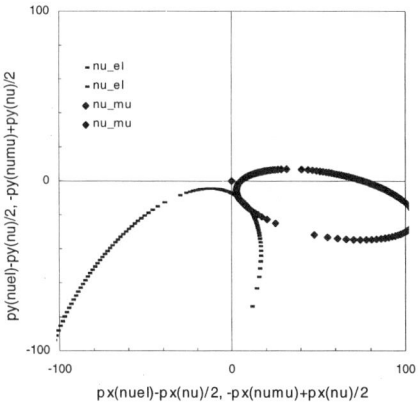

Figure 15. (a) Definition of the top-to-dilepton decay chain and description of the 0C fit. (b) $(p_y(\nu_e) - \frac{1}{2}p_y(\nu_e + \nu_\mu))$ vs. $(p_x(\nu_e) - \frac{1}{2}p_x(\nu_e + \nu_\mu))$ or $(p_y(\nu_\mu) - \frac{1}{2}p_y(\nu_e + \nu_\mu))$ vs. $(p_x(\nu_\mu) - \frac{1}{2}p_x(\nu_e + \nu_\mu))$ (GeV/c) for 1993 analysis of D0 event 58976-417 with $m_t = 146$ GeV/c^2. The intersections of the ellipses are the solutions.

Weighting the top mass solutions

In the high resolution limit, the general weight is

$$w(m_t) = \frac{d^{14}\sigma(t\bar{t})}{\sigma_{\text{vis}}(t\bar{t})\, do_1 \cdots do_{14}},$$

where "σ" reflects the decay as well as the production matrix elements. For any event with a fixed jet assignment, w is invariant to the choice of observables $\{o_i\}$. The visible cross section divisor $\sigma_{\text{vis}}(t\bar{t})$ removes the *a priori* bias favoring low m_t.

Since $d\sigma(t\bar{t})/d\text{LIPS}$ rather than $d\sigma(t\bar{t})/d\{o_i\}$ is calculated, w implicitly contains the Jacobian factor

$$J = \left|\frac{\partial \text{LIPS}}{\partial\{o_i\}}\right|.$$

J favors solutions for which a large LIPS volume maps into a small observable phase space volume. Typically, for a given set of observables, J varies widely with m_t.

Components of the general top mass solution weight

The general weight may be factored into three parts:

1. $d\sigma^{\text{prod}}/\sigma^{\text{prod}}_{\text{vis}}\, d\,\text{LIPS}$ [σ^{prod}=production cross section only]
2. $dP(E_e*|m_t)/dE_e*$ [*=top CM, one factor for each lepton]
3. density of states = $|\partial\,\text{LIPS}/\partial\{\text{observed variables}\}|$

Below is shown factor 1. Using the 2 leading jets as b's yields the 2 solutions shown in the high mass contour, converging at ~145 GeV. One solution's weight falls faster with top mass than the other. (The low mass contour uses the now defunct third jet as one of the b's.)

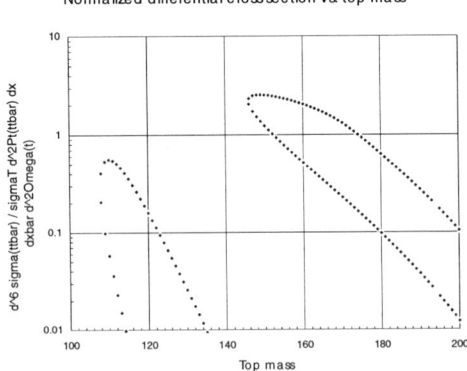

Figure 16. (a) Definition of the general weight for top mass solutions. (b) $d^6\sigma^{\text{prod}}(t\bar{t}) / (\sigma^{\text{prod}}_{\text{vis}}\, d^2\mathbf{p}_T(t\bar{t})\, dx\, d\bar{x}\, d^2\Omega^*(t))$ vs. m_t for 1993 analysis of D0 event 58976-417 (* refers to the $t\bar{t}$ CM). This is the first factor in the general weight.

Components of the general weight (cont'd)

Below is shown the product of the two second factors

$dP(E_e^*|m_t) / dE_e^*$ [*=top CM] .

The variation with top mass is only slight.

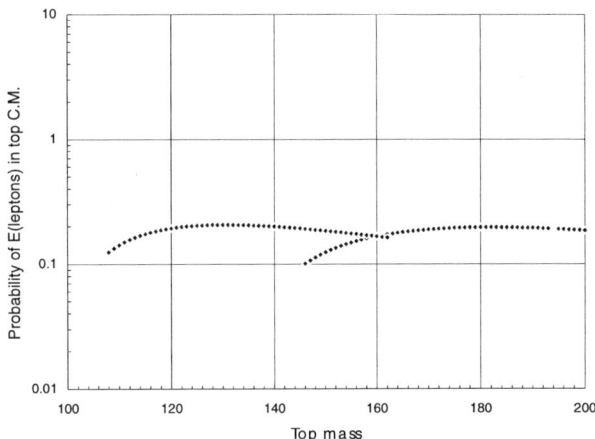

Probability of E(leptons) in top C.M. vs. top mass

Components of the general weight (cont'd)

The third factor

density of states = $|\partial \text{ LIPS}/\partial \{\text{observed variables}\}|$

has the strongest variation with top mass.

The relative normalization between the solutions involving the third jet (low mass contours) and only the two leading jets (high mass contours) depends on the choice of observable variables. But for a fixed set of jets the density of states is independent of this choice.

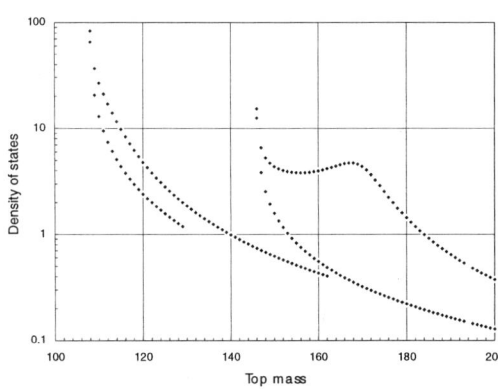

Density of states vs. top mass

Figure 17. (a) $(dP(E_e^*|m_t)/dE_e^*) \cdot (dP(E_\mu^*|m_t)/dE_\mu^*)$ (arb. units) vs. m_t, where * refers to the t or $\bar{t}$ CM, for 1993 analysis of D0 event 58976-417. This is the second factor in the general weight.
(b) $|dx\, d\bar{x}\, d^2\Omega^*(t)\, d^2\Omega^*(b)\, d^2\Omega^*(\bar{b})\, d^2\Omega^*(e)\, d^2\Omega^*(\mu) / (d^3\mathbf{p}_b\, d^3\mathbf{p}_{\bar{b}}\, d^3\mathbf{p}_e\, d^3\mathbf{p}_\mu)|$ (arb. units) vs. m_t, where * refers to the rest frame of the indicated particle's parent and $d^3\mathbf{p}_i \equiv d^2\mathbf{p}_{Ti} d\eta_i$, for 1993 analysis of D0 event 58976-417. This is the third factor in the general weight.

Approximations to the general weight for top solutions

For high-statistics Monte Carlo tests, w is cumbersome to compute. Simplifying it does not grossly degrade the m_t resolution, which is dominated by other effects such as measurement error, jet combinatorics, and gluon radiation.

DØ uses two methods with independent simplified weights. Data and Monte Carlo are always treated symmetrically, so both methods are unbiased. They turn out to be comparably efficient as well.

Following Dalitz and Goldstein, Method 1's weight drops the Jacobian and substitutes the PDF product $q\bar{q}$ for the production cross section. Method 1's weight is

$$w_1(m_t) = A(m_t)\, q(x)\, \bar{q}(\bar{x})\, P(E_e^{\text{CM of }t})\, P(E_\mu^{\text{CM of }\bar{t}}),$$

where the P's are decay probability densities. D-G's technique is extended by introducing $A(m_t)$, a function chosen empirically to cancel the m_t bias, and by fitting the measured rather than an assumed value of $\mathbf{p}_\perp(t\bar{t})$.

Method 2's approximation to the general weight

Method 2's weight is the complement of Method 1's. Here the production and decay factors are dropped in favor of a "neutrino phase space" approximation to the Jacobian.

Fixing the measured jet and lepton momenta, the two ν's are stepped through their expected distributions in η. Each pair $\eta_m \eta_n$ yields a 0C fit, which is given a weight w_{mn} based on compatibility between fit and measured missing E_T (which is not used by the fit). The weight $w_2(m_t)$ is the sum of the w_{mn}'s.

One finds that the product $w_2 w_1$ is not significantly more efficient than w_2, so one chooses to keep the two weights complementary.

Figure 18. DØ's approximations to the general weight, (a) for Method 1, (b) for Method 2.

Method 2 average weight curve

As a check, one can apply Method 2 to a sample of perfect-resolution parton-level Monte Carlo events, considering only the correct pairing of leptons to jets.

Below are the average weight curves for 10 000 such events at two widely separated top masses. Clearly the peaks follow the generated masses.

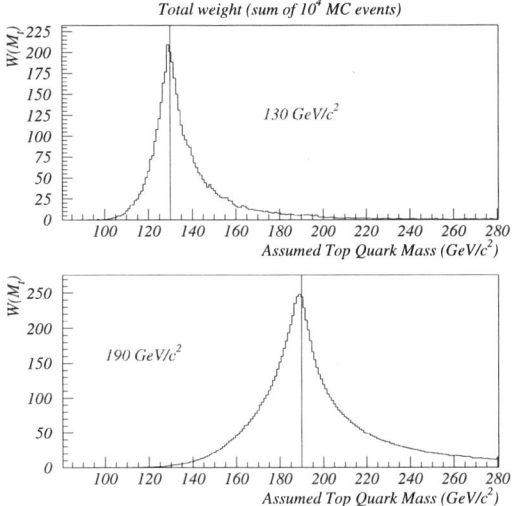

Common aspects of Methods 1 and 2

Weights aside, Methods 1 and 2 share the following features:

- Because the experiment is not in the high resolution limit, the measured parameters of each event are smeared many times within the experimental resolution. The weights from from each smear are summed to get $w(m_t)$ for that event.

- If a third jet is present, DØ (A) assumes that it is an initial state gluon, or (B) assigns a probability to the hypothesis that it is a final state gluon which contributes to m_t. Ensemble studies show that, for events in which m_t^A is quite different from m_t^B, it helps to average the two, which DØ does.

- DØ then sums the weights for the 2 different jet assignments and the 0, 2, or 4 different solutions per assignment.

- The m_t which maximizes the summed weight is taken to be the single reconstructed top mass for each event.

Figure 19. (a) Method 2 weight *vs.* m_t, average of curves for 10 000 perfect-resolution parton-level MC events, with each event's curve normalized to the same area, for generated top masses of 130 and 190 GeV/c^2. (b) Common features of Method 1 and Method 2 analyses.

Method 1 event likelihoods

Shown are Method 1's top mass likelihoods for D0's five dilepton candidates.

Contours labeled "2j" ("3j") use the just described method A (B) for assigning the third jet. "2j" assumes that the third ranking jet is ISR, "3j" calculates weight curves for all six possible ISR/FSR assignments and sums them.

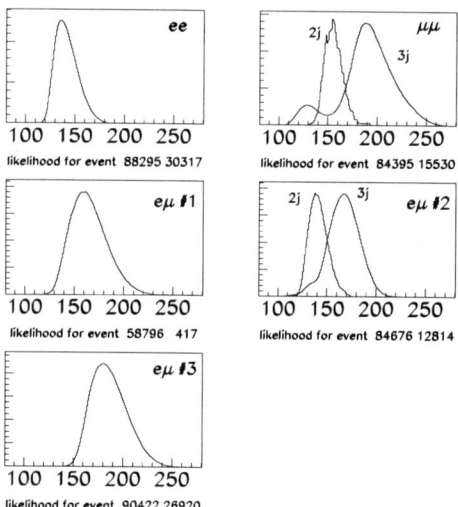

Method 2 event likelihoods

The event likelihoods for Method 2 are remarkably similar, considering that its solutions are weighted completely differently from those of Method 1.

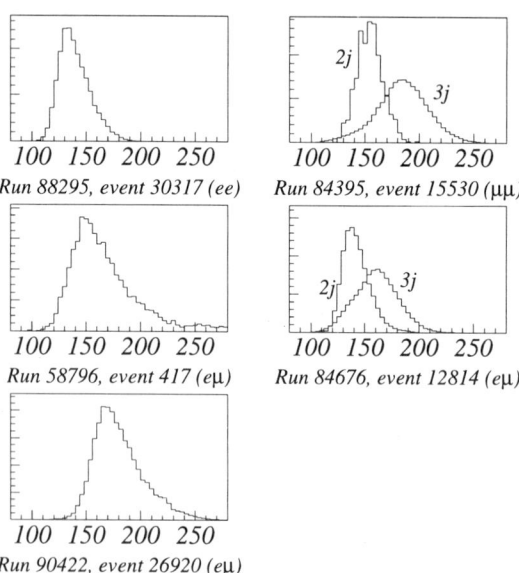

Figure 20. (a) Method 1 weight vs. m_t for five top-to-dilepton candidate events. (b) Same for Method 2.

Common aspects of DØ's dilepton and lepton+jets top mass analyses

Top mass reconstruction methods aside, DØ's dilepton and lepton+jets top mass analyses share the following procedures:

- Templates in m_t are constructed for various top masses and backgrounds.

- Maximum likelihood fits are made to m_t and the expected number $\langle n_s \rangle$ of signal and $\langle n_b \rangle$ of background events in the sample. The dilepton analyses use the external constraint on $\langle n_b \rangle$ (with errors) provided by the counting experiment.

- Quoted errors are based not on the likelihood curves, but rather on studies of ensembles of Monte Carlo experiments in which n_s and n_b as well as the event kinematics are allowed to fluctuate. This tends to increase the errors.

Method 1 mass resolution functions for $e\mu$ channel

Method 1's reconstructed top mass templates for background and various generated top masses are shown.

The peak follows the generated mass except at 130 GeV/c^2, where the H_T cut biases the apparent top mass of the events which pass it.

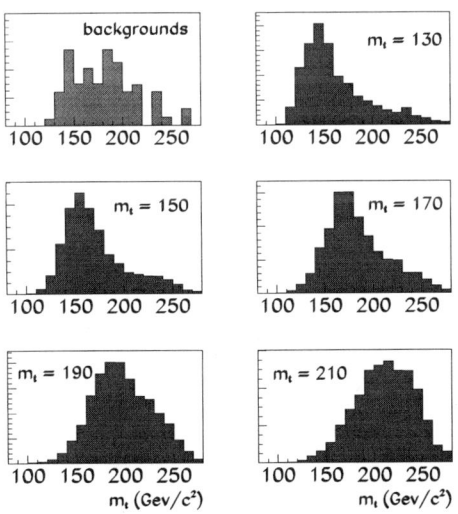

Figure 21. (a) Common features of DØ dilepton and lepton+jets top mass analyses. (b) Method 1 mass resolution functions for $e\mu$ channel background and five top masses.

Method 2 mass resolution functions for $e\mu$ channel

Method 2 takes an average within a 16 GeV/c^2 peak window on the distribution of weight vs. m_t to define the best m_t.

The reconstructed top mass templates below are similar to those of Method 1. The tails on the high side are less pronounced, but the peak does not move quite as rapidly with the generated mass.

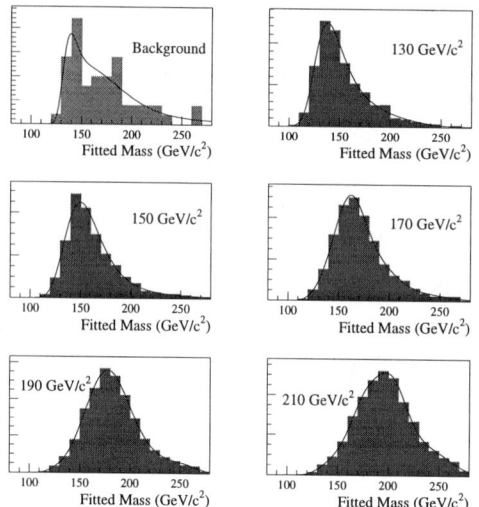

Method 1 likelihood fitting and ensemble test

At top masses where Monte Carlo exists, Method 1 evaluates $\ln L$. To the five points nearest the maximum a parabola is fit, whose peak is taken to be the central value.

The statistical error is based on ensemble tests exemplified below. The error interval is defined as the smallest interval which includes 68% of the ensembles.

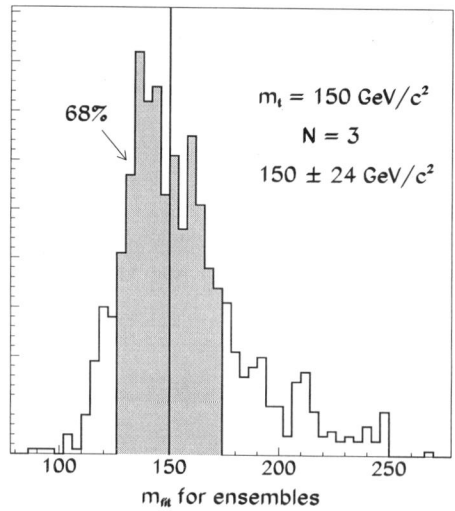

Figure 22. (a) Method 2 mass resolution functions for $e\mu$ channel background and five top masses. (b) Method 1 maximum likelihood top mass for an ensemble of MC experiments each consisting of 3-$e\mu$-event mixtures of 150 GeV/c^2 top and background.

Method 1 result (3 $e\mu$ events)

For the $e\mu$ channel, with 3 events *vs.* a background of 0.36 ± 0.09, the Method 1 result is $m_t = 158 \pm 24(\text{stat}) \pm 10(\text{syst})$ GeV/c^2. The statistical error is from the ensemble tests just described, and the systematic error remains to be discussed.

Results are available either from the $e\mu$ or from the $ee+e\mu+\mu\mu$ channels, and from either Method 1 or 2. D0 quotes the above dilepton top mass because of the smaller backgrounds and because it has used Method 1 longer.

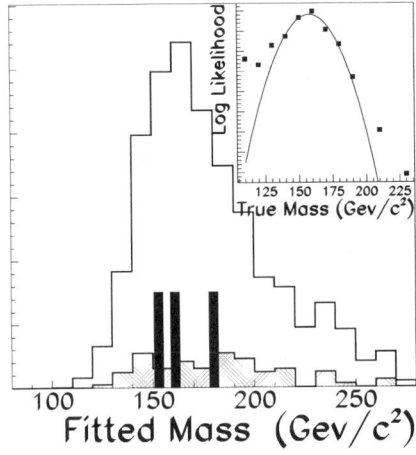

Method 1 result (5 $ee+e\mu+\mu\mu$ events)

For all three dilepton channels, with 5 events *vs.* a background of 1.6 ± 0.4, the Method 1 result is

$$m_t = 151 \pm 21(\text{stat}) \pm 10(\text{syst}) \text{ GeV}/c^2.$$

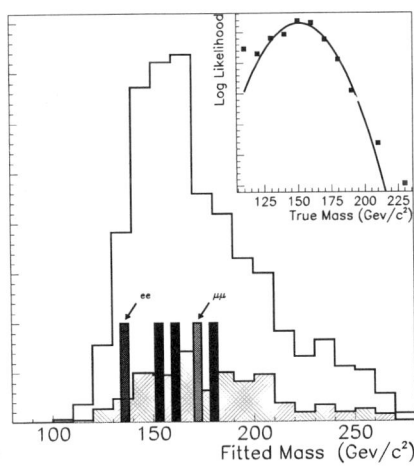

Figure 23. (a) Method 1 fitted mass for 3 $e\mu$ candidates. This is the basis for DØ's quoted result for its top mass analysis in the dilepton channel. (b) Method 1 fitted mass for 5 ee, $e\mu$, and $\mu\mu$ candidates.

Method 2 result (3 $e\mu$ events)

In method 2, the reconstructed templates are parametrized continuously *vs.* generated top mass and an unbinned maximum likelihood fit is performed. The result is

$$m_t = 157 \pm 23(\text{stat}) \pm 9(\text{syst}) \text{ GeV}/c^2.$$

Again the statistical error is taken from the results of ensemble tests (yet to be discussed for Method 2). (I note that the likelihood-curve statistical error is ±15 GeV/c^2.)

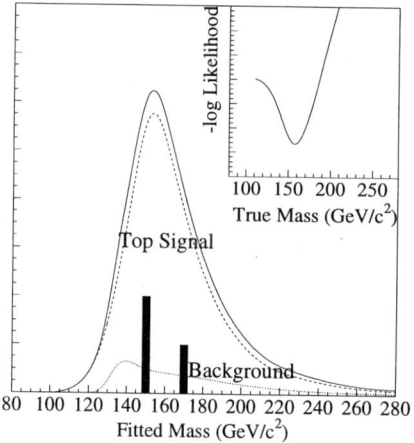

Method 2 result (5 $ee+e\mu+\mu\mu$ events)

When the five events in all three dilepton channels are analyzed, the Method 2 result is

$$m_t = 157^{+20}_{-32}(\text{stat}) \pm 9(\text{syst}) \text{ GeV}/c^2.$$

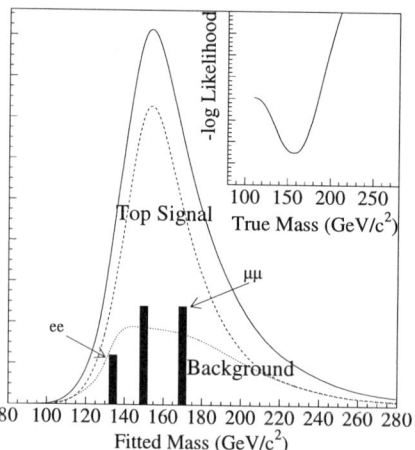

Figure 24. (a) DØ Method 2 fitted mass for 3 $e\mu$ candidates. (b) Method 2 fitted mass for 5 ee, $e\mu$, and $\mu\mu$ candidates.

Method 2 ensemble tests

The Method 2 statistical errors are based on ensemble tests exemplified by those below. Again, the error interval is the smallest range which includes 68% of the ensembles.

The median maximum likelihood mass tracks the generated mass except at 230 GeV/c^2, near the end (280 GeV/c^2) of the range in m_t within which the weights are computed.

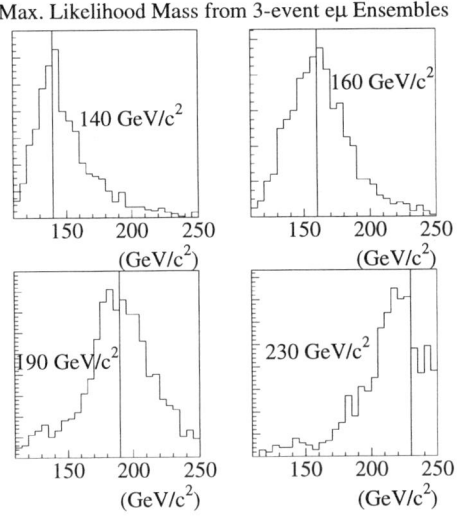

Max. Likelihood Mass from 3-event $e\mu$ Ensembles

Comparison to CDF dilepton top mass analysis

D0's cuts for its dilepton top mass analysis are the same as previously described for its cross section analysis.

CDF's cuts also are unchanged, except for the extra cut $H_T > 160$ GeV. This reduces the sample from 10 to 8 events and the expected background from 2.0 to 1.1 ± 0.3.

Summarized are the differences between D0 and CDF cuts:

Dilepton top mass analysis parameter	D0	CDF
no. of leptons	2	2
no. of isolated leptons	2	1 or 2
lepton charges	--	opposite
E_T of lepton (GeV)	>15 or 20	>20
$|\eta|$ of electron	<2.5	<1.0
$|\eta|$ of muon	<1.7 or 1.0	<1.1
missing E_T (GeV)	>20 or 25	>25
E_T of jets (GeV)	>20	>10 (raw)
$|\eta|$ of jet	<2.5	<2.0
H_T of jets + (e in $e\mu$, e_1 in ee, 20 in $\mu\mu$)	>120	--
H_T of jets + leptons + missing E_T (GeV)	--	>160
no. of candidates	5	8
expected no. of top events (top mass)	3.14 (180)	<4.2 (175)
calculated background	1.6 ± 0.4	1.1 ± 0.3
no. of $e\mu$ candidates	3	7
calculated $e\mu$ background	0.36 ± 0.09	0.76 ± 0.21

Figure 25. (a) Method 2 maximum likelihood top mass for an ensemble of MC experiments each consisting of 3-$e\mu$-event mixtures of 140, 160, 190, and 230 GeV/c^2 top and background. (b) Table comparing DØ and CDF event selection criteria for dilepton top mass analysis.

CDF top mass result for dilepton channel

Given a perfect MC model, any parameter which is correlated with m_t can be used to measure m_t with some efficiency.

CDF plots the total energy of each of an event's two highest E_T jets for data and a mixture of MC top and background.

The result of a binned maximum likelihood fit is $m_t = 159^{+24}_{-22}(\text{stat}) \pm 17(\text{syst})$ GeV/c^2, where the statistical error is taken from the likelihood curve.

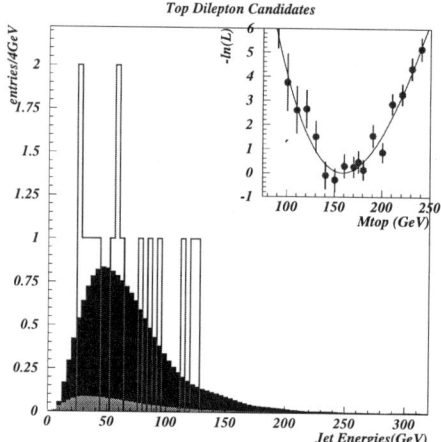

Comparison of CDF and D0 systematic errors on top mass in dilepton channel

The D0 systematic errors due to jet scale are smaller because both of its methods for reconstructing the top mass are not strongly correlated with the jet energies.

D0's smaller systematic errors due to background shape reflect the low background in their $e\mu$ channel.

	D0 Method 1 (GeV)	D0 Method 2 (GeV)	CDF (GeV)
central top mass	158	157	159
statistical error	±24	±23	+24 −22
Systematic error source:			
jet energy scale	5	5	<16
top event generator	5	5	3
structure functions	--	--	3
MC statistics	5	5	--
background shape	4	3	8
systematic error	10	9	17
total error	26	25	~29

Figure 26. (a) Jet energy spectrum for CDF dilepton candidates, 160 GeV/c^2 MC top events, and background events, yielding CDF result for top mass analysis in the dilepton channel. (b) Table comparing DØ and CDF systematic errors for dilepton top mass analysis.

Conclusions -- top mass in the dilepton channel

In the dilepton channel, the top mass is measured to be

$$(\text{D0}) \ m_t = 158 \pm 24(\text{stat}) \pm 10(\text{syst}) \ \text{GeV}/c^2,$$

$$(\text{CDF}) \ m_t = 158^{+24}_{-22}(\text{stat}) \pm 17(\text{syst}) \ \text{GeV}/c^2.$$

Though the statistical errors are large, the results from three different methods and two different data sets are in close agreement.

Figure 27. Conclusions for Lecture II.

LECTURE III. TOP MASS IN THE LEPTON + JETS CHANNEL

Events with one isolated lepton are the principal source of information on the top quark mass. I begin with a brief description of the 2C fit to m_t which is possible in this channel, followed by a short discussion of the difficulties in choosing both the jets to fit and the quarks to which they should be assigned. After summarizing each experiment's event selection criteria, I report the CDF analyses in this channel at the full level of detail allowed by the material available to me, including results from samples with both one and two b tags. I conclude the CDF based discussion with a brief account of their top mass analysis in the all jets channel.

After summarizing DØ's March 1996 top quark mass result in the lepton+jets channel, I describe a recent advance in the technique for separating top signal from background. This advance promises to improve substantially the precision of DØ top mass analysis in this final state.

The CDF material for this lecture was drawn from a May 96 talk by Rolli[20]. The DØ transparencies are from Ref. [19].

III. Top mass in the lepton+jets channel

Top mass reconstruction and combinatorics
Event selection
The following is for CDF:
Backgrounds
Top mass result and statistical errors
Jet energy scale and other systematic errors
Lepton + 4 jet sample with 2 b tags
W mass
Top mass and errors
SVX tagged all jets top mass
The following is for DØ:
March 1996 lepton + 4 jet top mass
Top likelihood discriminant cut

Reconstructing the top mass in the lepton + 4 jet final state

With one neutrino unmeasured rather than two, fits to lepton+jets final states are 2C (with m_t fit) rather than 0C (with m_t assumed) in the dilepton case.

The two constraints are

$$m(b_{\text{jet1}} W(\to l\nu)) = m(b_{\text{jet2}} W(\to \text{jet3 jet4}))$$
$$m(W(\to \text{jet3 jet4})) = m^W_{\text{pole}} .$$

One does not measure $p_z(\nu)$; it is determined by the requirement

$$m(W(\to l\nu)) = m^W_{\text{pole}} .$$

CDF's and 3 of DØ's fitting routines minimize the χ^2 constructed from the inverse measurement-error matrix; a "fast fitter" also used by DØ minimizes a χ^2 based on the above constraint equations.

Figure 28. (a) Outline of Lecture III. (b) 2C fit to the lepton+jets final state.

Combinatorics in the lepton + 4 jet final state

There are 12 possible jet assignments (6 for b tagged events). Usually the fitted m_t varies strongly with reassignment of b_{jet1} (4 or 2 permutations), and less strongly with reassignment of b_{jet2} when b_{jet1} is fixed (3 permutations or 1). Also, for a fixed jet assignment, often there are local χ^2 minima for each of two solutions for $p_z(\nu)$.

Minimizing χ^2 does yield the best fit to a fixed permutation, but for typical measuring errors the lowest χ^2 permutation is not usually correct. Also, initial and final state gluon radiation frequently cause the four highest E_T jets not to correspond to the four quarks to which one wishes to fit.

These combinatoric uncertainties make necessary, as in the dilepton case, a statistical determination of m_t in which the relation between true and measured top mass varies in a complex way from event to event.

Top mass analysis in the lepton + 4 jet channel

In the lepton+jets channels, DØ applies the same basic cuts used in its cross section analysis. A 4th jet is required in the tagged channels (4 jets are always required in the untagged channels). The 4th jet fully satisfies normal cuts, including $|\eta| < 2$ and $E_T > 15$ GeV.

In the untagged channels, for the top mass analysis first reported in March 1996, DØ applied an early version of its top likelihood cut in lieu of the cut on A and H_T made in its cross section analysis. This cut will be discussed later.

For lepton+jets top mass analysis, CDF also requires a 4th jet, but looser cuts are imposed than on the jets used for cross section analysis. Jet 4's $|\eta|$ may extend to 2.4 and its raw E_T may fall to 8 GeV. As in the cross section analysis, a b tag (SVX or SLT) is required.

Both experiments impose a χ^2 cut on the best solution to a full 2C fit using the four highest E_T jets. DØ requires $\chi^2<7$ and CDF requires $\chi^2<10$. When a jet is b tagged, it is required to be fit as one of the b jets.

For its lepton+jets top mass first reported in March 1996, DØ analyzed 30 events satisfying these requirements, *vs.* a calculated background of 17.4±2.2.

CDF bases its lepton + 4 jet top mass analysis on 34 events *vs.* a calculated background of $6.4^{+2.0}_{-1.5}$.

Figure 29. (a) Combinatoric and radiative uncertainties in the 2C fit. For a discussion of gluon radiation in top quark production and decay, see Orr, Stelzer, and Stirling.[21] (b) CDF and DØ event selection criteria for lepton+jets top mass analysis.

CDF backgrounds

Backgrounds in lepton + jets analysis:

The dominant source is W + multijets production

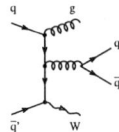

Rejection is obtained with b-tag

The background to the b-tagged events is calculated from the observed tag excess and measured b-tag efficiency.

Major sources:

- non W-events, WW, WZ, Drell-Yan, $Z \to \tau\tau$.
  Derived from combination of data and MC.

- mistagged events.
  Parametrized from generic jet data.

- $Wb\bar{b}$, $Wc\bar{c}$, Wc.
  Determined from MC and scaled to the observed number of W events in a given jet multiplicity.

CDF top mass for b tagged lepton + 4 jet events

With the background constrained to the calculated value within its error, CDF's fit yields

$$m_t = 175.6 \pm 5.7(\text{stat}) \pm 7.1(\text{syst})\ \text{GeV}/c^2 .$$

With this background constraint, the fit background is 6.3 ± 1.7 events.
The statistical error is taken from the likelihood curve (inset).

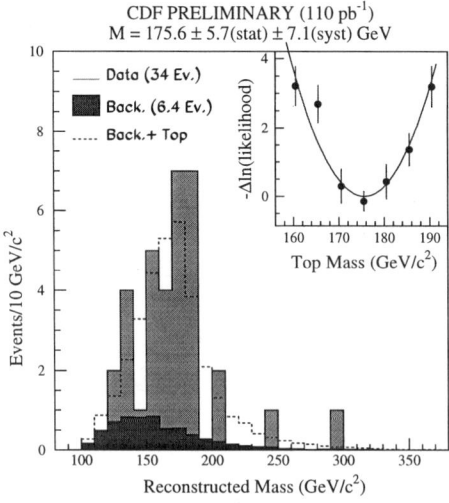

Figure 30. (a) CDF backgrounds (transparency from Rolli[20]). (b) Reconstructed top mass for data, 175 GeV/c^2 top MC, and background, yielding CDF result for top mass from lepton+jets final state.

Ensemble test of stat. error on CDF top mass for b tagged $1 + 4$ jets events

CDF repeats the fit for 1000 pseudo experiments consisting each of 34 events with mean background of 6.4 events. Below is the distribution of central masses and of statistical errors on the mass from the likelihood curves.

These distributions provide two other estimates (which CDF does not quote) of its top mass statistical error (nominally 5.7 GeV/c^2). One is the median value, 7.1 GeV/c^2, of the likelihood errors found in the 1000 pseudo experiments. The other is the interval, ±6.6 GeV/c^2, within which 68% of the 1000 central values fall.

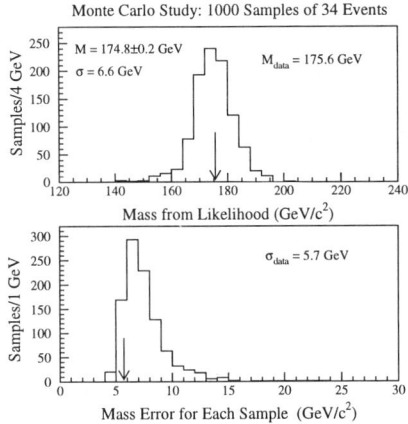

CDF systematic errors

Systematic uncertainties on top mass measurement in b-tagged W+≥ 4 jets

CDF Preliminary

Systematics	Value GeV	Value (%)
Jet E_T scale	3.1	1.8
Soft Gluon Effects	1.9	1.1
Different Generators	0.9	0.6
Hard Gluon Effects	3.6	2.1
Fit Configuration	2.5	1.4
b-tagging Bias	2.3	1.3
Background Spectrum	1.6	0.9
Likelihood method	2.0	1.1
Monte Carlo statistics	2.3	1.3
Total	7.1	4.0

Final result:

$$M_{top} = 175.6 \pm 5.7 \text{ (stat)} \pm 7.1 \text{ (syst) GeV}/c^2$$

Figure 31. (a) Maximum likelihood top mass, and error in top mass from likelihood curve, for 1000 MC samples of 34 CDF events consisting of a mixture of 175 GeV/c^2 MC top events and background, the latter having a mean of 6.4 events. (b) Systematic errors on CDF lepton+jets top mass (transparency from Rolli[20]).

CDF jet energy corrections

Jet Energy Corrections

- Mass fit requires going back from the jet to the parton energy.
- several sources of uncertainties in reconstructing the quark 4-momentum
- several sets of corrections on the raw jet P_T, derived using MC events.

$$P_T = P_T^{raw}(R) \times f_{rel} \times f_{abs}(R) - UE(R) + OC(R) \quad (1)$$

- f_{rel}: η response, relative to central region;
- $f_{abs}(R)$: absolute energy scale correction;
- UE(R) takes into account the underlying event;
- OC(R): jet energy outside the cone of radius R.

Corrections derived for a topology with single jets and then *corrected* to the top topology with a further set of corrections (AA corrections).

On each of these corrections we have a systematic uncertainty that is going to propagate to the fitted top mass.

Systematic error on jet energy scale due to CDF calorimeter effects

Plotted is the fractional systematic uncertainty in jet energy scale (solid line) as a function of true jet E_T attributed to the CDF calorimeter itself. The contributions are:

Uncertainty in detector uniformity (dot), arising from η dependence (±2%) and drifts [±1% (1A) or ±2% (1B)].

Underlying event uncertainty (dash) (30% of correction).

Uncertainty in p_T dependence of calorimeter response (dot dash), arising in part from fragmentation uncertainty.

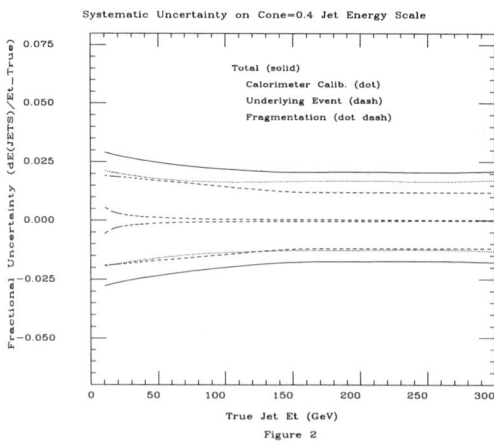

Figure 32. (a) CDF jet energy corrections (transparency from Rolli[20]). (b) Fractional systematic uncertainty in jet energy scale *vs.* true jet E_T ascribed to various effects in the CDF calorimeter.

Systematic error on CDF jet energy due to soft gluon radiation

Uncertainties in the out-of-cone jet energy are studied using the fraction F of $\Delta R<0.4$ jet energy within $0.4<\Delta R<1$.

Plotted vs. jet p_T is the fractional difference in F between HERWIG simulated and W+1 jet data. Z+1 jet and b pair data are also studied.

The jet energy scale error from this source ranges from 5.6% at 8 GeV to 1.4% at 150 GeV.

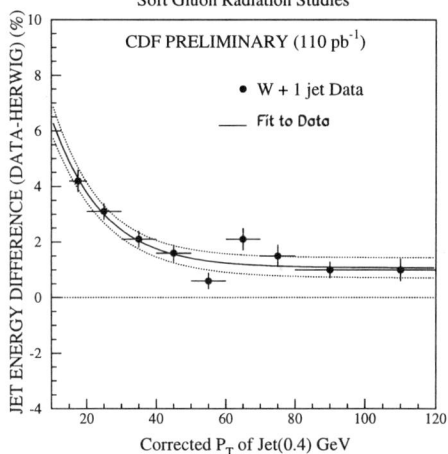

Systematic error on CDF top mass due to hard gluon radiation

For HERWIG top Monte Carlo at $m_t = 170$ GeV/c^2, among the four highest E_T jets 55% of the events have at least one "gluon jet" which is separated by $\Delta R>0.4$ from the nearest parton.

Since the top mass fit always assumes that the four highest E_T jets correspond to the four quarks in the lepton+jets final state, these "gluon jets" confuse the fit and shift the reconstructed top mass.

Varying the percentage of events with "gluon jets" among the four highest E_T jets from 25% to 85% causes the mean reconstructed top mass to vary by 1 GeV/c^2 and its error by 3.5 GeV/c^2.

The top mass systematic error from this source is taken to be the sum in quadrature of these two variations, or 3.6 GeV/c^2.

Figure 33. (a) % difference in F between $W + 1$ jet data and HERWIG, vs. jet p_T, where F is the fraction of $\Delta R < 0.4$ jet energy in the annulus $0.4 < \Delta R < 1$. (b) Systematic error on CDF top mass due to hard gluons.

CDF lepton + 4 jet sample with 2 *b* tags

In addition to the standard SVX and SLT tags, CDF also considers as loosely *b* tagged those jets which have less than 5% probability to be prompt according to SVX criteria. A subset of the lepton + 4 jet sample requires 2 *b* tags, one of which is usually loose.

Plotted for such events is the dijet invariant mass of pairs of jets which contain 0, 1, or 2 *b* tags. The untagged pairs cluster in the *W* mass region 60-100 GeV/c^2.

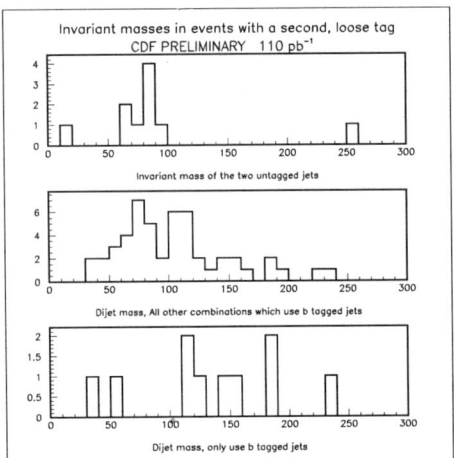

CDF lepton + 4 jet sample with 2 *b* tags (cont'd)

Below is compared the same untagged dijet mass spectrum with the expectation from top 175 + non-top background.

Within the 60-100 GeV/c^2 mass window the expected non-top background is only 0.4 ± 0.1 events.

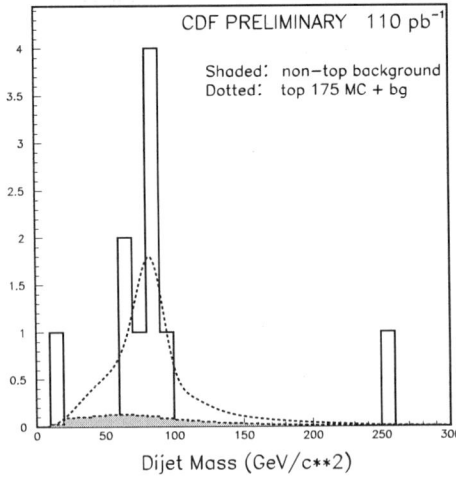

Figure 34. (a) Mass of CDF dijets in lepton + jets events with two *b* tags, for pairs of jets which have 0, 1, or 2 *b* tags. (b) Same for pairs of jets with 0 *b* tags, compared to 175 GeV/c^2 top MC and background.

CDF top mass for lepton + 4 jet events with 2 b tags

Applying the W mass window cut and including a ninth event with 3 b tags yields the top mass spectrum plotted below.

A fit similar to that for the singly tagged sample yields

$$m_t = 174.8 \pm 7.6(\text{stat}) \pm 5.6(\text{syst}) \text{ GeV}/c^2 \,,$$

where of course the two results are not statistically independent.

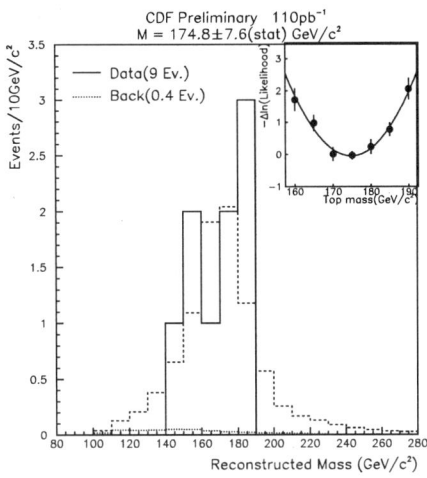

CDF systematic errors on double tagged top mass

benefit from smaller combinatoric and background errors:

Systematic uncertainties in top mass measurement on 2 b-tagged W+≥ 4 jet events

CDF Preliminary 110 pb^{-1}

Systematics	Value GeV	Value (%)
Jet E_T scale	2.9	1.7
Soft Gluon Effects	1.7	1.0
Different Generators	0.9	0.5
Hard Gluon Effects	3.6	2.1
Fit Configuration	0.9	0.5
b-tagging Bias	2.0	1.1
Background Spectrum	0.1	< 0.1
Likelihood method	0.6	0.3
Monte Carlo statistics	1.0	0.6
Total	5.6	3.2

Final result:

$$M_{top} = 174.8 \pm 7.6 \text{ (stat)} \pm 5.6 \text{ (syst) GeV}/c^2$$

Figure 35. (a) Reconstructed top mass for double b tagged data, 175 GeV/c^2 top MC, and background, yielding CDF result for top mass from double b tagged lepton+jets final state. (b) Systematic errors on CDF top mass from double b tagged lepton+jets events (transparency from Rolli[20]).

CDF top mass for SVX tagged all jets events

Requiring 6 jets otherwise satisfying the same cuts as in the cross section analysis, except that minimum H_T is reduced from 300 to 200 GeV, CDF observes 142 SVX tagged all jets events. The 109 ± 7 event background is calculated using a per-jet SVX tag rate parametrization.

Without an unmeasured neutrino, the final state allows a 5C rather than a 2C top mass fit as in the lepton + 4 jet final state. CDF makes a 3C fit, avoiding use of missing E_T.

Below is plotted the 3C fit mass compared to background + 175 GeV/c^2 top.

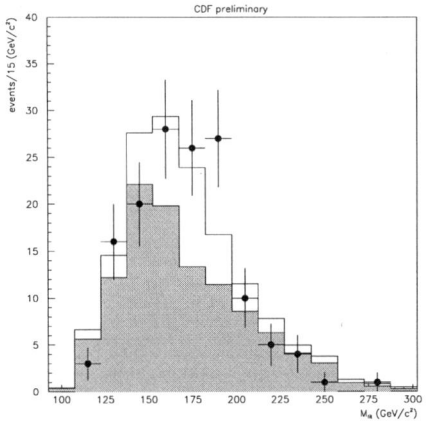

CDF top mass for SVX tagged all jets events (cont'd)

Substituting a top mass of 185 for 175 GeV/c^2 in the previous plot represents the data better by ~1σ.

Of the 142 events plotted, 28 have a second b tag.

The 6.5% background rate error is dominated by uncertainties in kinematics (5%), tag rate parametrization (3%), and run conditions (2.3%).

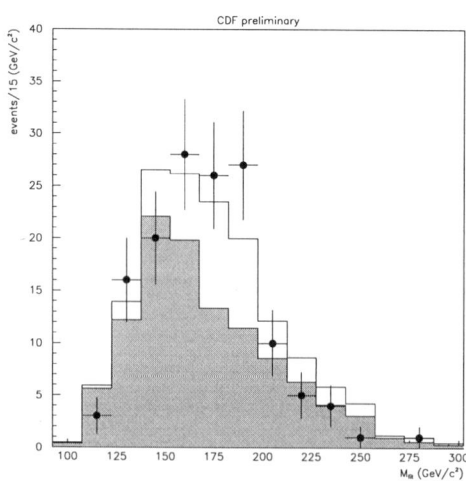

Figure 36. (a) 3C fit top mass for CDF all jets data and for 175 GeV/c^2 MC top + calculated background. (b) Same with 175 replaced by 185 GeV/c^2 MC top.

CDF top mass for SVX tagged all jets events (cont'd)

Plotted is the maximum likelihood fit with background constrained within errors to the calculated value.

The cubic fit shown yields $m_t = 187 \pm 8(\text{stat}) \pm 12(\text{syst})$ GeV/c^2.

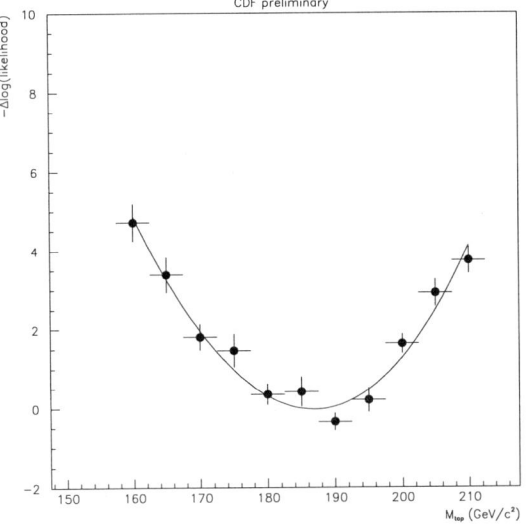

CDF systematic error for all jets top mass

Systematic uncertainties on top mass measurement from multijets sample

CDF Preliminary

Systematics	Value (%)
Jet E_T scale	+2.9 -2.1
Soft Gluon Effects	±1.6
Different Generators	+0.3 -1.0
Hard Gluon Effects	±4.3
Fit Configuration	±2.1
b-tagging Bias	< 0.1
Background Spectrum	-1.5
Likelihood method	+0.4 -2.4
Monte Carlo statistics	±0.3
Total	+6.6 -6.3

Figure 37. (a) $-\ln(\text{likelihood})$ vs. m_t for fit to CDF all jets data. (b) Systematic errors on top mass from CDF all jets events (transparency from Rolli[20]).

D0's March 1996 top mass for lepton + 4 jet events

D0's spectrum in 2C fit top mass plotted below is based on 30 candidates *vs.* a calculated background of 17.4 ± 2.2.

The curves are from a background constrained unbinned maximum likelihood fit which used a continuous parametrization of the top mass templates. Its result is $m_t = 170 \pm 15(\text{stat}) \pm 10(\text{syst})$ GeV/c^2, where the statistical error is based on ensemble tests.

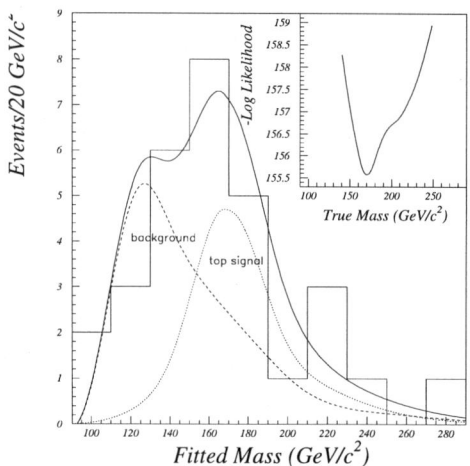

Ensemble test for D0's March 1996 lepton + 4 jet top mass

The just mentioned statistical error is based on 1000 Monte Carlo experiments of 30 events each in which the expected background is 17.4. Their best fit top masses are plotted.

The fit to a gaussian has an rms of 15 GeV/c^2, which is taken to be the statistical error. If instead it were assigned using the likelihood curve on the previous plot, the error would be smaller.

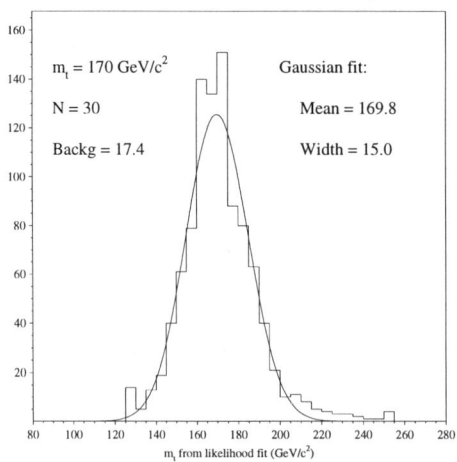

Figure 38. (a) Reconstructed top mass for data, 170 GeV/c^2 top MC, and background, yielding DØ's March 1996 result for top mass from lepton+jets final state. (b) Maximum likelihood top mass for 1000 MC samples of 30 DØ events consisting of a mixture of 170 GeV/c^2 MC top and background, the latter having a mean of 17.4 events.

D0 systematic errors in lepton + 4 jet top mass

Jet scale error
The jet energy scale uncertainty is estimated from Z+jets data to be ±(4% + 1 GeV). Ensemble tests are used to propagate this to a top mass error. The result is ±7 GeV/c^2.

Other errors
±6 GeV/c^2 -- variations among Monte Carlo generators and jet definitions.
±3 GeV/c^2 -- background shape.
±3 GeV/c^2 -- likelihood fitting method.
±1 GeV/c^2 -- Monte Carlo statistics.

Total systematic error: 10 GeV/c^2

Progress in methods for top mass analysis in the lepton + 4 jet channel

To suppress background with respect to top without unduly distorting either mass spectrum, DØ has developed a "top likelihood discriminant". Its method is:

- Identify kinematic variables which discriminate top from background with high reconstructed top mass ($m_t > 150$), without significantly biasing the m_t spectrum of either top or bkgnd.

$$\text{missing } E_T \text{ near cut}$$
$$\mathcal{A} \propto \text{least eigenvalue of } \mathcal{P} \text{ tensor}$$
$$H'_{T2} \equiv \frac{H_T - E_T^{j1}}{H_\parallel}$$
$$K'_{T\min} \equiv \frac{(\min \text{ of } 6\, \Delta\mathcal{R}_{jj}) \cdot E_T^{\text{lesser } j}}{E_T^W}.$$

E_T^W is the scalar E_T of the leptons, and $H_\parallel$ is the scalar $|p_z|$ of the jets and leptons. (For its lepton+jets top mass analysis first presented in March 1996, DØ used a different definition of the last two variables.)

Figure 39. (a) Systematic errors in DØ lepton+jets top mass analysis. (b) Kinematic variables used to form DØ's top likelihood discriminant.

Top likelihood discriminant (cont'd)

- Parametrize the normalized ratio of distributions
$$\ln \text{top}/\text{bkgnd} \equiv \ln \mathcal{L}_i(v_i)$$
for each variable v_i. For each event form
$$\ln \mathcal{L} \equiv \sum_i \omega_i \ln \mathcal{L}_i$$
where ω_i is a constant weight.

- Taking into account the correlations of the $\ln \mathcal{L}_i$ with m_t and with each other, choose the ω_i to optimize S/N while requiring a null correlation of $\ln \mathcal{L}$ with m_t.

- Form and cut on the top likelihood discriminant
$$P_{\text{top}} \equiv \frac{\mathcal{L}}{1+\mathcal{L}} \ .$$

- Apply a light cut on $H_{T2} \equiv H_T - E_T^{j1}$ to control the shape of the bkgnd spectrum at very low m_t.

- Neural networks trained on the same variables provide a similar discriminant.

Distributions of top likelihood variables and discriminant

Shown are distributions in 3 variables for top signal (HERWIG 180) and background events with high (>150 GeV/c^2) reconstructed top mass, normalized to equal area. The last panel plots the top likelihood discriminant for all events. The cut is at the arrow.

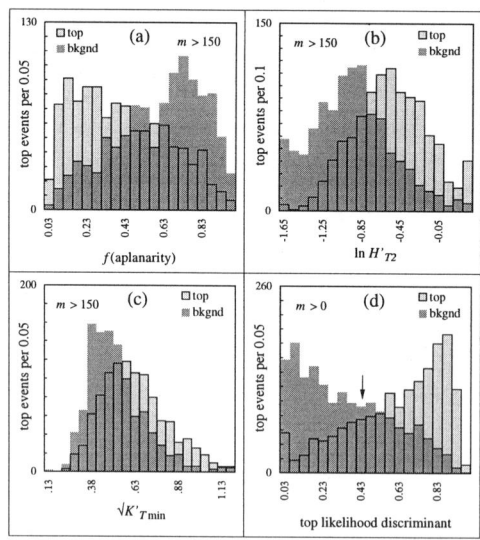

Figure 40. (a) Definition of DØ's top likelihood discriminant. (b) Distributions of 180 GeV/c^2 top signal and background in 3 variables used to form DØ's top likelihood discriminant, and in the discriminant itself.

Top mass templates after top likelihood cut

After DØ's top likelihood cut and the extra cut $H_{T2} = H_T - E_T(\text{jet 1}) > 70$ GeV, the peak of the fast fitted mass template continues to follow the generated mass while the background peak remains well separated.

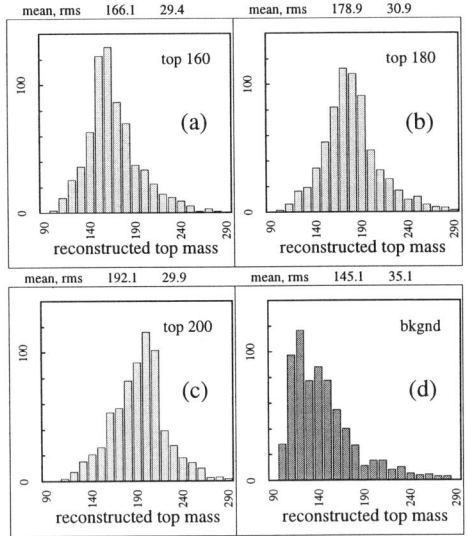

Expected effect of top likelihood discriminant cut on DØ's top mass distribution

The top likelihood and H_{T2} cuts are applied only to untagged events.

Shown is the expected effect on the full mass analysis sample's reconstructed top mass after applying the top likelihood cut and the requirement $H_{T2} > 90$ GeV. Top 180 signal (yellow) and background (blue) are plotted for Monte Carlo events with typical relative normalization.

More than 3/4 of the top signal and less than 1/4 of the background are retained, while the top and background peak masses remain distinct.

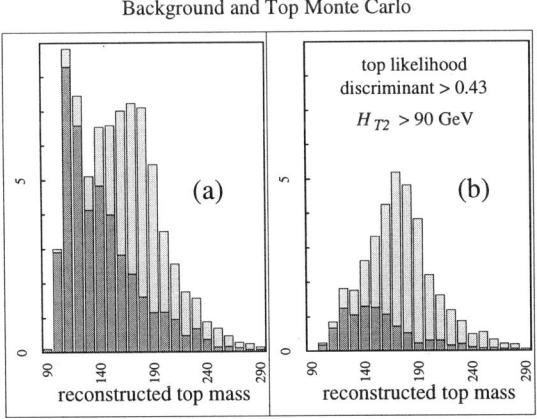

Figure 41. (a) Fast fitted top mass for background and for 160, 180, and 200 GeV/c^2 MC top after DØ's top likelihood cut. (b) Expected distribution of fast fitted top mass for a typical mixture of 180 GeV/c^2 MC top and background before and after DØ's top likelihood cut.

Conclusions -- top mass in the lepton + jets channel

In the b tagged lepton + 4 jet channel, CDF measures the top mass to be
$$m_t = 175.6 \pm 5.7(\text{stat}) \pm 7.1(\text{syst}) \text{ GeV}/c^2.$$

A doubly b tagged subset of this sample yields essentially the same mass with a larger statistical and a smaller systematic error.

In the SVX tagged all jets channel, based on a sample heavily dominated by background which is subtracted using a per-jet SVX tag rate parametrization, CDF obtains
$$m_t = 187 \pm 8(\text{stat}) \pm 12(\text{syst}) \text{ GeV}/c^2.$$

D0's March 1996 top mass in the lepton+4 jet channel is
$$m_t = 170 \pm 15(\text{stat}) \pm 10(\text{syst}) \text{ GeV}/c^2.$$
Application of a recently developed top likelihood cut should significantly reduce the background and the statistical error in this measurement.

Figure 42. Conclusions for Lecture III.

LECTURE IV. W MASS

The W boson mass has been measured to an accuracy of ± 370 MeV/c^2 by UA2[22], and to ± 180 MeV/c^2 by CDF[23] based on their 1992-3 data set. The methods used for these analyses are well understood and have been described thoroughly, particularly in the CDF publication. In this last lecture I present the results of a new W mass measurement by the DØ collaboration to an accuracy of ± 170 MeV/c^2 based on 1994-6 data, and separately to ± 270 MeV/c^2 based on the 1992-3 sample. The material is drawn from recent presentations by E. Flattum[24] and K. Streets.[25]

After touching on the motivation for precisely determining the W mass (for a serious discussion see G. Altarelli's contribution to this volume), I describe the basic method, the selection of events, and their modeling. Energy calibration is the key to this analysis, so I devote some effort to explaining the principal calibration issues. I conclude with an error summary, a comparison with previous results, and a rough and unofficial world average W mass.

IV. W mass

Motivation
Method for W mass measurement
Event selection and backgrounds
Fast Monte Carlo model
Electron angle and energy calibration
Recoil momentum calibration
Recoil momentum resolution
Removing underlying event from electron energy
Effect of recoil energy on electron efficiency
Luminosity dependence
Error summary
Other recent W mass measurements
Combining W mass measurements

Motivation

In the Standard Model at tree level the W mass is determined by the precisely measured parameters

$$m_Z = 91.1884(22) \text{ GeV}$$
$$G_\mu = 1.16639(2) \times 10^{-5} \text{ GeV}^{-2}$$
$$\alpha^* = 1/128.87(12),$$

where G_μ is the Fermi coupling constant measured from the muon lifetime and corrected for purely electromagnetic loops, and α^* is the fine structure constant evaluated at $q^2 = m_Z^2$. The W mass is given by

$$m_W = m_Z \cos\theta_W = 79\,957 \text{ MeV}/c^2,$$

where θ_W is the weak mixing angle defined by

$$\sin^2(2\theta_W) = (4\pi\alpha^*/\sqrt{2}) / G_\mu m_Z^2.$$

Beyond tree level, the W mass is shifted by the factor

$$(1-\Delta r)^{-1/2}$$

by a loop diagram involving the t and b quarks and another involving the Higgs boson. Δr is proportional to $(m_t/m_W)^2$ or $\ln(m_H/m_W)$ in the infinite m_t or m_H limit.

Figure 43. (a) Outline for Lecture IV. (b) Motivation for precise measurement of the W mass.

Motivation (cont'd)

For a given Higgs mass, the W and top masses must lie on a band like the ones exhibited below. Superimposed are the CDF and D0 W mass measurements from Run 1A and the CDF and D0 top mass results as of March 1996. Thus a precise measurement of m_W and m_t not only tests the Standard Model but also constrains the Higgs mass.

This presentation will be devoted to D0's new W mass measurement using the 1994-6 data from Run 1B.

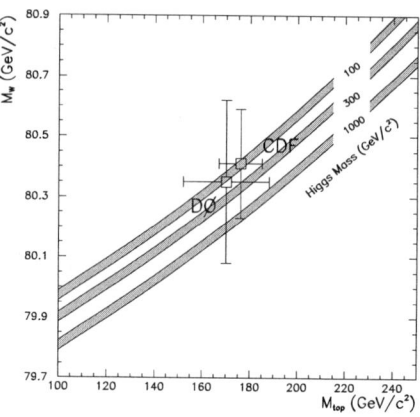

Method for W mass measurement

D0 compares the W mass to the precisely known Z mass, as measured in the same detector at the same time.

Plotted is the central-calorimeter ee mass spectrum from Run 1B data (points, 1562 events) and MC for Z+background (curve). The line near the axis is the background.

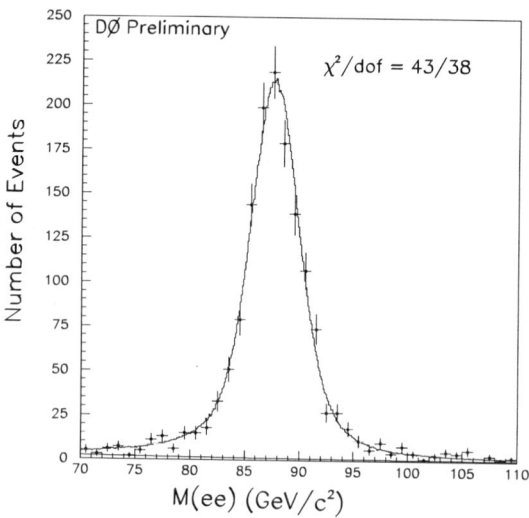

Figure 44. (a) Allowed bands in W vs. top mass for various Higgs masses, showing Tevatron W masses from 1992-3 data and top masses as of March 1996. (b) DØ dielectron mass in Z region, used to calibrate the W mass measurement.

Method for W mass measurement (cont'd)

Since the longitudinal component of the ν momentum cannot be deduced from transverse momentum balance, the W invariant mass can be calculated only in two dimensions.

This is the W transverse mass $m_T(W)$ plotted below (32 856 events from Run 1B). The D0 data show a Jacobian peak at the W mass.

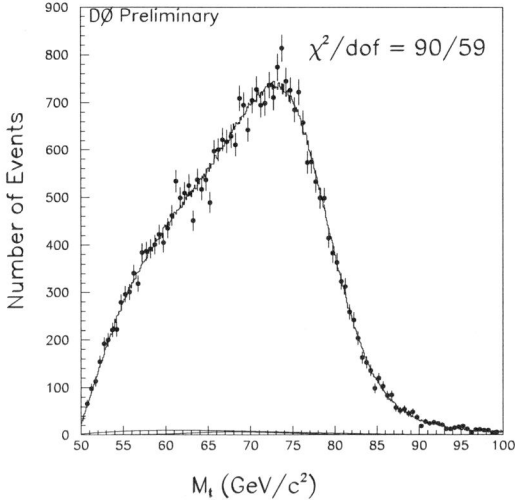

Method for W mass measurement (cont'd)

Because the W is produced with a p_T which is not large compared to its mass, its daughter e and ν also exhibit a Jacobian peak in E_T at $\sim m(W)/2$.

Shown is the electron E_T spectrum from the same D0 data. Though it can be measured without a hadron calorimeter, the shape is sensitive to the details of W production.

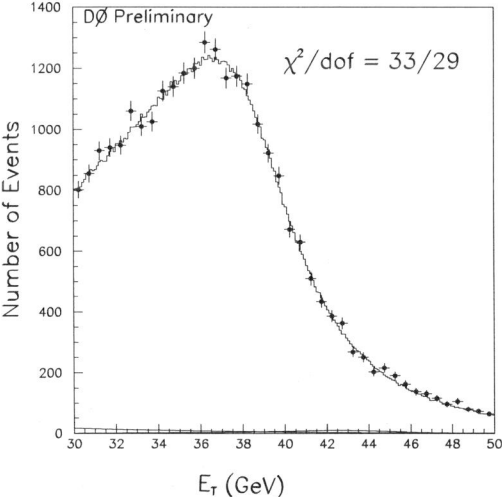

Figure 45. (a) Transverse mass of e and missing E_T (main input to the W mass determination). (b) Transverse energy of e (for auxiliary W mass measurement).

Method for W mass measurement (cont'd)

Lastly, $E_T(\nu)$ = missing E_T is plotted below. This variable has the advantages of neither $m_T(W)$ nor $E_T(e)$ -- it is sensitive to $p_T(W)$ and it requires the measurement of hadronic energy.

D0 bases its W mass measurement on $m_T(W)$, at present reserving $E_T(e)$ and $E_T(\nu)$ for cross-checks.

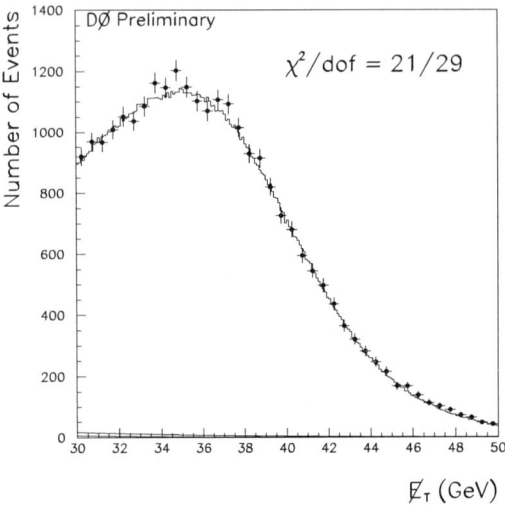

Event selection and backgrounds

The data just shown are drawn from Run 1B (76 pb^{-1}) and survive the cuts tabulated below.

W backgrounds are from QCD multijet events with misidentified electrons or from $Z \to ee$ events with one electron not identified and with mismeasured $p_T(\nu)$.

> QCD background is measured by studying the $p_T(\nu)$ distribution of events with a normal and with a flawed electron signature.
>
> Z background is estimated from MC tuned in $p_T(\nu)$ to match the data.

Z backgrounds are from QCD multijet events with misidentified electrons and/or direct photon production.

> The combined effects of the backgrounds and the tails from Drell-Yan and $Z\gamma^*$ interference are derived from the lineshape fit to the Z data.

D0 cut	$W \to e\nu$	$Z \to ee$
missing E_T (GeV)	>25	--
no. of electrons	1	2
no. of isolated electrons	1	2
E_T of electron (GeV)	>25	>25
$|\eta|$ of electron	<1.1	<1.1
p_T of W (GeV/c)	<30	--
m (transverse or invariant) (GeV)	50-110	70-110
fit region	60-90	70-110
D0 electron identification		
	EM fraction, isolation, track match	
	shower shape consistent with test beam electrons	

Figure 46. (a) Missing E_T (for auxiliary W mass measurement). (b) Backgrounds and tabulated event selection criteria.

Fast Monte Carlo model

The curves just shown are obtained from a fast MC program written by D0.

W and Z production use the Ladinsky-Yuan cross section currently constrained by fitting D0's Run 1A $p_T(Z)$ spectrum. The allowed variation is shown below.

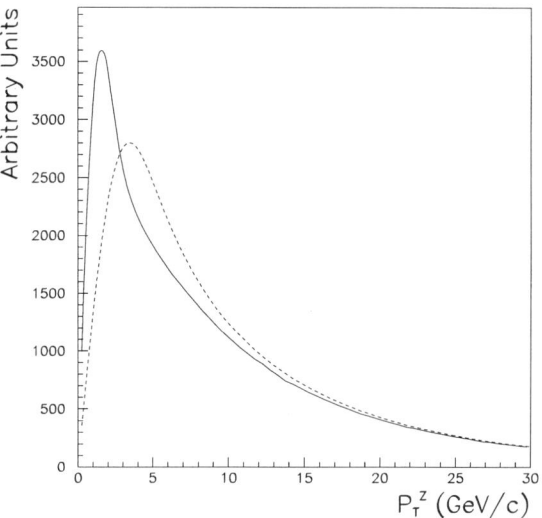

Fast Monte Carlo model (cont'd)

W's produced by quark-antiquark annihilation are required by V–A to be 100% polarized along the direction of the antiquark. This correlates the lepton charge and its direction with respect to the beam. Usually the antiquark comes from the antiproton beam; in sea-sea interactions with ~20% likelihood, it comes equally from either beam.

To explore the sensitivity of fitted W mass to parton density variations, CDF's W charge asymmetry data (below) were employed. Parton densities CTEQ3M' and CTEQ3M'' were derived from CTEQ3M after a coherent increase and decrease, respectively, of 1σ in the data points, and the variations in fitted mass were studied.

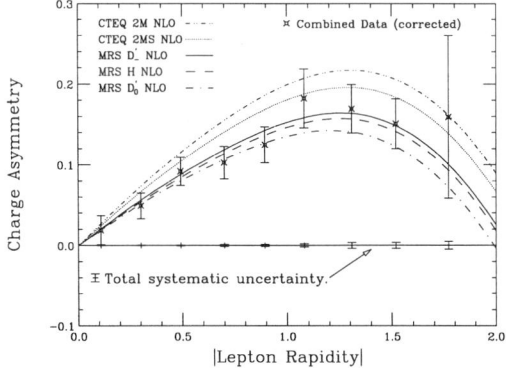

Figure 47. (a) Transverse momentum of Z calculated by Ladinsky and Yuan[26], showing the effect of the variation in the parameters of the calculation which is allowed in the fast MC. (b) Lepton charge asymmetry with respect to beam direction in W decay, measured vs. lepton $|\eta|$ by CDF[27], and used to restrict the allowed variation in parton density functions.

Fast Monte Carlo model (cont'd)

D0 uses the MRSA parton densities to measure the W mass.

Five other parton densities (MRSH, MRSD', CTEQ2M, CTEQ3M, GRV H0) yield mass shifts from -20 ± 15 to $+47\pm15$ MeV.

Excellent agreement with the RESBOS triple-differential W cross section generator is achieved.

Radiative corrections use Berends et al.'s prescription with a minimum E_γ of 50 MeV. Effects of radiative corrections on the electron and recoil energies are calculated by putting the e and γ through a GEANT simulation.

The combined systematic error in fitted W mass due to variation of $p_T(W)$ and parton densities is 65 MeV.

Electron angle and energy calibration

D0 measures the electron angle by connecting the centroids of the electron shower in the calorimeter (at ~92 cm radius) and the electron track segment in the central driftchamber (at ~62 cm). This avoids complications from multiple interaction points.

The electron position resolution in the calorimeter is ~8 mm. It was studied extensively in the test beam and is modeled with the help of full GEANT Monte Carlo.

The central driftchamber track centroid resolution is ~3 mm along the axial (delay line) direction. It was calibrated using collider and cosmic ray muons.

The systematic error on $m(W)$ due to uncertainty in electron angle determination is 40 MeV.

In the test beam, for two calorimeter modules, the measured electron energy was related to the true beam energy by a constant slope factor α and offset δ. Above 10 GeV the fractional deviation from this form was <0.3%.

The Z peak already has been shown. Mainly it determines α, but some information on δ is available from the variation of electron energy in asymmetric decays. The Z peak also fixes the electromagnetic energy resolution, which contributes a systematic error in the W mass of 30 MeV.

Low energy calibration points mainly determine δ. These are obtained from π^0 and J/ψ decay.

Figure 48. (a) Characteristics of the fast MC which simulates W production and decay in the DØ detector. (b) Calibration of electron energy and angle.

Electron energy calibration (cont'd)

A "symmetric mass" related to the π^0 invariant mass is constructed by measuring the opening angle between two converted photons, summing their (spatially unresolved) energy deposits in the calorimeter, and assuming the decay to be symmetric.

Shown is the background subtracted symmetric mass distribution. The line is the MC simulation.

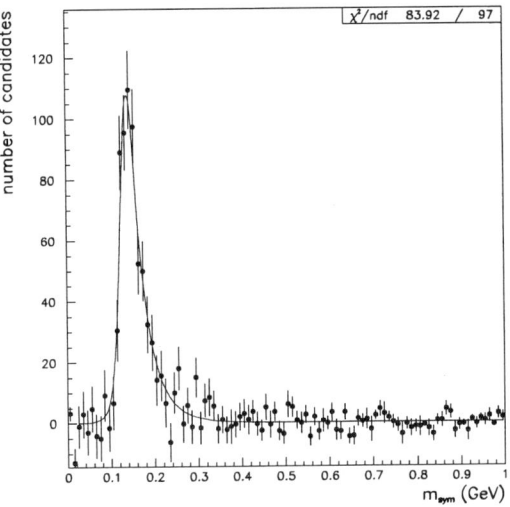

Electron energy calibration (cont'd)

The J/ψ invariant mass provides D0's other low-energy electromagnetic energy calibration point.

The Run 1A J/ψ signal below (histogram) has a significance of 5σ. The background (points) is obtained by pairing energy clusters in the calorimeter which have no associated electron tracks and subjecting them to the same cuts as the signal events. The curve is the fit.

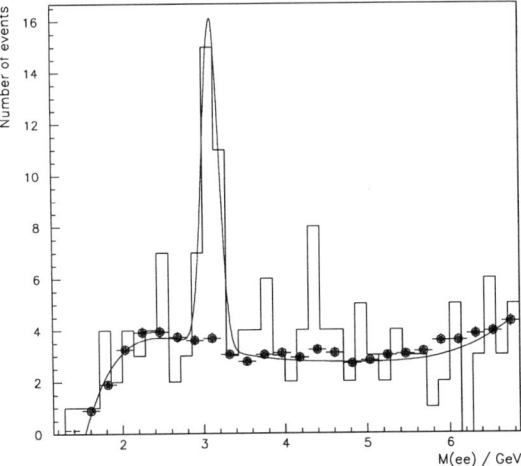

Figure 49. (a) "Symmetric mass" of two converted photons, showing a π^0 peak. (b) Dielectron mass showing a J/ψ peak. Both are used to anchor the electron energy calibration at the low end.

Electron energy calibration (cont'd)

The 68% confidence level constraints on α and δ from the J/ψ (wide band), π^0 (narrow band), and Z data (large ellipse) combine to give the small ellipse, which yields

$$\alpha = 0.9537 \pm 0.0009, \quad \delta = -0.158 \pm 0.015 \, ^{+0.030}_{-0.210} \text{ GeV}.$$

$m(W)$ and $m(Z)$ are the same within 12%, so the error in δ propagates only weakly to the ratio $m(W)/m(Z)$. The resulting scale error on $m(W)$ from α and δ is 80 MeV, including allowance for nonlinearity.

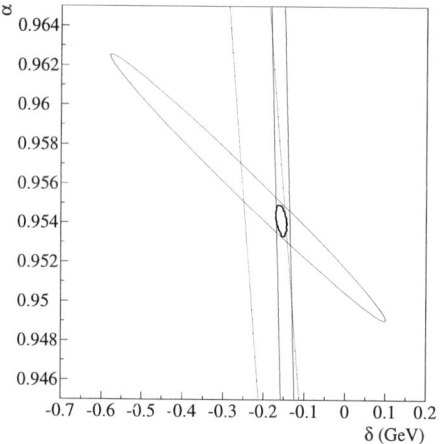

Recoil momentum calibration

The response of the calorimeter to the (hadronic) recoil transverse momentum p_T^{recoil} ($= u_T$) is calibrated with $Z \to ee$ decays using the recoil component p_η ($= u_\eta$) along the bisector of the electron transverse momenta as shown.

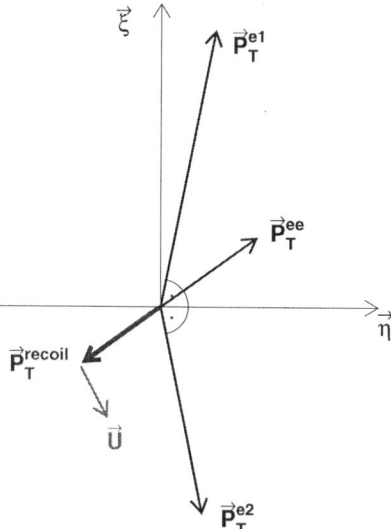

Figure 50. (a) 68% confidence limits on α vs. δ, where $E_{meas} = \alpha \, E_{true} + \delta$ for electrons. (b) Coordinate system in the transverse plane used for calibrating the hadronic jet response by measuring the energy recoiling against $Z \to ee$.

Recoil momentum calibration (cont'd)

The sum of the Z and raw recoil momenta along this bisector is plotted vs. the projected Z momentum only. Its nonzero value and constant slope fix the hadronic recoil response of the calorimeter to be 0.810 ± 0.016 of the ideal value.

The systematic error in $m(W)$ due to hadronic energy scale is 30 MeV.

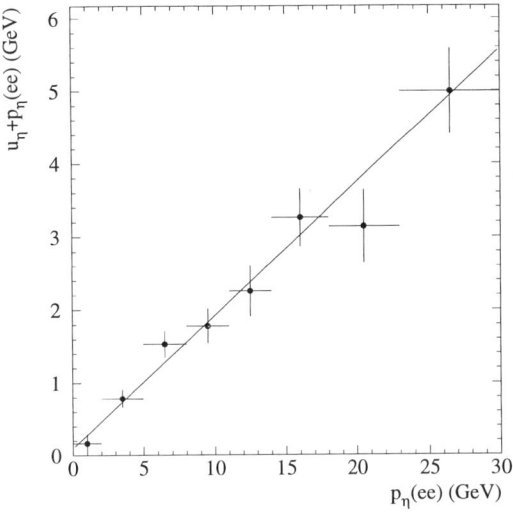

Recoil momentum resolution

The dispersion of these same data fixes the fractional energy resolution of the hard component of the hadronic recoil energy to be $0.04 \oplus (0.50 \pm 0.15)/\sqrt{u_T}$. The points below are data and the histogram is MC.

The soft (azimuthally symmetric) component of the hadronic recoil momentum is modeled using the missing p_T from minimum bias events with the same mean number of interactions as the W sample.

The syst. error in $m(W)$ due to uncertainty in recoil momentum resolution is 45 MeV.

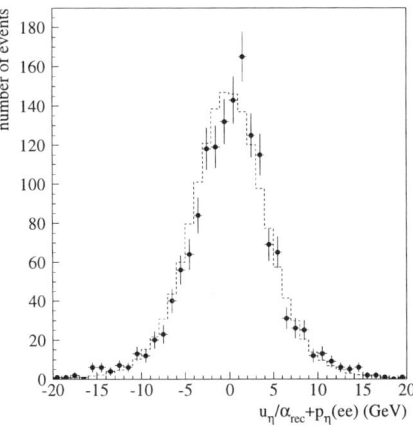

Figure 51. (a) Sum of dielectron and *raw* recoil momentum projected along the transverse dielectron bisector, *vs.* projected dielectron momentum alone. (b) Sum of dielectron and *corrected* recoil momentum projected along the transverse dielectron bisector.

Removing underlying event from electron energy

The 5×5 tower ($\Delta\eta \times \Delta\phi = 0.25$) region centered on the electron is not included in the calorimeter measurement of the recoil transverse momentum u_T. Both u_T and $p_T(e)$ must be corrected for this effect.

The component of u_T which contributes significantly to the W transverse mass is $u_\|$, the projection of u_T along $p_T(e)$.

The correction $\Delta u_\|$ is computed from the W events themselves by opening other 5×5 tower windows in regions away from the electron which would also satisfy the electron isolation requirement.

The mean $\Delta u_\|$ is 460 ± 25 MeV with a modest dependence on luminosity and $u_\|$ itself.

Uncertainty in underlying event removal contributes a systematic error on $m(W)$ of 30 MeV.

Effect of recoil energy on electron efficiency

The D0 electron trigger and identification require isolation. This causes electrons in events in which the recoil transverse momentum u_T lies along $p_T(e)$ to be detected less efficiently than when they are opposed.

The electron selection efficiency vs. $u_\|$ is estimated by superimposing GEANT simulated electrons onto the W signal sample. Monte Carlo which incorporates this estimate compares well to the data in the distribution of $u_\|$ shown below.

Uncertainties in efficiencies contribute a systematic error on $m(W)$ of 20 MeV.

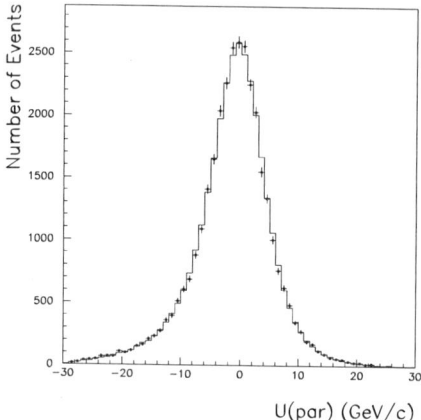

Figure 52. (a) Procedure for removing underlying event energy from electron, and its uncertainty. (b) Correction for electron inefficiency due to isolation requirement in electron trigger and offline identification.

Luminosity dependence of fit to $m_T(W)$

The fit to the $m_T(W)$ distribution yields a W mass of 80 380 ± 70(stat) MeV.

The χ^2 probability for this fit is 0.6%, due largely to 2 (2) points with χ>2.5 (χ<2.5) spread over the 60-90 GeV fitting window (the result is stable against varying the window). The Kolmogorov-Smirnov probability is 95%.

When the data are divided into four luminosity regions, the fits to $m(W)$ and $m(Z)$ (below) are consistent with being independent of luminosity. If the highest luminosity point were dropped, $m(W)$ would rise by 70 MeV.

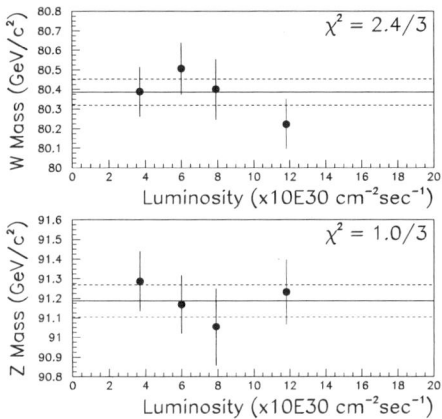

Luminosity dependence of fits to $p_T(e)$ and $p_T(\nu)$

As cross-checks, the fit to the $p_T(e)$ distribution shown earlier yields a W mass of 80 275 ± 95(stat) ± 180(sys), and the fit to $p_T(\nu)$ yields 79 930 ± 120 ± 270 MeV.

Removing common systematic errors and comparing to the main fit using $m_T(W)$, $m(W)$ is low by 0.7σ (48% probable) for the $p_T(e)$ fit and by 2.2σ (3%) for $p_T(\nu)$.

Shown is the luminosity variation of the cross-check fits. The top (bottom) plot has the $p_T(e)$ ($p_T(\nu)$) fits. At present, since the largest deviation below occurs in the highest luminosity bin, we assign a systematic error based on the effect of dropping that bin.

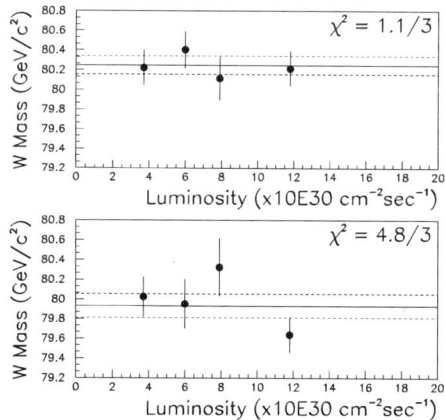

Figure 53. (a) Result of fitting W mass to electron + missing E_T transverse mass, and the luminosity dependence of the fit. (b) Results of fitting W mass to electron E_T and to missing E_T, and their luminosity dependences.

Error summary

The DØ central values and errors are preliminary. In particular, the Run 1B luminosity error is expected to disappear or diminish as further studies of luminosity dependence are completed.

	Run 1B (MeV)	Run 1A (MeV)
central W mass	80 380	80 350
statistical error	70	140
Systematic error source:		
$p_T(W)$ and parton densities	65	65
recoil momentum resolution	45	90
electron angle calibration	40	50
EM energy resolution	30	70
underlying event removal	30	35
efficiencies	20	30
radiative decays	20	20
backgrounds	15	35
W natural width	10	20
non-uniformities	10	10
bias in fit	5	5
luminosity dependence	70	--
systematic error	130	160
energy scale error	80	160
total error	170	270

Other recent W mass measurements

Experiment	m(W) (MeV)	stat	syst	scale	total error
CDF 90	79 910	292	227		390
UA2 92	80 360	220	297		370
CDF 95 (1A)	80 410	118	136		180
*DØ 96 (1A)	80 350	140	160	160	270
*DØ 96 (1B)	80 380	70	130	80	170
	*preliminary				

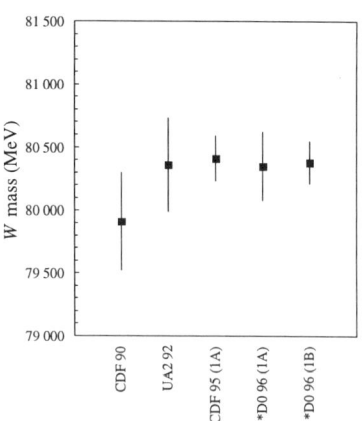

Figure 54. (a) DØ W mass error summary table. (b) Table and plot of recent W mass measurements.

Conclusion -- W mass

Combining W mass measurements:

To obtain a rough world average, not defended by any experimental collaboration, working group, or data compilation group, I have taken the error common to the two D0 96 preliminary results to be 100 MeV. I have assumed that the error common to the D0 96 average and the CDF result is 70 MeV (mainly $p_T(W)$ and parton densities). I have ignored errors common to any other group of measurements.

With these coarse assumptions, the average of the W mass measurements in the previous table is

$$m(W) = 80\,350 \pm 130 \text{ MeV}/c^2 .$$

Figure 55. Conclusion for Lecture IV: rough and unofficial world average of W mass measurements.

Acknowledgments

I would like to express my thanks to Tom and Barbara Ferbel for their warm and expert organization of the Advanced Study Institute, to the other lecturers from whom I learned a great deal, and to the students whose enthusiasm reaffirms my confidence in the future of experiments in fundamental physics. For help in preparing these lectures, I am grateful to Mark Kruse, Sandra Leone, and Simona Rolli of CDF, and to Jim Cochran, Eric Flattum, Uli Heintz, Meenakshi Narain, Scott Snyder, Kathy Streets, and Erich Varnes of DØ.

The DØ and CDF groups thank the staffs at Fermilab and the collaborating institutions for their contributions to the success of this research, and acknowledge support from the Department of Energy, National Science Foundation, and the A.P. Sloan Foundation (U.S.). DØ acknowledges support from the Commisariat à L'Energie Atomique (France), Ministries for Atomic Energy and Science and Technology Policy (Russia), CNPq (Brazil), Departments of Atomic Energy and Science and Education (India), Colciencias (Colombia), CONACyT (Mexico), Ministry of Education and KOSEF (Korea), and CONICET and UBACyT (Argentina). CDF is grateful for support from the Instituto Nazionale di Fisica Nucleare (Italy), the Ministry of Science, Culture, and Education (Japan), the Natural Sciences and Engineering Research Council (Canada), and the Grainger Foundation.

REFERENCES

1. F. Abe et al., Observation of the top quark in $p\bar{p}$ collisions, *Phys. Rev. Lett.* 74:2626 (1995).
2. S. Abachi et al., Observation of the top quark, *Phys. Rev. Lett.* 74:2632 (1995).
3. S. Abachi et al., Study of diboson production at DØ, submitted to 28th Intl. Conf. on HEP, Warsaw (1996); M.L. Kelly, Study of the $WW\gamma$ and WWZ interactions at DØ, in M.L. Kelly, G. Landsberg, J.N. Tarazi, A.M. Narayanan, A.V. Kotwal, and S. Rajagopalan for the DØ Collaboration, DØ papers on electroweak physics submitted to DPF '96, FERMILAB-Conf-96/236-E (1996).
4. G. Landsberg, Recent DØ results on $Z\gamma$ production, in M.L. Kelly, G. Landsberg, J.N. Tarazi,

A.M. Narayanan, A.V. Kotwal, and S. Rajagopalan for the DØ Collaboration, DØ papers on electroweak physics submitted to DPF '96, FERMILAB-Conf-96/236-E (1996).

5. See http://www-d0.fnal.gov/~strovink/ for links to these transparencies.
6. M. Narain, Top production at DØ, to appear in "Proc. Les Rencontres de Physique de la Valée d'Aoste", La Thuile (1996), FERMILAB-Conf-96/192-E (1996).
7. M.C. Kruse, Production and decays of top quark pairs in the single lepton and dilepton channels at CDF, to appear in "Proc. XIth Topical Conference on $p\bar{p}$ Collider Physics", Abano Terme (Padova), Italy (1996), FERMILAB-Conf-96/238-E (1996).
8. S. Leone, New observations of top at CDF, to appear in "Proc. XIth Topical Conference on $p\bar{p}$ Collider Physics", Abano Terme (Padova), Italy (1996), FERMILAB-Conf-96/195-E (1996).
9. G.L. Kane and S. Mrenna, Do about half the top quarks at FNAL come from gluino decays?, to appear in *Phys. Rev. Lett.*, ANL-HEP-PR-96-43, UM-TH-96-06 (1996), hep-ph/9605351.
10. E. Laenen, J. Smith, and W.L. Van Neerven, Top quark production cross section, *Nucl. Phys.* B321:254 (1994).
11. E. Berger and H. Contopanagos, The perturbative resummed series for top quark production in hadron reactions, *Phys. Rev.* D54:3085 (1996).
12. S. Catani, M.L. Mangano, P. Nelson, and L. Trentadue, The top cross section in hadronic collisions, *Phys. Lett.* B378:329 (1996).
13. R. Akers *et al.*, Search for a scalar top quark using the OPAL detector, *Phys. Lett.* B337:207 (1994).
14. S. Abachi *et al.*, Search for light top squarks in $p\bar{p}$ collisions at 1.8 TeV, *Phys. Rev. Lett.* 76:2228 (1996).
15. R.M. Barnett and L.J. Hall, Squarks in Tevatron dilepton events?, to appear in "Proc. Annual Meeting of the APS Division of Particles and Fields, DPF '96", Minneapolis (1996), hep-ph/9609313 (1996).
16. R.H. Dalitz and G. Goldstein, Analysis of top-antitop production and dilepton decay events and the top quark mass, *Phys. Lett.* B287:225 (1992).
17. K. Kondo *et al.*, Dynamical likelihood method for reconstruction of events with missing momentum. III. Analysis of a CDF high P_T $e\mu$ event as $t\bar{t}$ production. *J. Phys. Soc. Jpn.* 62:1177 (1993).
18. M. Strovink, Top quark search with the DØ detector, "Proc. Intl. Europhysics Conf. on HEP", Marseille, France (1993) (eds. J. Carr and M. Perrottet, Editions Frontières, 1994), 292.
19. M. Strovink, DØ top quark mass analysis, to appear in "Proc. XIth Topical Conference on $p\bar{p}$ Collider Physics", Abano Terme (Padova), Italy (1996), FERMILAB-Conf-96/336-E (1996).
20. S. Rolli, Top mass measurement at CDF, to appear in "Proc. XIth Topical Conference on $p\bar{p}$ Collider Physics", Abano Terme (Padova), Italy (1996), FERMILAB-Conf-96/141-E (1996).
21. L.H. Orr, T. Stelzer, and W.J. Stirling, Gluon radiation and top quark physics, to appear in "Proc. Annual Meeting of the APS Division of Particles and Fields, DPF '96", Minneapolis (1996), UR-1477, hep-ph/9609354 (1996).
22. J. Alitti *et al.*, An improved determination of the ratio of W and Z masses at the CERN $p\bar{p}$ collider, *Phys. Lett.* B276:354 (1992).
23. F. Abe *et al.*, Measurement of the W boson mass, Phys. Rev. D52:4784 (1995).
24. E.M. Flattum, A preliminary measurement of the W boson mass using $W \to e\nu$ decays at DØ, to appear in "Proc. XIth Topical Conference on $p\bar{p}$ Collider Physics", Abano Terme (Padova), Italy (1996), FERMILAB-Conf-96/199-E (1996).
25. S. Abachi *et al.*, Measurement of the W boson mass, submitted to *Phys. Rev. Lett.*, FERMILAB-Conf-96/177-E, hep-ph/960711.
26. G. Ladinsky and C.-P. Yuan, The nonperturbative regime in QCD resummation for gauge boson production at hadron colliders, *Phys. Rev.* D50:4239 (1994).
27. F. Abe *et al.*, The charge asymmetry in W boson decays produced in $p\bar{p}$ collisions at $\sqrt{s} = 1.8$ TeV, *Phys. Rev. Lett.* 74:850 (1994).

ADVANCES IN TRACKING CHAMBERS: THE MICRO-STRIP GAS CHAMBER AND THE MICROGAP CHAMBER

R. Bellazzini[1] and M. A. Spezziga[1,2]

[1] Istituto Nazionale di Fisica Nucleare
Via Livornese 582/a
56010 S. Piero a Grado (PI), Italy
[2] Scuola Normale Superiore
P.za dei Cavalieri 7
56100 Pisa, Italy

1. Introduction

The study of Electroweak symmetry breaking is the next step to be done for the comprehension of elementary particles and interactions. The Standard Model has shown itself as the best description available up to now of fundamental processes, but its correctness will not be proved until the symmetry breaking sector is checked. The further one goes with high energy experiments, the harder it is to build and operate a suitable experimental apparatus, and the Large Hadron Collider (LHC) with its detectors ATLAS and CMS is going to be the most difficult task ever carried out in this field. The physics channels to be investigated have very small cross sections and branching ratios, so that very high luminosity (10^{34} cm^{-2} s^{-1}) and centre-of-mass energy (14 TeV) are needed to detect them, and the constraints on the detectors are most stringent.

The central tracking system of CMS is one of the most delicate parts of the apparatus, the nearest to the interaction point and therefore the most exposed to a hard environment. As will be shown in the following, the core of CMS is composed by semiconductor devices and MicroStrip Gas Chambers (MSGC). The latter is a relatively new detector, one of the best performing gas detectors in terms of speed, spatial and energy resolution. In addition, its granularity and radiation hardness make it a good choice for this task.

In this paper an overview of latest studies on MSGC and its capabilities in view of its application to LHC is presented. Many of these results come from a simulation of MicroStrip Gas Chambers and Micro Gap Chambers (MGC, another gas detector on the style of the MSGC) that has been developed in the last few years and is continually (though slowly) extended and improved. The results of this simulation have been used to understand the principles of operation of this new kind of detector, and it is used to give a contribution to the simulation of the full tracker system.

The first two sections are a brief description of CMS (section 3) and of its physics goals (section 2). In particular the MSGC is described in section 3.3. Section 4 is a review of results coming from the simulation of a *single* detector.

2. The physics of CMS

In this section, the main physics goals of CMS are shortly outlined. As mentioned in the introduction, electroweak symmetry breaking is the first subject that the LHC is going to study, both from the classical (Standard Model) point of view and from alternative ones. These two matters are thus the subject of sections 2.1 and 2.2. Section 2.3 just remembers that with an apparatus like CMS it is possible to aim at different objectives, both frontier physics and more "softly" typical cross section, mass width and lifetime measurements, with precision and richness of data that other detectors cannot achieve.

2.1. The Higgs boson

The Higgs boson is needed in the Standard Model to justify the masses of fermions and gauge bosons, still having a gauge invariant theory. The Lagrangian that accounts for this model may be written as:

$$\mathcal{L} = \mathcal{L}_{gauge} + \mathcal{L}_{SB} + \mathcal{L}_f$$

with

$$\mathcal{L}_{gauge} = -\frac{1}{4}\mathbf{F}_{\mu\nu}\cdot\mathbf{F}^{\mu\nu} - \frac{1}{4}F_{\mu\nu}F^{\mu\nu} + i\overline{\Psi}_L \not{D}\Psi_L + i\overline{\psi}_R \not{D}\psi_R$$

being the $SU(2)_L \times U(1)_Y$ symmetric Lagrangian of massless gauge bosons and massless fermions. The symbols are:

$$F^{\mu\nu} = \partial^\mu X^\nu - \partial^\nu X^\mu$$
$$\mathbf{F}^{\mu\nu} = \partial^\mu \mathbf{W}^\nu - \partial^\nu \mathbf{W}^\mu + g\mathbf{W}^\mu \times \mathbf{W}^\nu$$
$$D^\mu = \partial^\mu - ig\mathbf{W}^\mu\cdot\mathbf{T} - ig'X^\mu Y$$

$\mathbf{W}$ and X represent the gauge fields of respectively the $SU(2)_L$ and $U(1)_Y$ symmetries, their linear combinations

$$W_\mu^- = \frac{1}{\sqrt{2}}(W_{1\mu} - iW_{2\mu})$$
$$W_\mu^+ = \frac{1}{\sqrt{2}}(W_{1\mu} + iW_{2\mu})$$
$$Z_\mu = W_{3\mu}\cos\theta_W - X_\mu\sin\theta_W$$
$$A_\mu = W_{3\mu}\sin\theta_W + X_\mu\cos\theta_W$$

being the physical W, Z and photon. The fields Ψ_L and ψ_R represent respectively SU(2)$_L$ fermions doublets and singlets, **T** and Y are the generators of SU(2)$_L$ and U(1)$_Y$. The symmetry-breaking term

$$\mathcal{L}_{SB} = (D_\mu \Phi)^\dagger (D_\mu \Phi) - \lambda \left(\Phi^\dagger \Phi - \frac{1}{2} v^2 \right)^2$$

introduces a SU(2)$_L$ doublet made of four scalar fields

$$\Phi = \frac{1}{\sqrt{2}} \begin{pmatrix} w_1 + iw_2 \\ h + iw_3 \end{pmatrix}$$

Those denoted by w are the goldstone bosons and are "gauged away" when a particular (*unitary*) gauge is fixed. The remaining field is the Higgs boson. It couples with the W's, Z and γ through the first part of $\mathcal{L}$, containing the covariant derivative D^μ, and with the fermions through

$$\mathcal{L}_f = g_u \overline{\Psi}_L \psi_{Ru} \Phi + g_d \overline{\Psi}_L \psi_{Rd} \tilde{\Phi} + \text{h.c.}$$

where $\tilde{\Phi} = i\sigma_2 \Phi^*$, and ψ_{Ru}, ψ_{Rd} are the right SU(2)$_L$ singlets, corresponding to each component of the left doublet $\overline{\Psi}_L$.

By these couplings, fermions and gauge bosons acquire their masses. Coupling constants result proportional to masses, so that Higgs production from and decays to heavy quarks and gauge bosons are favoured. In fact the most interesting mechanisms for Higgs production in LHC are those reported in fig. 2.1.

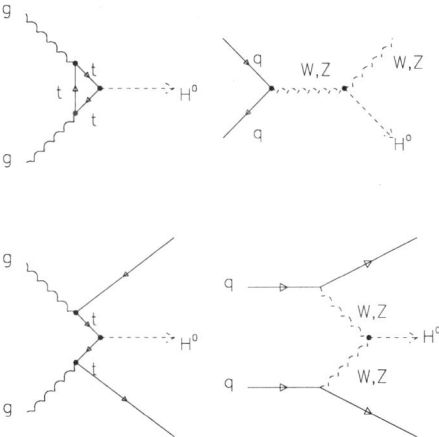

Figure 2.1: *Higgs production mechanism at hadron colliders.*

The mass of the Higgs is in principle unknown. However a lower experimental limit has been set [1] around 64 GeV, while upper limits come from theoretical considerations, in particular the violation of unitarity by partial amplitudes occurring for $M_h \approx 1\text{TeV}$, and indirect effects of the Higgs mass on low energy experiments [2], which suggest that the Higgs mass lies in the range $40 < M_h < 450$ GeV.

The favourite decay channels depend on the Higgs mass. In the lowest range $(80 \lesssim M_h \lesssim 130 \text{ GeV})$ $H \to \gamma\gamma$ should be the cleanest one, with a signature given by two isolated photons at energy ≈ 50 GeV. For this channel a very good electromagnetic calorimeter is needed, with excellent energy and angular resolution, since the precision of the mass measurement is given by (θ is the angle between the two photons directions, E_i their energies)

$$\frac{\sigma_M}{M} = \frac{1}{2}\left[\frac{\sigma_{E1}}{E_1} \oplus \frac{\sigma_{E2}}{E_2} \oplus \sigma_\theta \cot\left(\frac{\theta}{2}\right)\right]$$

with $\oplus$ meaning the quadratic sum. Figure 2.2 reports the calculated mass plot for four different values of the Higgs mass exploiting the $\gamma\gamma$ channel and the signal significance contours for 10^5 pb^{-1} integrated luminosity, that is the cross section times branching ratio needed for the discovery of Higgs with a given significance (N_s/N_B) as a function of Higgs mass. The dashed line represent the same quantity for the SM Higgs, so that, for example, it can be seen that after about one year of LHC running at the foreseen luminosity an Higgs can be discovered with a significance greater than 10 σ in the range between 110 and 140 GeV.

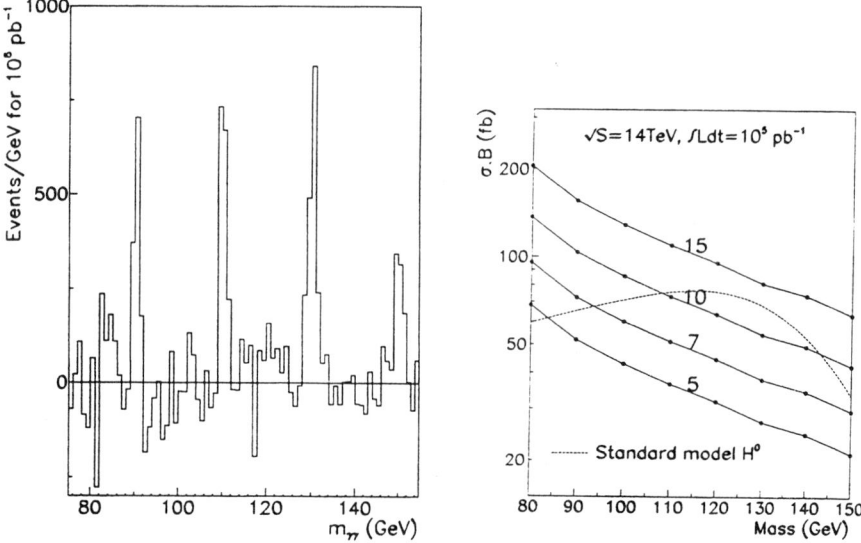

Figure 2.2: *Left: background-subtracted 2γ mass plot for 10^5 pb^{-1} with signals at 90, 110, 130 and 150 GeV in the PbWO$_4$ calorimeter. Right: Signal significance contours for 10^5 pb^{-1} taken at high luminosity.*

In the range between 130 and 800 GeV, the favourite channel is $H \to ZZ^*, ZZ \to 4l^\pm$, that gives as a signature a peak in the four lepton invariant mass distribution. This mass region can be divided into two subregions discriminated by the $2M_Z$ threshold, below which the Z production is small, as well as the Higgs mass width, so to require a good lepton energy and momentum resolution. Above this threshold the event rate increases and above $\sim$300 GeV the Higgs width becomes larger than the

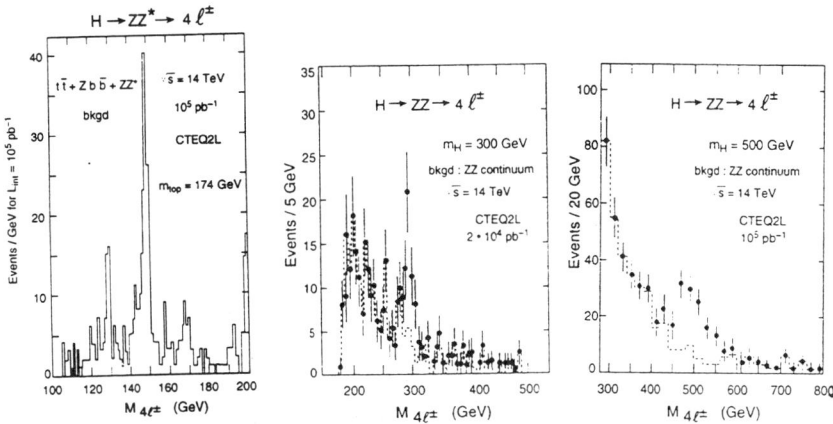

Figure 2.3: *The four lepton mass distributions for $H \to ZZ^* \to 4l^{\pm}$ and signals at 130, 150, 170 GeV (left), and $H \to ZZ \to 4l^{\pm}$ and signals at 300 GeV with 2×10^4 pb^{-1} (centre) and at 500 GeV with 10^5 pb^{-1}.*

instrumental resolution. In fig. 2.3 the four lepton mass distributions for Higgs lying in this range discovered through these channels.

Over 500 GeV, the channel $H \to ZZ \to ll\nu\nu$ gives a signature consisting of two high p_t leptons and high missing transverse energy. This channel is affected by a strong background. It can be reduced at high energy (> 700 GeV) by requiring forward energetic jets coming from the WW or ZZ fusion production, the second graph illustrated in fig. 2.1. In fig. 2.4 an example of the signal.

Going up with the energy other channels become interesting. These are $H \to WW \to l\nu jj$ and $H \to ZZ \to lljj$. Their limits are a strong background that must be carefully studied to be reduced, and the broadness of the Higgs resonance. In fig. 2.5 there is an example of mass distribution of such a decay.

2.2. "Non resonant" models

The mechanism mentioned at the beginning of the previous section is not the only possibility of Electro-weak Symmetry Breaking (ESB). Furthermore, it is affected by the *"naturalness problem"*, that is the need, in the renormalisation procedure, to fix the cancelling parameters with a precision of one part to 10^{17}. This is not a problem from a mathematical point of view, but seems to be very "unnatural" physically speaking.

Some models, based on low energy Lagrangians, non renormalizable but $SU(2)_L \times U(1)_Y$ symmetric, have been developed and their issues concerning production and decays of vector bosons and existence of new ones, have been computed. A simple idea of this kind of procedures is given by the linear sigma model with the (scalar) potential

$$V = \frac{\lambda}{4} \left(\sigma^2 + \pi^2 - F_\pi^2 \right)^2 .$$

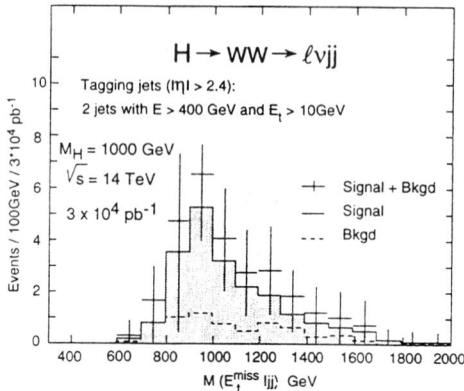

Figure 2.4: $H \to ll\nu\nu$ signal for $M_h = 200$ GeV and 10^4 pb^{-1}

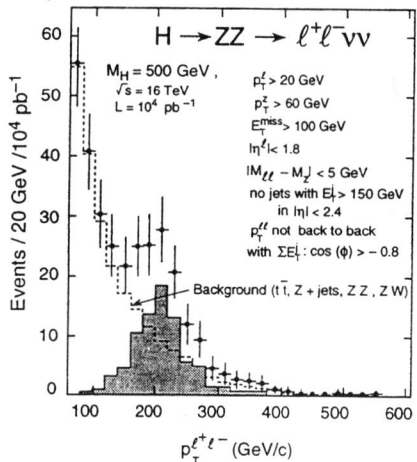

Figure 2.5: Higgs signal in $H \to WW \to l\nu jj$ final state for $M_h=1$ TeV with 3×10^4 pb^{-1}.

The Lagrangian containing this potential has a $SU(2)_L \times SU(2)_R$ symmetry, that is broken down to $SU(2)_V$ when a minimum of the potential is chosen as $\langle \sigma \rangle = F_\pi$ and the singlet field is redefined as $\sigma \to \sigma + F_\pi$, in the usual mechanism. The triplet of π's are the goldstone bosons of the model and the mass of σ is $\sqrt{2\lambda} F_\pi$. When this quantity is taken to infinity, σ can be thought no more as a physical particle, but as a parameter of the theory and the constraint $\sigma = \sqrt{F_\pi^2 - \pi^2}$ must be imposed (e.g. [3]). This identity may be developed in powers of π^2/F_π^2 and substituted in the kinetic term

$$\mathcal{L}_{kin} = \frac{1}{2}(\partial\sigma)^2 + (\partial\pi)^2$$

thus obtaining an infinite series of interaction terms of increasing order [4]. The leading ones can be used to predict the interaction amplitudes of the goldstone bosons of the theory, which in this simple case are the pions, but in more advanced models are for example the longitudinal components of W's and Z, through the "low energy theorems". Each term of the low energy Lagrangian is proportional to a coefficient indicated as α_n, that may be experimentally measured if the sensitivity of CMS is good enough. In fig. 2.6 the mass distributions are shown of some WZ scattering and production processes for given values of some of the α parameters, compared with the Standard Model QCD background. These results are foreseen after 3×10^5 pb^{-1}[5].

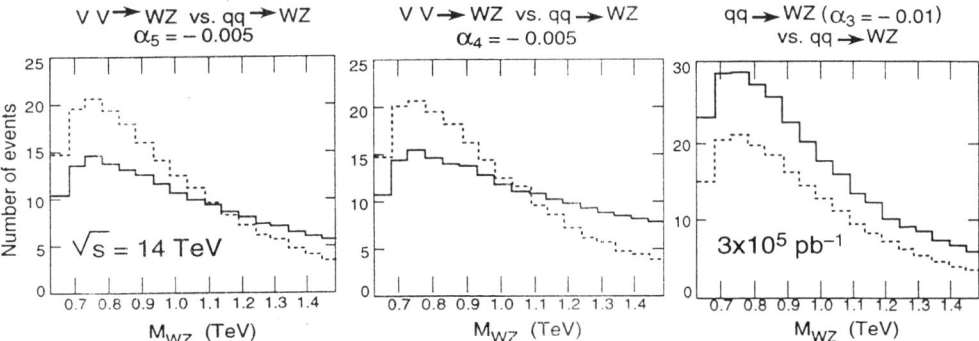

Figure 2.6: *WZ mass distributions in the non resonant EChL model for WZ final states from WZ fusion (solid lines in the first two figures), and from $q\bar{q}$ annihilation (solid line in the third figure). Dashed lines are distributions from $q\bar{q}$ annihilation background.*

The general treatment of these models is a part of the theory of non linear representations [6] which deals with the spontaneous breaking of a global symmetry group G down to a local H⊂ G. G and H can be chosen in many different ways (see e.g. [3, 6, 7, 8, 9, 10]), provided SU(2)$_L\times$ U(1)$_Y$ ⊂ H. For each pattern of symmetry breaking, a different set of new goldstone bosons arises, which couple with the W's and the Z and, through them, to the fermions. The detection of these states or of effects on fermions interactions due to their existence, is among the possibilities of CMS.

2.3. Other searches

More theories have been developed to face the naturalness problem and other problems arising in the Standard Model. Among these, supersymmetry foresees other kinds of Higgs bosons. In the simplest model there are a charged one, H$^\pm$, and three neutral ones, h, H, A. One of them can be light enough to be easily seen in CMS.

Supersymmetry establishes a correspondence between each particle and its "SUSY" (SUper SYmmetric) partner, whose spin differ by 1/2. Thus "sleptons" ($\tilde{l}$), and "squarks" ($\tilde{q}$) will exist with spin 0; "gluinos" ($\tilde{g}$), "W-inos" ($\tilde{W}$), "Z-inos" ($\tilde{Z}$), and "photinos" ($\tilde{\gamma}$) with spin 1/2. It is supposed the existence of a $R = -1$ quantum number for superparticles, while common particles would have $R = 1$. The possible conservation of R would imply the production of SUSY particles in pairs and the stability of the lightest one. All of them should be massive, to justify the fact that none has been found up to now, and the lightest one could be the photino with mass ≥ 0.5 GeV.

CMS will search for these particles as well as for (at least) the lightest SUSY Higgs. Some of SUSY Higgs decays are similar to SM Higgs ones: h, H$\to \gamma\gamma$; h, H$\to$ZZ; h, H, A$\to \tau\tau \to$ l$^\pm$+hadrons+X; H$^\pm \to \tau\nu_\tau$; h, H, A$\to \mu\mu$ and more.

Other important fields of particle physics that CMS should be able to investigate are the B-physics, which is thought to be a more sensitive ground to CP violation than K mesons experiments, and heavy ions (from O-O to Pb-Pb) physics with interesting phenomena of gluon-plasma production.

Furthermore, study of new gauge bosons with masses up to 4.5 TeV, detailed study of top physics, search for compositness of quarks and leptons etc. are foreseen, besides more traditional measurements of "soft physics".

For all the processes mentioned in this section and many more, details and references can be found in [11].

3. The CMS project

Purposes, layout, detailed characteristics of CMS are described in [12]. We will only quickly survey in the first two paragraphs the general layout of CMS, with a little more space to the central tracker, where the MSGC's will be used. More details and figures (among which those presented here) can be found in the quoted references.

3.1. General layout

The physics outlined in the previous chapter leads to the production of several new particles, that must be identified by their decay products, especially Z, W, charged leptons and photons. Z's and W's will be mainly detected by their leptonic decays, because hadronic channels will be affected by a strong background. Small cross sections and branching ratios force the use of a very high luminosity, with important drawbacks on the detector structure.

An example of process to be studied by CMS is the detection of a Higgs boson by its decay to two Z's, which in turn go to pairs of electrons or muons. The leptons must be detected up to rapidity values of $\eta = 2.5$, with $p_t \geq 5$ GeV. It is necessary a very good p_t resolution ($\leq 1\%$ under 100 GeV) for this and all decays having as an intermediate state Z's and W's. However, if the Higgs mass is 90 and 130 GeV, its most important channel is two photons, and a very good energy resolution in the

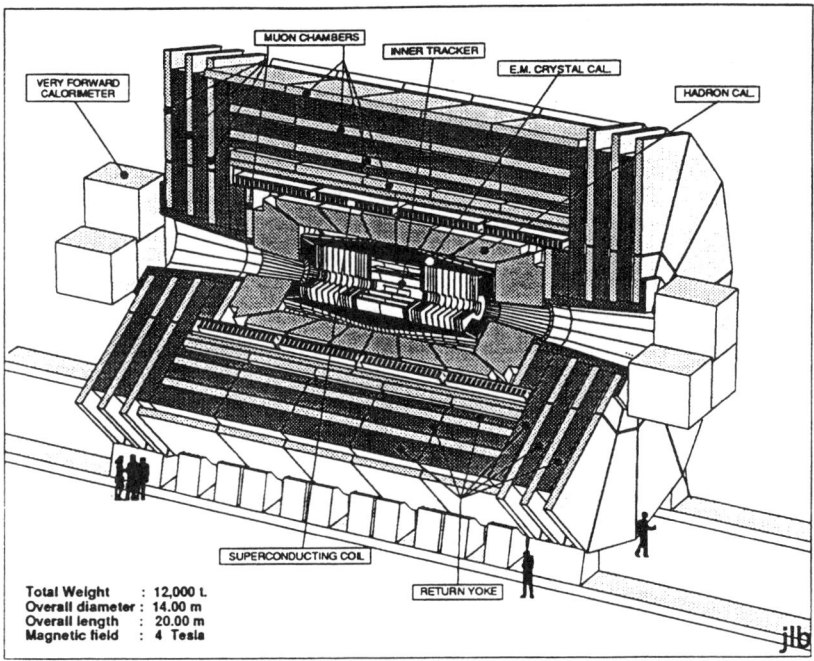

Figure 3.1: *Layout of the Compact Muon Solenoid [12].*

Electromagnetic Calorimeter (ECAL) is needed (< 1% at 50 GeV). The resolution of the Hadron Calorimeter (HCAL) is not required to be as good. For a precise measurement of the momentum of leptons, and especially muons, a high magnetic field and a good tracking system (both central and outer) is needed. That is the reason why CMS has been thought as a system to be developed around a powerful solenoidal magnet, with several muon stations, and from here its name. In figure 3.1 there is an open view of CMS, which can be divided into two different symmetries, the *barrel* and the *end caps*. In both symmetries can be found, from outside inward, alternating muon chambers and iron plates, that provide both absorber material and the return yoke for the magnetic field, the hadron calorimeter, the electromagnetic calorimeter, the central tracker. In the barrel, between muon stations and calorimeters, are the magnet and a further thin hadron calorimeter (*tail catcher*), between ECAL and tracker there is a preshower (not well distinguishable from the figure), while after the end caps there is a *very forward calorimeter* for very low p_t's. The whole structure is 21.6 meters long (very forward calorimeters excluded), 14.6 meters wide, 14500 tonnes heavy.

The peculiar geometry of magnet and return yoke, allow a double measurement of the muon momentum, first in the central tracking system, under the influence of the strong uniform field (4 T) inside the solenoid, then in the muon chambers among plates, where the flux is quantised and inverse. This provides two independent measurements of the momentum.

The assembly of CMS must take into account the need of easy insertion in sub-

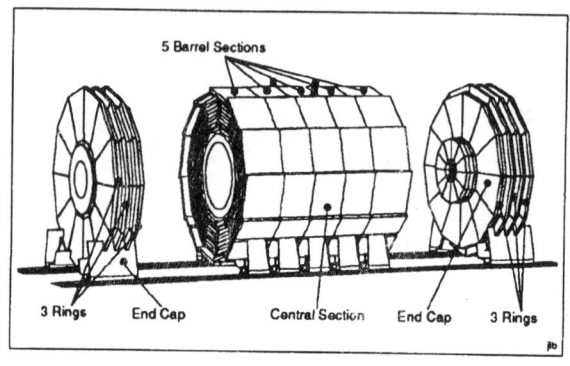
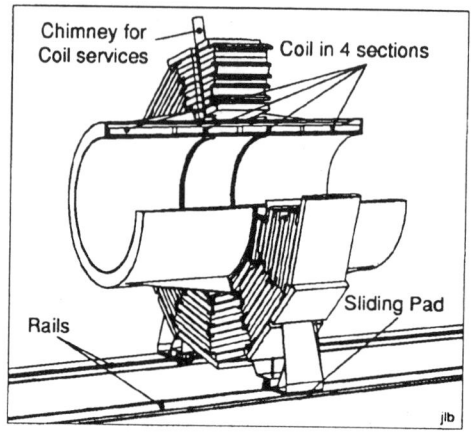

Figure 3.2: *Division of CMS into five sections of the barrel and two end caps.*

sequent times of different components of the detector (single chambers, calorimeter and tracker units, cooling pipes, electronic connections etc.) and of maintenance and modifications. In fig. 3.2 an open view of CMS is shown, with five sections and three end caps sustained by supports sliding on rails parallel to the beam. The supports have hydraulic mechanisms for alignment and adjustment to the beam. The central section holds the structure of the magnet. The major problems of this system are the heat dissipation and the mechanical stress due to weight and magnetic forces. The cryogenic system is liquid helium based.

The muon detection system is represented in fig. 3.3 and is made of four layers (or *stations*) interleaved with iron plates. It is also divided in five sections along the beam direction and in twelve sectors around the barrel, each covering $30°$. The stations are not perfectly aligned so to avoid cracks and dead points. The detectors that will constitute the muon chambers are special kinds of drift tubes and drift chambers.

The ECAL most difficult performance concerns the detection of the decay $H \to \gamma\gamma$. As already mentioned, this is the most interesting channel, though not the most probable one, if the Higgs mass is around 100 GeV. Photons direction and energy must be measured with very good precision. For this purpose, detectors made of lead tungstate crystals ($PbWO_4$) have been developed and will take place in positions

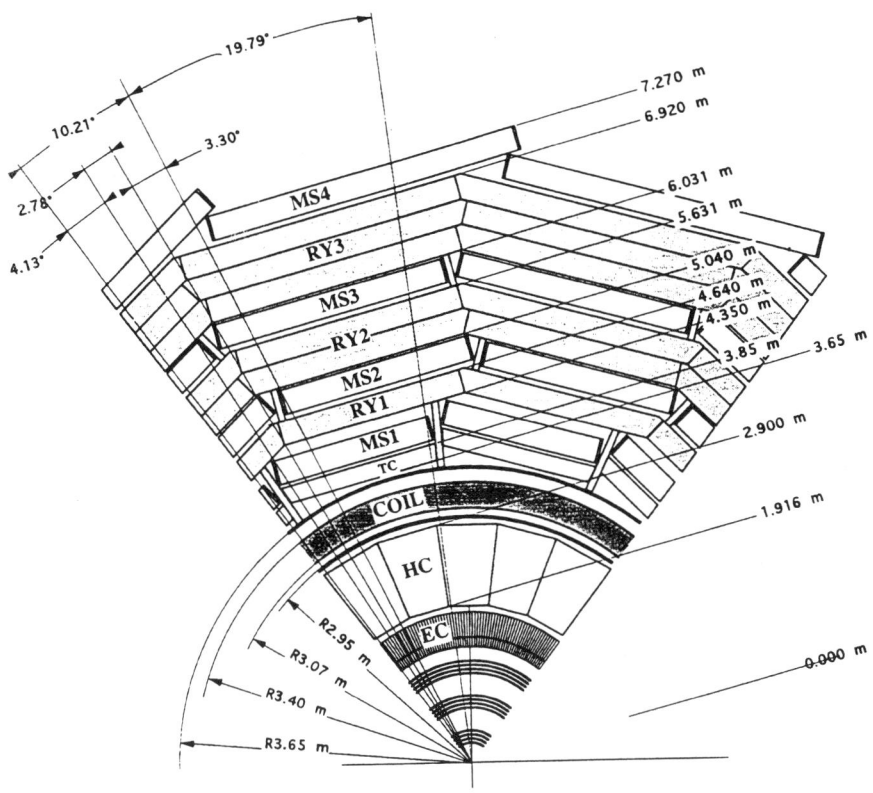

Figure 3.3: *Sector of CMS showing the position of muon chambers.*

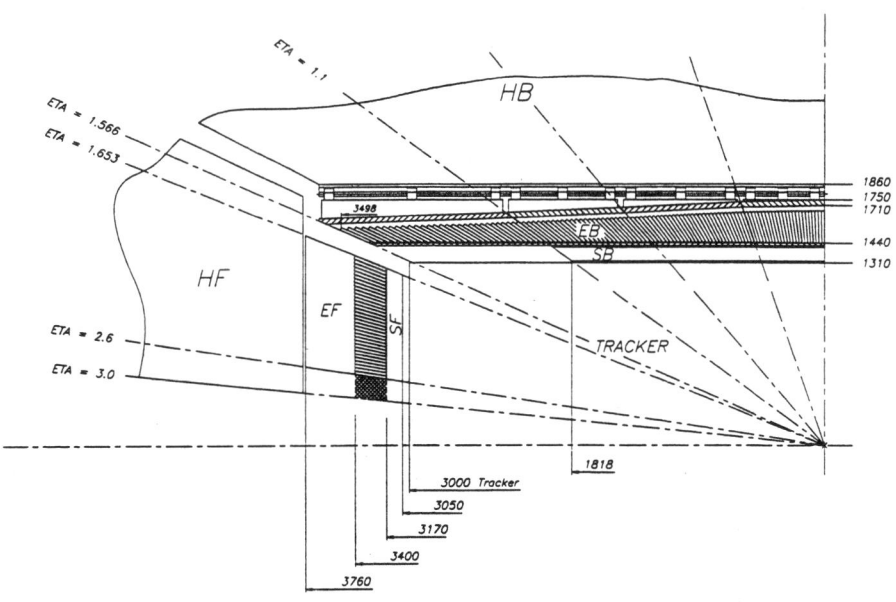

Figure 3.4: *The layout of the barrel and end cap $PbWO_4$ electromagnetic calorimeter [12].*

shown in fig. 3.4, approximately pointing towards the interaction point. A preshower with Si wafers will be between the tracker and the ECAL, separated from the latter by insulating material and a thermal screen, to keep them at different working temperatures.

The hadron calorimeter will be composed by alternating copper layers and plane scintillators, divided in square cells each containing wavelength shifter fibres for collection and transmission of the light signal. The surface of the cells varies with the radial position keeping the solid angle constant.

For the Very Forward Calorimeter, two options are being examined, the first being arrays of parallel plate chambers working in the avalanche mode, which would give the radiation hardness required for this difficult region of the detector. The second option consists in quartz fibres, embedded in iron or copper, that will give a signal by Cherenkov effect. This would make them only sensitive to relativistic particles.

3.2. The tracking system

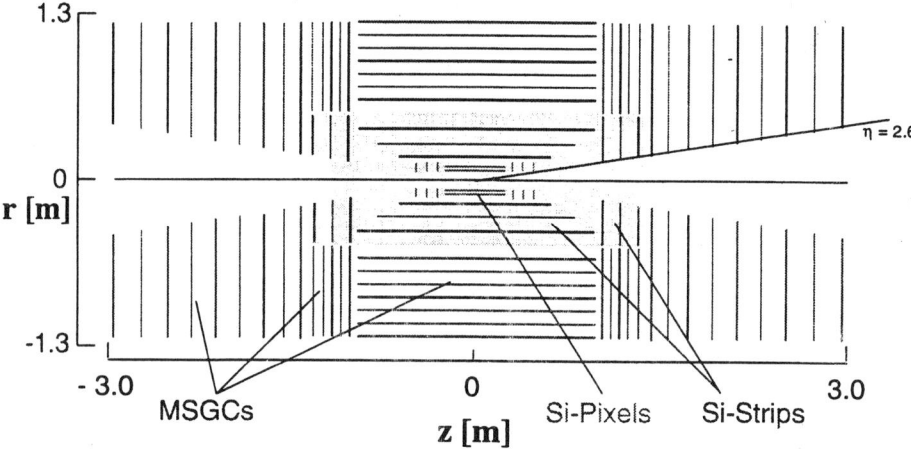

Figure 3.5: *Schematic of inner tracker. Thick lines denote double-sided readout.*

The tracker must provide a momentum resolution of $\Delta p_t/p_t \approx 0.1 p_t$ (p_t in TeV) for isolated charged leptons at low rapidities with efficiency better than 95%. In latest designs, a vertex detector near to the interaction point has been added to allow b-tagging. The high track density in the central region of CMS requires the use of high granularity detectors. Silicon detectors and MSGC's have been chosen for this reason and for their good resolution and speed. Fig. 3.5 shows the central tracker frame, that is composed of three detector kinds: the silicon pixel detectors immediately around the interaction point, the silicon micro-strip detectors in the following layers and MSGC's in the outer part.

The silicon pixel detector must help vertex reconstruction, improving the three dimensional resolution and especially that in the z position, which is rather poorly determined by other detectors. Pixels shape will be square, $125 \times 125 \mu m^2$, each row staggered with respect others, to improve the resolution, and not tilted with respect to the magnetic field, to take advantage of the Lorenz angle effect.

The silicon micro-strip detectors will form three layers in the barrel and three in the forward part, with 50 μm pitch between strips and some of them with small angle stereo strips for two-dimensional read out. Both barrel and forward structures will present some overlapping detectors to avoid dead regions.

For both pixel and micro-strip silicon detectors the problem of radiation damage will be serious. It will be necessary to keep them at lower than ambient temperature ($\approx 0^oC$) and change the bias voltage with time. They will undergo type inversion, the pixel detector very soon, the micro-strip detectors after some years.

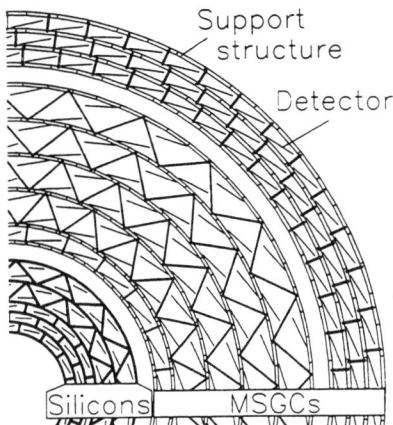

Figure 3.6: *Barrel wheel structure.*

The outer part of the barrel tracker will be formed by twelve layers of MSGC's, five of them double sided. The forward region will contain five layers at the same z position as the forward silicon micro-strip detectors and nine layers beyond. In the barrel, the micro-strip detectors (both MSGC and silicon) will not be disposed in circles but in spiral arms, to allow introduction of services (cooling, gas, cables...). This will be simplified by the fact that they are all tilted to compensate the Lorenz angle effect. This last peculiarity will also introduce a slight asymmetry in the resolution between negative and positive charged particles, that will bend either in the same or contrary direction with respect to the tilt angle of detectors. In fig. 3.6 the plot of the frame containing the barrel detectors.

For the forward part, the option of "open" design is also being considered, where the MSGC are positioned in rings without walls between each MSGC and its neighbour, and all in a common gas path.

MSGC's must also be cooled down during operation, but only to keep them at uniform ambient temperature. A thermal shield will be needed between the two detector types (gaseous and semiconductor). The cooling will be indirect, connecting each detector module to the cooling pipes by means of a mesh of conductive material, and direct, with water pipes applied to the substrate and a continuous flow of cool gas through the whole tracking volume. The cold gas will enter the region from near the beam pipe and flow outwards following the spiral structure.

From experimental tests [13] it turns out that the Lorentz angle will not be a problem for the MSGC's used in the barrel (the forward region should not be affected by this problem because of the different orientation of the electron drift lines with respect to the magnetic field). Tilting the module by the same amount as the Lorentz angle will recover the original resolution for normal tracks. Aging has also been tested (see e.g. [14]) and seems to be completely uninfluent for many years of CMS running, if clean materials are adopted.

3.3. The MicroStrip Gas Chamber

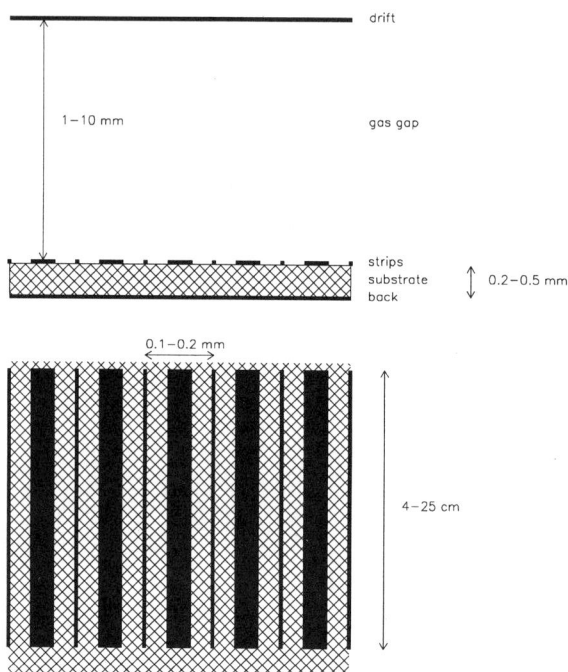

Figure 3.7: *Layout of the simplest model of MSGC.*

The MicroStrip Gas Chamber has been built in several versions, different for proportions and materials, after the model illustrated in figure 3.7. The principle is to miniaturize the wire chamber (MWPC) cell, replacing the wires with thin aluminium strips, fixed to an insulating support, for example plastic or glass, to prevent the electrostatic forces from distorting or breaking them. This way, their width can be reduced down to few microns as well as the distance among them. The strips acting as anodes are thinner (2 to 10 μm) than cathodes (70 ÷ 90μm). Each electrode is about 50μm far from its neighbour, so that the distance between an anode and the next (*pitch*) goes from 100 to 200μm.

Over the plane supporting these electrodes there is a volume filled with gas, closed on the upper side by a metal foil, which forms the electric drift field for the electrons and called for this reason the *drift cathode*. Its distance goes from 1 to 10mm. The bottom surface of the insulating plate, whose thickness is usually $200 \div 500\mu m$, may be covered with a further metallic layer that forms the lower cathode or *back*. A voltage on this electrode may have a secondary, sometimes beneficial, effect on the electric field allowing a control of the charging phenomenon, that is the gathering of ions on the insulating substrate (see later on).

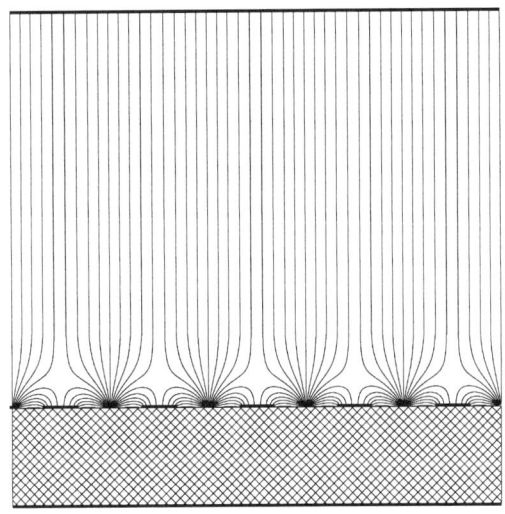

Figure 3.8: *The electric field in the MSGC.*

A typical bias configuration for the different electrodes consists of some hundreds volts between anode and cathode (for example 0 volt on anodes and −500 on cathodes) and a negative voltage on the upper cathode (≈ -2000 V). It is convenient to put to ground the read out electrodes, so to avoid cumbersome decoupling capacitors between the high voltage and the read out electronics. The electric field in the volume of the detector thus assumes the shape of figure 3.8, that divide the gas filled volume in a drift region, of uniform and weak field ($\approx 1 \div 10$ kV/cm), and another of *multiplication*, where the field lines coming from the drift and side cathodes converge on anodes. In this region (fig. 3.9) the electric field intensity reaches some hundreds of kV/cm.

The operation of this detector is similar to that of a wire chamber: particles that pass through leave a trail of ionised molecules whose electrons are brought towards the

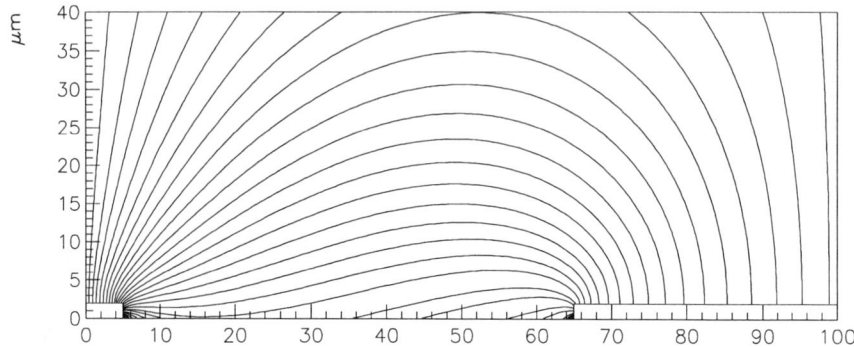

Figure 3.9: *Electric field in the multiplication region between anode (down left) and side cathode (right).*

multiplication region by the drift field. Here the field gets strong enough to trigger an avalanche that produces some thousands of free electrons for every one in the track. They are immediately collected by the anode, while some of the ionised molecules are pushed towards the drift cathode and (many more) towards the side cathodes, where they are collected. It is mainly in the collection of charges that the MSGC proves its superiority with respect to the MWPC, because the distances that the ions must cover to reach the side cathodes are extremely reduced and thoroughly included in the high field region. This offers the important advantage of a fast charge collection with very short signal shaping times and clean up of the multiplication region from space charge soon after the event. As a result this detector can work correctly at very high irradiation rates: the best versions can face a flux of 10^7 particles/mm$^2\cdot$s [15].

More merits that make this detector proper for use in CMS are its radiation hardness, low cost, excellent resolution (it can localise a single track with 30 μm resolution while two tracks are discriminated with 250 μm precision [16]) and high granularity. High granularity implies low occupancy ($\approx 1\%$ as requested by pattern recognition and track reconstruction algorithms.

The parameters of the model described up to here may be modified with more or less important consequences on the detector performances. Thinner anodes create stronger fields in smaller volumes, with improvement of energy resolution [17, 18]. Anode sizes have however a lower limit in the strip impedance and therefore in the signal-to-noise ratio.

Another element on which modification can be applied is the substrate. The support of surface electrodes must guarantee the necessary insulation among anode, cathode and back. However it is by now sure that a perfectly insulating substrate is a negative condition. Detailed studies on some materials in use for the support, especially some kinds of glass and plastic [19, 20], lead to the conclusion that chambers built on substrate with a comparatively low resistivity (down to $10^9\Omega\cdot$cm) yield a better gain stability both in time and with increasing particle flux. The surface of the "conductive" substrate drains more easily the electric charges that are not collected by the electrodes, with less disposition to *charging*. This is an important problem which consists in the piling up of charged particles (almost solely ions produced in the avalanche) on the dielectric surface between anode and cathode. Ideally, every ion produced during the avalanche process should be collected by the side or drift cathode, more or less quickly, following the electric field lines, which should all and always end on one of the charge collecting electrodes. In practice this situation is never accomplished and we always have some field lines which tend to bring some of these charges to the substrate surface. Charges which are not removed are sources of an electric field which is superimposed to the original one and which modifies in time the gain and the rate capability of the detector. A typical measurement of the gain variation is reported in fig. 3.10 [21], while in fig. 3.11 a simulation is shown of the electric field before and after charging, of gain as a function of time and of the surface charge altogether deposited between electrodes.

To avoid this behaviour some methods have been experienced, like the use of slightly conductive substrates, therefore able to drain partly or totally the charges which gather over them. The conductivity of the substrate is obviously limited by the leakage current and by the heat production and noise that it introduces.

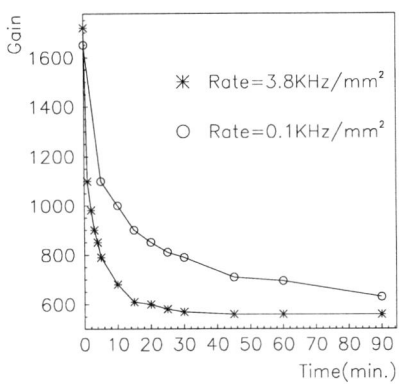

Figure 3.10: *Variation of the gain in time due to the substrate charging [21].*

To be able to use cheap but low resistance materials, some methods have been thought to give the substrate a high conductivity, but confined on a thin layer on the surface of the dielectric support(fig. 3.12). One of them consists in the implantation of boron ions, that by their impact create modification in the dielectric structure, leading to an increase in the material conductivity, but only down to the depth that they reach, usually of the order of a tenth of a micron. A second way is to put on the surface of the detector (either before or after the electrodes deposition) a layer of conductive material, like germanium, with a thickness that can be easily tuned. A job of some tenth of a Ångström has proved itself deep enough to have a satisfying conductivity. Actually, the introduction of a non uniform conductivity in the dielectric volume leads to a change in the electric field also in the active region of the detector. The electric field in this case results as in fig. 3.13. Consequences of this

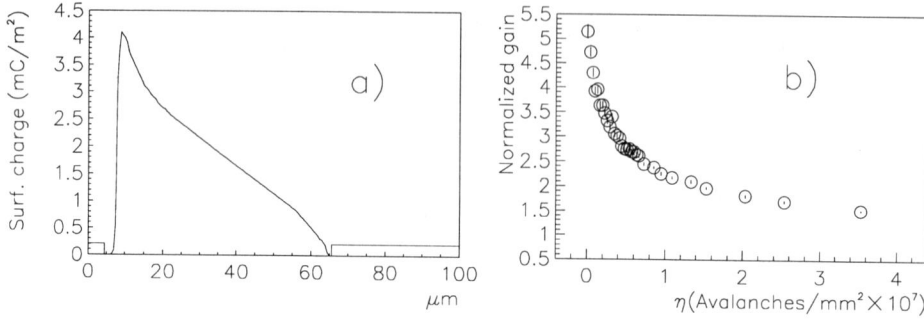

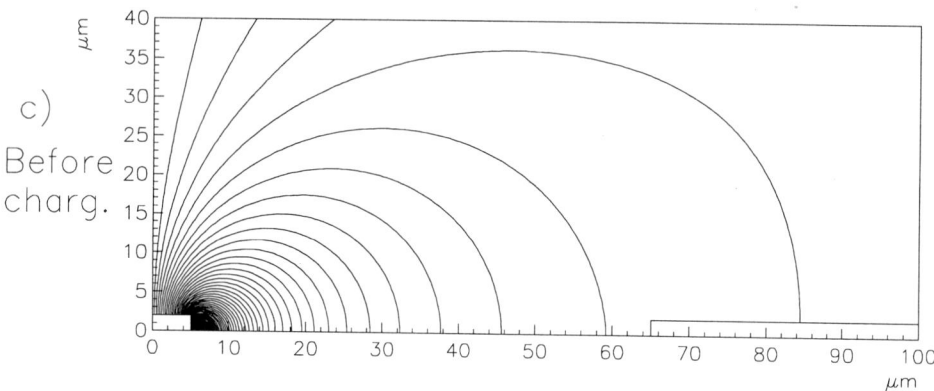

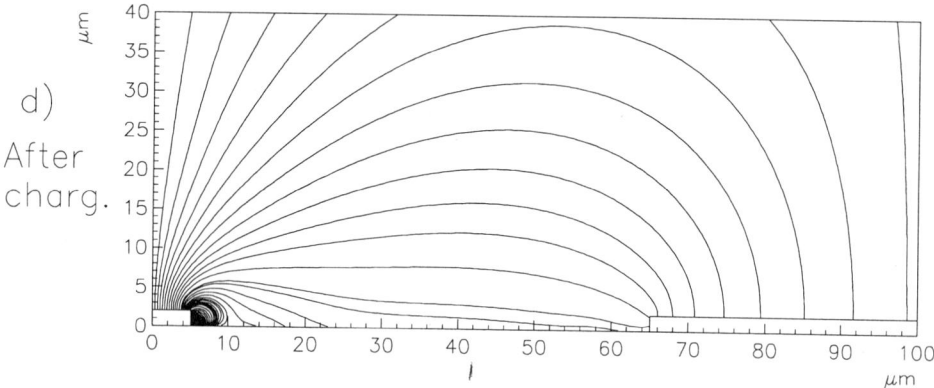

Figure 3.11: *Simulation of charging: a) surface charge deposited on substrate, b) gain as a function of time, c) electric field before charging, d) electric field after charging.*

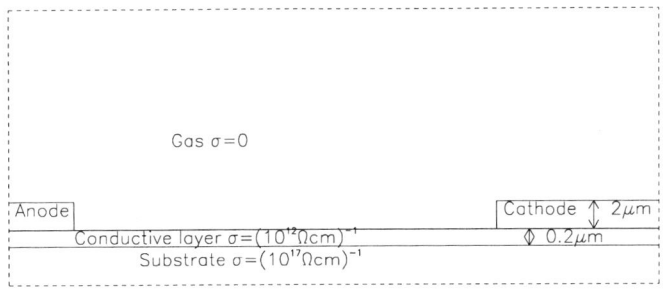

Figure 3.12: *MSGC with the substrate surface covered with a slightly conductive material (not on scale)*

field configuration are the almost total absence of charging (because of the repulsive field near to the anode) and the decrease of the detector gain, due to the weakness of the electric field over the anode. Presently all detectors in use in Pisa exploit this technique.

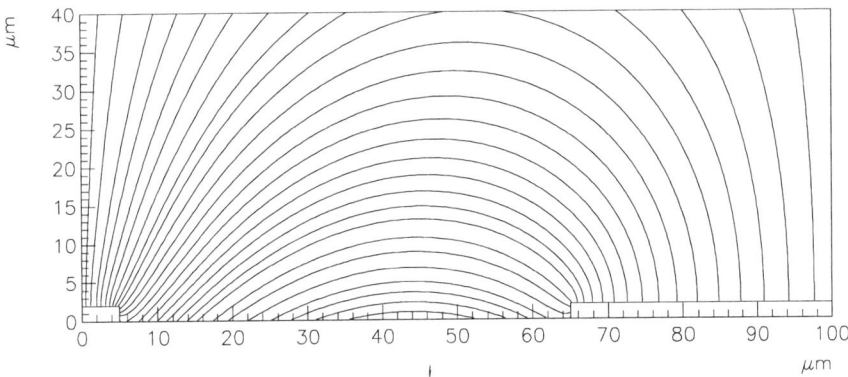

Figure 3.13: *The electric field in the MSGC with conductive layer.*

To push away from the surface the avalanche ions, the back electrode can be used too, connecting it to the anodic voltage, so to create a repulsive field for ions. The effect of this choice is indicated in fig. 3.14.

The classic MSGC is able to find the position of the event only along one coordinate, normal to strips on their plane. Information on the second coordinate can be obtained making a second set of strips directly on the same detector, either inclined by a small angle with respect to the first one or normal, depending on the choice of

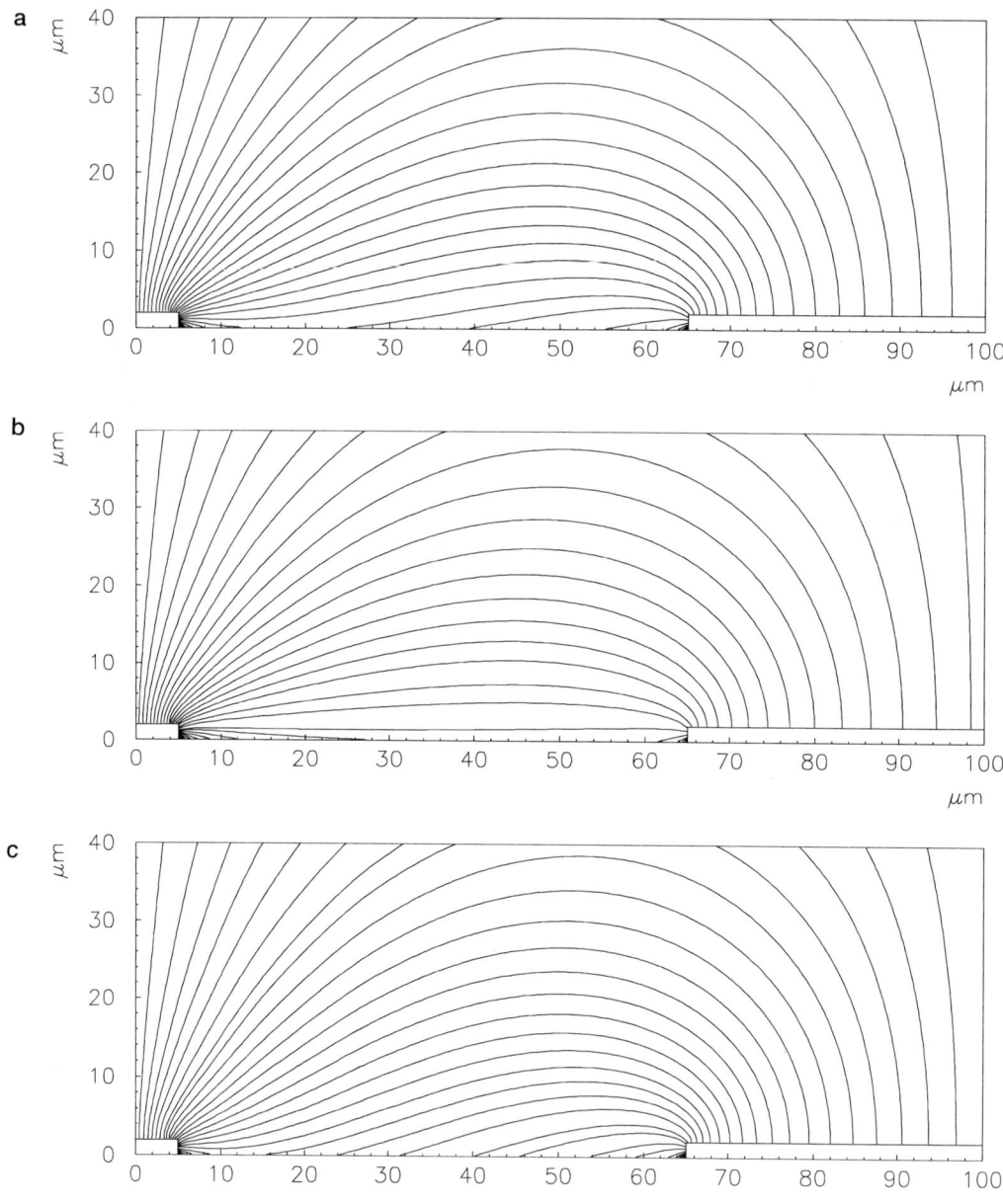

Figure 3.14: *The change of the field lines on the surface when changing the voltage on back: a) 0 V (as the anode), b) −500 V (as cathode), c) +500 V.*

reading both coordinates from the same side of the detector or from two contiguous ones.

This kind of MSGC has actually carried out and put in operation with excellent results [15], using as a plane to dispose the series of normal electrodes the back one (fig. 3.15), suitably covered by an insulating substrate. This electrode indeed did not have up to now an essential function in the detector operation, and can be divided in substructures of suitable shapes and sizes for reading the position out, for instance small squares, rings or pie sectors, or just strips normal to the surface ones.

It is necessary that the plane of these electrodes be situated as near as possible to the surface for avalanche ions to induce on them a sensitive signal. In fig. 3.16 the fraction of charge induced on back is reported as a function of the dielectric thickness that divides it from the surface, as it results from the simulation that will be described in the following.

However in this way the lower electrode strongly affects the electric field. If the insulating substrate is of uniform conductivity, this problem can be extremely troublesome: a voltage on back near to the cathodic one induces a gain instability due to charging, while anodic-like voltages make inefficient the multiplication field near to the anode. The problem can be solved by the above mentioned layer of higher conductivity, that damps the influence of the electric field created by the back in the gas. The effect is clearly visible in fig. 3.17 that compares electric fields obtained putting on back negative (left column) or null voltages (right) with respect to the anode. The three rows correspond to different values of the ratio σ_1/σ_2 between the conductivity of the resistive layer σ_1 and that of the dielectric support σ_2. This ratio has proved itself to be the key quantity to parametrise this effect.

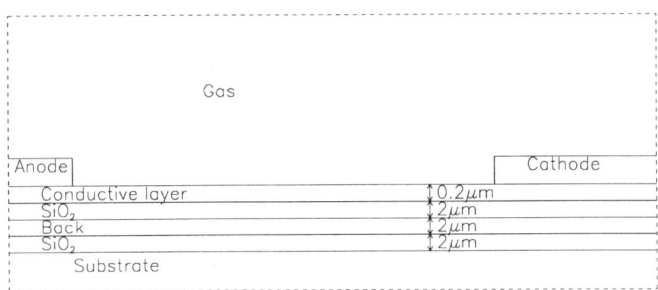

Figure 3.15: *Two dimensional MSGC: the substrate (not necessarily insulating) is covered successively by a layer of SiO_2, one of aluminium, again SiO_2, and a slightly conductive thin layer ($\approx 10^{12}\Omega \cdot cm$), (not on scale).*

We can remark that for $\sigma_1/\sigma_2 > 10^5$ the electric field in the gas is almost completely independent of the presence of the back. The second row is instead clearly affected by the variation of the voltage on the back, while the effect of this variation is really strong on the electric fields of the third row, which represents the complete absence of the conductive layer ($\sigma_1/\sigma_2 = 1$).

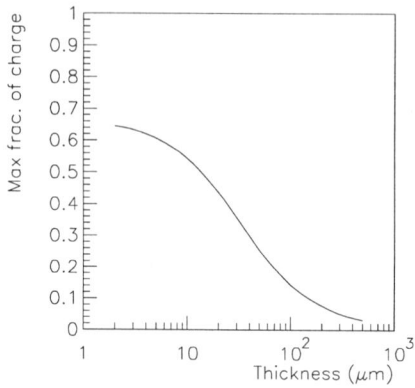

Figure 3.16: *Maximal value of of the signal as a function of the substrate thickness, that is of the distance of back from surface.*

This situation is confirmed by table 1, which presents a simulation of the gain of all these configurations, in the same order of columns and rows of figure 3.17. These values concern the *initial* gains, because in fact some of them are strongly affected by charging.

The two dimensional detectors built in Pisa have shown performances equivalent to the classic MSGC in energy resolution, gain, rate capability. An idea of two dimensional spatial resolution is given by fig. 3.18, where a drilled mask has been used to screen an X-ray beam incident on the detector.

4. The simulation of the single MSGC

The simulation of the single MSGC comes in different steps. As it is well known, the detection of a particle starts with ionization of gas molecules along its track. Primary electrons are occasionally produced with high kinetic energy (delta rays) and may further ionise, creating secondary electrons. For simplicity we will often call both kinds "primary electrons" to distinguish them from electrons created in the avalanche. Primary electrons, then, drift and diffuse towards the multiplication region and produce the avalanche. Electrons are collected and so are ions, inducing the electric signal on anodes and cathodes. The aim of the simulation of single MSGC is to reproduce all these processes, so to have a control as complete as possible on the response of the detector and to know what may be expected from a given modification of the structure and of the working conditions (gas, bias voltages...). This knowledge may be exploited to design new structures or to obtain better results from existing ones, especially in view of their application in CMS.

The first stage of the simulation is to compute the shape and intensity of the electric field inside the detector active area, because this will decide the motion of charges and will be responsible for charging of the substrate, gain, signal intensity and many more important quantities.

Secondly, the avalanche process is simulated, by means of interaction cross sections of electrons with gas molecules, and the gain and position of ionizations around the anode is found.

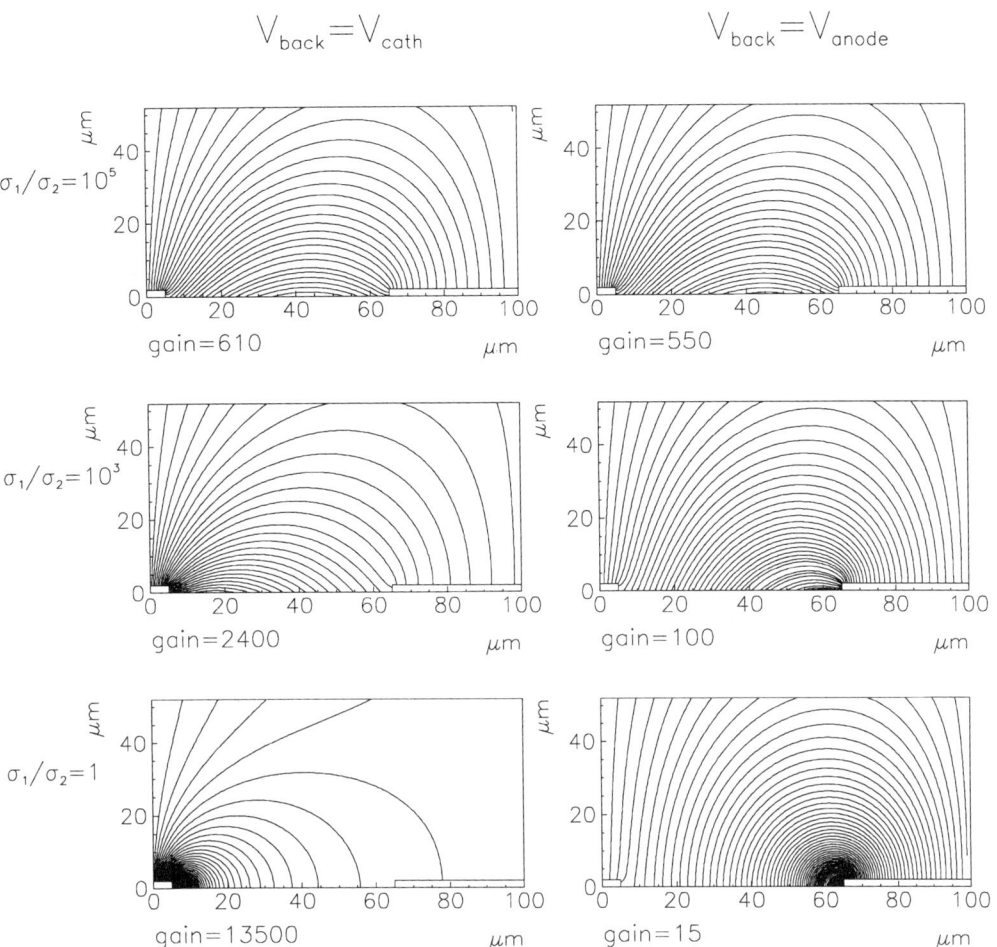

Figure 3.17: *How the electric field in the MSGC with conductive layer changes as a function of the ratio between the conductive layer resistivity and the insulating one.*

TABLE 1

GAIN		
	$V_{back} = V_{cathode}$	$V_{back} = V_{anode}$
$\sigma_1/\sigma_2 = 10^5$	610	550
$\sigma_1/\sigma_2 = 10^3$	2400	100
$\sigma_1/\sigma_2 = 1$	13500	15

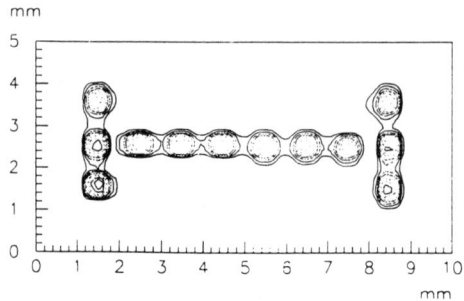

Figure 3.18: *Image of a mask with a two dimensional MSGC to prove the spatial resolution [15].*

The signal is then computed, knowing the motion of ions after the avalanche, and their electrostatic induction (weighting field) on the different electrodes of the detector.

The main reference for these "basic" parts of the simulation is [22]. The knowledge coming from all that, is then gathered to find the overall response of the detector to minimum ionising particles (m.i.p.), in terms of resolution, timing accuracy, efficiency and so on. This part is, so to say, the summary of the whole simulation of the single detector.

These stages are better described in the first four sections. The next four present briefly some applications, which are also open problems to be fully studied in future.

4.1. Electric field

The electric field of the MSGC is computed by a commercial program called ELECTRO, by the IES company, Canada. Some of the results obtained by this program have already been shown in the figures of previous sections. The electric field changes as a function of bias voltages and substrate material, and the shape of the electric field can, for example, give a first hint about the possibility that a given configuration is subject to charging. Dangerous fields, in this sense, are those that show field lines starting from the anode and ending on the substrate. An example is shown in fig. 4.1. On the contrary, fields like that of fig. 3.13 are free from charging, at least for ions (perhaps not electrons). One of the merits of this kind of computations is to have shown for the first time how the MSGC with a conductive layer on the top of the substrate can avoid the charging, even when the total leakage current is very low. Charges are not drained through but repelled from the substrate.

The electric field is then used as a basis for the following parts of the simulation.

4.2. Avalanche

In figure 4.2 the program flow is reported. It simulates the electron behaviour in the detector and the avalanche process. Input data are the initial velocity and position

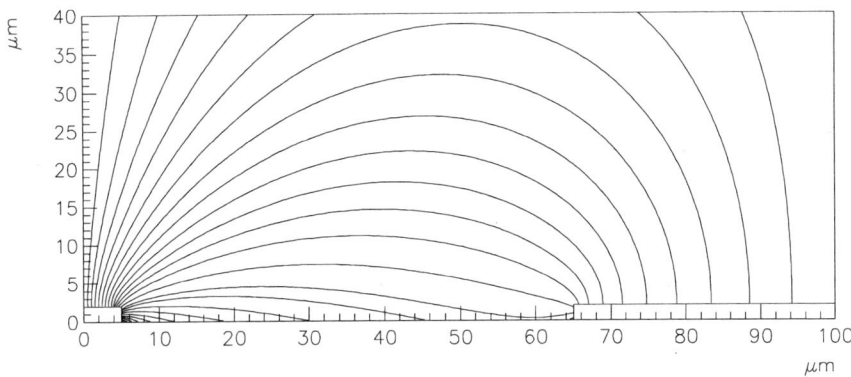

Figure 4.1: *Field lines near to the dielectric surface in an MSGC subject to charging*

of the electron and, obviously, the electric field into the detector. The electron is subject to the field acceleration for a small step with respect to its mean free path, and subsequently the probabilities of interaction with gas molecules are computed, namely elastic collision, molecule excitation and ionization, by means of cross sections kindly provided by S. Biagi, from Imperial College, London. Starting from these probabilities, by a random number extraction it is decided whether and which one of these interaction has to be simulated. Then the electron new velocity is computed according to the following rules:

- In case of *elastic collision* the electron energy is conserved, since the maximum energy loss by a mass m (electron) particle interacting with a mass M (molecule) is $2m/M$, less than numeric errors made by the simulation. The new direction is assigned with random isotropic distribution, to save computer time.
- In case of molecule *excitation* the electron energy is reduced by the excitation energy and its new direction is randomly assigned.
- In case of *ionization* the ionization energy is subtracted from electron energy and the left energy is randomly shared between the outgoing electrons. Both directions are randomly assigned.
- In absence of interaction the new velocity is computed accordingly to the electric field acceleration along the step.
- A small probability of attachment may also be taken into account.

The new position is computed, and the loop is repeated until the electron reaches the detector borders, usually on the anode. When needed the whole process can be again performed for every electron created in ionising collisions. At every electron's step the interesting quantities (like energy, cross sections, time, velocity and position after ionization) are computed and possibly recorded.

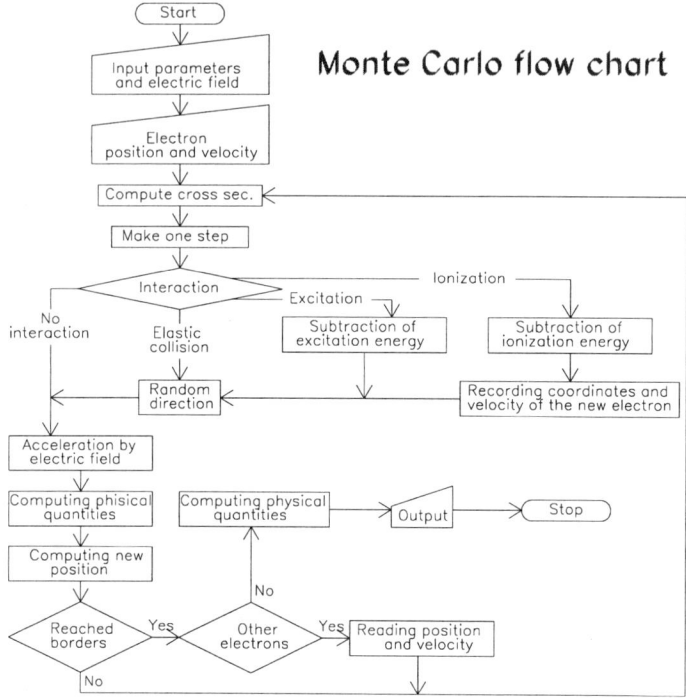

Figure 4.2: *Simplified flow chart of the monte carlo.*

The main approximations of this procedure are:

- Photo-ionization is not taken into account.
- Isotropically random direction after each interaction is assumed.
- The electric field is not modified by the space charge created in the avalanche.

These approximations seem not to be very bad, as it turns out from comparison with experimental data of results that should mainly be affected.

In figure 4.3 the path of an electron released at $\approx 100\mu m$ from electrodes is plotted.

With respect to [22] this part of the simulation has been improved in the following points:

- Introduction of attachment (negligible).
- Better cross sections with more excitation levels and higher energies.
- Introduction of more gases beside $Ar(90\%)$-$CH_4(10\%)$.

One of the results is that experimental quantities like drift velocities and diffusion coefficients now fit better the simulation over a wide range of electric field intensities and gas kinds, as can be seen in figures 4.4 to 4.6.

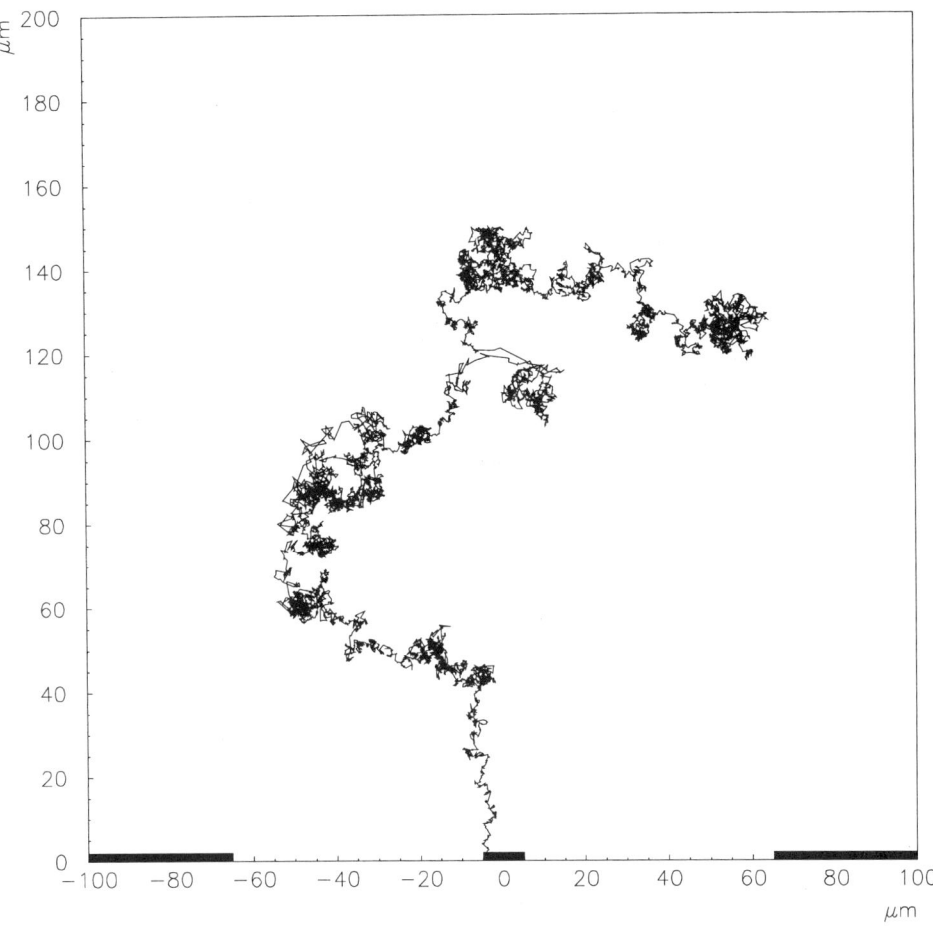

Figure 4.3: *The path of an electron inside the detector.*

Fig. 4.7 shows an avalanche in terms of the ionization points.

Some interesting quantities that can be directly extracted from this part of the simulation are, for example, the gain and its fluctuations as a function of voltages or detector shapes or materials. The behaviour with respect to the charging has already been mentioned (fig. 3.11). In fig. 4.8 another application: the fraction of charge collected by the side cathode (the rest will go to the drift cathode) as a function of anode-to-cathode voltage, with comparison to experiment.

4.3. Signal

The computation of the signal induced on every electrode of the detector can be carried out by the following steps:

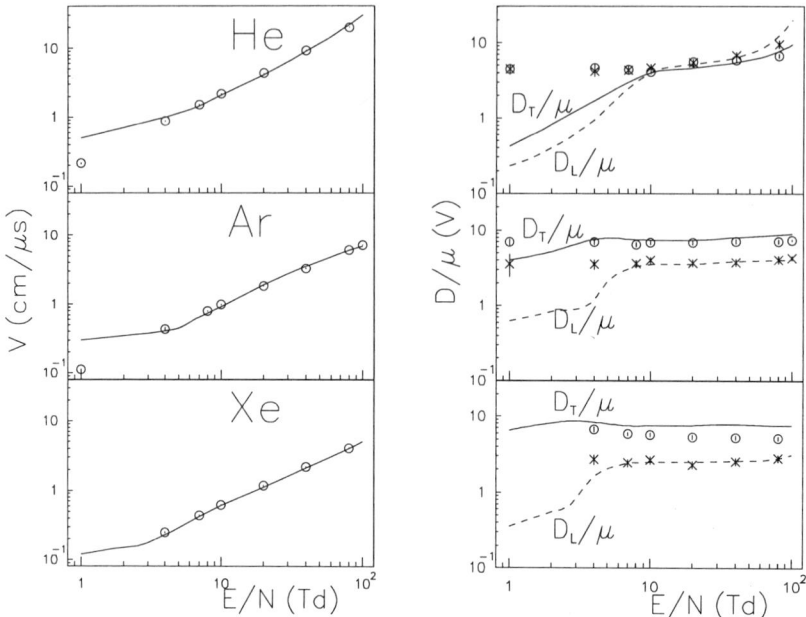

Figure 4.4: *Comparison of drift velocities and diffusion coefficients as a function of the reduced electric field obtained with the Monte Carlo (circles and crosses) and from experimental data (continuous line) [23].*

- find the electric field $\mathbf{E}(\mathbf{x})$ in the detector,
- define the starting points of ions and electrons, which should be the ionization points of the avalanche and can be obtained by the simulation described above,
- solve the equation of motion of the charges (for example, $\mathbf{v}_j(t) = \mu \mathbf{E}[\mathbf{x}_j(t)]$ where j refer to the j-th electron)
- compute the weighting field $\mathbf{E}_w(\mathbf{x})$ of the investigated electrode [26],
- use the following formulae to compute the current induced on the electrode at time t and/or the charge from time t_1 to t_2 (V_w is the weighting voltage):

$$i_k(t) = \frac{dQ_k}{dt} = -\sum_{j=1}^{m} q_j \frac{dV_w}{d\mathbf{x}_j} \cdot \frac{d\mathbf{x}_j}{dt} = \sum_{j=1}^{m} q_j \mathbf{E}_w[\mathbf{x}_j(t)] \cdot \mathbf{v}_j(t)$$

$$\Delta Q_k = \int_{t_1}^{t_2} i_k(t)\, dt = \sum_{j=1}^{m} q_j \{V_w[\mathbf{x}_j(t_1)] - V_w[\mathbf{x}_j(t_2)]\}$$

By this method it is possible to have the time development of the signal for every electrode in every detector configuration. As an example, see figure 4.9, that shows the signal of a single avalanche in a *thick* and *thin* MSGC, where thick and thin refer to the substrate, which in terms of the signal mean the distance that divide the back from the detector surface. This signal can be parametrised by a function of time to

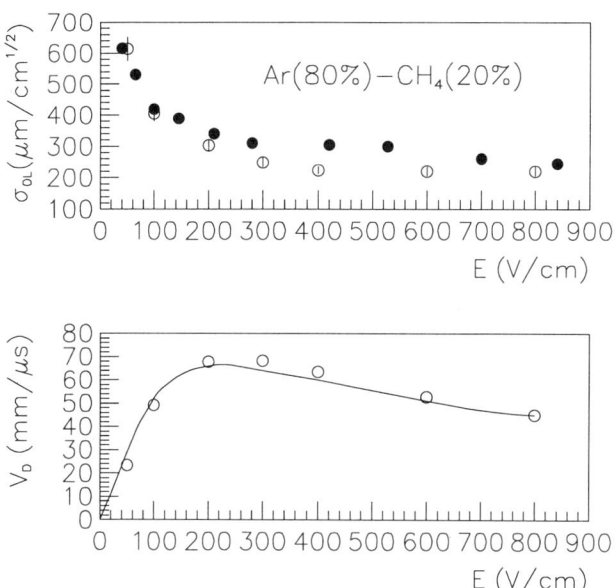

Figure 4.5: *Comparison of drift velocity and longitudinal diffusion in Ar(80%)-CH$_4$(20%) obtained with the Monte Carlo (hollow circles) and experiment (continuous line and full circles)[24].*

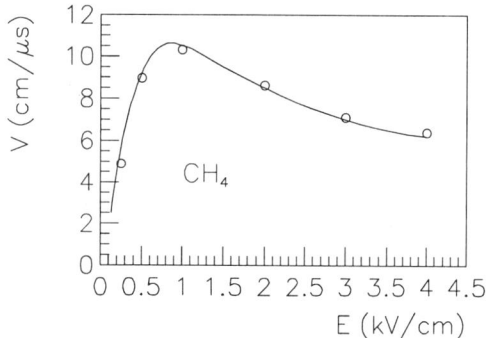

Figure 4.6: *Comparison of drift velocity in CH$_4$ obtained with Monte Carlo (circles) and experiment (line)[25].*

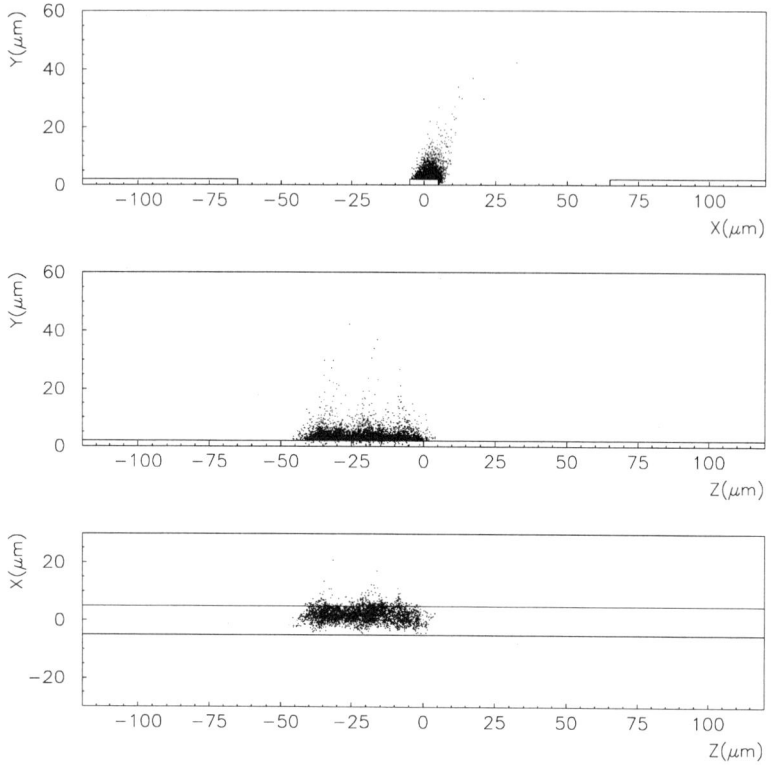

Figure 4.7: *An avalanche seen from three orthogonal directions.*

be implemented as a fast algorithm in the simulation of m.i.p. signal (see below). A function that fits well the shape of the signal of a thick MSGC is:

$$Q(t) = eA_\infty - \frac{eA_\infty - eA_0}{\left(1 + \frac{\mu t}{t_0}\right)^m}$$

$$I(t) = eA_0\delta(t) + \frac{I_0}{\left(1 + \frac{\mu t}{t_0}\right)^{m+1}}$$

eA_∞ represents the total charge collected (e is the elementary charge), eA_0 the charge induced by the electrons created in the avalanche, which are collected almost instantaneously (less than 0.5 ns), and the δ-like part of the current signal corresponds to this component, μ is the mobility of ions in the gas, and $I_0 = me\mu(A_\infty - A_0)/t_0$. A_∞, A_0, t_0, and m must be fitted to the signal shape and depend on the configuration. Typical values may be: $A_\infty = 1$, $A_0 = 0.15$, $t_0 = 4$ns, $m = 0.4$, $\mu = 1$cm^2/Vs, $I_0 = 13.6$pA.

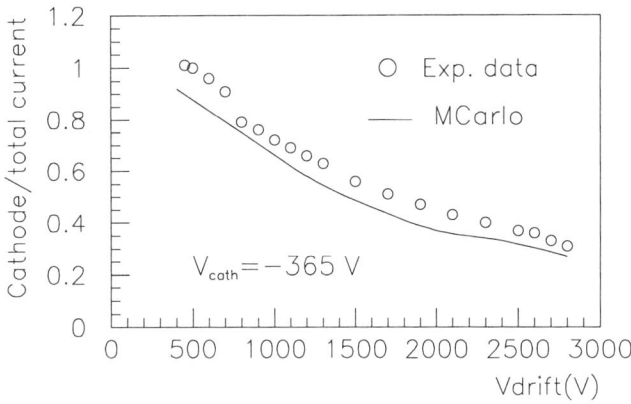

Figure 4.8: *The fraction of charge collected by the cathode as a function of the drift voltage, according to our monte carlo (continuous line) and experimental results (circles).*

4.4. Response to Minimum Ionising Particles

The "single avalanche" is the avalanche triggered by one electron, that may be similar to what is obtained for example by a point-like ionization (photon). A m.i.p. would yield a different shape, coming by the superposition of single avalanche signals, each belonging to one of the primary electrons of its track, since they reach the multiplication region at different drift times, according to the point where they are produced.

Examples of such current and charge signals are shown respectively in figures 4.10 and 4.11. They are obtained following these steps:

- Randomly distribute primary electrons along the track of the particle. This is done by a Poisson assignment of ionization clusters and a parametrisation of the cluster size distribution. The important parameters are the total number of primary electrons per unit length and the number of clusters per unit length, and depend on the gas (assuming a negligible dependence on the kind of m.i.p.).
- Let primary electrons drift to the respective anodes, where they are multiplied. Drift velocities, diffusion coefficients and drift field intensity are needed for this task. The first two quantities depend on the gas and on the field. The last can be found by the simulation of section 4.1 or by the simple formula [22]:

$$E_d = \frac{V_d - V_a}{G} - \frac{w_c + w_s}{P} \frac{V_c - V_a}{G}$$

where V_d, V_a and V_c are respectively the drift, anode and cathode voltage, G is the gas gap, w_c and w_s are the cathode width and its distance from the neighbour anode, P is the pitch.

- Randomly assign the gain of each primary electron, that is the size of the avalanche that it triggers. This number is extracted by a (usually Polya) distribution whose important parameters are its average and RMS. These may be obtained again from one of the previous stages of the simulation, the avalanche.

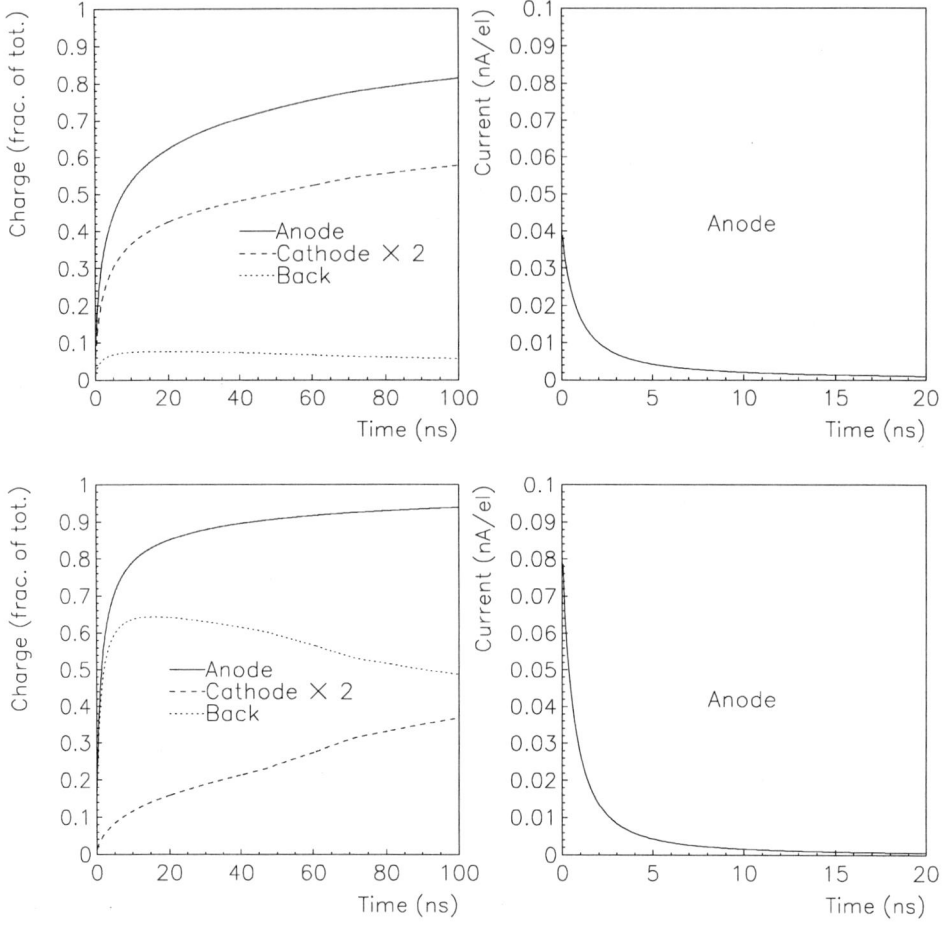

Figure 4.9: *On the left the charge signal, on the right the current signal of a classic MSGC (upper row) and a thin MSGC (bottom row)*

- Finally, the signal coming by each single avalanche is summed up with starting times corresponding to the drift time of each primary electron. The function seen in the previous section may be used for this end.

The simulation of m.i.p.'s detection does not stop here. The quantities that we are interested in are for example detection efficiency, resolution, dependence from gas, angle of incidence of the particle, signal processing algorithms etc. For this purpose more work is due after we get the signal as described above, consisting in shaping the signal, adding random noise and cross talk among strips, setting thresholds and so on, and all relevant parameters must yet be chosen or fitted to experimental data.

As an example, fig. 4.12 shows a comparison of an experimental Landau distribution with the monte carlo result. The fact that it does not fit perfectly indicates

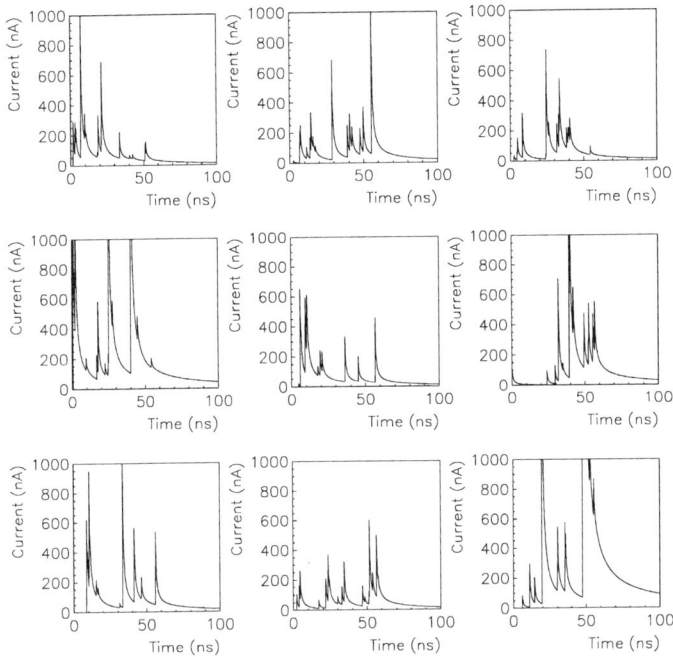

Figure 4.10: *Examples of current signal from a minimum ionising particle*

that the work is not yet fully satisfying and will be hopefully completed in future.

4.5. The problem of edges

The electric field is not always as regular as that in fig. 3.8. Indeed in that figure the detector is supposed to be infinite in two dimensions: horizontally and normally to the plane of the page. Let us forget this last coordinate and consider the field in fig. 4.13 which shows how the electric field may be distorted near to the border of the detector. The grey box on the left of the upper picture represents the frame that closes the gas region and holds the drift cathode. It is usually in some plastic and possibly polarizable material. This very field belongs to a given configuration, other shapes can be obtained as we will see later.

The edge problem is made up of several effects:
- *Distortion of detected position and resolution loss.* The effect is clear from the behaviour of drift lines. They will bring primary electrons to cells different from the one exactly below, and, in some cases, spread them on many cells instead of one or two, as usually happens. The consequence is that the detector will detect the particle systematically more to left or to right and with a poorer resolution.
- *Loss of efficiency.* This is due to two reasons. In configurations where the drift field lines are directed to the frame, instead than away from it, some of the primary electrons will be lost on the dielectric surface. This will subtract part of the charge and may cause the signal not to cross the threshold. The same can happen because of the spreading of primary electrons on more cells.

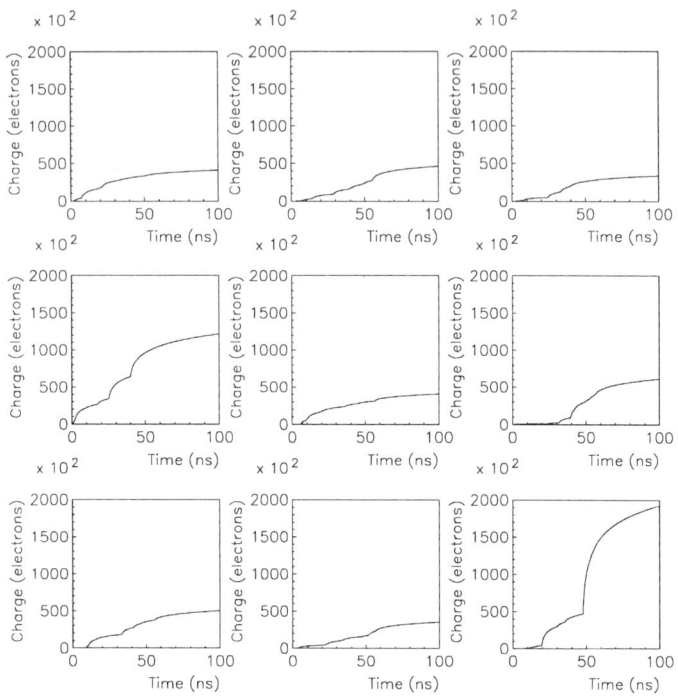

Figure 4.11: *Examples of charge signal from a minimum ionising particle*

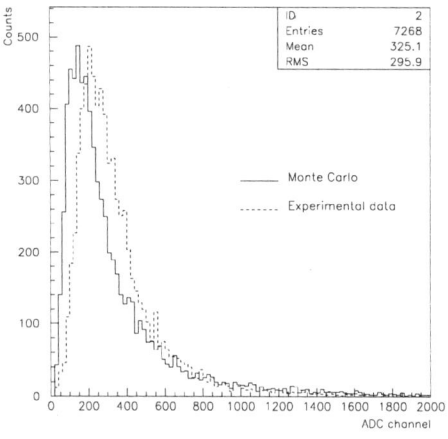

Figure 4.12: *Comparison of a Landau distribution obtained by our monte carlo and experimental data*

- *Gain modulation.* The electric field in the multiplication region of the cells closer to the border may be stronger or weaker than that of the average cell. Even a small difference in the field can lead to a sensitive difference in the gain, and in

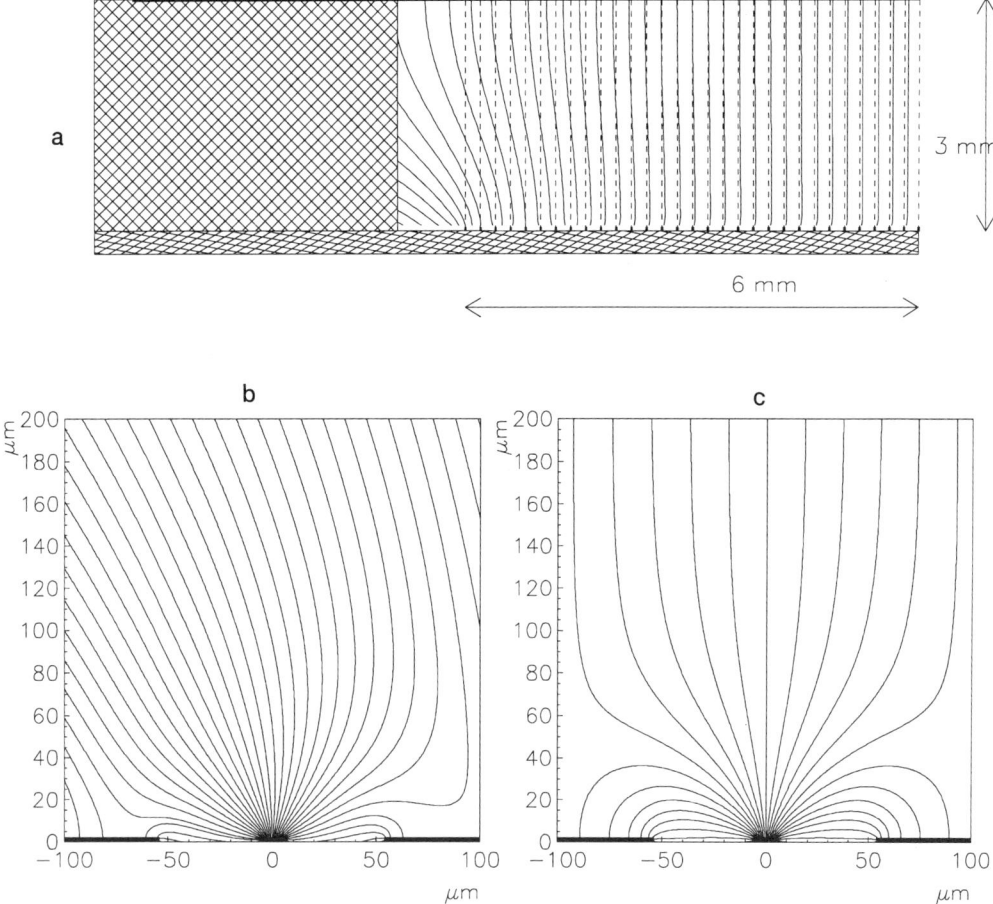

Figure 4.13: *The electric field near to the border of the MSGC. a) Large vision, hatched lines represent cell limits. b) Zoom on the leftmost anode. c) Zoom on a cell in the middle of the detector, for comparison.*

the behaviour of this area of the detector.

- *Spark probability.* When the electric field is reinforced on cathodes near to the edge, the probability of sparking increases, limiting the maximal bias voltage attainable and the gain. If a non standard field on the edge is unavoidable, a weaker one is to be preferred instead of a stronger one.

- *Influence of external fields.* The environment where the MSGC must work may be very complicated from the electrostatic point of view. Cables and other detectors may be very close to each other, and some of their parts can be put at very different voltages (for instance the drift cathode of a neighbour MSGC may be close to the back of another MSGC). The middle region of a MSGC is usually not much affected by this problem, because it is almost totally screened by the drift cathode on the upper side and by the grid of anode and cathode strips on the bottom side. The border region is however more open to these influences, and two MSGC's with different neighbour conditions would behave differently.

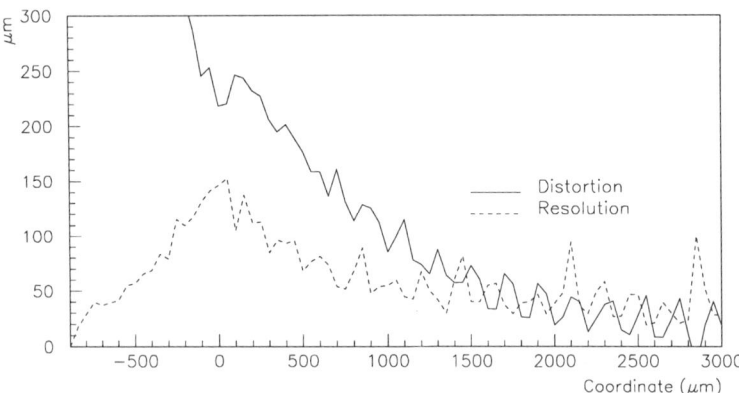

Figure 4.14: *Distortion and resolution of the detector as a function of the distance of the track from the border (the origin is on the first anode, the border of the frame is at position $-900\mu m$).*

We can investigate these problems more precisely with the aid of the simulation. Fig. 4.14 shows, for the field of figure 4.13, the distortion, that is the difference between the detected and the real coordinate of a particle track incident normally to the plane of the MSGC, and the resolution of the detector as a function of the (real) coordinate of the track. A modulation of both quantities can be noticed, with the pitch as a period. This effect is not due to the border and continues in the middle of the MSGC. Some peaks in the graph of the resolution, that are taller than average, are due to a bug in the algorithm that has already been found and will be soon corrected.

The influence of external voltages can be seen in figure 4.15, where the lower field differs from the upper one because of a plane (for example the drift cathode of a neighbour MSGC) put at the drift cathode voltage, which is located 3 mm below. Notice for example that field lines starting exactly over the leftmost anode end their path on the fourth in the lower detector, on the second in the upper one. This behaviour can be corrected suitably screening the internal area, for example by a back plane (see fig. 4.16) that unfortunately, as is well known, introduces stray capacitances and noise.

In the last-but-one detector the drift cathode ended just at the beginning of the gas gap. This may be a useful modification in view of a reduction of the edge field to a more regular shape. Another idea may be the introduction of a "guard strip", that is a wide strip (e.g. 500 μm) beyond the last cathode, which can be put at anode or cathode voltage. Fig. 4.17 shows some of these settings. More interesting is however the use of the coated substrate, i.e. the substrate covered with a more conductive thin layer. This is known to screen the active area of the detector from external influences and gives the fields of fig. 4.18 or fig. 4.19, according to the voltage on the guard strip. Now the surface under the frame tends to the same voltage of the leftmost strip,

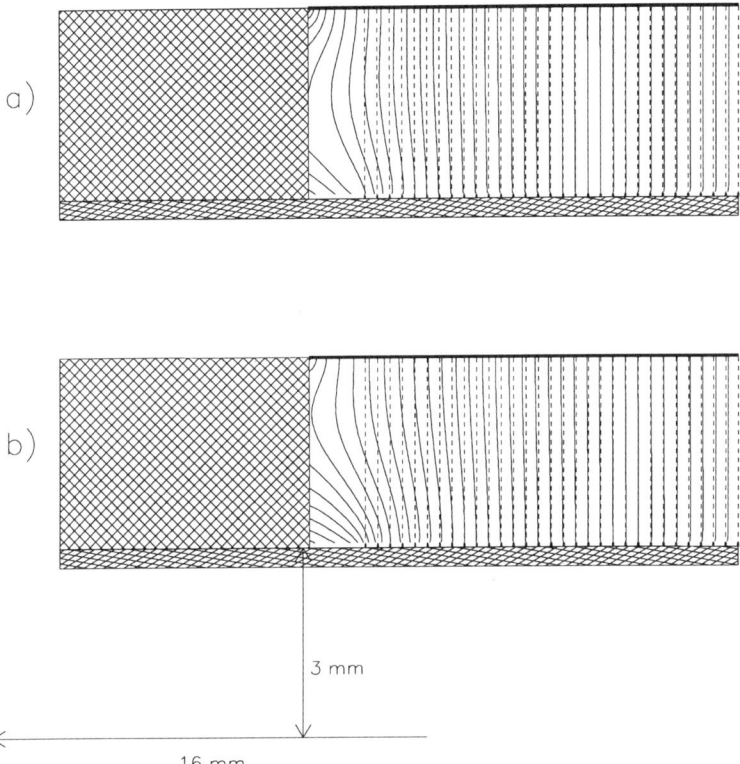

Figure 4.15: *Influence of a cathode 3mm away from the bottom side of an MSGC. a) without cathode, b) with cathode*

that in this case is the guard strip. In absence of the guard strip, the leftmost strip is usually the last cathode. Under each picture of the field there is the plot of distortion and resolution. The last two configurations offer also a weaker gain modulation, as shown in fig. 4.20 where the gain of the last three strips is indicated and compared with the same for the field of fig. 4.13.

These configurations seem to behave better than the previous ones, and the coated MSGC is nowadays almost the standard choice, so no particular care should be taken to avoid edge problems, beside this one.

The fundamental defect of this analysis of the edge effect, is not to take into account the charging of the plastic frame. For this, a study like that of section 3.3 in the case of the substrate charging is required, which is a dynamical process and takes some time. It will be one of my next concerns.

4.6. Bunch Crossing Identification

The central tracking system of CMS will be formed by a core of solid state strip detectors and by outer MSGC layers. As components of the same system, and due to the similar structure, several problems concerning their use in CMS are common

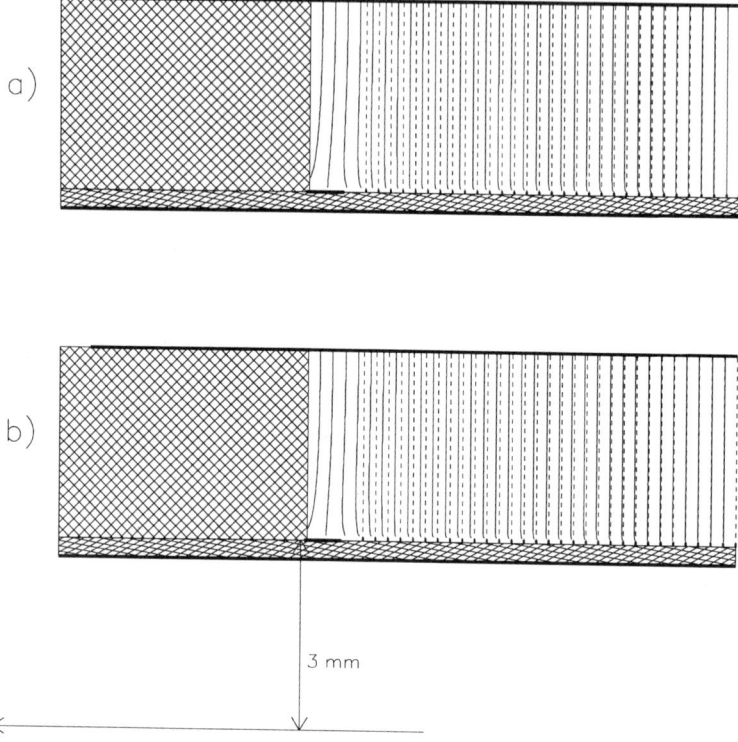

Figure 4.16: *The same as fig. 4.15, but for an MSGC with a back plane.*

to both kind of detectors. As an example, the high channel density imposes a read out electronics with low power consumption, while to assign each event to the bunch crossing in which it has been produced, a time resolution of the order of the *bunch crossing interval* (BCI) is needed, that is 25 ns. Furthermore, the signal provided by both detector types has more or less the same size and this suggests to use for the MSGC read out the same circuitry which has been built for the silicon devices [27]. The possibility to use an already developed technique would give quite a few practical and economical advantages. In particular the method of *signal deconvolution*, already successfully applied to semiconductor devices, is thought to be adapted to MSGC.

The deconvolution method [28,29] originates from two opposed requirements of CMS. The first one is a fast signal, which singles out without ambiguity the parent bunch crossing. The second one is the need of a limited heat and noise production, which on the contrary requires not very fast amplifiers. The method consists of renouncing an amplifier with shaping times of the order of BCI, obtaining a relatively slow signal upon which the deconvolution is then performed. This process (fig. 4.21) allows one to go back to the initial signal shape, i.e. its shape at the output of the detector and before integration, therefore to the instant of the event development, but only for triggered or selected events. The method is based on the read out of

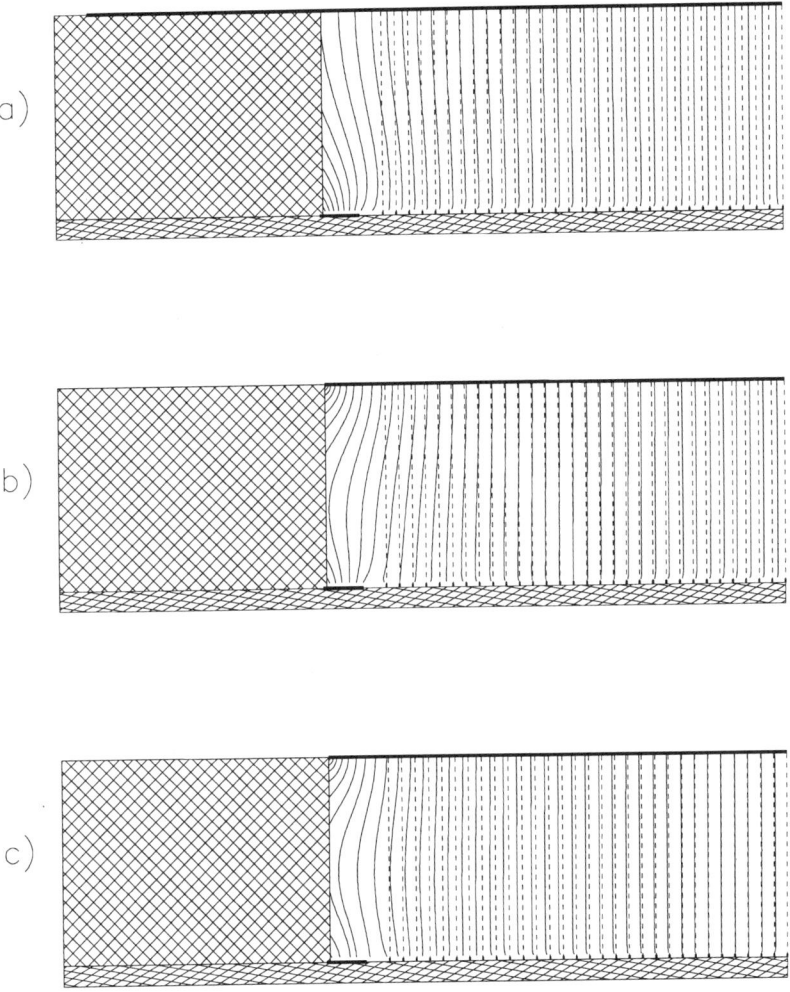

Figure 4.17: *Three configurations: a) long drift cathode, guard strip at side cathode voltage, b) short drift, guard strip at anode voltage, c) short drift, guard strip at side cathode voltage.*

three signal values separated by sampling times less than the shaping time, and on their weighted sum in a suitable way as to reproduce the original signal. The task is performed by a circuitry already designed and tested, which dissipates less than 0.1 mW per channel. In practice, all this is equivalent to a signal integration with shorter shaping times, with in addition some advantages concerning noise and power dissipation.

The biggest problem to be solved to adapt this technique to MSGC, is that it works well if the signal supplied by the detector shows a single pronounced peak, as it happens in semiconductor detectors. The MSGC signal is made, as has been shown,

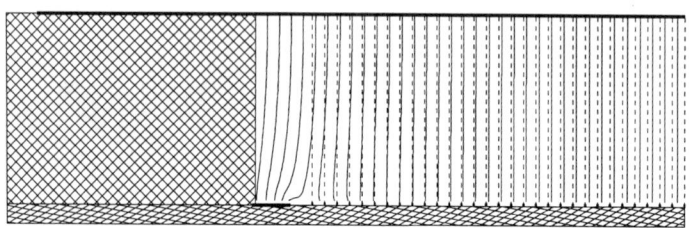

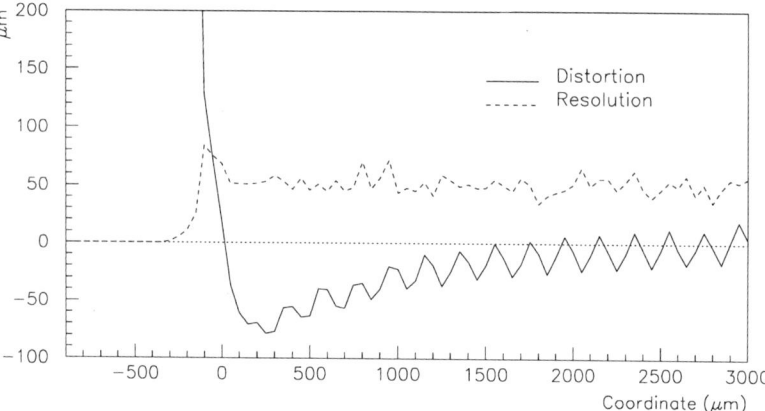

Figure 4.18: *Electric field and plots of distortion and resolution for a coated MSGC with guard strip at anode voltage.*

of a series of peaks spread along a 50÷100 ns long interval and is then difficult to fix the particle passage with a precision of less than two BCIs. A more sophisticated study of the method (for example a four times sampling instead of three), according to the peculiar MSGC signal, is then needed

4.7. The Micro-Gap Chamber

Another application of the simulation is the study of other detectors beside the MSGC. This is useful both before the implementation of the structure, for a useful design, and afterwards, to understand and improve it. This has been done, for example, with the Micro-Gap Chamber, a new device, similar in some respects to the MSGC, that provides a very good performance in terms of speed, spatial and energy resolution [30].

The structure of the detector is shown in fig. 4.22, where we recognise (from bottom up):

- The detector substrate (quartz), which acts only as a mechanical support. Any substrate material compatible with the silicon processing lines could be used as well (silicon, glass, sapphire, etc.).

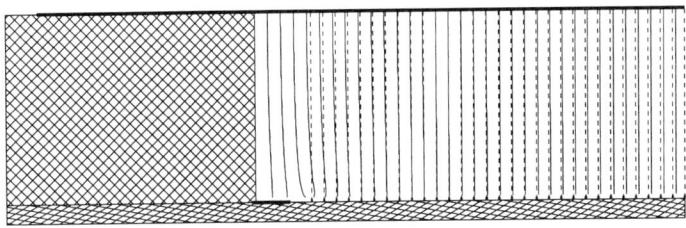

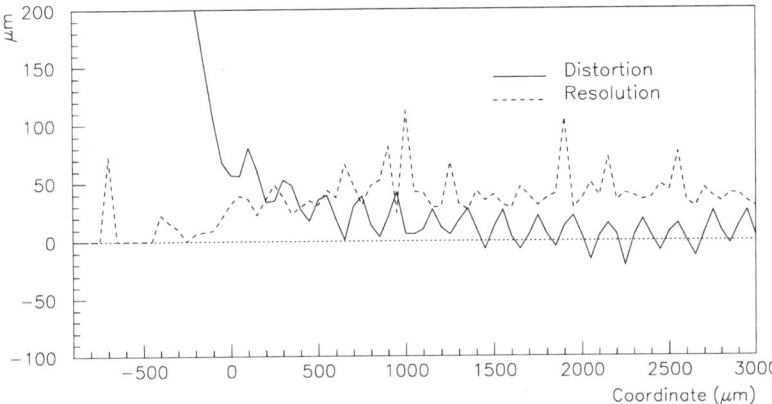

Figure 4.19: *Electric field and plots of distortion and resolution for a coated MSGC with guard strip at cathode voltage.*

- The first Aluminium film (Metal 1, 1 μm thick) deposited onto the underlying quartz substrate. This metal layer is the positive charge collecting electrode (the cathode). The Al film can be patterned by standard photo-lithography and dry etching techniques to provide a measurement of the coordinate along the anode strips (2-D read-out) or to provide the trigger;
- The 2 μm thick insulating strips obtained by patterning an inter-metal oxide deposited on Metal 1 by a plasma enhanced chemical vapour deposition technique (PECVD). These insulating strips set the anode-cathode distance and are only a few microns wider than the overlying anode micro-strips. A new technique, called *self alignment* can provide insulating strips exactly as wide as the anode. In both cases practically no insulating material is left exposed, so that charging of the substrate cannot occur.
- The second Aluminium film (Metal 2, 2 μm thick) deposited onto the oxide strips and on which the anode micro-strips (5-9 μm wide, 100-200 μm apart) are engraved with a plasma etching technique. The anode micro-strips lie on and run parallel to the oxide micro-strip beneath. The side cathode strips of the MSGC do not exist anymore.
- The usual gas gap ($\approx$ 3 mm) and drift electrode (of course not to scale).

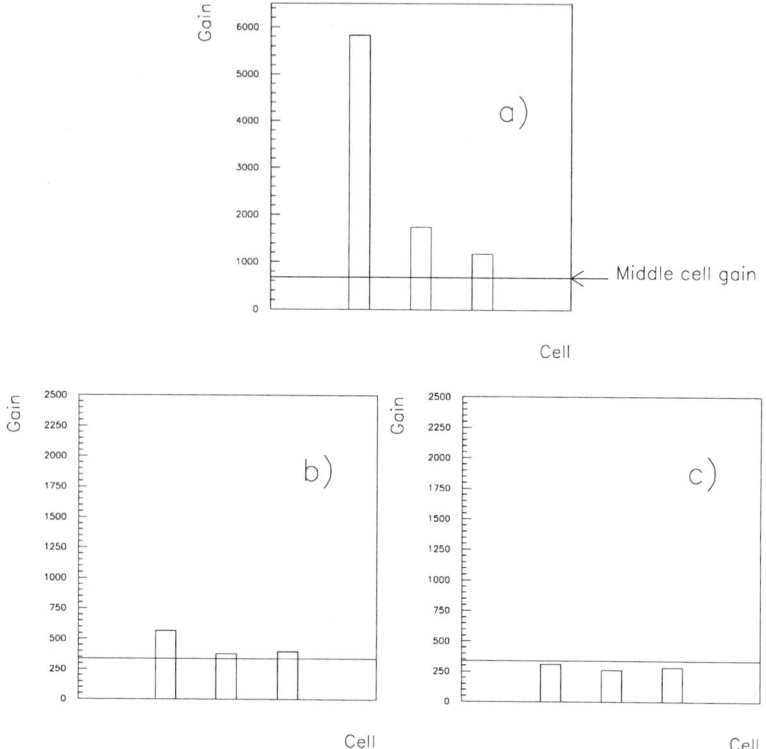

Figure 4.20: *Gain of the first three cells. a) Simple substrate, no guard strip, b) coated substrate, no guard strip, c) coated substrate, guard strip at anode voltage.*

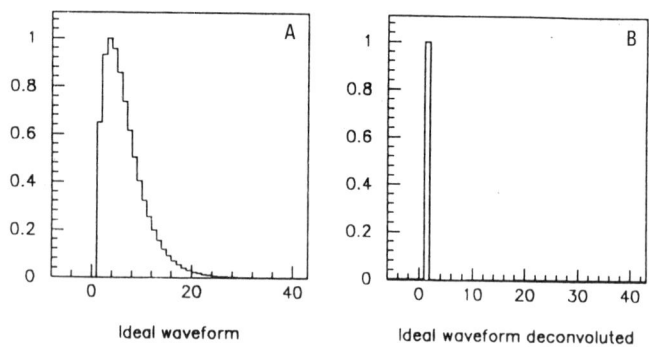

Figure 4.21: *A signal of ideal shape, produced by a CR-RC shaper with equal time constants (a) and the result of the deconvolution (b). The time unit is the* Bunch Crossing Interval *(BCI=25 ns)* [28].

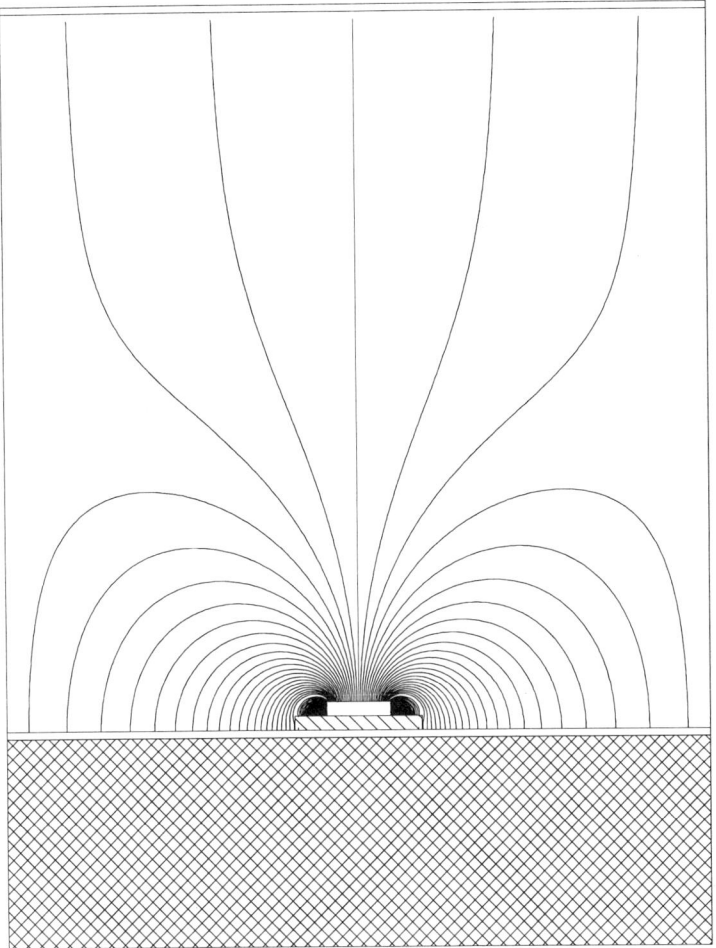

Figure 4.22: *The Micro Gap Chamber.*

The MGC collects extremely quickly the avalanche charge, since the large cathode area is practically under the multiplication region and the electric field intensity is high even for comparatively low voltages. The last property assures a good gas amplification even for voltages at which the MSGC would not yet gain. The extreme proximity of the cathode (a few microns) imposes however a greater fraction of quencher in the gas mixture and prevents from raising the voltage to values typical of standard MSGC's. This drawback is overcome by the high signal speed, see fig. 4.23, which provides an equivalent or larger signal than MSGC even at lower gains. The energy resolution has been measured until now as being approximatively equal to the best MSGC's.

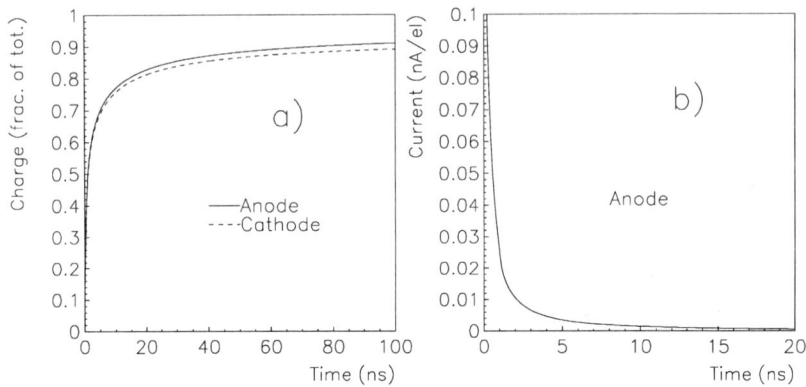

Figure 4.23: *The signal of a single avalanche in the MGC. a) charge signal, b) Current signal.*

The very short charge collecting time ($\approx$ 20 ns) allows a very high rate capability ($\approx 10^7$ mm^{-2}s^{-1}). From this point of view the MGC behaves almost like a solid state device, with speed respectively one and three orders of magnitude greater than MSGC and common wire chambers. The high fraction ($\approx$ 80%) of the avalanche charge detected also on the back strips allows to use the MGC as a 2-dimensional device. As such it will be used in CMS to measure the z-coordinate of the event vertex. In this case a small angle stereo arrangement of the strips will be implemented [31].

4.8. Pixel detectors

While the Micro Gap Chamber has been actually built, some detectors have only been simulated. One of them is a *pixel detector*, i.e. a detector whose aim is to give a (two-dimensional) point-like information of the track position. For this purpose, a structure like that in fig. 4.24 has been imagined and simulated. a, b, and c are free parameters on which the gain and the collection efficiency depend. The problem of collection efficiency comes from the fact that the anode strip of MGC and MSGC is now reduced to a "button" and some of the primary electrons diffuse towards the cathode, which is anyway a positive electrode with respect to the drift cathode, instead than to the anode.

In table 2 a, b and c for 17 models are listed and a name ranging from R1 to R17 is assigned to each of them. Other characteristics common to any of them are:

- distance among pixels (pitch): 200μm,
- cathode thickness: 2μm,
- gas gap: 3mm,
- gas: Ne(50%)-DME(50%),
- anode voltage: 0 V,
- drift voltage: 3000 V,
- cathode voltage at which the maximum efficiency and the gain are calculated: 500 V,

- cathode voltage at which the minimum efficiency is calculated: 300 V.

The results are in figs. 4.25 and 4.26

In the simulation the structure has been approximated by a rotationally symmetric one (instead of hexagonal or squared). The general effect of raising the button height is to lower the gain when it crosses about 5μm, in some cases even before and to improve charge collection. The same is obtained decreasing the anode-to-cathode distance. Some models behave differently from this trend. Many of them show a good performance from the point of view of these quantities.

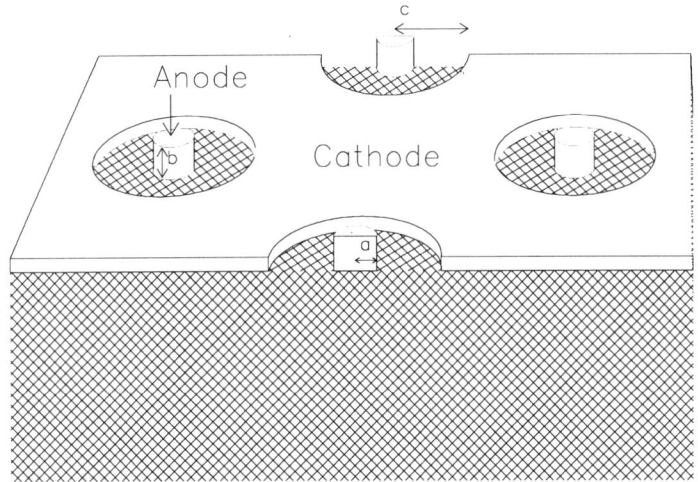

Figure 4.24: *The pixel detector*

5. Conclusion

The MSGC is a device for particle detection designed about eight years ago. The MGC has been built less than three years ago. Though they work on the same principles of other gas detectors, now well known, their microscopic structure and the presence of the substrate give new problems, and several characteristics of their behaviour are not yet very clear. Laboratory studies can do a lot to answer many questions, but they would be much more effective if supported by a theoretical prediction of phenomena taking place inside the detector. This could lead to order them

TABLE 2

	a	b	c		a	b	c
R1	10	2	50	R10	5	10	25
R2	10	10	50	R11	10	2	25
R3	10	20	50	R12	20	2	50
R4	5	10	50	R13	20	10	50
R5	5	2	50	R14	20	20	50
R6	10	10	60	R15	10	2	40
R7	10	10	25	R16	10	10	40
R8	10	20	25	R17	10	20	40
R9	2	2	25				

in a more complete and constructive scheme. Since both detectors are still in their infancy this knowledge was not developed up to now.

The aim of this work has then been an attempt to fill this lack, providing a tool to foresee and explain some features of MSGC and MGC behaviour.

6. References

[1] The ALEPH Collaboration – *"Mass limit for the Standard Model Higgs Boson with the full LEP I ALEPH data sample."* – CERN PPE/96-079

[2] J. Ellis et al. – *"The top quark and Higgs boson masses in the Standard Model and the MSSM."* – Phys. Lett. B**333**(94)118

[3] R. Casalbuoni et al. – *"Physical implication of possible $J = 1$ bound states from strong Higgs."* – Nucl. Phys. B**282**(1987)235

[4] A. C. Longhitano – *"Low-energy impact of a heavy Higgs boson sector."* – Nucl. Phys. B**188**(1981)118

[5] A. Dobado et al. – *"Learning about the strongly interacting symmetry breaking sector at LHC."* – CMS TN/94-276

[6] M. Bando et al. – *"Non linear realization and hidden local symmetries."* – Phys. Rep. **164**(1988)217

[7] R. Casalbuoni et al. – *"Effective weak interaction theory with a possible new vector resonance from a strong Higgs sector."* – Phys. Lett. B**155**(1985)95

[8] R. Casalbuoni et al. – *"Vector and Axial Vector bound states from a strongly interacting electroweak sector."* – Int. Jou. of Mod. Phys. A**5**(1989)1065

[9] R. Casalbuoni et al. – *"Symmetries for vector and axial-vector mesons."* – Phys. Lett. B**349**(1995)533

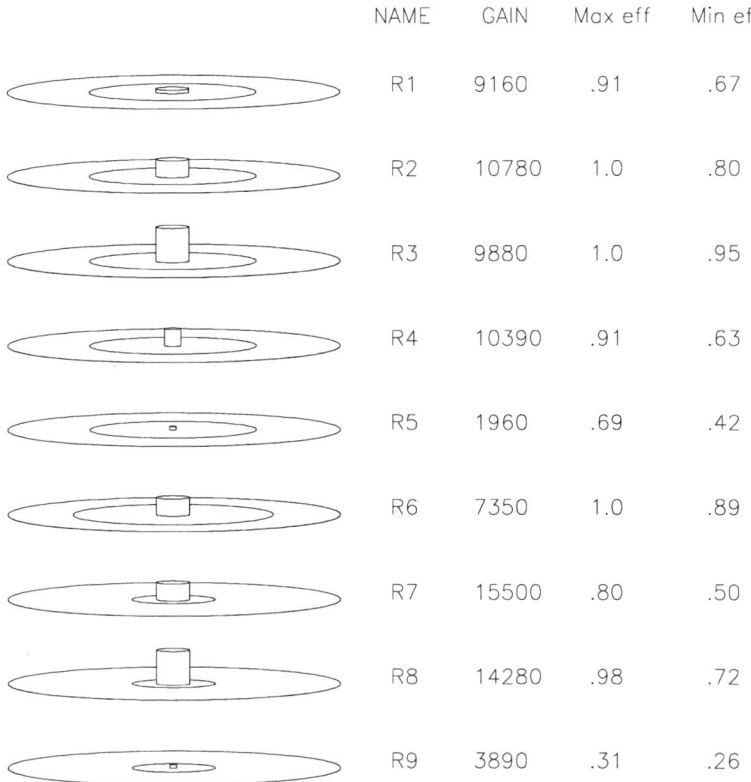

Figure 4.25: *Results of the pixel simulation (I).*

[10] R. Casalbuoni et al. – *"Low energy strong electroweak sector with decoupling."* – Phys. Rev. D**53**(1996)5201
[11] European Committee for future accelerators – *"Large Hadron Collider workshop proceedings."* – CERN 90-10, ECFA 90-133
[12] The CMS Collaboration – *"CMS Technical Proposal."* – CERN/LHCC 94-38,LHCC/P1
[13] F. Angelini et al. – *"Behaviour of microstrip gas chamber in strong magnetic field."* – Nucl. Inst. and Meth. A**343**(1994)441
[14] F. Angelini et al. – *"A thin, large area microstrip gas chamber with strip and pad read-out."* – A**336**(1993)106
[15] F. Angelini et al. – *"A MSGC with true two-dimensional and pixel read out."* – Nucl. Inst. and Meth. A**323** (1992) 229
[16] F. Angelini et al. – *"The microstrip gas chamber."* – Nucl. Phys. B **23-A** (1991) 254

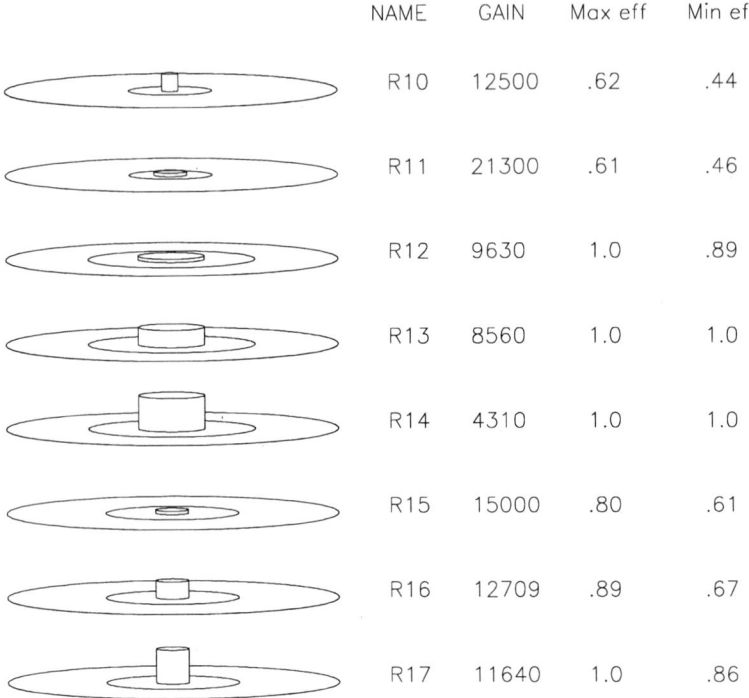

NAME	GAIN	Max eff	Min eff
R10	12500	.62	.44
R11	21300	.61	.46
R12	9630	1.0	.89
R13	8560	1.0	1.0
R14	4310	1.0	1.0
R15	15000	.80	.61
R16	12709	.89	.67
R17	11640	1.0	.86

Figure 4.26: *Results of the pixel simulation (II).*

[17] A. Oed et al. – *"Substratum and layout parameters for microstrip anodes in gas detectors."* – Nucl. Inst. and Meth. **A310**(1991)95

[18] H. Sakurai et al. – *"Dependence of energy resolution on anode diameter in xenon proportional counters."* – Nucl. Inst. and Meth. **A313**(1992)155

[19] R. Bouclier et al. – *"Performance of gas microstrip chambers on glass and plastic supports."* – Nucl. Inst. and Meth. **A232**(1992)240

[20] R. Bouclier et al. – *"Performance of gas microstrip chambers on glass substrata with electronic conductivity."* – Nucl. Inst. and Meth. **A332**(1993)100

[21] S. Biagi et al. – *"Initial investigations of the performance of a microstrip gas avalanche chamber fabricated on a thin silicon dioxide substrate."* – Nucl. Inst. and Meth. **A323**(1992)258

[22] R. Bellazzini et al. – *"Electric field, avalanche growth and signal development in MSGC's and MGC's."* – La Rivista del Nuovo Cimento, vol. 17, num. 12, 1994

[23] J. L. Pack et al. – *"Longitudinal electron diffusion coefficients in gases: Noble gases."* – J. Appl. Phis. **71**(1992)5363

[24] F. Piuz – *"Measurement of the longitudinal diffusion of a single electron in gas mixtures used in proportional counters."* – Nucl. Inst. and Meth. **205**(1983)425

[25] B. Jean-Marie et al. – *"Systematic measurement of electron drift velocity and study of some properties of four gas mixtures."* – Nucl. Inst. and Meth. **159**(1979)213

[26] V. Radeka – *"Low noise techniques in detectors."* – Ann. Rev. Nucl. Part. Sci., **38**(1988)217

[27] R. Sachdeva et al. – *"Fast electronic readout of microstrip gas chambers."* – Nucl. Inst. and Meth. **A348**(1994)378

[28] S. Gadomski et al. – *"The deconvolution method of fast pulse shaping at hadron colliders."* – Nucl. Inst. and Meth. **A320**(1992)217

[29] N. Bingeforts et al. – *"A novel technique for fast pulse shaping using a slow amplifier at LHC."* – Nucl. Inst. and Meth. **A326**(1993)112

[30] F. Angelini et al. – *"The Micro Gap Chamber."* – Nucl. Inst. and Meth. **A335**(1993)69

[31] R. Bellazzini et al. – *"The MicroGap Chamber: a new detector for the next generation of high energy, high rate experiments."* – Nucl. Inst. and Meth. **A368**(1995)259

EXPERIMENTAL TESTS OF THE ELECTROWEAK THEORY AT LEP AND SLD

Alain BLONDEL
L.P.N.H.E. Ecole Polytechnique 91128 Palaiseau FRANCE
Mailing address: CERN, Geneva, Switzerland

Abstract

Experiments in Electroweak Physics try to answer two main questions: i) can one see New Physics beyond the Standard Model (SM)? ii) what is the Higgs boson mass? The first question is addressed by improved measurements of fermion couplings, both in Charged Currents, from τ physics in particular, and in Neutral Currents, where hopes were high with the R_b, R_c "puzzle". The measurements of R_c have been entirely revisited and are now in agreement with SM. A new, precise measurement of $R_b = 0.2158 \pm 0.0014$, agrees also with SM; the world average is now within two standard deviations of SM. This "normality" is corroborated by the agreement of the total hadronic width with SM expectations using the new world average of $\alpha_s(M_Z^2)$. A 3 standard deviations discrepancy now appears for the b quark NC coupling asymmetry. As for the Higgs boson, it has not been found yet, so its mass is the only free parameter left in the electroweak fits. Its indirect determination from radiative corrections benefits from improvements in several ingredients: i) the top and W masses from the Tevatron; ii) new Z line shape scan; iii) new asymmetries from SLD and LEP and iv) further refinements in neutrino scattering. The Higgs boson mass comes out to be

$$M_H = 150^{+150}_{-80} \text{ GeV}; \quad M_H < 540 \text{ GeV at 95\% C.L.}.$$

Introduction: what is to be tested?

Family Universality

The first unavoidable consequence of gauge theories, on the basis of which the Electroweak Theory [1] is constructed, is family universality: if there is one group, there has to be the same coupling for all fermions in similar multiplets. In the case of Charged Currents (CC), this is modulated for quarks by mixing, and it could be so for leptons as well. Since the number of families is three, following the measurement of the number of light neutrinos at LEP, mixing must be described by a 3x3 unitary matrix. Universality is violated by fermion masses and could therefore receive small radiative corrections. Unexpected violations in CC could signal further families. In both Neutral Currents (NC) and CC, violations could arise from new particles with badly broken family symmetry, coupling to known fermions, through radiative corrections.

Universality of Weak couplings

The choice of $SU(2)_L \times U(1)$ for gauge group and the particular assignment of fermions in its multiplets was imposed by the properties of CCs. Parity violation, for instance, is completely ad-hoc. Given this, however, the relationship between NC chiral couplings of the various fermions is strictly imposed:

$$\begin{aligned} g_{Vf} &= (g_{Lf} + g_{Rf}) = I^3_{Lf} - 2Q_f \sin^2\theta_w \\ g_{Af} &= (g_{Lf} - g_{Rf}) = I^3_{Lf}. \end{aligned} \quad (2)$$

These relations must hold for all fermions with the **same** value of the weak mixing angle $\sin^2\theta_w$. Again, symmetry between the various fermions is broken by masses, and these relations might receive small radiative corrections. More sizeable violations would point to either i) a new gauge group (Z' etc..), or ii) to the existence of new particles affecting radiative corrections in a non-universal way.

Universality of NC and CC couplings

The overall coupling strength of CC and NC weak currents is uniquely predicted in terms of the weak mixing angle and the electric charge:

$$g = e/\sin^2\theta_w, \quad g_{NC} = e/\sin^2\theta_w \cos^2\theta_w. \quad (3)$$

These relations have to be valid for the same value of $\sin^2\theta_w$ as that of equation 2. In the SM, these relations are affected by radiative corrections in the photon, W and Z propagators, but these should be universal, e.g. independent of the fermions involved.

Symmetry Breaking

All the above considerations are consequences of the gauge structure of the theory. The fact that W and Z, but also fermions, have masses requires symmetry breaking [2]. This is achieved most economically by introducing one doublet of scalar fields, of which, after absorption of all additional degrees of freedom, only the neutral Higgs boson remains as physical particle. The first consequence of this minimal scheme is the relation between the W and Z masses:

$$\sin^2\theta_w = 1 - \frac{M_W^2}{M_Z^2}. \quad (4)$$

This relation is specific of symmetry breaking with Higgs doublets. It can be different for different Higgs structure. It is affected by radiative corrections that differ for the W and Z propagators, most particularly those involving $SU(2)$ breaking, such as the top-bottom mass difference.

The mechanism of symmetry breaking is certainly the least satisfactory part of the Standard Model. It is the area where most conceptual difficulties arise, not the least because radiative corrections to the Higgs boson propagator send its mass to infinity, unless new physics is invoked to cancel these divergences. Nevertheless, this minimal scheme provides us with a renormalizable theory [3], allowing calculation of radiative corrections and predictions for most electroweak observables at a level of precision of a few 10^{-4}.

Radiative Corrections

The free parameters of the theory are
- g, g', the Higgs boson mass, M_H, and its vacuum expectation value ¡v¿.
- all fermion masses and mixing angles,
- α_s,

where g, g' are the $SU(2)_L$ and $U(1)$ couplings and $<v>$ is the Higgs vacuum expectation value. In practice they are replaced by the three best measured electroweak quantities available, namely α, G_F, M_Z.

Every electroweak observable can be calculated in the MSM from the above set of parameters. The largest sources of uncertainties come from:
- the light quark masses, which will be discussed in Sec. .
- the top quark mass, which enters in the W and Z propagators, and in the $Z \to b\bar{b}$ vertex correction. Until the top quark observation [5] in 1994, its mass was unknown and precision measurements were measuring it indirectly, very successfully indeed [4]. Now M_t is a known input parameter. The CDF [6] and D0 [7] collaborations have provided a combined value [8] of 175 ± 6 GeV.
- the strong coupling constant $\alpha_s(M_Z^2)$ enters in hadronic observables, in particular the hadronic Z width and other related quantities. There has been much progress in the determination of $\alpha_s(M_Z^2)$ recently [9, 10] in particular with the new measurement of Λ_{QCD} [11] and the new theoretical studies on several fronts, such as the Lattice QCD calculations of charmonium and bottomonium levels, as well as the τ hadronic partial width. The World average now stands as $\alpha_s(M_Z^2) = 0.118 \pm 0.003$, where the error takes into account possible positive correlations between experiments.
- technical precision: this uncertainty pertains to the treatment of higher order corrections. The corrections related to the QCD corrections have been calculated for the inclusive hadronic width up to 3^d order. Top loop corrections are corrected to order $\alpha\alpha_s$ and even partly to order α^2. These errors are at the level of a few 10^{-4} for most Z peak observables and the W mass.
- the only free parameter left is the Higgs boson mass. Its determination has become the main goal of the precision measurements. The calculation of radiative corrections involving the Higgs mass suffers uncertainties at the level of a few 10^{-4}.

Not all electroweak observables are equally sensitive to all corrections. Table 1 gives the measured values, the SM expectations and sensitivity of the predictions to the various effects mentioned above.

Universality in Charged Current

Tests of universality in hadronic charged current processes are reviewed by Gibbons [13]. Here we concentrate on the leptonic couplings, which were reported by

Videau [15]. New data come from e^+e^- annihilation into τ pairs, both by CLEO and ARGUS in the $b\bar{b}$ threshold region, and on the Z peak. The available statistics are large, $3 \cdot 10^6$ τ pairs in CLEO, 10^5 τ pairs per experiment at LEP. Improved tests of the Lorentz structure of τ decays were reviewed by Stroynowski [16, 17, 18]. Tests of family universality in the leptonic CCs are as follows.

1. Comparison between the leptonic branching ratios $\tau \to e\nu_e\nu_\tau = 0.17817(74)$ and $\tau \to \mu\nu_\mu\nu_\tau = 0.17338(85)$. Correcting for the phase space difference of 0.97256 one obtains the coupling ratio:
$$\frac{g_\mu}{g_e} = 0.9997 \pm 0.0032$$

2. The τ partial widths into electron or muon are obtained from the above branching ratios and from the average τ lifetime, $\tau_\tau = 290.34 \pm 1.15$. They can be compared with the muon lifetime. Partial widths scale with the fifth power of the decaying lepton mass, so the comparison uses the τ mass, $\mathbf{m}_\tau = 1777.00^{+0.30}_{-0.27}$ MeV. Electron-muon universality is assumed. This gives:
$$\frac{g_\ell}{g_\tau} = 1.0010 \pm 0.0025$$

3. The most precise test of electron-muon universality comes from the ratio of $\pi \to e\nu_e$ versus $\pi \to \mu\nu_\mu$, from which one gets:
$$\frac{g_\mu}{g_e} = 1.0012 \pm 0.0016.$$

Because this involves coupling to a spin 0 pion, this test probes charged current with a longitudinal W, contrary to the tests 1 and 2 above which involve a transverse W.

4. Comparison of the $\tau \to \pi\nu_\tau$ decay with $\pi \to \mu\nu_\mu$ directly compares the τ and μ couplings to a longitudinal W, again, giving:
$$\frac{g_\tau}{g_\mu} = 1.0105 \pm 0.0066$$

5. The same test with $\tau \to K\nu_\tau$ versus $K \to \mu\nu_\mu$ gives:
$$\frac{g_\tau}{g_\mu} = 0.977 \pm 0.035$$

Table 1. Summary of the present experimental errors, physics sensitivity and theoretical uncertainties in SM predictions for the main Electroweak observables. The SM values are given for BHM [12], with $M_t = 175$ GeV, $M_H = 300$ GeV, $\alpha_s = 0.118$.

Physics sensitivity of Electroweak observables

Observable	ICHEP'96 value (error)	SM	Sensitivity to:							
			M_t(GeV) 175 ± 6	M_H(GeV) 60	1000	α_s ±0.003	$\alpha(M_Z^2)^{-1}$ 128.89(9)	(m_b) 4.7 ± 0.3	Higher orders
Γ_Z(MeV)	2494.6(2.7)	2493.4	+1.5	+4.2	-5.3	1.7	0.7	0.2	0.6
$\Gamma_{\ell\ell}$(MeV)	83.91(0.11)	83.92	+0.06	+0.11	-0.14	.02	–	–	0.02
$\sigma_{had}^{peak,0}$(pb)	41508(56)	41476	+3	-4	+4	-16	2	2	2.
$R_\ell \times 10^3$	20778(29)	20741	-1.8	+15	-13	21	4	2	4.
$\sin^2\theta_w^{eff} \times 10^4$	2316.5(2.4)	2319.2	-2.0	-8.5	+6.9	0.05	2.3	–	1.
$R_b \times 10^4$	2178(11)	2158	-2	-0.4	0.	0.05	0.2	0.8	1.
M_W(MeV)	80356(125)	80320	+37	+103	-97	1	14	–	10.

The evolution with time of these comparisons is interesting, especially of test 2. This uses the τ mass, which has shifted four years ago with the measurement at BES [14], by 2 σ. It uses also the τ life-time, of which all measurements have shifted coherently by also 2 σ since 1992. This is traced back to a collective lifetime bias for large tracking errors. This was reduced with the introduction and use of silicon vertex detectors. This story kindly reminds us that detector related systematic errors are not always uncorrelated between experiments....

QED

Muon magnetic anomaly

A new measurement of the muon magnetic anomaly $(g_\mu - 2)$, BNL experiment E821, is soon to begin [19]. Because of the muon large mass, $(g_\mu - 2) = (116\,592\,300 \pm 840)\,10^{-11} \simeq \frac{\alpha}{2\pi}$ is more sensitive to weak corrections than that of the electron, with which it is compared. Weak corrections amount to $151\,10^{-11}$, far below the current experimental precision from the CERN $g-2$ experiment, $840\,10^{-11}$. The hope is to reduce the experimental error to $40\,10^{-11}$.

The technique is really beautiful. The following numbers are humbling for those who triumph from the LEP measurement of the Z mass. The technique is to store monochromatic, polarized (they come from pion decays) muons in a ring of 7 m radius. The rate of decay electrons above a given energy threshold provides a measure of the orientation of muon polarization as a function of time, from which $g-2$ is extracted. The muon momentum is "magic", 3.094 GeV: at that momentum, spin precession in electrostatic fields vanishes. This trick allows beam steering with electrostatic quadrupoles, so that the precession frequency is independent of the exact muon trajectory. The B field integral around the ring must still be homogeneous to 10^{-6} for all trajectories, which requires an excruciating amount of shimming. The field is monitored to 10^{-7} by NMR probes placed in situ or circulating on a little train around the ring (more on trains later). This procedures relates the field to the proton magnetic factor g_p, which in turn is related to that of the muon by the hyperfine line of muonic atoms, and ultimately to that of the electron by the hyperfine line of hydrogen. First beams are expected early 1997.

The usefulness of the result is potentially upset by the uncertainty in the hadronic vacuum polarisation, presently $154\,10^{-11}$. The on-going or foreseen efforts to remeasure the cross sections $e^+e^- \to$ hadrons should allow improvement down to $60\,10^{-11}$ [21]. The physics reach of such measurement is, for several scenarios of new physics, equivalent to that of LEP2 or even LHC [22].

Hadronic Vacuum polarization

When relating measurements performed at different energy scales, and if the relation involves the QED coupling constant – as was the case for the muon $g-2$ in relation with that of the electron which *defines* $\alpha_{QED}(0)$ to $4\,10^{-8}$ – then one has to know the running of α_{QED} from one scale to the other. The running of α to the Z mass scale is given by

$$\alpha(M_Z^2) = \frac{\alpha(0)}{1 - \Delta\alpha(M_Z^2)} \quad ; \quad \Delta\alpha(M_Z^2) = -\Pi_{\gamma\gamma} \tag{5}$$

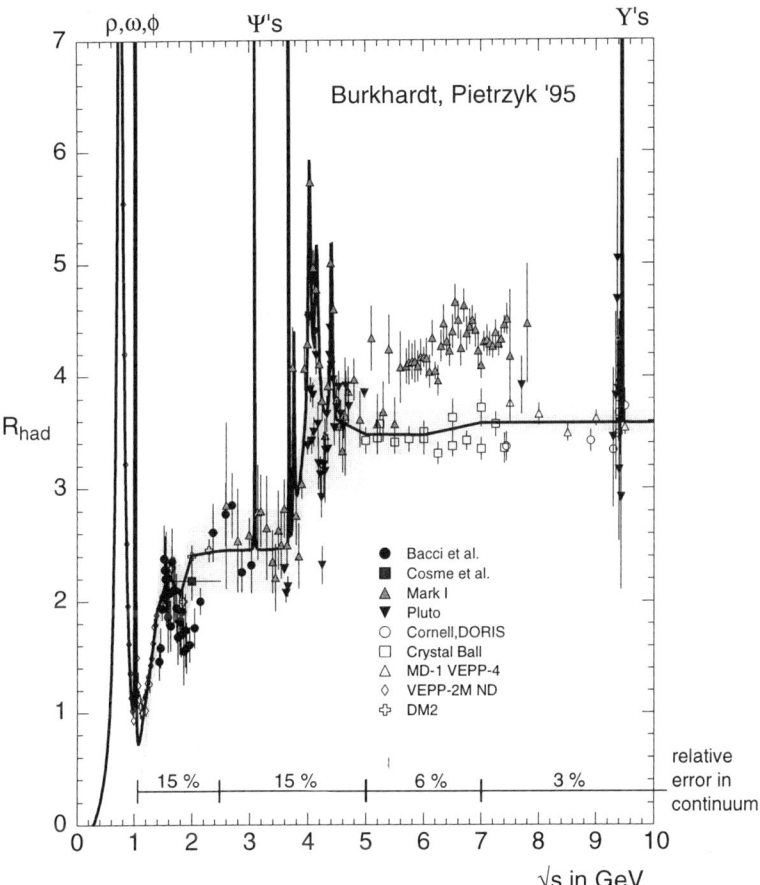

Figure 1. *Measurements of hadronic cross sections used in the evaluation [25] of $\alpha(M_Z^2)$. The band shows the correlated uncertainty assigned to each energy region.*

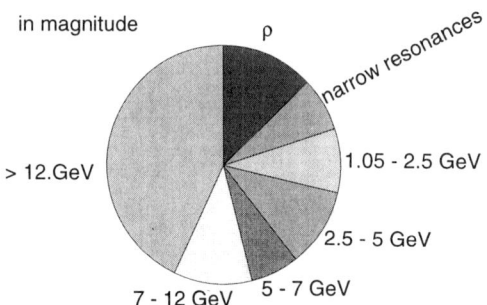
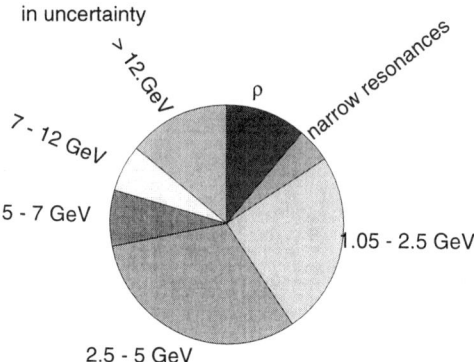

Figure 2. Relative contribution of each energy region to the hadronic vacuum polarization at the Z peak [25].

The contributions of leptons to the photon self-energy $\Pi_{\gamma\gamma}$ is well calculated, but that of quarks must be obtained from measurements. The method so far has been to use the measurement of $e^+e^- \to$ hadrons cross sections, R_{had} by the dispersion integral:

$$\Re(\Pi_{\gamma\gamma}^{had}) = \frac{\alpha s}{3\pi} \mathcal{P} \int \frac{R_{had}(s')}{s'(s'-s)} ds' \qquad (6)$$

This integral has been evaluated by several groups [23, 24, 20, 25] with consistent results. The value used here is [20, 25] $\alpha(M_Z^2) = 1/(128.89 \pm 0.09)$.

The data are shown in Fig 1. Error analysis, Fig. 2, shows a dominant contribution from the region between 1.5 and 5 GeV center-of-mass. Uncertainties in the measured cross sections in this region are as large as 15%. Many of these data are rather old, and/or this measurement was not the design goal of the detector. In most cases the uncertainty comes largely from acceptance corrections. The hadronic final state at low energies is not so well known, and cuts involving, for instance, at least two charged tracks, require large and badly known extrapolation. Above charm threshold, a complication arises when the charm channels open successively with rapidly varying characteristics. More modern detectors with better solid angle coverage and hermeticity should have no problem in improving the errors. Such measurements are planned at the low energy end in Novosibirsk and, for the very important range from 2.5 to 5 GeV, in BES. DAPHNE or the B-factories should have their say, too.

At this moment, examination of table 1, reveals that the most sensitive observable for Higgs mass determination is $\sin^2\theta_w^{\text{eff}}$, and that the uncertainty due to $\Delta\alpha(M_Z^2)$ in the prediction of this quantity is equal to the experimental error.

Re-measurement of the hadronic cross sections has as much weight as all measurements of $\sin^2\theta_w^{\text{eff}}$ put together!

The high energy measurements

A synopsis of the measured quantities

The building blocks of electroweak physics at the Z are measured cross sections for various final states, forward-backward and polarization asymmetries. Assuming that Z and photon exchange are the only processes that occur, they can all be expressed in terms of the chiral couplings, eq. 2. The $Z \to f\bar{f}$ partial width is given by:

$$\Gamma_f = \frac{\alpha}{6\sin^2\theta_w \cos^2\theta_w} M_Z (g_{Lf}^2 + g_{Rf}^2). \qquad (7)$$

The total width is the sum over all open channels. Within very good limits, only the fermions of the first three families, with the exception of the top quark, contribute to the cross section.

Around the Z pole, the photon exchange is only a correction to the Z-channel, which dominates the cross section and can then be written as:

$$\sigma_f = \frac{12\pi(\hbar c)^2}{M_Z^2} \frac{s\Gamma_e\Gamma_f}{(s-M_Z^2)^2 + s^2 \frac{\Gamma_Z^2}{M_Z^2}}. \qquad (8)$$

Forward-backward asymmetries or polarization asymmetries are sensitive to the chiral coupling asymmetry:

$$\mathcal{A}_f \equiv \frac{g_{Lf}^2 - g_{Rf}^2}{g_{Lf}^2 + g_{Rf}^2} = \frac{2g_{Vf}g_{Af}}{g_{Vf}^2 + g_{Af}^2}. \qquad (9)$$

For unpolarized beams the forward-backward asymmetry is:

$$A_{FB}^{(f)} \simeq \frac{3}{4} \mathcal{A}_e \mathcal{A}_f. \tag{10}$$

For the tau lepton, the polarization of the final state fermion is measurable, as a function of polar angle. For unpolarized beams:j

$$\mathcal{P}_\tau(\cos\theta) \simeq -\frac{\mathcal{A}_\tau + \frac{2\cos\theta}{1+\cos^2\theta}\mathcal{A}_e}{1 + \frac{2\cos\theta}{1+\cos^2\theta}\mathcal{A}_e\mathcal{A}_\tau} \tag{11}$$

from which one can derive both $\mathcal{A}_e$ and $\mathcal{A}_\tau$.

Interesting observables are obtainable if longitudinal beam polarization is available. The Left-Right asymmetry of Z production [26, 27]

$$A_{LR} = \frac{\sigma_L - \sigma_R}{\sigma_L + \sigma_R} \simeq \mathcal{A}_e, \tag{12}$$

and the forward-backward polarized asymmetry [28]:

$$A_{FB}^{\text{pol}(f)} = \frac{(\sigma_{L,F} - \sigma_{R,F}) - (\sigma_{L,B} - \sigma_{R,B})}{(\sigma_{L,F} + \sigma_{R,F}) + (\sigma_{L,B} + \sigma_{R,B})} \simeq \frac{3}{4}\mathcal{A}_f. \tag{13}$$

The values of Neutral Current couplings and their sensitivity to $\sin^2\theta_W^{\text{eff}}$ are given in table 2.

f	I_{3f}	Q_f	g_{Af}	g_{Vf}	$\mathcal{A}_f$	$\frac{\partial \mathcal{A}_f}{\partial \sin^2\theta_W^{\text{eff}}}$
ν	1/2	0	1/2	1/2	1	0
e	-1/2	-1	-1/2	-0.04	0.16	-7.9
u	1/2	2/3	1/2	0.19	0.69	-3.5
d	-1/2	-1/3	-1/2	-0.35	0.94	-0.6

Table 2. *Numerical values of quantum numbers, Neutral Current couplings, chiral coupling asymmetry $\mathcal{A}_f$ and sensitivity of $\mathcal{A}_f$ for the four types of fermions. The value of $\sin^2\theta_W^{\text{eff}}$ is 0.23.*

A strategy of tests and radiative effects

Besides QED radiative effects (emission of real or virtual photons) which are conceptually straightforward, LEP observables are sensitive to electroweak (propagator or vertex) radiative effects. Electroweak corrections are sensitive [29] to heavy, yet undiscovered, particles, such as the top quark or the Higgs boson, in an inclusive way. There are four [30, 31, 32, 33, 34, 35] main radiative effects at the Z pole:

- The running of the QED coupling constant $\alpha(q^2)$ from $q^2 = 0$ to $q^2 = M_Z^2$.

- The isospin-breaking loop corrections to the W and Z propagators. They are absorbed conveniently in the ρ parameter, $\rho = 1 + \Delta\rho$.

- The running of the Z self-energy, absorbed in the parameter Δ_{3Q}.

Table 3. *Synopsis of precision Electroweak measurements at the Z energy scale. R_ℓ is defined as $R_\ell \equiv \Gamma_{\text{had}}/\Gamma_{\ell\ell}$, where $\Gamma_{\ell\ell}$ refers to the partial width into a pair of massless charged leptons.*

Quantity	Main Technologies	Physics Outputs	Relative Precision
line shape			
M_Z	**Absolute energy scale** relative cross sections line shape fit (QED rad. corr.)	input	$2 \cdot 10^{-5}$
Γ_Z	**Relative energy scale** relative cross sections line shape fit (QED rad. corr.)	$\Delta\rho$	$1.1\ 10^{-3}$
$\sigma_{\text{had}}^{\text{peak},0}$	**Absolute cross sections**	$N_\nu \cdot \frac{\Gamma_{\text{inv}}}{\Gamma_{\ell\ell}}$ universality	$1.2\ 10^{-3}$
$R_\ell \equiv \frac{\Gamma_{\text{had}}}{\Gamma_{\ell\ell}}$	lepton, hadron event selection	universality $f(\alpha_s, \sin^2\theta_w^{\text{eff}}, \delta_{vb})$	$1.\ 10^{-3}$
M_W	**absolute energy scale** in pp collisions: structure functions	Δr^{ew}	$1.5\ 10^{-3}$
$R_b \equiv \frac{\Gamma_b}{\Gamma_{\text{had}}}$	b-**tagging**	δ_{vb}	$5 \cdot 10^{-3}$
$R_c \equiv \frac{\Gamma_c}{\Gamma_{\text{had}}}$	c-**tagging**	universality	3%
asymmetries		$\sin^2\theta_w^{\text{eff}}$	$1.1\ 10^{-3}$

- The $Z \to b\bar{b}$ vertex correction.

One more parameter, Δr^{ew}, is necessary for the W mass. The propagator corrections modify equation 2 by an overall scaling factor $\sqrt{\rho}$ and a global change of $\sin^2\theta_w$, in an universal way. Non-universal corrections are small and – with the notable exception of the $Z \to b\bar{b}$ vertex – insensitive to heavy physics. Furthermore, all Z-pole asymmetries with unpolarized beams and the most precise asymmetry with polarized beams are proportional to the electron coupling $\mathcal{A}_e$, while the sensitivity to $\sin^2\theta_w$ of hadronic asymmetries is contained in the $\mathcal{A}_e$ term (see eq. 10 and table 2). It is therefore convenient to express all asymmetry measurements at LEP in terms of the effective weak mixing angle [36] defined as:

$$\sin^2\theta_w^{\text{eff}} \equiv \frac{1}{4}\left(1 - \frac{g_{Ve}}{g_{Ae}}\right) \quad (14)$$

where the ratio $\frac{g_{Ve}}{g_{Ae}}$ is extracted from pole asymmetries. This definition absorbs vertex corrections for leptons, but not for quarks. See [31, 32, 37] for various avatars of the concept. This definition of $\sin^2\theta_w^{\text{eff}}$ and the $\overline{\text{MS}}$ one [38] are very close [35, 39].

The relations between LEP observables, the Fermi constant G_F and the QED running constant can be written in terms of these universal electroweak corrections $\Delta\rho, \Delta_{3Q}, \Delta r^{\text{ew}}$ and δ_{vb} as [35, 34, 40, 41]:

$$M_Z^2 = \frac{\pi\alpha(M_Z^2)}{\sqrt{2}G_F(1+\Delta\rho)(1+\Delta_{3Q})\sin^2\theta_w^{\text{eff}}\cos^2\theta_w^{\text{eff}}};$$

$$\begin{aligned}
\Gamma_{\ell\ell} &= \frac{G_F M_Z^3}{24\sqrt{2}\pi}[1+\Delta\rho]\left[1+(\frac{g_{V\ell}}{g_{A\ell}})^2\right](1+\frac{3}{4}\frac{\alpha}{\pi}); \\
\Gamma_b &= \Gamma_d(1+\delta_{vb}); \\
M_W^2 &= \frac{\pi\alpha(M_Z^2)}{\sqrt{2}G_F(1-\Delta r^{ew})(1-\frac{M_W^2}{M_Z^2})}.
\end{aligned} \qquad (15)$$

In the SM, δ_{vb} depends on M_t only (quadratically), $\Delta\rho$ depends on M_t (quadratically) and on M_H (logarithmically), while Δ_{3Q} has a logarithmic dependence on both M_t and M_H, and thus is relatively more sensitive to M_H. The leading terms are:

$$\begin{aligned}
\Delta\rho &\simeq \frac{\alpha}{\pi}\frac{M_t^2}{M_Z^2} - \frac{\alpha}{4\pi}ln\frac{M_H^2}{M_Z^2} \\
\Delta_{3Q} &\simeq \frac{\alpha}{9\pi}ln\frac{M_H^2}{M_Z^2} \\
\delta_{vb} &\simeq -\frac{20}{13}\frac{\alpha}{\pi}\left(\frac{M_t^2}{M_Z^2} + \frac{13}{6}ln\frac{M_t^2}{M_Z^2}\right)
\end{aligned} \qquad (16)$$

More complete expressions have been calculated and implemented in computer codes [42].

Table 3 summarises the main observables, their physics output and the most critical technique involved. With this specific choice, these observables are almost uncorrelated, from both points of view of statistical and systematic errors. Table 4 gives an overview of the progress in these measurements over the past year.

Table 4. *Changes over the last year in precision Electroweak measurements and in important additional inputs.*

Quantity	1995 value	1996 value	Main improvements
line shape			
M_Z	91.1884(22)	91.1863(20)	prelim results from 1995 scan, LEP energies
Γ_Z	2.4963(32)	2.4946(27)	idem
$\sigma_{\text{had}}^{\text{peak},0}$ (nb)	41.488(78)	41.508(56)	new σ_{Bhabha} calculation
$\rightarrow N_\nu$	2.991(16)	2.989(12)	
$R_\ell \equiv \frac{\Gamma_{\text{had}}}{\Gamma_{\ell\ell}}$	20.788(32)	20.778(29)	statistics, improved hadronic selection
$\rightarrow \Gamma_{\ell\ell}$	83.93(14)	83.94(11)	all of above
W mass			
M_W	80.26(16)	80.356(125)	new D0 measurement on run 1b
νN			
$1 - M_W^2/M_Z^2$	0.2257(47)	0.2244(42)	CCFR Update
Heavy Flavors			
$R_b \equiv \frac{\Gamma_b}{\Gamma_{\text{had}}}$	0.2219(17)	0.2178(11)	New measurement/technique ALEPH,SLD
			new measurement/update L3, DELPHI
			new inputs charm prod., gluon splitting...
$R_c \equiv \frac{\Gamma_c}{\Gamma_{\text{had}}}$	0.1540(74)	0.1715(56)	all NEW!
asymmetries			
$\sin^2 \theta_w^{\text{eff}}$:			
$A_{FB}^{(\ell)}$	0.23096(68)	0.23085(56)	increased statistics
$\mathcal{P}_\tau$	0.23218(95)	0.23240(85)	more stat. from DELPHI, OPAL final
$A_{FB}^{\text{pol}(\tau)}$	0.23250(110)	0.23264(96)	idem
A_{LR}	0.23045(50)	0.23061(47)	also includes lepton polarized asymmetries
leptonic $\sin^2 \theta_w^{\text{eff}}$	0.23106(35)	0.23114(31)	average of above
$A_{FB}^{(b)}$	0.23209(55)	0.23246(41)	new measurements from ALEPH, OPAL
			new QCD correction
$A_{FB}^{(c)}$	0.2318(13)	0.23155(112)	stat., new input
Q_{FB}	0.2325(13)	0.2320(10)	DELPHI update, ALEPH final
average $\sin^2 \theta_w^{\text{eff}}$	0.23143(28)	0.23165(24)	average of all above
other inputs			
$1/\alpha(M_Z^2)$	128.896(90)	128.896(90)	unchanged
M_t	180(12)	175(6)	new mass average of CDF and D0
$\alpha_s(M_Z^2)$	0.123(6)	0.118(3)	Shifted from LEP only to
			World average. New CCFR DIS data,
			New lattice calculations.

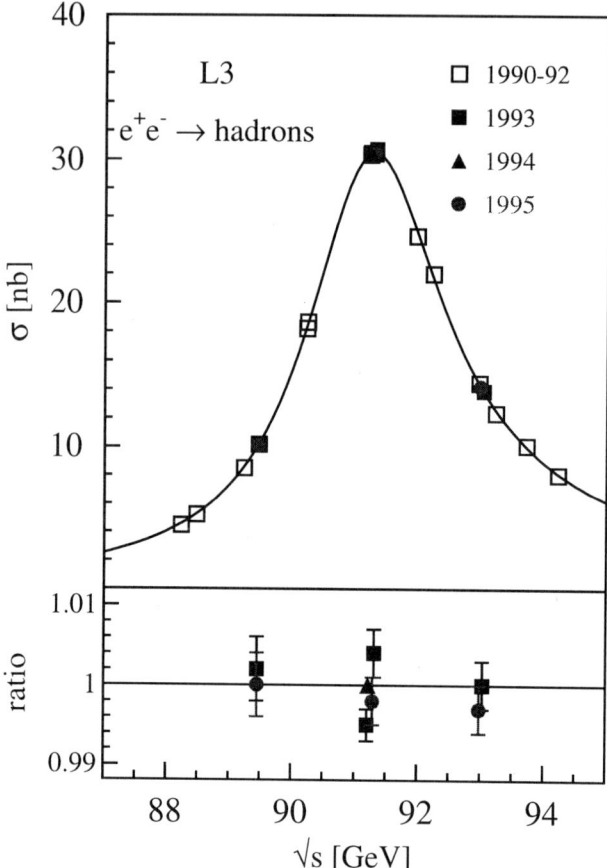

Figure 3. *The L3 hadronic cross sections as a function of center-of-mass energy. The solid line shows the result of the SM fit. The lower part of the plot shows the agreement with the fit for the data collected in 1993-1995.*

The measurement of hadronic cross sections around the Z peak, as in Fig. 3 allows determination of the Z mass and width, and of the peak hadronic cross section. Measuring in addition the leptonic cross sections allows determination of R_ℓ for each or all leptons combined. It has become customary to include the lepton forward-backward asymmetries in the analysis. The cross section measurement is the business of the experiments, the energy measurement is performed by a collaboration of machine experts, in particular the polarization team, and of volunteers from the experiments.

In 1995, a new scan of the Z line shape was performed, and the first results were

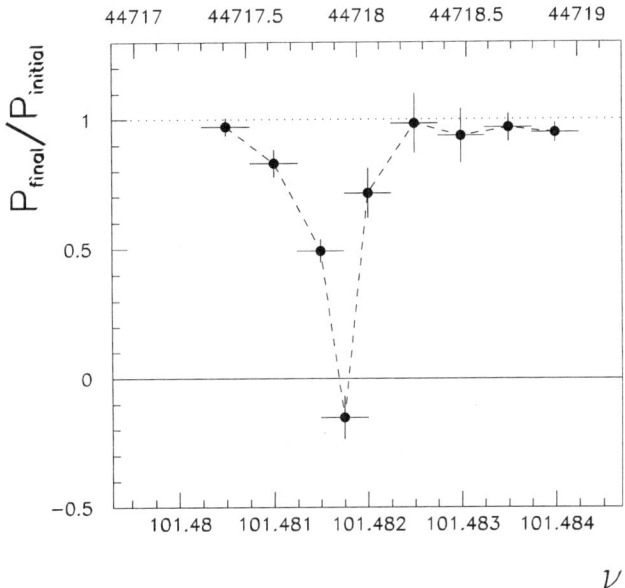

Figure 4. *Measurement of the width of the artificial depolarizing resonance, showing a width of 200 KeV.*

given for this conference by three of the LEP experiments.

LEP beam Energy

The LEP data were taken in 1990-1991 at seven different center-of-mass energies inter-spaced by 1 GeV from 88.25 GeV to 94.25 GeV. In 1992 and 1994 all data were taken at the Z pole. In 1993 and 1995 scans of the Z line shape were performed at the energies of 89.4, 91.2, 93 GeV, which were chosen i) to minimize the statistical error on Γ_Z, ii) to obtain beam polarization at all three points [43].

The corner stone of the LEP energy calibration is resonant depolarization [44]. Transverse spin Polarization builds up in a storage ring by the Sokolov-Ternov effect [45]. It has been observed in all storage rings where it has been searched for [46], and in LEP since 1990 [51]. Resonant depolarization has been used previously in e^+e^- machines, providing accurate measurements of the masses of the J/ψ, ψ', Υ, Υ', at VEPP4 in Novosibirsk [47, 48], at DORIS in Hamburg [49], at CESR in Cornell [50]. It was first performed in LEP in 1991 [44].

The intrinsic resolution of the method is better than 200 KeV [52], fig. 4. However, energy measurements require polarizing the beams, which is a delicate operation. So far, polarization has only been obtained reliably with separated beams, although dedicated experiments have achieved 40% polarization in colliding mode. In 1993, calibration was performed 25 times, at the END of physics fills (EOF), so as disturb machine operation as little as possible. 40% of the 1993 integrated luminosity was obtained during calibrated fills. In 1995 calibration was attempted at the END of every fill, with a success rate of 70%. Six times, calibration were also performed at the beginning of the fill (BOF), to verify that calibrating at EOF only would not introduce a bias.

The extrapolation to the whole scan data requires tracking in time the properties of the magnets, current, field, temperature, as well as the geometrical properties of the ring. The most spectacular source of energy variations comes from ground motion. Because of the strong focusing of LEP, these movements are amplified by a factor of nearly 10^4, so that a small expansion by $\pm 10^{-8}$ leads to energy shifts of 10 MeV. Among these are terrestrial tides [53], but also the effect on geometry of the Pays de Gex of such events as heavy rainfall and the level variations of lake Geneva. Nicely enough, these can be tracked by orbit measurements [55, 56]. These and other known sources of energy variations (such as temperature variations) are included in a model of LEP energy that is used to extrapolate the calibrations.

To monitor the unexplained energy jumps observed in 1993 during a long term experiment, it was decided to install NMR probes directly inside two of the LEP magnets. Reading of these NMRs revealed noise and jumps at all times except between midnight and 5:00 a.m.. A continuous rise of the magnetic field was observed corresponding to about 5 MeV beam energy. After investigation it turned out that the Train Grand Vitesse (TGV) was the origin of the problem. An experiment measuring the current flow on the LEP beam pipe showed very clearly the correlation with the passage of the TGV leaving Geneva at 16:40, Fig. 5.

A model of the rise was build using the two available NMRs. It was verified that the rise was consistent with the that observed during the six BOF/EOF experiments, and during the 1993 experiments. This findings have forced a revision of the 1993 energies, shifting the Z mass by more than 1 σ. The Z width measurement is robust against this effect, which biases energies below and above the peak in a similar way.

The 1995 scan was performed under delicate conditions due to i) opposite sign vertical dispersion due to the bunch-train mode of operation; this effect was anticipated and successfully tackled; ii) development of super-conducting cavities implying irregular acceleration patterns around LEP. Both effects lead to small systematic errors.

The LEP energy spread contributes to a widening of the apparent Z width (4 MeV) and must be corrected for. It was re-evaluated, using the bunch length measured in the experiments, and a direct measurement of the synchrotron frequency from its depolarizing satellites [57]. The energy corrections errors for 1993,1994,1995 are discussed by Wilkinson [58]. They result in $\Delta M_Z \approx 1.5$ MeV, $\Delta \Gamma_Z \approx 1.7$ MeV. The error on leptonic forward-backward asymmetries is $\Delta A_{FB}^{(\ell)} \approx 0.0005$ for each lepton species ($\ell = e, \mu, \tau$), the effect for electrons being anti-correlated with the others.

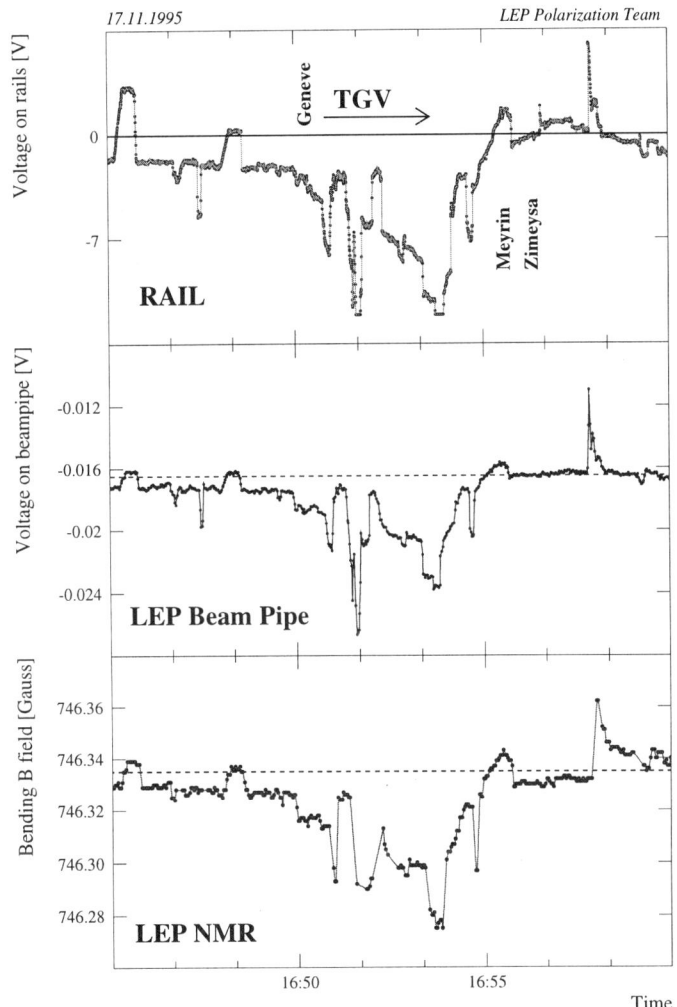

Figure 5. *The correlation between magnetic field in LEP, the current flowing in the beam pipe, and the passage of a train.*

Line shape results

The line shape results were presented by Gruenewald [59]. The total statistics and the systematic errors for the four LEP collaborations are given in Tables 5 Not all experiments have used the full 1995 data yet. Details of the individual analyses can be found in References [60, 61, 62, 63, 64].

Table 5. *The LEP statistics in units of 10^3 events used for the analysis of the Z line shape and lepton forward-backward asymmetries.*

		A	D	L3	O	LEP
$q\bar{q}$	'90-'91	451	357	416	454	1678
	'92	680	697	678	733	2788
	'93 prel.	640	677	646	646	2609
	'94 prel.	1654	1241	1307	1524	5726
	'95 prel.	739	584	311	–	1634
	total	4164	3556	3358	3357	14435
$\ell^+\ell^-$	'90-'91	55	36	40	58	189
	'92	82	70	58	88	298
	'93 prel.	78	74	64	82	298
	'94 prel.	190	129	127	184	630
	'95 prel.	80	67	28	42	217
	total	485	376	317	454	1632

All experiments are equipped with second generation luminosity monitors, based on high precision silicon calorimeters (ALEPH, OPAL, DELPHI) or silicon tracker (L3). Experimental systematics on the luminosity measurement are in the range $0.7 - 1 \; 10^{-3}$. Further studies of the hadronic selection in ALEPH and L3 have led to a reduction of systematic errors to below 10^{-3} ($0.5 \; 10^{-3}$ in L3). The dominant systematic on the hadronic cross sections remains the theoretical error on the Bhabha cross section. The most recent calculations give an error of 0.11%[66] (ALEPH and DELPHI still use a slightly older estimate of 0.16%[67]). As a consequence the peak cross section has substantially improved, with repercussions on the leptonic width and on the determination of α_s.

Systematic errors on the purely leptonic final states remain somewhat higher, within $2 - 8 \; 10^{-3}$ for the cross sections and $1 - 6 \; 10^{-3}$ on the asymmetries. They are therfore larger than the statistical error on the global averages. The only common systematic error considered for asymmetries, besides the LEP energy errors described earlier, is that coming from the t-channel subtraction in the e^+e^- channel. This should still improve for the final publications.

The results of the fits to the measured cross sections and asymmetries are given in table 6. The data from the different experiments are consistent. The τ asymmetry is higher than that of the other leptons (and than the SM global fit prediction of 0.0159 as well) by 2 σ. The contribution of correlated systematic errors is large only for M_Z and Γ_Z (1.5 MeV and 1.7 MeV from LEP energies), and for σ_h^0 (0.045 nb error from the Bhabha cross section). In all other quantities, experimental errors, assumed uncorrelated, dominate.

Table 6. *Average line shape and asymmetry parameters from the four LEP experiments, without and with the assumption of lepton universality.*

Parameter	Average Value
m_Z(GeV)	91.1863±0.0020
Γ_Z(GeV)	2.4946±0.0027
σ_h^0(nb)	41.508±0.056
R_e	20.754±0.057
R_μ	20.796±0.040
R_τ	20.814±0.055
$A_{FB}^{0,e}$	0.0160±0.0024
$A_{FB}^{0,\mu}$	0.0162±0.0013
$A_{FB}^{0,\tau}$	0.0201±0.0018

With Lepton Universality:	
R_ℓ	20.778±0.029
$A_{FB}^{0,\ell}$	0.0174±0.0010

From these measurements one can derive the leptonic partial width for each lepton species or for an average massless lepton, the hadronic partial width, and the invisible width, Tab. 7. These quantities are extracted from the previous ones and contain no additional experimental information.

Table 7. *Partial Z decay widths, derived from the results of the 9-parameter and 5-parameter fits. In the case of lepton universality, $\Gamma_{\ell\ell}$ refers to the partial Z width for the decay into a pair of massless charged leptons.*

Without Lepton Universality:	
Γ_{ee} (MeV)	83.96±0.15
$\Gamma_{\mu\mu}$ (MeV)	83.79±0.22
$\Gamma_{\tau\tau}$ (MeV)	83.72±0.26

With Lepton Universality:	
$\Gamma_{\ell\ell}$ (MeV)	83.91±0.11
Γ_{had} (MeV)	1743.6±2.5
Γ_{inv} (MeV)	499.5±2.0

The W mass

Measurements at hadron coliders

So far, the discovery [68, 69] and studies of W bosons have taken place in $p\bar{p}$ collisions, where single W production is possible. Large samples of single W's decaying into $e\nu_e$ and $\mu\nu_\mu$ have been used to measure the W mass [70]. A new preliminary measurement of the W mass was reported by the D0 collaboration [71], based on the full statistics of run 1b. D0 used exclusively electrons.

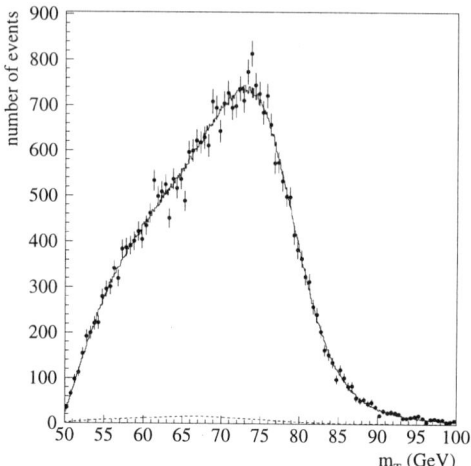

Figure 6. *The D0 transverse mass distribution for run 1b data. The line shows the result of the fit. The χ^2 is 90 for 59 degrees of freedom, but originated from a few points and not from the shape of the distribution. The mass-sensitive region from 60 to 90 GeV contains 27040 events.*

In $p\bar{p}$ collisions, a single W event decaying leptonically appears as a high P_T^ℓ lepton accompanied by missing transverse momentum P_T^ν from the neutrino. The latter is obtained by summation of the visible transverse momentum. Since any amount of longitudinal momentum can be missing in the beam pipe, the W mass is determined using the transverse mass distribution:

$$m_T^2 = 2P_T^\ell P_T^\nu (1 - \cos\phi^{e\nu})$$

where $\phi^{e\nu}$ is the angle between the projections of the electron and missing p_T on the plane perpendicular to the beam axis. This distribution is shown on figure 6 for the new D0 data. This is in essence a jacobian peak, and the W mass information is contained mostly in the region between 75 and 85 GeV. The fit is performed on the range 60-90 GeV, which contains 27040 events, giving a statistical accuracy of 70 MeV.

The energy calibration and the resolution function are both obtained by measuring the Z mass in the same sample. Possible non-linear effects are calibrated by using $\psi \to e^+e^-$ and $\pi^0 \to \gamma\gamma$ decays. This error contributes 80 MeV to the D0 measurement.

The exact shape of this distribution is sensitive to many effects, but two are particularly serious:
i) modeling of the missing P_T resolution. This depends to a large extent on the correct description of the recoil momentum. QCD is able to predict the recoil at a partonic level in case of interactions involving only one parton, but this is spoilt by spectator effects, multiple interaction, or worse, rate-dependent pile-up. These efects can be calibrated by study of the recoil against identified Z decays, so this error of 95 MeV should scale down with increasing statistics.

ii) because the limited angular acceptance, the shape of the transverse mass distribution depends on the longitudinal momentum of the W. This in turn depends on the exact shape of structure functions. This can be constrained by the forward-backward asymmetry in W^+, W^- production, which has been measured by CDF [72]. This constraint is not unambiguous, and this error will decrease more slowly. This structure function uncertainty of 65 MeV is the better part of the common error used in the combination. Fig. 9 summarises the measurements of the W mass at the proton colliders.

W mass at LEP

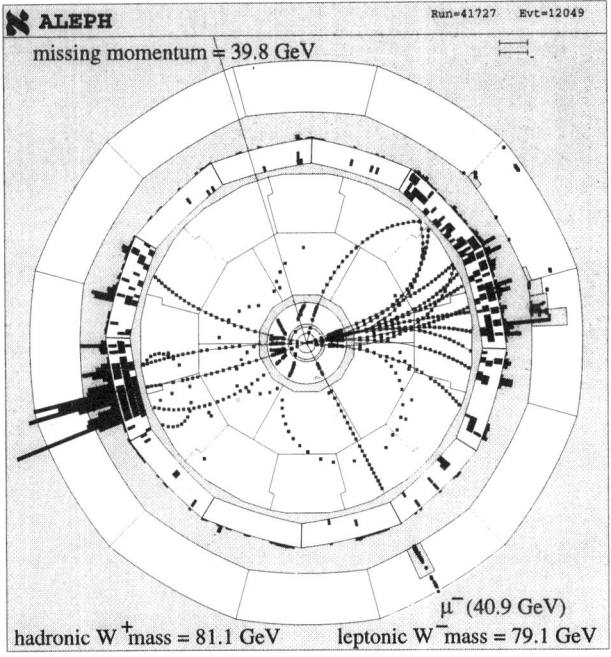

Figure 7. An ALEPH event with a W pair decaying to $\mu\nu qq$

Production of pairs of W bosons is one of the major goals of the LEP2 programme. At LEP2, the measurement of the W mass should improve significantly, thanks to the possibility to detect W's through all decay modes, and thanks to the precise knowledge of the center-of-mass energy. The cross section at W-pair threshold, at the optimum energy of $2.M_W + 0.5$ GeV, provides a sensitive measure of the W mass, with very little model dependence [73]. In June 1996, the LEP center-of-mass energy reached the W pair threshold of 161 GeV with the adjunction of superconducting RF cavities. A first data set of 10 pb^{-1} took place between end of June and 15 August 1996. The results from the LEP collaborations were presented on october 8 at CERN [74, 75].

The cross section is rather low, 3.5 pb, and the measurement consist in selecting as many W pairs as possible. Events where one or both of the Ws decay semileptonically are very clean. A semileptonic one is shown in figure 7. Since both W's are nearly at rest, the leptons and quarks are in a narrow energy range around 40 GeV. Selections of 60% efficiency in the purely leptonic channel and 70% in the semileptonic channel

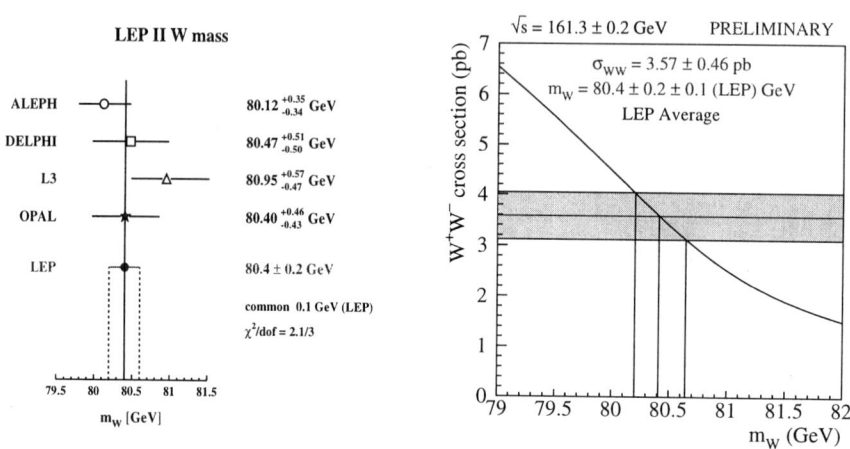

Figure 8. *Measurements of the W mass at the kinematic threshold from LEP. This result was not available in ICHEP96 and is not used in the global fits.*

(including τ decays), have been obtained, for background contamination of 10% or less. The purely hadronic channel is more difficult, and selection efficiency of 50% for a background of 30% is obtained. A multidimensional analysis based on several event shapes and multiplicities of particles and jets has to be applied. The systematic error on the total cross sections are dominated by the uncertainty in the background rejection for this channel. From the cross section, the W mass is inferred as shown in figure 8, where the results from the LEP experiments are summarized.

In october 1996, LEP started running above the W pair threshold, at 172 GeV. At these energies, the W mass has to be reconstructed from its decay products, as described in [73]. First results should be available early 1997. If LEP is able to accumulate 500 pb^{-1} per experiment, precision on M_W could reach 25 MeV.

Deep inelastic neutrino scattering

Neutral Currents were discovered in neutrino scattering more than 20 years ago, but the measurement of the ratio of Neutral Current to Charged Current deep inelastic cross sections still provides a meaningful test of the Standard Model [76, 77]. The CCFR collaboration keeps improving the methods and is planning for a large improvement at nuTeV [78]. It is well known that this measurement can be turned into a measurement of $\frac{M_W}{M_Z}$. The remaining sensitivity to the top mass is completely negligible, and that to the Higgs mass is less than ten times smaller than the experimental precision. This would not be true if, for instance, running of the W and Z self energies were pathological. Comparison between the W mass obtained from neutrino experiments and the world average shows that this is not the case.

The present World average value of $\sin^2 \theta_w = 1 - \frac{M_W^2}{M_Z^2}$ from neutrino scattering is 0.2244 ± 0.0047, or equivalently a value of $M_W = 80.22 \pm 0.21$ GeV.

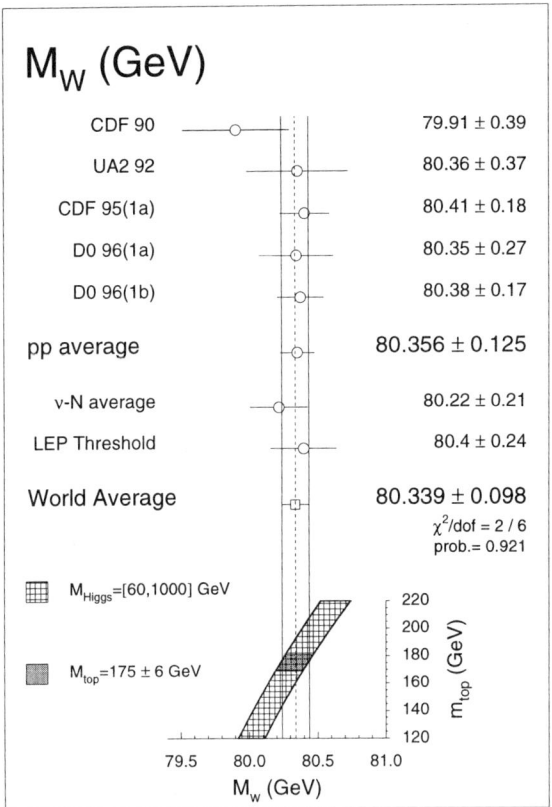

Figure 9. *Measurements of the W mass. The LEP average represents the results from the 161 GeV run, and is not used in the global fits.*

Polarization asymmetries at the Z peak

Polarization at SLD

The measurement of A_{LR} at SLD [79] is certainly one of the cleanest possible. Thanks to the development of an electron gun with a GaAs strained cathode, SLD is able to accelerate electrons with 80% longitudinal polarization at a rate of 120 Hz, with the possibility of reversing the spin at each pulse. The polarization is measured with a precision of $\pm 0.67\%$. The comparison of total cross sections for the two different helicities gives A_{LR}:

$$A_{LR} = \frac{1}{P_e}\frac{\sigma_L - \sigma_R}{\sigma_L + \sigma_R}$$

The final result for the 1992-1995 data, based on 20 000 Zs with 22% polarization 50 000 Z with 63% polarization and 100 000 Zs with 77% polarization, is:

$$A_{LR} = 0.1543 \pm 0.0039$$

which leads to
$$\sin^2\theta_w^{\text{eff}} = 0.23060 \pm 0.0050$$

The forward-backward polarised asymmetries have also been measured for the three species of leptons, allowing the final state couplings to be measured:

$$\mathcal{A}_e = 0.202 \pm 0.038 \quad (17)$$
$$\mathcal{A}_\mu = 0.102 \pm 0.033 \quad (18)$$
$$\mathcal{A}_\tau = 0.190 \pm 0.034 \quad (19)$$
$$<\mathcal{A}_\ell> = 0.148 \pm 0.016 \quad (20)$$

These can be averaged with the measurement of A_{LR} to give:
$$\sin^2\theta_w^{\text{eff}} = 0.23061 \pm 0.00047$$

Systematic errors are below 0.0001. There are very few possible loopholes in this measurement and one should resist the temptation to set it aside because it differs by two standard deviations from the LEP average. The possible e^+ polarization and beam beam effects will be cross-checked in the coming run, but they are expected to be exceedingly small. SLD is approved to run into 1998, for a total of 500 000 Zs, and should reach a precision on $\sin^2\theta_w^{\text{eff}}$ of $\Delta\sin^2\theta_w^{\text{eff}} \simeq 0.0002$.

τ Polarization

OPAL has now published [84] a final result comprising 1991 to 1994 data; DELPHI [83] has a new preliminary analysis of the 1993-1994 data, and L3 [85] of 1994 data. All experiments have therefore analysed the data up to 1994, with the exception of ALEPH for which data until 1992 only have been published.

The tau polarization is extracted from the reconstructed kinematic parameters of the decay particles as shown by Rougé [80, 81]. This is a one dimensional problem for $\tau \to e\nu_e\nu_\tau, \tau \to \mu\nu_\mu\nu_\tau$ and $\tau \to \pi\nu_\tau$. for the $\tau \to \rho\nu_\tau$ decay, the τ helicity affects the distributions of both τ and ρ decay angles, in a way that depends on the $\pi^-\pi^0$ mass. This set of 3 observables, $\{\xi\}$, defines the final state. The probability density functions for the ± 1 helicity, $W^\pm(\{\xi\})$, are used to build an optimal variable $\omega(\{\xi\}) = (W^+ - W^-)/(W^+ + W^-)(\{\xi\})$, and fit the τ polarization. For the a_1, the decay is defined by six variables. Full use of the 6 dimensional density function makes the a_1 channel more sensitive than the leptonic one. Distributions for e, μ, π, ρ are shown in figure 10. By analyzing the polarization as a function of polar angle, eq. 11, one can derive the coupling asymmetry for both τ, $\mathcal{A}_\tau$ and electron $\mathcal{A}_e$. The LEP averages of these couplings are nicely consistent with universality and can be averaged:

$$\mathcal{A}_\tau = 0.1401 \pm 0.0067 \quad (21)$$
$$\mathcal{A}_e = 0.1382 \pm 0.0076 \quad (22)$$
$$\text{average } A_\ell = 0.1393 \pm 0.0050. \quad (23)$$

R_b and R_c

Precision measurements using the b quark, i.e. the isopartner of the heaviest fermion top, probe new physics that could have escaped the more precise tests using

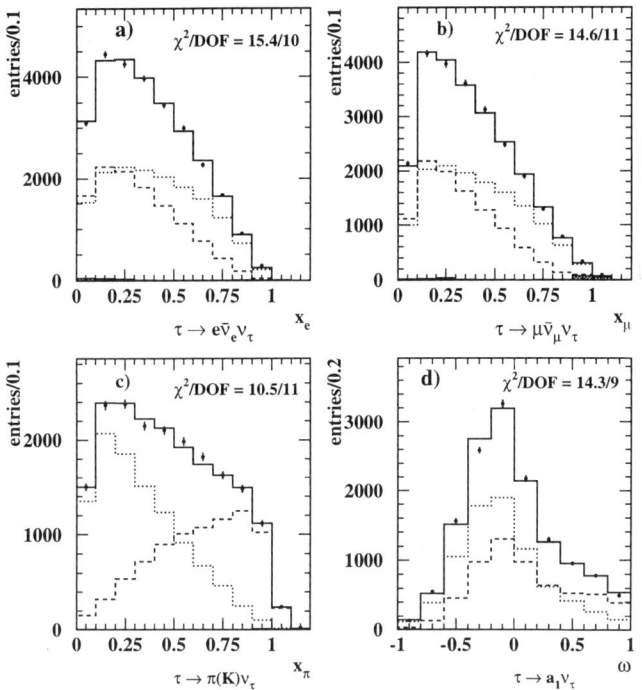

Figure 10. *Extraction of the τ polarization in OPAL: a) $\tau \to \mu \nu_\mu \nu_\tau$; b) $\tau \to e \nu_e \nu_\tau$; c) $\tau \to \pi \nu_\tau$; d) $\tau \to \rho \nu_\tau$. The full line is the result of the fit, which is a linear sum of the components due to positive (dotted line) or negative (dashed line) τ helicities.*

light fermions and leptons in particular. In summer 1995, the measured decay rate of the Z boson in b quarks was 3.7 standard deviations away from the Standard Model (SM) value, while that for charm quarks was 2.5 standard deviations away, see Fig. 13. If one believed the systematic errors to gaussian the Standard Model was ruled out at 99.9% C.L.! Was this the long-sought new physics? Or could this be an experimental artifact? Let us start with R_c.

R_c

The charm quark is squeezed between b's and light quarks, and designing an efficient tag has been very difficult. For a long time, double tag was too costly in statistics to be worthwhile.

Four methods have been used to measure R_c:
1. lepton fits [92, 93, 95, 94]: this has been done by all experiments. After selection of hadronic events with an identified lepton, the distribution in the P_T, P_L plane is analysed to extract a signal of charm and beauty. Since background is large and statistics limited, double-tag is not a very good solution. one has to rely in this case on the charm semi-leptonic branching ratio $B_{c \to \ell} = 9.8 \pm 0.5\%$ obtained from ARGUS and CLEO. This input was cross-checked with limited accuracy by DELPHI by performing a purer tag on one hemisphere of the events, the D^* tag, and counting leptons on the other.

The result, $B_{c\to\ell} = 9.8 \pm 1. \pm 0.6\%$ agrees with the above. However, if one were to use the measured production rates of the various charmed particle species, folding with the known semileptonic branching ratios, one would get $B_{c\to\ell} = 9.1 \pm 0.5\%$ and R_c would increase by 7%. This latter solution was not used here.

2. Single exclusive charmed particle tag (OPAL [96], DELPHI [97]). The easiest charmed particle to tag is the D^*: the small $D^* - D$ mass difference is very well measured experimentally and leads to very clear signals. By selecting high momentum D^*, one can obtain a clean sample of charm events. Obtaining R_c from this method is sensitive both to fragmentation and to the probability that a charm quark fragments into a D^*, $P_{c \to D^*}$.

3. Reconstruction of inclusive charmed particles (OPAL [96], DELPHI [97]). By measuring the production of all possible charmed particles, one frees oneself from the need to know the probability for a charm quark to give specific charmed hadrons. Nevertheless, the method is still sensitive to decay branching ratios and acceptance corrections. Errors stemming from a branching ratios are well defined. However, momentum cuts are necessary to remove the background from $b\bar{b}$ events, which is mostly peaked at low momenta. This leads to strong sensitivity to fragmentation. Nevertheless, an important by-product of this method pioneered by OPAL has been a measurement of the relative production rates of individual charmed hadrons from charm quarks.

4. With increased statistics and the improvements in tracking brought about by the new vertex detectors, double tag of charm has become possible. Low transverse momentum pions sign a $D^* \to \pi D$ decay chain, without need to reconstruct the D decay, leading to higher efficiency than for the full D^* reconstruction. By combining this tag with a a full D^* tag (OPAL [96], DELPHI [97]), or by combining all possible combinations of fully reconstructed D^0, D^+, D^*, (ALEPH [98]), one can extract both R_c $P_{c \to D^*}$ more reliably. The knowledge of $P_{c \to D^*}$ in turn, has allowed use of the large sample of single D^* to extract R_c.

The value of $P_{c \to D^*}$, which was previously taken from lower energy data, $P_{c \to D^*} = 0.178 \pm 0.013$, is therefore now been measured at LEP to be $P_{c \to D^*} = 0.163 \pm 0.005$. Since the most powerful method, statistically, is the single D^*, R_c has a correlation of -60% with this number. This change in $P_{c \to D^*}$ has led to an increase in R_c of 7%.

All measurements presented at ICHEP96 were new or updated. All changes in the D decay branching ratios and charm production rate have been propagated consistently to the analyses of R_c and of the other heavy flavour Electroweak observables [86]. Furthermore, gluon splitting and momentum correlations are now properly accounted for. The result is shown in figure 11. All results agree now very well with each other – but this was always the case – and with the Standard Model prediction:

$$\text{ICHEP96} \quad R_c = 0.1715 \pm 0.0056 \qquad (24)$$
$$\text{SM} \quad R_c = 0.1723 \qquad (25)$$

R_b

The art of experimenters is to design efficient and self-calibrating b-tags. There are three main sources of information to design a tag: i) the presence of large transverse momentum leptons (lepton tag); ii) the presence of tracks that miss significantly the primary vertex, indicating the presence of a heavy object with large multiplicity decaying with a long lifetime (lifetime tag); iii) consistency of the event shape with the

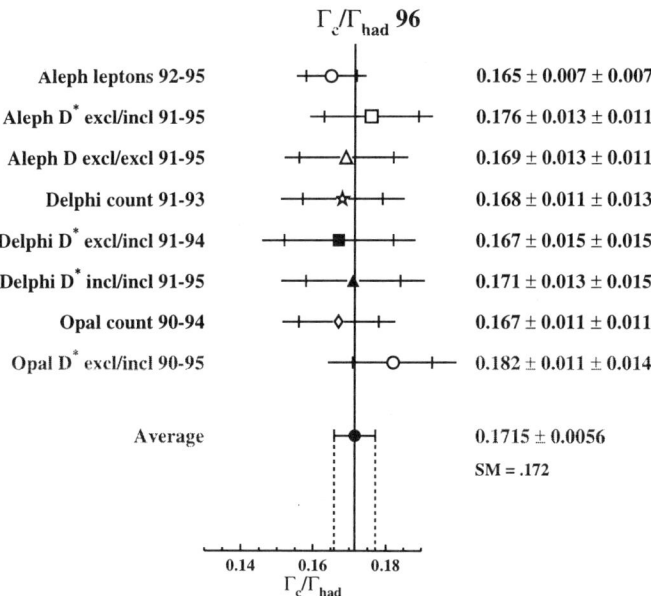

Figure 11. *Measurements of R_c at LEP. All measurements are new or updated since last year. The χ^2 of the average is 1 for 7 d.o.f..*

presence of a heavy boosted object (event shape tag). The lifetime tag is by far the most efficient.

The tagging efficiencies are subject to the large uncertainties concerning both the decays of b-hadrons, and their production. They can be calibrated by the double tagging technique: since b quarks are produced in pairs, one can take events where one b is tagged and count which fraction of these have the second b tagged to extract the tagging efficiency (ϵ_b). If N_t is the number of events with one tag and N_{tt} the number of events with two tags, then one gets two equations:

$$\frac{N_t}{2N_{had}} = \epsilon_b R_b; \qquad \frac{N_{tt}}{N_{had}} = \epsilon_b^2 R_b. \qquad (26)$$

which can be solved for ϵ_b and R_b. In practice, things are not so simple. Equations 26 are only valid in absence of backgrounds and assume explicitly that in a $Z \to b\bar{b}$ event, the probabilities to tag the two bs are independent. They have to be modified for these effects:

$$\begin{aligned}
\frac{N_t}{2N_{had}} &= \epsilon_b R_b + \epsilon_c R_c + \epsilon_{uds} R_{uds}; \\
\frac{N_{tt}}{N_{had}} &= C_b \epsilon_b^2 R_b + \epsilon_c^2 R_c + \epsilon_{uds}^2 R_{uds},
\end{aligned} \qquad (27)$$

where $C_b \simeq 1 - \rho$, ρ being the correlation of tagging efficiency between the two hemispheres in a $Z \to b\bar{b}$ event, while $\epsilon_c, \epsilon_{uds}$ are the efficiency of the tag for charm and u, d, s events respectively.

Both hemisphere correlations and background tag efficiencies are obtained from Monte Carlo. This is where systematic errors originate, and they dominate the measurement errors. A collective mistake of 2% in the hemisphere correlations could explain the 1995 anomaly.

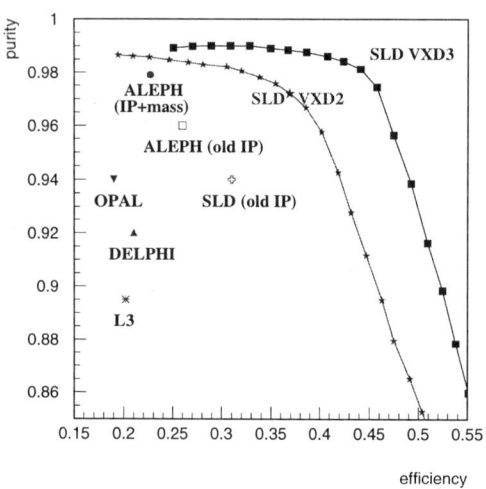

Figure 12. *Comparison of tagging efficiencies and purities at LEP and SLD.*

New results were presented at this conference by L3 [89] and DELPHI [90], with techniques similar to those applied until 1995. A novelty appeared in spring 1996, [87] when the requirement that the tracks that miss the primary vertex, or make a secondary vertex, have an invariant mass larger than that of a charmed hadron. This is very efficient at cutting charm away and ensures a better purity for a given efficiency – or vice-versa. SLD has recently achieved up to 37% efficiency with a purity of 97% [88]. ALEPH has developed a similar method, with somewhat lesser performance, due evidently to the fact that the beam spot is much smaller at SLD then at LEP, and that the vertex detector of SLD is closer to the beam. This technique is easy at SLD, already harder for ALEPH, and requires silicon vertex detector readout in both (r,ϕ) and (r,z) planes. Figure 12 compares the various tagging efficiencies, highlighting the gain obtained by this method. SLD obtains a precision similar to that of the LEP experiments with 20 times less statistics.

The main improvement however came from the measurement presented at this conference by ALEPH [91], which by itself is more precise than the previous world average. I refer to the contribution of Tomalin [91] for a detailed explanation of the method and will outline here the main progress with respect to previous measurements and possible remaining loopholes.

1. The measurement combines the several tags outlined above: lepton, event shape, life-time/mass tag. This increases considerably the overall efficiency, which, in the double tag technique, leads directly to a similar improvement in the statistical precision.

2. One of the tags is very pure, the lifetime tag with an invariant mass requirement.

3. The analysis uses background tags (a c tag and a uds tag). This combined with the

previous point allows a complete determination of the background contamination in the other tags, and to keep under control the charm background.
4. hemisphere correlations were reduced as much as possible, by reconstructing a different primary vertex in each hemisphere of each event.
5. The full LEP1 statistics are used.

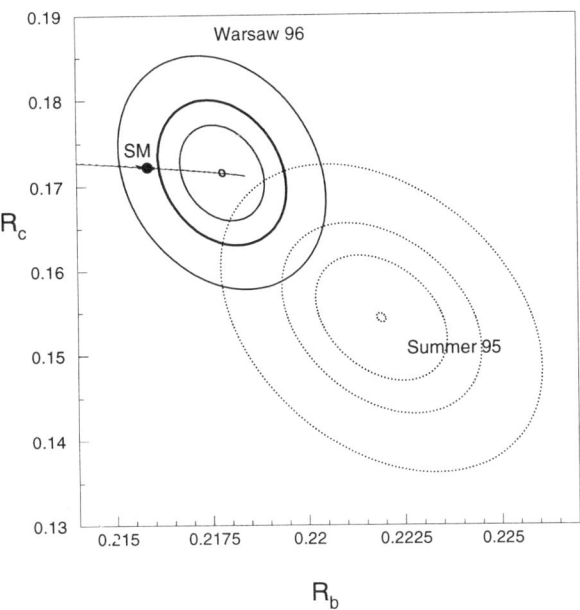

Figure 13. *Curves of constant χ^2 in the R_b, R_c plane from LEP-SLD data in summer 96, and comparison with the corresponding contours in 1995. These curves show the minimum and increases by +1, +2.3, +6, corresponding to one standard deviation (i.e. 39% C.L. in two dimensions), 68%, and 95% C.L. contours, if systematic errors were assumed to be gaussian. The Standard Model prediction is indicated by the dark point, and the thick line around it indicates the range allowed for $M_t = 175 \pm 6$ GeV. The thin continuation indicates the variations allowed in models affecting R_b alone, such as the Minimal Supersymmetric Model.*

The situation can then be summarized as follows:

$$\text{Summer 95}: R_b = 0.2219 \pm 0.0017$$
$$\text{ICHEP96}:$$
$$\text{L3} \quad R_b = 0.2185 \pm 0.0028 \pm 0.0032$$
$$\text{DELPHI} \quad R_b = 0.2176 \pm 0.0028 \pm 0.0027$$
$$\text{SLD} \quad R_b = 0.2149 \pm 0.0032 \pm 0.0021$$
$$\text{ALEPH} \quad R_b = 0.2161 \pm 0.0009 \pm 0.0011$$

$$\text{W.A.} \quad R_b = 0.2178 \pm 0.0011$$

$$\text{SM} \quad R_b = 0.2158 \pm 0.0003$$

The new and old contour plots in the R_b, R_c plane are compared in figure 13. One can legitimately be surprised by such a large change. The change is due mostly to the new ALEPH measurement. New input on gluon splitting and charm production fractions change the result by only -0.0005. Although there are good reasons to think that the new measurement is better methodologically, it is almost impossible to identify one single reason why there could have been a mistake before. Charm background is obviously a suspect, and so are hemisphere correlations, but my ALEPH colleagues assure me that the old analysis performed on the full data set would have given the same central value....

Studies are continuing. It should be emphasized that not all question marks concerning the measurement of R_b have vanished. In particular the question of fragmentation induced b-hadron momentum correlations remains particularly worrying since the tag efficiency is essentially proportional to b-hadron momentum. Contrary to the correlations induced by gluon radiation, which are in principle well simulated, this source of errors is difficult to control. Our best hope, since LEP is not taking more Z peak data anymore, is that SLD will be able to collect many Zs.

Hadron asymmetries

Asymmetry for b-quarks at LEP

Hadron asymmetries at the Z pole are larger and more sensitive to electroweak effects than the leptonic ones. The best measured is the b-asymmetry. Its value is about 10%, so that the relative precision achievable is 10 times lower than the precision on R_b for the same statistics. Although one has to estimate the direction and charge of the original quarks in the $Z \to b\bar{b}$ process, the demand on systematic errors related to knowledge of the purity are correspondingly less important than for R_b. In addition, the selection efficiency does not enter in the measurement of an asymmetry. Consequently, analyses are optimized differently than for the R_b measurement, emphasis being placed on statistical errors rather than on systematic precision. At the end the relative precision on $A_{FB}^{(b)}$ is 2.5%, while it was 0.5% on R_b.

The asymmetry is measured by to principal techniques.

– **Lepton tag**, where the high P_T lepton gives the sign of the b-quark jet to which it is closest. Typical efficiencies of 5% for a purity of 80% are obtained. Corrections are necessary for $B - \bar{B}$ mixing, for background (mostly charm decays), and for the assignment of the lepton to the b-jet. The selection procedure with high momentum cuts on the lepton leads to an interplay with QCD effects. To first order the effect of gluon emission is to smear the direction of the final state quarks, leading to $A_{FB}^{(b)} = A_{FB}^{(b)\,0}(1 - \alpha_s/\pi)$. The selection of high momentum leptons reduces this effect by about $(50 \pm 25)\%$. This question is under further investigation, the correction should really be applied by each experiment.

– **Lifetime tag**, as for R_b, complemented with charge tagging by means of **jet charge**. This technique was first developed in ALEPH [99, 100], and performed also by OPAL [101] and DELPHI [102]. The charges Q_F and Q_B of two (Forward and Backward) hemispheres of each event are measured,

$$Q_{F,B} = \frac{\sum_{F,B} p_{\|i}^\kappa q_i}{\sum_{F,B} p_{\|i}^\kappa} \tag{28}$$

where q_i is the charge of particle i, $p_{\|i}$ its momentum projected on the thrust axis. The optimal value of the parameter κ is around $\kappa = 0.5$. From Q_F and Q_B one can form
- the charge flow $Q_{FB} = Q_F - Q_B$;
- the total charge $Q = Q_F + Q_B$;
- the product $Q_F \cdot Q_B$;

The average over a pure selected sample would be $\langle Q_{FB} \rangle = \delta_b \cdot A_{FB}^{(b)}$ while $\langle Q_F \cdot Q_B \rangle \simeq -\delta_b^2/4$, where δ_b is the average charge difference between an hemisphere containing a b or a $\bar{b}$. The method is thus largely self-calibrating. It was verified by Monte Carlo studies that this method automatically accounts for $B - \bar{B}$ mixing and gluon emission, further QCD corrections are unnecessary. Jet charge and lepton tagging lead to similar statistical precision.

In addition, $A_{FB}^{(b)}$ is also measured [93, 103], with more limited precision using low momentum D^*, as a by-product of the measurement of $A_{FB}^{(c)}$ by that technique.

The physics input is similar to that of the measurements of R_c, R_b and the measurement is treated in the same global heavy flavours fit [86]. The new results from OPAL and ALEPH both reduce the asymmetry. After corrections for initial state radiation and photon exchange, the pole asymmetry is obtained:

$$A_{FB}^{(b)\,0} = \frac{3}{4}\mathcal{A}_e\mathcal{A}_f = 0.0979 \pm 0.0023 \qquad (29)$$

which is 1.8 standard deviations lower but consistent with the SM best fit of 0.1022.

Asymmetry for c-quarks at LEP

Similarly as for the measurement of R_c the measurement of the charm asymmetry is performed using a lepton tag (all experiments [92, 93, 94, 95]) or a D^* tag (ALEPH [104], DELPHI [93], OPAL [103]). The result of the heavy flavours global fit is:

$$A_{FB}^{(c)} = \frac{3}{4}\mathcal{A}_e\mathcal{A}_c = 0.0733 \pm 0.0048 \qquad (30)$$

which is consistent with the SM fit expectation of 0.0733.

Polarized Asymmetries for b and c-quarks at SLD

This was described by D. Falciai [105]. Techniques are similar to those used at LEP, with in addition a fast Kaon tag for charge tagging of b-events. The asymmetry, eq. 13, is much larger than with unpolarized beams, so that the relative contribution of statistical errors is smaller while that of systematic errors remains the same at SLD as in LEP. The b and c Polarized asymmetries of SLD were treated by the same global fit as the heavy flavours results of LEP; the results are

$$\mathcal{A}_b = 0.863 \pm 0.049 \qquad (31)$$
$$\mathcal{A}_c = 0.625 \pm 0.084. \qquad (32)$$

Light quark asymmetries

The average charge flow in the inclusive samples of hadronic Z decays is related to the forward-backward asymmetries of individual quarks:

$$\langle Q_{FB} \rangle = \sum_{\text{quark flavors}} \delta_f A_{FB}^{(f)} \frac{\Gamma_f}{\Gamma_{\text{had}}}, \qquad (33)$$

where δ_f, the *charge separation*, is the average charge difference between the quark and antiquark hemispheres in an event: $\delta_f = \langle Q_f - Q_{\bar{f}} \rangle$. The forward backward asymmetries, see equation 10, are all positive, but the signs of the charge separations are different. Typical values, for $\kappa = 1$ are $\delta_u = 0.406 \pm 0.008, \delta_d = -0.229 \pm 0.009, \delta_s = -0.329 \pm 0.005, \delta_c = 0.211 \pm 0.010, \delta_b = -0.208 \pm 0.004$, as obtained in the recently published ALEPH [106] final analysis. This results in a large cancellation, already at parton level. The b charge separation was obtained as by-product of the b asymmetry by jet-charge. The c charge was obtained using hemispheres opposite a fast D^* tag. The u, d, s charge separations cannot be obtained directly and were extracted from the JETSET fragmentation model. Systematic errors are dominated by uncertainties in the baryon production mechanism. Using another fragmentation Model such as HERWIG led to a consistent result with larger systematic uncertainties, In the final result the JETSET error estimate was used.

OPAL [110] and DELPHI [109] have produced preliminary numbers with similar analyses, where the c quark charge separation is, however, still obtained from Monte Carlo.

The results are expressed in the Standard Model as measurements of $\sin^2 \theta_{\rm w}^{\rm eff}$:

	$\sin^2 \theta_{\rm w}^{\rm eff}$ $\times 10^4$:	
ALEPH	2320	$\pm 8_{\rm stat} \pm 7_{\rm exp} \pm 8_{\rm frag}$
DELPHI	2311	$\pm 10_{\rm stat} \pm 10_{\rm exp} \pm 10_{\rm frag}$
OPAL	2326	$\pm 12_{\rm stat} \pm 4_{\rm exp} \pm 13_{\rm frag}$
LEP	2320	± 10

The method has clearly reached a systematic limit.

More on light quarks

This section summarizes attempts made to tag light quarks in Z decays and measure their couplings. Light quark couplings are best measured from Deep Inelastic Neutral Current scattering of neutrinos and antineutrinos [10]. Measuring them at LEP is interesting because of the very different energy, but difficult.

By selecting events with a fast $K^{\pm}$ and Λ an enriched sample of signed $s\bar{s}$ events can be obtained. This is possible in DELPHI thanks to the particle identification provided by the Ring Imaging Cerenkov [111]. Fast K^0 and neutrons, detected in the hadron calorimeter, provide an unsigned tag for $s\bar{s}$ and $d\bar{d}$ events. This can be combined with jet charge to measure the down-quark asymmetry. The measured asymmetries [112] are consistent with the SM value 0.0937:

$K^{\pm}$:	0.118	$\pm 0.031_{\rm stat.} \pm 0.016_{\rm syst.}$
Λ :	0.135	$\pm 0.055_{\rm stat.} \pm 0.037_{\rm syst.}$
K^0, n :	0.111	$\pm 0.031_{\rm stat.} ^{+0.068}_{-0.054}$ syst.

Triggered by the 1995 discrepancy in R_b, R_c, ALEPH [113] has performed a measurement of the light quark rate, R_{uds}. This is done by selecting events on the basis of three main variables in each hemisphere of hadronic Z decays: i) the momentum of the highest momentum particle, which is higher for light quarks than heavy ones; ii) the absence of lifetime, which anti-tags heavy quarks; iii) the absence of very low transverse momentum particles that would tag the presence of a $D^* \to D\pi$ decay chain. The tagging efficiencies are constrained by the double tagging technique. A tagging efficiency of 28% for light quarks is obtained, with a 10% efficiency for charm and 2.5% efficiency for bs. The analysis obtains $R_{uds} = 0.614 \pm 0.014$, that compares well with the SM expectation of 0.612 but is not precise enough to exclude the value that would result from the 1995 R_b, R_c values, $R_{uds} = 0.624$.

A more ambitious programme of individual light quark tagging has been attempted by OPAL [114]. Here the hemisphere tags include normal heavy flavour tags (lifetime and D^*), while light quark tags use the presence of a high momentum stable hadron $\pi^\pm, K^\pm, p, K^0, \Lambda$. The measurement of the large number of single and double tags for each tag leads to a large number of non-linear coupled equations. Like in the Q_{FB} analyses, symmetries in fragmentation help by limiting the number of unknown parameters. A solution is found if one further assumes equality of d and s couplings:

$$\begin{aligned}
\{R_d = R_s\} &= 0.230 \pm 0.010 \pm 0.010 \\
R_u &= 0.159 \pm 0.019 \pm 0.021 \\
\{A_{FB}^{(d)} = A_{FB}^{(s)}\} &= 0.067 \pm 0.034 \pm 0.011 \\
R_u &= 0.034 \pm 0.067 \pm 0.028
\end{aligned}$$

These results are strongly correlated. It is clear that very precise measurements of individual light quarks couplings cannot be obtained from Z decays.

Measurements of the effective weak mixing angle

The different values of $\sin^2 \theta_w^{\text{eff}}$ measured from lepton and quark asymmetries and τ polarization at LEP and A_{LR} at SLC are summarized on fig. 14. The measurements presented above average to:

$$\sin^2 \theta_w^{\text{eff}} = 0.23165 \pm 0.00024. \tag{34}$$

The χ^2 of 12.8 for 6 degrees of freedom is acceptable but not great. Two measurements contribute to it: A_{LR} and $A_{FB}^{(b)}$, being 2.2 and 1.9 s away from the average. The global consistency of Electroweak measurements will be discussed below. New data from SLAC and final $A_{FB}^{(b)}$ values from LEP are awaited eagerly. This discrepancy between the two most precise determinations of $\sin^2 \theta_w^{\text{eff}}$ will be emphasized further when extracting the couplings of the b quark.

It is remarkable that the experimental precision of ± 0.00024 is comparable to the uncertainties in the prediction of $\sin^2 \theta_w^{\text{eff}}$ for a given Higgs boson mass coming respectively from $\alpha(M_Z^2)$, (± 0.00023) and from the top quark error ± 0.00020. In order to improve the Higgs mass determination, one now must improve all three of them.

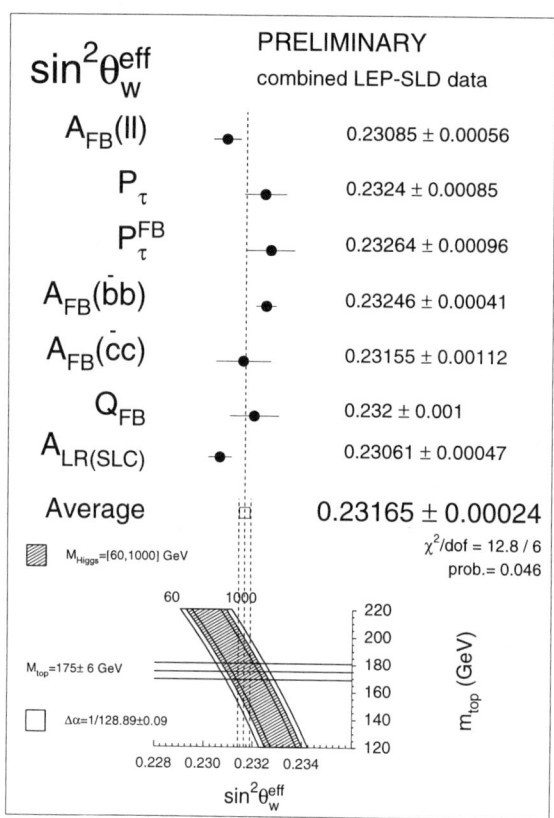

Figure 14. Summary of measurements of $\sin^2 \theta_w^{\text{eff}}$ from the forward-backward asymmetries of leptons, τ polarization, inclusive quarks, heavy quarks asymmetry and the SLC polarization asymmetry. Also shown is the SM prediction as a function of M_t.

Analysis of Electroweak Measurements

Lepton Couplings and Universality

The couplings of the leptons can be extracted from the measurements of purely leptonic observables, $\Gamma_e, \Gamma_\mu, \Gamma_\tau, A_{FB}^{(e)}, A_{FB}^{(\mu)}, A_{FB}^{(\tau)} A_{FB}^{pol(\tau)} \mathcal{P}_\tau$, and A_{LR}. The results of the fit are shown on table 8. Lepton universality is well verified and will be assumed in the following.

Table 8. Lepton couplings $g_{V\ell}$ and $g_{A\ell}$ extracted from leptonic asymmetries and τ polarization, showing the validity of lepton universality. The neutrino coupling, extracted from the invisible width assuming three light neutrino species with equal couplings, is also shown.

	$g_{V\ell}$	$g_{A\ell}$
e	-0.03828 ± 0.00079	-0.50119 ± 0.00045
μ	-0.0358 ± 0.0030	-0.50086 ± 0.00068
τ	-0.0367 ± 0.0016	-0.50117 ± 0.00079
ℓ	-0.03776 ± 0.0062	-0.50108 ± 0.00034
ν	0.5009 ± 0.0010	0.5009 ± 0.0010

b and c couplings

While the SLD polarized forward-backward asymmetries, eqs. 32 and 31, provide direct measurements of the chiral coupling asymmetries $\mathcal{A}_c, \mathcal{A}_b$, the LEP asymmetries for c and b quarks measure the product of them by $\mathcal{A}_e$. Using the same purely leptonic asymmetries as above one finds: $\mathcal{A}_e = 0.1466 \pm 0.0033$ if one uses LEP data alone, or $\mathcal{A}_e = 0.1500 \pm 0.0025$ if one includes also the SLD data. This allows to extract the parameters $\mathcal{A}_c, \mathcal{A}_b$ from LEP data as well, as shown in Tab. 9.

Table 9. Determinations of the quark coupling parameters $\mathcal{A}_b$ and $\mathcal{A}_c$ from LEP data alone (using the LEP average for $\mathcal{A}_e$), from SLD data alone, and from LEP+SLD data (using the LEP+SLD average for $\mathcal{A}_e$) assuming lepton universality.

	LEP ($\mathcal{A}_e = 0.1466 \pm 0.0033$)	SLD	LEP+SLD ($\mathcal{A}_e = 0.1500 \pm 0.0025$)	SM
$\mathcal{A}_b$	0.890 ± 0.029	0.863 ± 0.049	0.867 ± 0.022	0.935
$\mathcal{A}_c$	0.667 ± 0.047	0.625 ± 0.084	0.646 ± 0.040	0.667

The charm quark coupling is in nice agreement with the SM expectation of 0.667. The b coupling is in disagreement with the SM by three standard deviations. If one correlate this with the 1.8 σ discrepancy in R_b, it is fair to say that the experimental situation is far from confirming nicely the SM couplings for the b quark. This might be a real effect, but it could also be that we are again plagued by the great difficulty in performing measurements with heavy flavours. It should also be stressed that the most precise measurements of b asymmetries are preliminary. A little patience is probably advisable before a firm conclusion can be drawn.

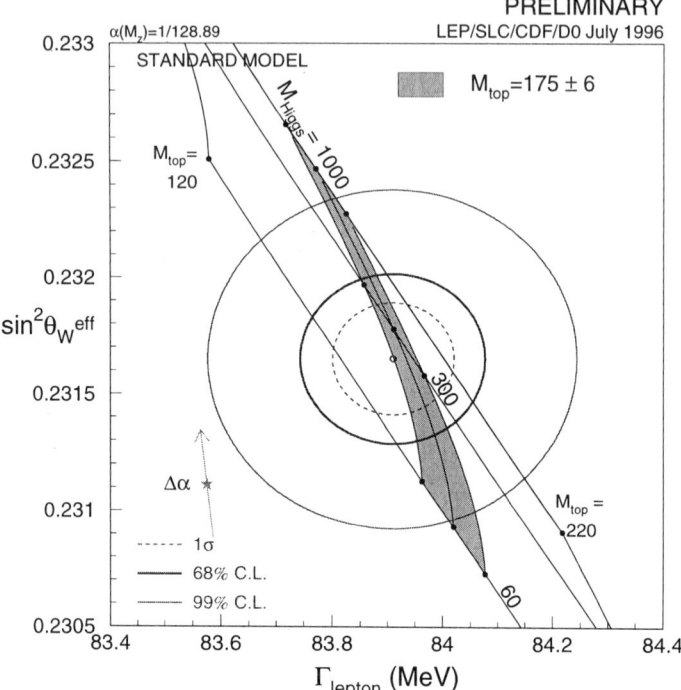

Figure 15. Contours of constant χ^2 for $\sin^2\theta_w^{\text{eff}}$ versus $\Gamma_{\ell\ell}$. The SM predictions as a function of M_t and M_H are shown. The star indicates the predictions of the SM, if the only electroweak correction applied is the running of $\alpha(M_Z^2)$. The arrow indicated on the star shows the influence of the error on $\alpha(M_Z^2)$ on the predictions.

Determination of the top and Higgs masses

The agreement of LEP data with the standard model can be well seen in Figure 15, where the leptonic width of the Z is plotted against the Standard Model prediction. The agreement is striking for these purely leptonic variables which concentrate all information on line shape and asymmetries from LEP. From this graph one can see the considerable change brought about by the determination of the top quark mass: while the lines corresponding to different higgs masses are too close to be distinguished by the data without external input on the top mass, the determination from the Tevatron allows constraints to be placed.

Good agreement also hold for the W mass. This could be seen in Fig. 9. The agreement with $\sin^2\theta_w^{\text{eff}}$ and $\Gamma_{\ell\ell}$ can be seen if one extracts from these three measurements the radiative corrections defined in eq. 15, or equivalently [34] $\Delta\rho \equiv \epsilon_1$ and $\Delta_{3Q} \equiv -\epsilon_3/\cos^2\theta_w$. Here one assumes that the correction specific to the W mass ϵ_2 takes the SM value. The bands in the ϵ_1, ϵ_3 plane are shown in Fig. 16. The three bands overlap on an area which is smaller than the SM prediction itself.

The quality of the SM comparison is highlighted when one fits the top and higgs

Table 10. *Summary of measurements included in the combined analysis of Standard Model parameters. The total errors in column 2 include the systematic errors listed in column 3. The determination of the systematic part of each error is approximate. The SM results in column 4 and the pulls in column 5 are derived from the Standard Model fit including all data with the Higgs mass treated as a free parameter. The systematic errors on m_Z and Γ_Z contain the errors arising from the uncertainties in the LEP energy only.*

		Measurement with Total Error	Systematic Error	Standard Model	Pull
	$\alpha(m_Z^2)^{-1}$	128.896 ± 0.090	0.083	128.907	-0.1
a)	LEP line-shape				
	m_Z(GeV)	91.1863 ± 0.0020	0.0015	91.1861	0.1
	Γ_Z(GeV)	2.4946 ± 0.0027	0.0017	2.4960	-0.5
	σ_h^0 (nb)	41.508 ± 0.056	0.055	41.465	0.8
	R_ℓ	20.778 ± 0.029	0.024	20.757	0.7
	$A_{FB}^{0,\ell}$	0.0174 ± 0.0010	0.007	0.0159	1.4
	τ polarisation: $\mathcal{A}_\tau$	0.1401 ± 0.0067	0.0045	0.1458	-0.9
	$\mathcal{A}_e$	0.1382 ± 0.0076	0.0021	0.1458	-1.0
	b and c quark results:				
	R_b	0.2179 ± 0.0012	0.0009	0.2158	1.8
	R_c	0.1715 ± 0.0056	0.0042	0.1723	-0.1
	$A_{FB}^{(b)}$	0.0979 ± 0.0023	0.0010	0.1022	-1.8
	$A_{FB}^{(c)}$	0.0733 ± 0.0049	0.0026	0.0730	0.1
	+ correlation matrix				
	$(\langle Q_{FB} \rangle) \sin^2 \theta_W^{eff}$	0.2320 ± 0.0010	0.0008	0.23167	0.3
b)	SLD				
	$\sin^2 \theta_W^{eff}$ (A_{LR})	0.23061 ± 0.00047	0.00014	0.23167	-2.2
	R_b	0.2149 ± 0.0038	0.0021	0.2158	-0.2
	$\mathcal{A}_b$	0.863 ± 0.049	0.032	0.935	-1.4
	$\mathcal{A}_c$	0.625 ± 0.084	0.041	0.667	-0.5
c)	$p\bar{p}$ and νN				
	m_W (GeV) ($p\bar{p}$)	80.356 ± 0.125	0.110	80.353	0.0
	$1 - \frac{M_W^2}{M_Z^2}$ (νN)	0.2244 ± 0.0042	0.0036	0.2235	0.2
	M_t (GeV) ($p\bar{p}$)	175 ± 6	4.5	172	0.5

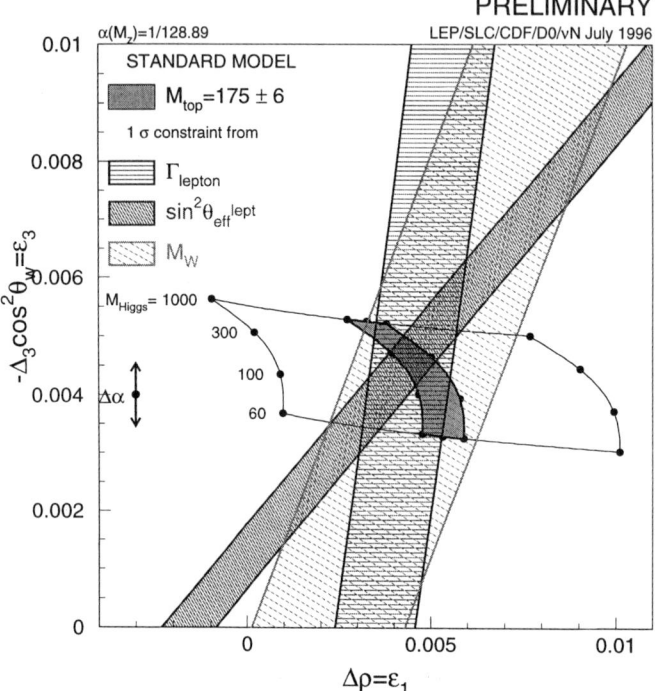

Figure 16. One standard deviation bands in the ϵ_1, ϵ_3 plane corresponding to the measurement of $\sin^2\theta_w^{\text{eff}}$, $\Gamma_{\ell\ell}$, M_W. The SM predictions as a function of M_t and M_H are shown. The arrow shows the influence of the error on $\alpha(M_Z^2)$ but this time on the experimental bands.

masses to the precision data. The most precise electroweak measurements are displayed in table 10, produced under the same format each 6 month by the LEP electroweak working group. Since the apparent agreement or disagreement can depend on the choice of observables or combinations thereof, the stability of the format ensures statistical reliability. Out of these 20 measurements, only one is more than two standard deviations away from the SM best fit, and six out of 20 at more than one standard deviation. This is exactly what is expected from gaussian distributed errors. The 3 σ effect in $\mathcal{A}_b$ appears to result from dividing one measurement which is high by another which is low. I conclude that the agreement of all observables with the SM is exactly as one would expect.

Various fits can be performed to these data. A fit with $M_t, \alpha_s(M_Z^2)$ as free parameters (not including the measured values of $M_t, \alpha_s(M_Z^2)$) yields

$$\begin{aligned}
M_H &= 60: & \chi^2 &= 17.8 \\
M_t &= 158 \pm 7 & \alpha_s(M_Z^2) &= 0.119 \pm 0.003 \\
M_H &= 300: & \chi^2 &= 20.4 \\
M_t &= 177 \pm 7 & \alpha_s(M_Z^2) &= 0.121 \pm 0.003
\end{aligned}$$

$$M_H = 1000: \quad \chi^2 = 23.9$$
$$M_t = 193 \pm 7 \quad \alpha_s(M_Z^2) = 0.123 \pm 0.003$$

The number of degrees of freedom being 14. This is in very good agreement with the direct measurements of $M_t, \alpha_s(M_Z^2)$.

Including now the direct determination of M_t one can perform a fit to $M_H, \alpha_s(M_Z^2)$. The result is:

$$M_H = 150^{+190}_{-80} \text{GeV}; \quad \alpha_s(M_Z^2) = 0.1202 \pm 0.0032$$

Uncertainties in the calculation of radiative corrections lead to an uncertainty of typically $\Delta(\log M_H) = \pm 0.2$. The results of the fit are shown in Fig. 17, in the $M_H, \alpha_s(M_Z^2)$ plane, together with present experimental constraints.

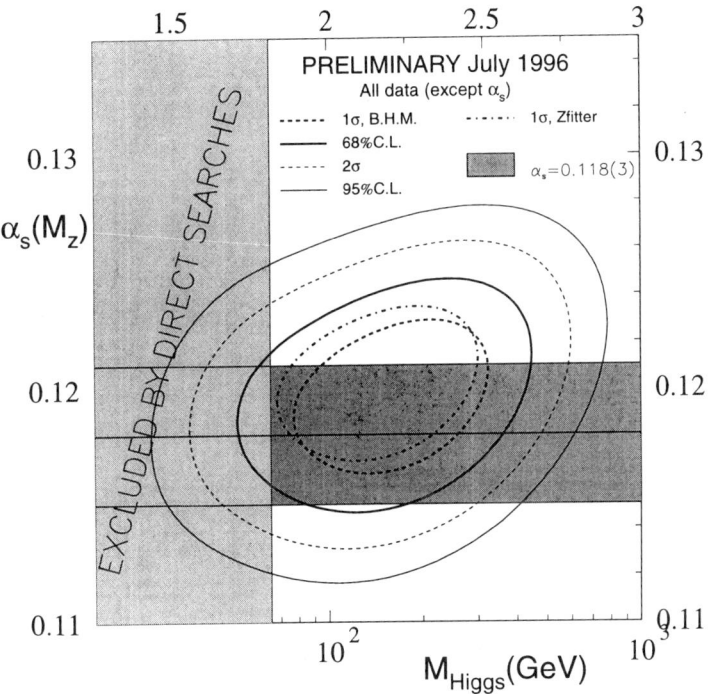

Figure 17. *Contours of constant χ^2 for M_H versus $\alpha_s(M_Z^2)$. The 1σ contours are shown for two different theoretical calculations to give a feel for the theoretical uncertainty.*

If one would also include the world average value of $\alpha_s(M_Z^2)$ in the fit, the following Higgs boson mass is obtained:

$$M_H = 140^{+150}_{-80} \text{GeV}; \quad \alpha_s(M_Z^2) = 0.1190 \pm 0.0022$$

The χ^2 curve is shown in Fig. 18. A two sigma upper limit on the Higgs mass can be derived:

$$M_H < 540 \text{ GeV at } 95\% \text{ C.L..} \tag{35}$$

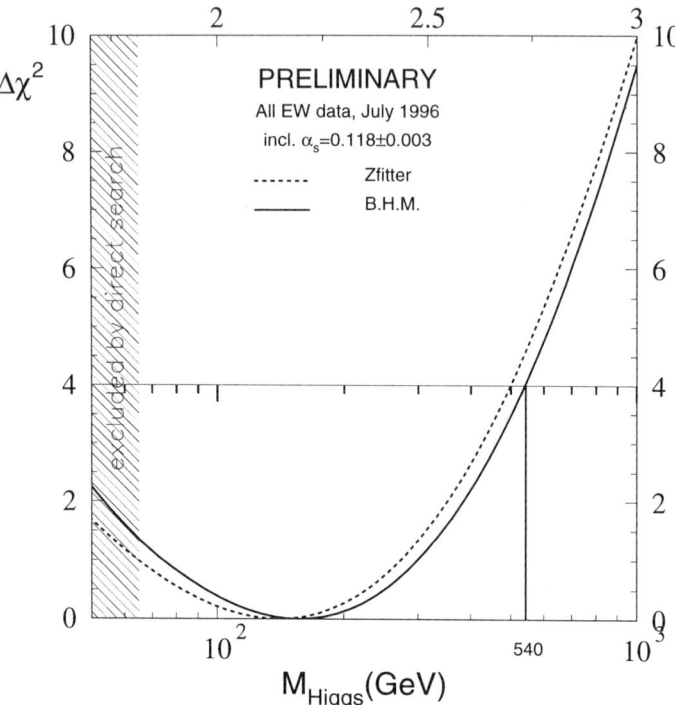

Figure 18. χ^2 curve for the Higgs boson mass in the minimal standard model fit to summer 1996 electroweak data. The theoretical uncertainty in the predictions is suggested by curves obtained using different codes [42].

R_b and $\alpha_s(M_Z^2)$

The disagreement of R_b with the standard model prediction has led to a variety of interpretations, which were summarized by Pokorski. Among those, several invoked additional radiative effects that would modify the b-partial width by an amount $\delta\Gamma_b$, but none of the others. This is the case for the Supersymmetric interpretations. Since there are 22% b events in the hadronic sample, an anomaly in R_b should show up as well as a discrepancy in R_ℓ. More precisely the hadronic width should be modified by $\delta\Gamma_{had} = \delta\Gamma_b$. This in turn would affect the value of $\alpha_s(M_Z^2)$ obtained from electroweak fits by an amount:

$$\delta\alpha_s(M_Z^2) = -4.\delta R_b \tag{36}$$

This has two possible consequences.

1. in the context of these scenarios of new physics that affect only R_b, one can use the fact that $\alpha_s(M_Z^2)$ is measured externally to express the electroweak measurements involving the hadronic width as "indirect" measurements of R_b, which one can compare with or average with the direct one:

$$\delta R_b = 0.0005 \pm 0.0011$$
$$\text{"indirect" } R_b = 0.2163 \pm 0.0011$$

$$\text{"direct" } R_b = 0.2178 \pm 0.0011$$
$$\text{"all LEP" } R_b = 0.2171 \pm 0.0008. \tag{37}$$

This constitutes the best estimate of R_b in this particular context. Note that the "indirect" evaluation does not support a large discrepancy with SM, and never has. It is as precise as the direct one.

2. Alternatively one can choose to believe the direct measurement of R_b and distrust the measurements of $\alpha_s(M_Z^2)$ (the reader has understood that I do not share this point of view). Then the modification of Γ_{had} leads to a different evaluation of $\alpha_s(M_Z^2)$:

$$\alpha_s(M_Z^2) = 0.112 \pm 0.006 \tag{38}$$

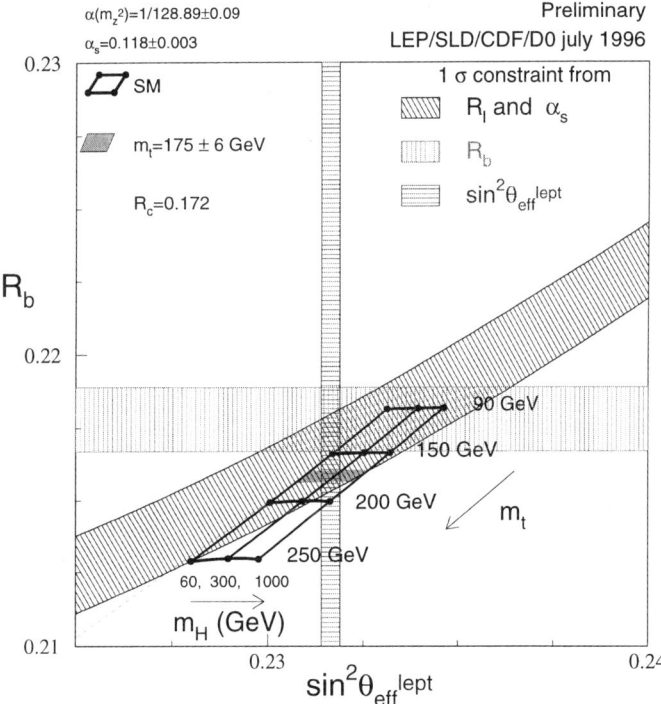

Figure 19. *Comparison of* $\sin^2 \theta_w^{eff}$ *and* $\frac{\Gamma_b}{\Gamma_{had}}$ *and the SM prediction. The constraint from* R_ℓ, *assuming* $\alpha_s = 0.118 \pm 0.003$, *is an oblique band.*

This discussion is illustrated by Fig. 19. Here the measured values of $\sin^2 \theta_w^{eff}$ and R_b (assuming the SM value for R_c) are plotted against each other. The measurement of R_ℓ is also indicated in this figure. If one uses the world average value $\alpha_s = 0.118 \pm 0.003$, the agreement of R_ℓ with the SM prediction is excellent.

Conclusions and Outlook

0. Precision electroweak physics is a lot of fun. It involves domains of knowledge as varied as fragmentation and beam polarization.

1. The tree level structure of the Standard Model is extremely well verified, at the level of below 10^{-3}. There is a 3σ problem with the b couplings, but it involves many preliminary numbers.

2. The radiative corrections to the SM are also very well verified: the prediction of the top quark exactly where it was to be found is a spectacular success. Consistency of all measurements implies

$$M_H = 150^{+150}_{-80} \text{GeV} \quad \alpha_s(M_Z^2) = 0.1190 \pm 0.0022 \tag{39}$$
$$M_H < 540 \text{ GeV at 95\% C.L.}.$$

3. LEP has now left the Z pole, and improvements will come from a number of analyses that remain at LEP, and from more data taken at SLC. Improvements of at most a factor 1.5 for $\sin^2 \theta_w^{\text{eff}}$ could be expected. The W mass measurement should improve down to the region $\Delta M_W = 25 - 50$ MeV.

4. Improving the indirect Higgs mass will require a combination of improved low energy hadronic cross sections, top quark mass determination and measurements of $\sin^2 \theta_w^{\text{eff}}$ and M_W. If these take place it might be possible to obtain a 20% measurement of M_H.

5. Enormous amount of work remains to understand and ascertain R_b. There is a lot that we do not know about b-hadron production in $Z \to b\bar{b}$ events!

Acknowledgements

Tom Ferbel is really someone special. It is a pleasure to congratulate him for such a smooth operation. St Croix has been quite an experience indeed and I had the pleasure to meet new interesting people, i.e. students. Precision measurements owe a lot to our colleagues from the CERN/SLAC/FERMILAB accelerator teams. My friends from the LEP polarization collaboration are thanked particularly. The LEP/SLC/Fermilab collaborations were very cooperative to give me all necessary informations. None of this work could have been made without the painstaking work of more than 100 colleagues performing precision calculations. Most numbers in this report are the results of the careful work of the LEP/SLD Electroweak working group.

REFERENCES

[1] S.L.Glashow, Nucl. Phys. 22(1961)579.
A.Salam, Proc. 8th Nobel Symposium, Aspenaegarden, Almqvist and Wiksell ed., Stockholm (1968)367.
S.Weinberg, Phys. Rev. Lett. 19(1967)1264; Phys. Rev. D5(1972)1412.

[2] P.W.Higgs, Phys. Rev. Lett. 12(1964)132; Phys. Rev. Lett. 13 (1964)508; Phys. Rev. 145(1966)1156;
F.Englert and R.Brout Phys. Rev. Lett. 13 (1964)240;
G.S.Guralnik, C.R.Hagen and T.W.B.Kibble, Phys. Rev. Lett. 13 (1964)585;
T.W.B.Kibble, Phys. Rev. 155(1967)1554.

[3] G.t'Hooft, Nucl. Phys. 33B(1971)173; Nucl. Phys. 35B(1971)167.

[4] B. Pietrzyk gave the last summary of top mass from precision measurements, before the actual evidence from Fermilab became public, at Moriond 1994 March 1994, that was $M_t = 174 \pm 11^{+17}_{-19}$ GeV.

[5] F.Abe et al, (CDF Coll.) Phys. Rev. Lett. 74 (1995) 2626;
D0 Collab., Phys. Rev. Lett. 74 (1995) 2632.

[6] J.Lys, CDF Coll., FERMILAB-CONF-96/409-E. Proceedings ICHEP'96, Warsaw,(1996).

[7] S. Protopopescu, D0 coll., Proceedings ICHEP'96, Warsaw,(1996).

[8] P.Grannis, presentation at ICHEP'96, Warsaw, PA07; M. Demarteau, memorandum to the LEP electroweak working group.

[9] M.Schmelling, Rapporteur's talk at ICHEP'96.

[10] R.M. Barnett et al., Phys. Rev. D54 (1996).

[11] CCFR Coll., Cont. ICHEP96 pa03-030, pa03-031.

[12] BHM: G.Burgers, W.Hollik and M.Martinez; Computer code available from M. Martinez.

[13] L.Gibbons, rapporteurs talk at ICHEP'96.

[14] BES coll., Phys. Rev Lett. 69(1992)3021;
Eric Soderstrom, 1994 APS Meeting, DPF, Albuquerque, NM, August 1994.

[15] H.Videau, presentation at ICHEP'96, Warsaw, PA07.

[16] R.Stroynowski, presentation at ICHEP'96, Warsaw, PA07.

[17] CLEO Coll., Michel parameters in Leptonic Tau Decays, cont. ICHEP96, Warsaw, Pa07-068;

[18] SLD Coll., A Measurement of the Tau Michel Parameters at SLD cont. ICHEP96, Warsaw, Pa07-065;
Tau Neutrino Helicity Measurement at SLD cont. ICHEP96, Warsaw, Pa07-065.

[19] B.L.Roberts, BNL E821 Coll., presentation at ICHEP'96, Warsaw, PA07.

[20] S.Eidelmann and F.Jegerlehner, Z. Phys. C67(1995)585.

[21] . D.H.Brown and W.A.Worstell, Phys. Rev. D54(1996)3237.

[22] A.Czarnecki, B.Krause, W.J.Marciano, [hep-ph] 9512369 - Electroweak corrections to the muon anomalous magnetic moment;
T. Kinoshita and W. J. Marciano, in Quantum Electrodynamics, T. Kinoshita ed., World Scientific, Singapore (1990)419.

[23] M.L.Swartz, SLAC-Pub-95-7001.

[24] A.D.Martin and D.Zeppenfeld, Phys. Lett. B345 (1994) 558.

[25] H.Burkhardt and B.Pietrzyk, Phys. Lett. B356(1995) 398.

[26] C.Y.Prescott, proc. 1980 Int. Symp. on High-Energy Physics with Polarized Beams and Polarized Targets, Lausanne, 1980, eds. C. Joseph and J. Soffer (Birkhäuser Verlag, Basel, 1981), 34. M.Böhm and W.Hollik, Nucl. Phys. B204 (1982), 45.

[27] B.W.Lynn and R.G.Stuart, Nucl. Phys. B253 (1985) 84.

[28] A.Blondel, B.W.Lynn, F.M.Renard and C.Verzegnassi, Nucl. Phys. B304 (1988) 438.

[29] M. Veltman, Nucl. Phys. B123 (1977) 89.

[30] B.W.Lynn, M.E.Peskin and R.G.Stuart, "Physics at LEP" CERN 86-02, (1986) 90.

[31] D.C.Kennedy and B.W.Lynn, Nucl. Phys. B322 (1989).

[32] M.Consoli and W.Hollik, in "Physics at LEP1", CERN 89-08 (1989) 7.

[33] M.E.Peskin and T.Takeuchi, Phys. Rev. Lett. 65 (1990) 964; V.A.Novikov, L.B.Okun, M.I.Visotsky, Nucl. Phys. B397 (1993) 35.

[34] G.Altarelli and R.Barbieri, Phys. Lett. B253 (1991) 161; G.Altarelli, R.Barbieri, S.Jadach, Nucl. Phys. B369 (1992) 3; Err. Nucl. Phys. B376 (1992); G.Altarelli et al, Nucl. Phys. B405 (1993) 3.

[35] A. Blondel, TASI 1991, Ellis, Hill and Lykken eds., world scientific (1992) 283; A.Blondel and C.Verzegnassi, Phys. Lett. B311 (1993) 346.

[36] The LEP collaborations, Phys. Lett. B276 (1992) 247.

[37] B.W.Lynn, SLAC-Pub 5077 (1989); D.Levinthal, F.Bird, R.G.Stuart and B.W.Lynn, CERN-Th 6094/91.

[38] $\overline{MS}$ scheme: S.Sarantakos, A.Sirlin and W.Marciano, Nucl. Phys. B217 (1983) 84; W.Marciano and A.Sirlin, Phys. Rev. D29 (1984) 75; A.Sirlin, Phys. Lett. B232 (1989) 123; A.Sirlin, Nucl. Phys. B332 (1990) 20.

[39] A.Olshevski, P.Ratoff and P.Renton, Z Phys. C60 (1993) 643; P.Gambino and A.Sirlin, Phys. Rev. D49 (1994) R1160.

[40] A.Blondel, F.M.Renard and C.Verzegnassi, Phys. Lett. B269 (1991) 419.

[41] A.Blondel, A.Djouadi, C.Verzegnassi, Phys. Lett. B 293 (1992) 253.

[42] Electroweak libraries: ZFITTER; D.Bardin et al, Z. Phys. C44 (1989) 493; Nucl. Phys. B351 (1991) 1; Phys. Lett. B255 (1991) 290 and CERN-TH 6443/92 (May 1992);
Computer programme ZFITTER, available from D. Y. Bardin.
BHM: G.Burgers, W.Hollik and M.Martinez;
M.Consoli, W.Hollik and F.Jegerlehner, in CERN 89-08 (1989) vol I 7 and G.Burgers, F.Jegerlehner, B.Kniehl and J.Kühn, ibid, 55. These computer codes have been upgraded by including the results of: B.A.Kniehl and A.Sirlin, DESY 92-102; S.Fanchiotti, B.A.Kniehl and A.Sirlin, CERN-TH.6449/92; R.Barbieri et al Phys. Lett. B288 (1992) 95; K.G.Chetyrkin, J.H.Kühn, Phys. Lett. B248 (1990)p. 359; K.G.Chetyrkin, J.H.Kühn and A.Kwiatkowski, Phys. Lett. B282 (1992) 221; J.Fleischer, O.V.Tarasov and F.Jegerlehner, Phys. Lett. B293 (1992) 437.
Report of the working group on precision calculations for the Z resonance, D.Bardin, W.Hollik, G.Passarino eds., CERN Yellow report 95-03 (1995).

[43] A.Blondel, 2d Workshop on LEP perf., CERN-SL/92-29 (DI) 339.

[44] L.Arnaudon et al, Phys. Letters B284 (1992) 431.

[45] A.A.Sokolov and I.M.Ternov, Sov. Phys. Doklady, 8 (1964) 1203.

[46] This effect was first seen in the Orsay ACO storage ring: R.Belbeoch et al, USSR Part. Accel. Conf. (1968) 129. For reviews, see B.W.Montague, Phys. Rep. 113 (1984); A.Blondel CERN-PPE/93-125 (1993); M.Böge, DESY 94-087.

[47] A.A.Zholentz et al, Phys. Lett. 96B (1980) 214.

[48] A.S.Artamonov et al, Phys. Lett. 118B (1982) 225.

[49] D.P.Barber et al, Phys. Lett. 135B (1984) 498.

[50] W.W.McKay et al, Phys. Rev. D29 (1984) 2483.

[51] L.Knudsen et al, Phys. Lett. B270 (1991) 97.

[52] L. Arnaudon et al, Z. Phys. C 66 (1995) 45; R. Assmann et al., Z. Phys. C 66 (1995) 567.

[53] L. Arnaudon et al, N.I.M. A357 (1995) 249.

[54] L.Arnaudon et al, CERN-SL/94-07 (BI) (1994).

[55] Investigation suggested to me by B.Richter and realized by: B.Jacobsen, SL-MD note 62, (1992).

[56] J.Wenninger, CERN-SL/94-14 (BI).

[57] E.Lancon and A.Blondel, LEP Energy Working Group note 96-07(1996).

[58] G.Wilkinson, presentation at ICHEP'96, Warsaw, PA07. LEP energy Groupnote 96-05 (1996).

[59] M.Gruenewald, presentation at ICHEP'96, Warsaw, PA07.

[60] ALEPH Coll., Z. Phys. C48(1990)365; Z. Phys. C53(1992)1; Z. Phys. C60(1993)71; Z. Phys. C62(1994)539; cont. ICHEP96, Warsaw, PA-07-069.

[61] DELPHI Coll., Nucl. Phys. B367(1991)511; Nucl. Phys. B417(1994)3; Nucl. Phys. B418(1994)403; DELPHI Note 95-62 PHYS 497, July 1995; DELPHI Note 96-118 CONF 65, cont. ICHEP96, Warsaw, PA-07-001.

[62] L3 Coll., Z. Phys. C51(1991)179; Phys. Rep. 236(1993)1; Z. Phys. C62(1994)551; L3 Note 1980, August 1996.

[63] OPAL Coll., Z. Phys. C52(1991)175; Z. Phys. C58(1993)219; Z. Phys. C61(1994)19; OPAL Physics Note PN166, Febuary 1995; OPAL Physics Note PN142, July 1994; OPAL Physics Note PN242, July 1996; cont. ICHEP96, Warsaw, PA07-015.

[64] The LEP collaborations, LEP electroweak working group and the SLD heavy flavour group, CERN-PPE/96-183(1996).

[65] S. Jadach, E. Richter-Wąs, Z. Wąs and B.F.L. Ward, Phys. Lett. B268 (1991) 253; Comput. Phys. Commun. 70 (1992) 305;
W. Beenakker and B. Pietrzyk, Phys. Lett. B304 (1993) 366.

[66] A.Arbuzov et al, Phys. Lett. B383(1996)238;
S.Jadach et al, CERN-TH/96-156 submitted to Comp. Phys. Comm.

[67] S. Jadach, E. Richter-Wąs, B.F.L. Ward and Z. Wąs, Phys. Lett. B353 (1995) 362.

[68] UA1 Coll., Phys. Lett. 122B (1983) 103

[69] UA2 Coll., Phys. Lett. 122B (1983) 476

[70] UA2 Coll., Phys. Lett. 276B (1992) 354;
CDF Coll., Phys. Rev. Lett. 65 (1990) 2243 and Phys. Rev. D43 (1991) 2070; FERMILAB-PUB-95/033-E;

[71] DØ Coll., M. Rijssenbeek at ICHEP '96, Warsaw, FERMILAB-CONF-96/365-E (1996).

[72] CDF coll., Phys. Rev. Lett. 74 (1995) 850.

[73] Working group on the W mass, in "Physics at LEP2", CERN 96-01 (1996) 141.

[74] OPAL Coll., CERN-PPE/96-141, to appear in Phys. Lett. B.

[75] Presentations by R.Miquel(ALEPH), W.DeBoer(DELPHI) and M.Pohl(L3), CERN seminar, 8 October 1996.

[76] H. Abramowicz et al. (CDHS Collaboration): Phys. Rev. Lett. 57(1986)298; A. Blondel et al., Z. Phys. C45(1990)361.

[77] J.V. Allaby et al. (CHARM Collaboration): Phys. Lett B177 (1986) p. 446, and Z. Phys. C36 (1987) p. 611.

[78] CCFR Coll., K. McFarland, cont. ICHEP96, Warsaw, Pa07-077.

[79] E.Torrence, presentation at ICHEP96, PA07. SLD coll., Phys. Rev. Lett. 73 (1994) 25.

[80] A.Rougé, Z. Phys. C48 (1990) 45.

[81] M.Davier et al, Phys. Lett. B306(1993)411.

[82] ALEPH Coll., Zeit. Phys. C69 (1996) 183.

[83] DELPHI Coll., Z. Phys. C67(1995)183; DELPHI 96-114 CONF 42, cont. ICHEP96, Warsaw, PA07-008.

[84] OPAL Coll., Z. Phys. C72 (1996) 365.

[85] L3 Coll., Phys. Lett. B341 (1994) 245; cont. ICHEP96, Warsaw, PA07-56; The 1994 data have been combined with the earlier data using 100% correlation of the systematic errors.

[86] LEP Heavy Flavours working group, LEPHF/96-01, ALEPH note 96-099, DELPHI 96-67 PHYS 627, L3 note 1969, OPAL Technical Note TN391.

[87] E. Etzion, in XXXI Rencontres de Moriond, March 1996, Electroweak interactions and unified theories, J.Tran Than Van Ed., Eds Frontieres (1996).

[88] SLD Collaboration, G. Crawford, ICHEP96, Warsaw.

[89] L3 Coll., cont. ICHEP96, Warsaw, PA05-049.

[90] DELPHI Coll., Z. Phys. C70 (1996) 531; cont. ICHEP96, Warsaw, (1996) PA01-061

[91] I. Tomalin, ALEPH Coll., ICHEP96, Warsaw, PA10-014 and PA10-015.

[92] ALEPH Coll., Z. Phys. C62(1994)179; Phys. Lett. B384 (1996)414; Cont. EPS-HEP-95 Brussels, **eps0404**,

[93] DELPHI Coll., Z. Phys C65(1995)569; Z. Phys C66 (1995) 341; DELPHI 95-87 PHYS 522;

[94] L3 Coll., Phys. Lett. B292(1992)454; Phys. Lett. B335 (1994) 542; L3 Notes 1449 (1993) , 1624,1625 (1994).

[95] OPAL Coll., Z. Phys. C70 (1996) 357; OPAL Physics Note PN226 cont. ICHEP96, Warsaw, 1996 PA05-007.

[96] OPAL Coll., Z. Phys. C67(1995)27; Z. Phys. C72(1996)1; cont. ICHEP96, Warsaw, PA05-011.

[97] DELPHI Coll., cont. ICHEP96, Warsaw, DELPHI 96-110 CONF 37, PA01-060.

[98] ALEPH Coll., cont ICHEP96, Warsaw, PA10-016.

[99] ALEPH Coll., Phys. Lett. B335 (1994) 99.

[100] ALEPH Coll., cont. ICHEP96, Warsaw, 1996 PA10-018.

[101] OPAL Coll., Z. Phys. C67 (1995) 365.

[102] DELPHI Coll., Z. Phys C65 (1995) 569; Z. Phys C66 (1995) 341; DELPHI 95-87 PHYS 522 (1995).

[103] OPAL Coll., CERN-PPE/96-101 (1996).

[104] ALEPH Coll., Z. Phys. C62 (1994) 1; Cont. EPS-HEP-95 Brussels, **eps0634**.

[105] D. Falciai, SLD coll., ICHEP96, Warsaw. SLD Coll., Phys. Rev. Lett. 74 (1995)2895; Phys.Rev.Lett.74 (1995) 2890; Nuovo Cim. A109 (1996) 663; Phys. Rev. Lett 75(1995) 3609; cont. ICHEP96, PA10-026.

[106] ALEPH Coll., Phys. Lett. B259(1991)377; Z. Phys. C71(1996)357.

[107] T. Sjöstrand, Comp. Phys. Com.82(1994)74.

[108] G. Marchesini et al, Comp. Phys. Comm. 67 (1992) 465.

[109] DELPHI Coll., Phys. Lett. B277(1992)371; DELPHI 96-19 PHYS 594.

[110] OPAL Coll., Phys. Lett. B294(1992)436; OPAL Physics Note PN195(1995).

[111] S.Anassontzis et al, Nucl. Inst. Meth. A323 (1992) 351.

[112] DELPHI coll., ICHEP 1994, GLS0232.

[113] A. Stacey, ALEPH coll., ICHEP96, Warsaw, PA10-017.

[114] J. Letts, P. Mättig, CERN-OPEN/96-012; J. Letts, OPAL Coll., ICHEP96, Warsaw, PA07-020.

THE PHYSICS OF MASSIVE NEUTRINOS

François Vannucci

LPNHE, Univ. Paris 7
4 place Jussieu Tour 33
75252-Paris, France

INTRODUCTION

Neutrinos originate from many varied sources. They were first detected at a nuclear reactor more than 40 years ago. Later on results were obtained at accelerators, then with solar atmospheric and supernova neutrinos. We also believe that the Big Bang has left relic neutrinos which fill the Universe at the level of $110/cm^3$/flavour. But nobody has invented yet a clear way to look for them.

After many years of experimental data, interactions of neutrinos are now well understood, and the emphasis of the field has moved to the search for effects of masses and/or mixings of neutrinos.

This domain is full of ups and downs, with claims which have not been confirmed. One remembers the stories of the 30 eV ν_e or of the 17 keV neutrino which lasted for eight years before their final demise. This points to the fact that neutrino physics remains difficult. But these upheavals make the field particularly interesting and subject to unorthodox ideas.

Neutrino physics is one of the favourite ways to reach beyond the Standard Model of electroweak interactions so successfully tested otherwise. Today we are confronted with several experimental facts which would be signs of the new physics but require a better understanding: solar and atmospheric deficits, LSND results.

Everybody has his own scale of belief or disbelief in these results. The solar neutrino deficit is becoming an evidence, while atmospheric and LSND results remain puzzles. Will they move to genuine discoveries or will they become mere anomalies? The future will tell. This series of lectures will review the various inputs into the field of massive neutrinos and will try to sketch some possible directions for the future.[1]

WHAT DO WE KNOW ABOUT NEUTRINOS?

Neutrinos are the electrically neutral leptons, among the fundamental constituents of matter which are usually represented in three families:

$$\begin{array}{ccc} d & s & b \\ u & c & t \\ e & \mu & \tau \\ \nu_e & \nu_\mu & \nu_\tau \end{array}$$

They only feel the weak force. This is why, although they are one billion times more abundant than the other elementary constituents, we take little notice of their presence. Invented by W. Pauli in 1930 to save the principle of energy conservation in β decays, the neutrino remained a "theoretical" particle for many years. The ν_e was finally detected at the Savannah River reactor in 1955.

The second neutrino ν_μ associated with the muon was discovered in 1962 at the Brookhaven Laboratory in what has become the prototype of all subsequent neutrino beams at accelerators: a high-energy, high-intensity proton beam impinges on a target, it is followed by a decay tunnel and a lot of shielding to stop all particles but the neutrinos.

A third type of neutrino ν_τ was hypothesized as soon as the τ lepton was discovered in 1975. This neutrino has not yet been seen experimentally.

How many neutrinos exist in nature? Will the series continue? The answer came in 1989 with the first results from SLC and LEP which measured the Z line shape.[2] Each new species adds 166 MeV to the width and its precise determination measures the number of neutrinos. The result is unambiguous: there exist three and only three light and stable neutrinos having standard couplings to the Z (Fig. 1).

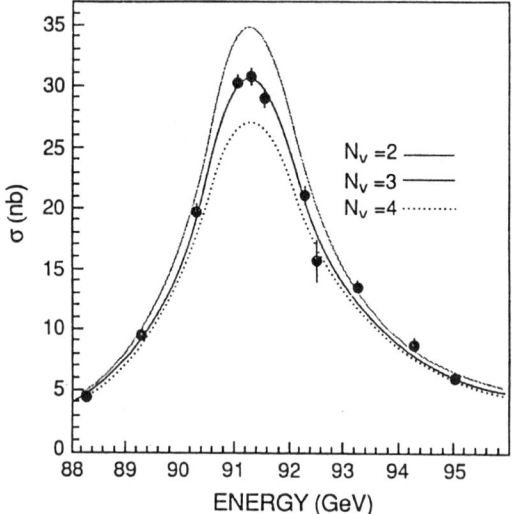

Figure 1. The Z^0 lineshape at LEP.

Twenty-five years of neutrino physics at accelerators have given a coherent picture of neutrino interactions with the discovery in 1973 of the neutral currents in the Gargamelle bubble chamber, then a precise measurement of the weak couplings. On another register, deep inelastic scattering of neutrinos has studied thoroughly the internal structure of the nucleons, and in particular has found evidence of the gluon content.

For the past ten years the emphasis has shifted from the study of interactions to the study of the intrinsic properties of the neutrinos, in particular masses and mixings. This shift has been coupled with more and more interest in detecting and understanding neutrinos of astrophysics origins, foremost those from the Sun. This cross-fertilization between these two separate frontiers of physics has been particularly fruitful in neutrino physics.

MASSIVE NEUTRINOS PHENOMENOLOGY

The usual mass term in a Lagrangian describing a spin 1/2 object is a Lorentz scalar bilinear of the form:

$$m\overline{\Psi}\Psi = m(\overline{\Psi}_L \Psi_R + \overline{\Psi}_R \Psi_L)$$

where Ψ_R and Ψ_L are the right-handed and left-handed field components, respectively.

A massless neutrino has only the left-handed field and the antineutrino has only the right-handed one (two-component neutrino). This is the case in the minimal standard model.

In order to construct a mass, one must have a right-handed neutrino field at one's disposal. There are two possibilities to achieve this:

- one adds "by hand" a R-handed component. This will give a Dirac mass-term. The neutrino now has four components like the electron. This is the solution favoured in grand unified theories which unify quarks and leptons;

- one can construct a Lorentz scalar using the C-conjugate partner of Ψ.

In our case this is the antineutrino. It will give a Majorana mass term. Such a term cannot appear for a particle other than the neutrino because it violates the conservation of electric charge. With the neutrino it violates (only) the conservation of lepton number, which is or is not a sacred law of nature.

An interesting experimental question arises here.

If neutrinos are massive, they do not propagate at the velocity of light. There exists a frame which goes faster. In such a frame the helicity reverses. Starting with a L-handed neutrino one ends up with a R-handed state. What is this final state? If neutrinos are Dirac particles, this new state is a sterile object, it does not interact. If neutrinos are Majorana, the new state is the antineutrino and it will give a μ^+ when interacting with matter. Unfortunately, it is difficult to test the idea since the amount of this new state is negligibly small in practical circumstances.

The most general mass term will have both Dirac and Majorana terms:

$$\overline{\Psi}_L M^D \Psi_R + 1/2 \overline{\Psi}_L M^L (\Psi_L)^C + 1/2 \overline{\Psi}_R M^R (\Psi_R)^C + \text{h.c.}$$

M^D, M^L and M^R are the so-called Dirac, left Majorana and right Majorana masses. The mass term can be rewritten in a matrix form:

$$\frac{1}{2}\overline{\Psi}\begin{pmatrix} M^L & M^D \\ (M^D)^T & M^R \end{pmatrix}\Psi^C + \text{h.c.}$$

where Ψ is a vector with all the possible L-handed chirality states.

The mass matrix has to be diagonalized in order to find the fields corresponding to the physical neutrinos. In general they will be of the Majorana type, they are their own antiparticles.

The popular see-saw model for masses,[3] assumes that there is no L-handed Majorana coupling. In this case the mass matrix simplifies as follows:

$$\begin{pmatrix} 0 & m \\ m & M \end{pmatrix}$$

where $m(M)$ refers to the Dirac (Majorana) mass. The m is presumably comparable to other Dirac masses. Diagonalizing the matrix, one finds two physical states for each generation. If $M \gg m$, the two physical masses will be

$$m_{\text{heavy}} \to M$$
$$m_{\text{light}} \to m^2/M .$$

In this framework, the neutrino which takes part in the traditional weak interactions is essentially the L-handed component of the light neutrino with a small admixture of the L-handed component of the heavy one. How large is M? Nobody knows the answer, but M is supposed to relate to the scale of "new physics" at which lepton numbers are violated, and this presumably happens at very high energies.

PHENOMENOLOGY OF OSCILLATIONS

Whatever the origin of the masses, the weak eigenstates ν_e, ν_μ and ν_τ have no reason to be identical to the mass eigenstates.

The weak eigenstates are by definition the states coupling to the W or the Z, associated to e, μ and τ, respectively.

The mass or energy eigenstates ν_1, ν_2 and ν_3 are eigenstates of the Hamiltonian:

$$id/dt|\Psi\rangle = H|\Psi\rangle = E|\psi\rangle .$$

They propagate as plane waves: $|\nu_1(t)\rangle = e^{-iE_1 t}|\nu_1(0)\rangle$.

The two sets of states are related by a unitary matrix, similar to the Cabibbo–Kobayashi–Maskawa of mixings among quarks.

$$\begin{pmatrix} \nu_e \\ \nu_\mu \\ \nu_\tau \end{pmatrix} = M \begin{pmatrix} \nu_1 \\ \nu_2 \\ \nu_3 \end{pmatrix} .$$

A weak state will be a superposition of mass eigenstates:

$$\nu_l = \sum_i U_{li}\nu_i .$$

This means that any physical process can be thought of as an incoherent sum over the various mass states allowed by phase space. The heavy Majorana states are included in the development.

Oscillations in Vacuum[4]

Let us first consider the case of two neutrino states, in order to simplify the demonstration. The mixing matrix in this case becomes a rotation matrix with a single mixing angle θ.

$$\nu_e = \cos\theta \nu_1 + \sin\theta \nu_2$$
$$\nu_\mu = -\sin\theta \nu_1 + \cos\theta \nu_2 \ .$$

Starting with a pure ν_μ state of momentum p at time $t = 0$,

$$\nu(t=0) = \nu_\mu$$

the state evolves at time t into:

$$\nu(t) = -\sin\theta e^{-iE_1 t}\nu_1 + \cos\theta e^{-iE_2 t}\nu_2 \ .$$

Projecting back onto the eigenstates of interactions, one finds the probability that the initial neutrino behaves as a ν_e

$$|\langle \nu_e|\nu(t)\rangle|^2 = 4\sin^2\theta \cos^2\theta \sin^2(E_1 - E_2)t/2 \ .$$

For relativistic neutrinos, one has:

$$E_1 - E_2 = (m_1^2 - m_2^2)/(2p)$$

and the probability of oscillations can be written:

$$P = \sin^2 2\theta \sin^2 \pi x/L$$

where x is the distance between production and detection, and L is the oscillation length:

$$L \text{ (km)} = 2.5 \ E \text{ (GeV)}/\delta m^2 (\text{eV}^2) \ .$$

There is oscillation between different weak neutrino states with a maximum amplitude $\sin^2 2\theta$ and a wavelength $L/2$. This is only possible if neutrinos mix ($\theta \neq 0$) and if $\delta m^2 \neq 0$. At least one neutrino must be massive.

While ν_1 and ν_2 verify by definition the equation:

$$i\frac{d}{dt}\begin{pmatrix}\nu_1\\ \nu_2\end{pmatrix} = \begin{pmatrix}E_1 & 0\\ 0 & E_2\end{pmatrix}\begin{pmatrix}\nu_1\\ \nu_2\end{pmatrix}$$

the ν_e and ν_μ states obey:

$$i\frac{d}{dt}\begin{pmatrix}\nu_e\\ \nu_\mu\end{pmatrix} = \begin{pmatrix}E_1\cos^2\theta + E_2\sin^2\theta & (E_1 - E_2)\sin\theta\cos\theta\\ (E_1 - E_2)\sin\theta\cos\theta & E_1\sin^2\theta + E_2\cos^2\theta\end{pmatrix}\begin{pmatrix}\nu_e\\ \nu_\mu\end{pmatrix}$$

and the oscillation comes from the non-diagonal terms in the matrix. Massless neutrinos are such that $E_1 = E_2$. In this case ν_e and ν_μ propagate independently.

This is a simplified description of the phenomenon of oscillations. It has been proven that a correct quantum mechanical treatment of the problem with considerations of wave packets gives the same result.[5]

Oscillations with Three Families

The mixing matrix is now 3×3 and can be parametrized with three angles and a phase which could be responsible for CP-violating effects.

The probability of oscillations between ν_μ and ν_e exhibits now three terms:[6]

$$P = (a+b-c)U_{e_1}^2 U_{\mu_1}^2 + (a-b+c)U_{e_2}^2 U_{\mu_2}^2 + (-a+b+c)U_{e_3}^2 U_{\mu_3}^2$$

with

$$a = 2\sin^2(1.27 \times \delta m_{21}^2/E)$$
$$b = 2\sin^2(1.27 \times \delta m_{31}^2/E)$$
$$c = 2\sin^2(1.27 \times \delta m_{23}^2/E) \,.$$

If there is hierarchy beween the three masses as would happen in the see-saw model, $m_3 \gg m_1, m_2$, then

$$\delta m_{31}^2 = \delta m_{32}^2 \gg \delta m_{21}^2 \quad \text{or} \quad b = c \gg a \,.$$

The probability becomes:

$$P = 2c U_{e_3}^2 U_{\mu_3}^2 \,.$$

The oscillation between the ν_e and the ν_μ has an oscillatory term driven by the third neutrino mass, and its amplitude depends on the mixings with the third neutrino. This is a kind of indirect oscillation.

One may then expect two different regimes:

- the $\nu_e\,\nu_\mu$ oscillation develops first, at small distances because of the ν_3 mass but corresponds presumably to a small amplitude,

- then the larger mixings $U_{e_1}^2 U_{\mu_1}^2$ and $U_{e_2}^2 U_{\mu_2}^2$ may take over at large distances due to the small term δm_{21}^2.

This double scenario could reconcile the results from solar and LSND experiments which both test the same physical channel but find very different parameters.

Oscillations in Matter[7]

In matter the neutrinos will interact. The propagation eigenstates now verify:

$$id/dt|\Psi\rangle = H|\Psi\rangle = (E+V)|\Psi\rangle \,.$$

The scattering probability enters as an average potential which adds a phase e^{-iVt} to the propagation.

This pseudopotential has the value:

$$V(x) = -2\pi\rho(x)f(0)/E$$

where $f(0)$ is the forward scattering amplitude and $\rho(x)$ is the density of scatterers.

If V is the same for the various neutrino species, it only adds a common phase to the propagation and does not change the pattern of oscillations.

But we know that V is different for ν_e or $\bar{\nu}_e$. Because of the presence of atomic electrons in matter and the absence of muons or tauons, ν_e and $\bar{\nu}_e$ can undergo the elastic scatterings of Figs. 2(b) and 2(c), which are forbidden for other neutrinos.

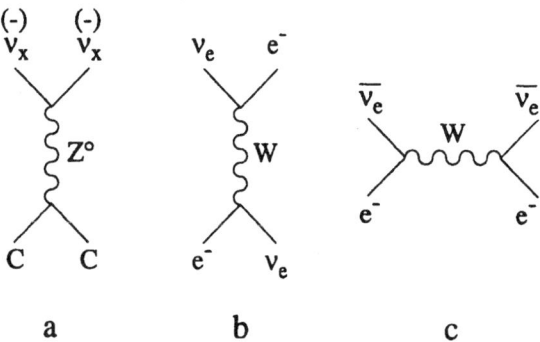

Figure 2. Elastic scattering of neutrinos in matter: (a) for all flavours; (b) only for ν_e; (c) only for $\bar{\nu}_e$.

This can be expressed in the following way, on the basis of the weak eigenstates which are the eigenvectors of interactions:

$$V\begin{pmatrix}\nu_e\\ \nu_\mu\end{pmatrix} = \begin{pmatrix}V + \sqrt{2}G\rho_e & 0 \\ 0 & V\end{pmatrix}\begin{pmatrix}\nu_e\\ \nu_\mu\end{pmatrix}$$

where V is the average potential common to all neutrinos.

The eigenstates of propagation in vacuum are ν_1 and ν_2, the eigenstates of interactions are ν_e and ν_μ. Neither are eigenstates of propagation in matter. To find these, we have to write down the complete mass matrix and diagonalize it.

The total Hamiltonian can be written on the basis of ν_e and ν_μ:

$$\begin{pmatrix} E_1\cos^2\theta + E_2\sin^2\theta + V + \sqrt{2}G\rho_e & (E_1 - E_2)\sin\theta\cos\theta \\ (E_1 - E_2)\sin\theta\cos\theta & E_1\sin^2\theta + E_2\cos^2\theta + V \end{pmatrix}.$$

The states propagating in matter are the eigenvectors of the previous matrix, or equivalently of the following reduced matrix:

$$\begin{pmatrix} \sqrt{2}G\rho_e & 1/2(E_1 - E_2)\sin^2\theta \\ 1/2(E_1 - E_2)\sin^2\theta & (E_2 - E_1)\cos^2\theta \end{pmatrix}.$$

The two states having well-defined energies in the medium can be expressed in the following way:

$$n_1 = \cos\theta_m \nu_e + \sin\theta_m \nu_\mu$$
$$n_2 = -\sin\theta_m \nu_e + \cos\theta_m \nu_\mu$$

with the mixing angle in matter θ_m given by the following expression:

$$\mathrm{tg}\,\theta_m = (E_2 - E_1)\sin^2\theta / [(E_2 - E_1)\cos^2\theta - \sqrt{2}G\rho_e].$$

If $\rho_e = 0$, then $\theta_m = \theta$ and we are back to the case of the propagation in vacuum. If ρ_e is very large, $\theta_m = 0$, ν_e and ν_μ are the propagating states.

In the general case, one can develop the propagating states on the $n_1 n_2$ basis:

$$|\nu(t)\rangle = \alpha(0)e^{-iE_1 mt}|n_1\rangle + \beta(0)e^{-iE_2 mt}|n_2\rangle .$$

The probability of oscillations becomes:

$$P = \sin^2 2\theta_m \sin^2 \pi x/L_m$$

where the mixing angle θ_m and the oscillation length L_m in matter are related to the values in vacuum θ and L by the following formulae:

$$\sin^2 2\theta_m = 1/[1 + (R-1)\cotg^2 2\theta]$$
$$L_m = L/[\sin^2 \theta \sqrt{1 + (R-1)^2 \cotg^2 2\theta}]$$

with R being the ratio $R = \rho_e/\rho_r$ where ρ_r is the resonant density

$$\rho_r = \delta m^2 \cos^2 \theta / (2\sqrt{2} G p) .$$

The variations of θ_m and L_m as a function of the density are shown in Fig. 3. They display a characteristic resonant behaviour.[8]

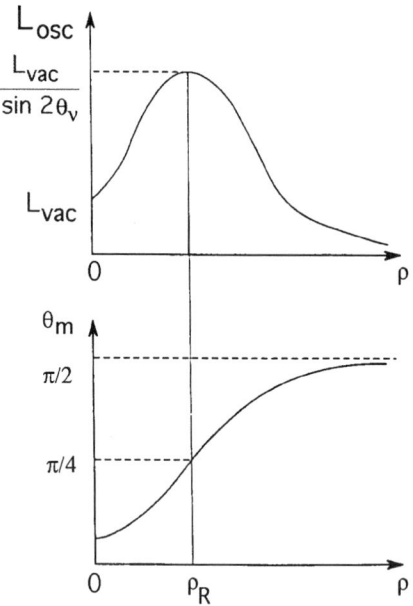

Figure 3. Variation of the oscillation length and of the mixing angle in matter as a function of the electronic density.

Even if θ is very small in vacuum, the mixing can be maximum $\theta_m = 45$ when the electronic density in the matter hits the resonant density. In this case there is complete disappearance of the initial flavour at the position $x = L_m/2$. The oscillation length is also maximum at resonance and reaches $L/\sin^2\theta$.

The value of the resonant density depends on the energy of the neutrinos. This resonant oscillation has been advocated to explain the problem of the solar neutrinos where neutrinos of intermediate energies seem to be almost completely depleted. We will return to this later.

Another possible origin of oscillations has been suggested in the effect of the gravitational field which may couple differently to the different flavours. This again would add new terms in the Hamiltonian, and the present experimental limits on oscillations have been used to set constraints on any violation of the equivalence principle.

DIRECT MEASUREMENTS

All of these discussions are relevant only if neutrinos have masses. But are neutrinos really massive? The present limits on masses (at 90% C.L.) are the following:

$$m(\nu_e) < 3.4 \text{ eV}$$
$$m(\nu_\mu) < 160 \text{ keV}$$
$$m(\nu_\tau) < 24 \text{ MeV}.$$

These results have been obtained respectively in ^{3}H, π and τ decays. Apart from ν_e the scales achieved here are still very large compared with the values suggested by astrophysics. These limits are called direct because they depend little on mixings, contrary to other methods sensitive to masses, in particular oscillations.

Tritium Decays

The decay

$$^3\text{H} \rightarrow {}^3\text{He}^+ + e^- + \nu_e$$

has been used for many years to set a limit on $m(\nu_e)$ because it is a transition with a minimum Q value of 18.6 keV. A non-zero mass of the neutrino would have a maximum effect.

The only experimental observable is the momentum of the detected electron. This is presented as a Kurie plot, which shows the counting rate divided by the momentum of the electron as a function of the electron energy. This distribution is fitted to the theoretical expression:

$$N/p_e \propto (E^0 - E_e)\sqrt{(E^0 - E_e)^2 - m^2(\nu_e)}.$$

The mass of the ν_e influences the shape of the electron spectrum near its upper endpoint where the electron takes all the available energy. A massive neutrino should deplete the end spectrum: the distribution should end at $E_0 - m(\nu_e)$ instead of E_0 the maximum energy.

The fraction of useful events that fall in the last bins of the spectrum is extremely small. This means that the source must be as powerful as possible.

For many years a Russian result[9] has claimed a non-zero mass of 30 eV (Fig. 4). The experiment is very difficult and many points were criticized: problem of the calibration peaks, thickness of the source, but mostly uncertainty in the molecular levels of the complex compound called valine $NH_3CH_3CHCOOH$ in which the tritium was implanted.

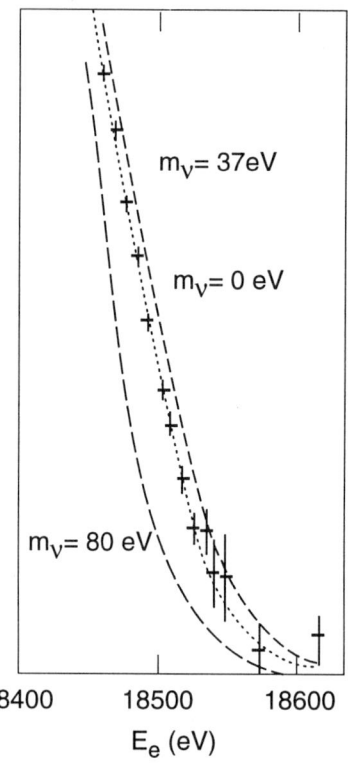

Figure 4. Kurie plot of the tritium decay in the 1980 ITEP experiment.

Many new experiments were started and the best results today come from both Mainz and Troitzk[10] which use a sophisticated solenoidal spectrometer with a retarding electric field and a magnetic gradient. This allows a very good resolution of 6 eV and excellent control of background. The source is made of gaseous molecular tritium.

But the quantity extracted in the measurement is $m^2(\nu_e)$, and the eight current results all find a negative m^2. This corresponds to an excess of events near the endpoint, instead of the expected deficit. Figure 5 shows the example of the Troitzk result.

Some unorthodox solutions have been proposed to explain this curious result: magnetic moment of the neutrino, tachyonic nature, interactions of fossil neutrinos. The answer could be more simple: the tritium is used in its molecular form and the spectrum is not known with enough precision. Given this problem it may be wise to adopt a conservative limit of 10 eV.

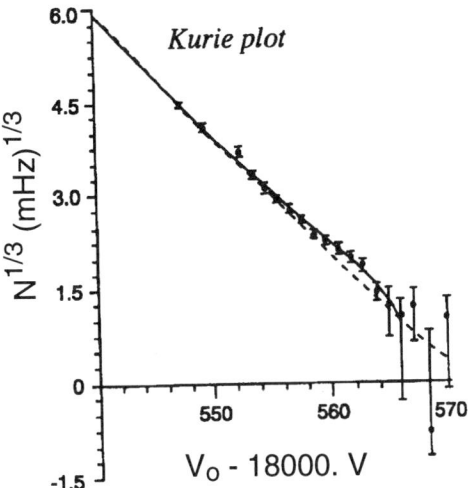

Figure 5. Integral Kurie plot measured in the Troitzk experiment.

π and τ Decays

A probe of $m(\nu_\mu)$ comes in the measurement of the μ momentum in π decays either at rest or in flight. The best limit of 160 keV at 90% C.L. obtained at PSI[11] comes from decays at rest. But it depends very much on the π mass for which there is still some uncertainty.

For $m(\nu_\tau)$ the technique is the study of hadronic decays of the τ where the ν_τ takes the minimum energy. Practically, the decay of choice is the one with 5 π, charged or neutrals in the final state. By studying the hadronic invariant mass of the 5 π system, Aleph at LEP puts the best limit of 24 MeV at 95% C.L.[12]

DOUBLE β DECAYS

Double β decays arise when single β decays are forbidden by energy conservation. Two neutrons are simultanously converted into two protons. This happens in 36 nuclei, and 11 different isotope decays have been observed so far. They agree reasonably well with theoretical predictions.

There are two very different types of double β decays corresponding to the two schemes sketched in Fig. 6.

The first type, the double β with emission of neutrinos has been studied for example in the case:

$$^{100}\text{Mo} \rightarrow {}^{100}\text{Ru} + e^- + e^- + \overline{\nu}_e + \overline{\nu}_e \ .$$

This decay, although very suppressed by phase space, is allowed in the Standard Model. The signal consists in the presence of two electrons of very low energy. The Nemo experiment[13] in the Fréjus underground laboratory uses a new technique to distinguish the signal from the background. The set-up consists of a light-tracking detector and compares natural Mo with only 9.6% of ^{100}Mo, and Mo enriched at 98.4%. Then the spectra are subtracted from one another. The result is shown in Fig. 7, and the lifetime for this process is already measured at better than 10^{21} years.

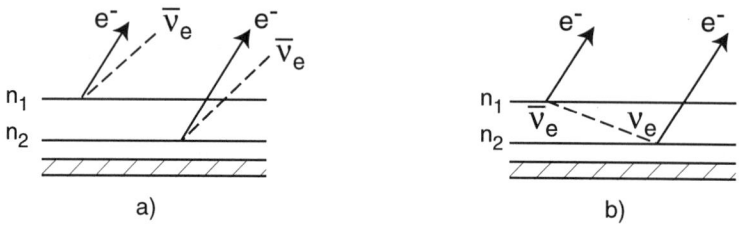

Figure 6. Graph of the double β decay: (a) with neutrinos; (b) without neutrinos.

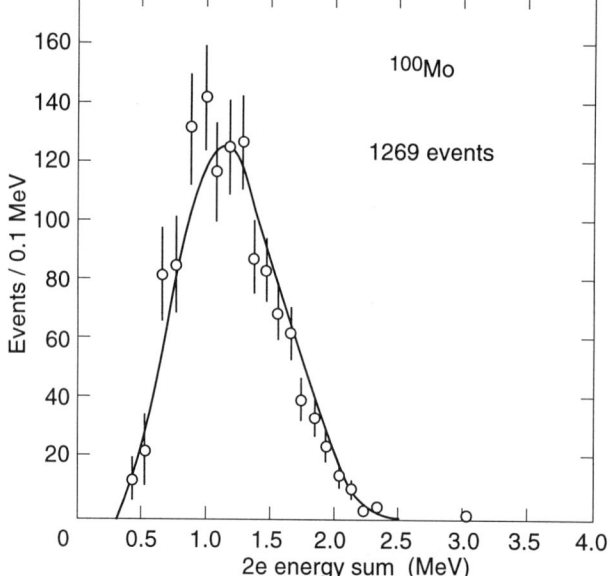

Figure 7. Energy spectrum in the double β decay with neutrinos of ^{100}Mo.

A much more interesting phenomenon is the double β decay without neutrinos. It is a lepton-violating process. As seen in Fig. 6, this corresponds to a right-handed $\overline{\nu}$ first emitted and then reabsorbed as a left-handed ν. This requires a massive Majorana particle.

The amplitude is proportional to an effective neutrino mass given by:

$$m_\text{eff} = \sum_i CP(\nu_i) m(\nu_i) U_{ei}^2 .$$

The best result comes from the study of:

$$^{76}\text{Ge} \rightarrow {}^{76}\text{Se} + e^- + e^-$$

by the Heidelberg–Moscow group[14] in the Gran Sasso laboratory.

The germanium is at the same time the source and the detector. This maximizes efficiency and the calorimetric measurement is very precise.

Figure 8 shows the spectrum in the region of interest. The signal would show up as a peak at 2.038 MeV. With about 20 kg of crystal the limit on lifetime is:

$$t_{1/2} > 5.1 \times 10^{24} \text{ years}.$$

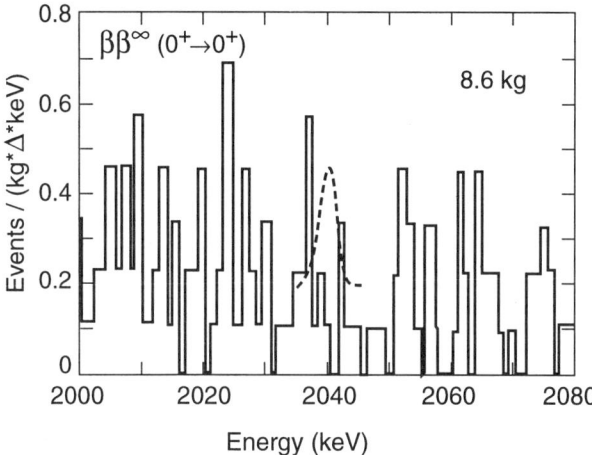

Figure 8. Energy spectrum of the Heidelberg–Moscow experiment.

This can be expressed as a limit on the effective mass:

$$m_{\text{eff}} < 0.68 \text{ eV}.$$

This limit should be taken with some care, because it does not directly apply to a mass eigenstate, and because it is difficult to determine the nuclear matrix elements which are necessary to extract the result. In fact with the same data other analyses find limits closer to 2 eV.

In any case the double β decay searches are still on their way to improvement and the aim is to reach in the near future the 0.1 eV level with various isotopes.

DECAYS OF NEUTRINOS

If neutrinos are massive, they will decay. Two classes of decays exist depending on the mass of the initial neutrino.

Decays into Charged Particles

As soon as the mass of a neutrino is above 1.2 MeV, it will decay into:

$$\nu' \to \nu_e + e^+ + e^-.$$

The corresponding graph is shown in Fig. 9(a). This graph is reminiscent of muon decay, and it is calculated in the same way, except for a mixing of the heavy neutrino with the electron which is not unity.

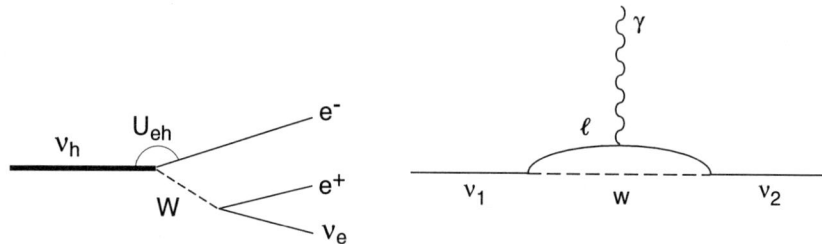

Figure 9. Decays of neutrinos: (a) decay emitting $e^+ e^-$; (b) radiative decay.

The lifetime comes out to be:

$$t \, [s] = 2.2 \times 10^{-6} \left(\frac{m_\mu}{m_\nu}\right)^5 \frac{1}{|U_{he}|^2}$$

where the relation to muon decay is evident. If the mass is higher, other channels open up:

$$\mu^- e^+ \nu \qquad \mu^- \mu^+ \nu \qquad e^- \pi^+ \cdots.$$

A search for such signatures is relatively easy since the signal appears as a V^0 arising in a decay volume properly instrumented intercepting a neutrino beam.

As explained before, there is a component of heavy neutrino in any beam, and a negative search for decay signatures, puts a limit on the mixings of a heavy neutrino with charged leptons as a function of its mass.

Figure 10 shows the present limits[15] obtained for masses ranging from 20 MeV to 45 GeV.

The large masses were looked for in $e^+ e^-$ collisions.

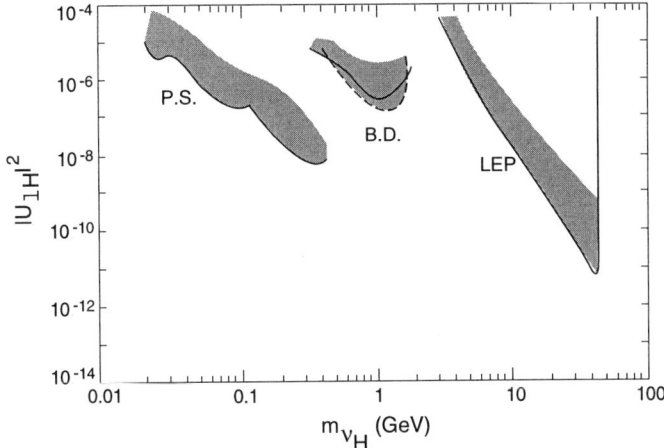

Figure 10. Limits on the mixings of heavy neutrinos in the range between 10 MeV and 45 GeV.

Radiative Decays

The decay:

$$\nu_2 \to \nu_1 + \gamma$$

is always possible, even for tiny masses, and corresponds to the graph of Fig. 9(b). This is a GIM suppressed mechanism and the corresponding lifetime is extremely small. The decay width has been calculated to be:[16]

$$\Gamma_v = (7 \times 10^{43} \text{ s})^{-1}(m_2/1 \text{ eV})^5(1-x^2)^3(1+x^2)|U_{l1}U_{l2}|^2$$

with $x = m_2/m_1$.

But recent papers have discussed the possibility of very large enhancements in a medium.[17] Just as the resonant oscillation (MSW effect) arises in matter, a coherent effect on atomic electrons can give rise to a decay width in matter much reduced:

$$\Gamma_m/\Gamma_v = 34 \times 10^{23} m_2/E(N_e/10^{23} \text{ cm}^{-3})^2(1 \text{ eV}/m_2)^4 \ .$$

There are also considerations of catalysis of the radiative decay in a strong electromagnetic field,[18] and stimulated conversion of neutrinos in sending a beam through an RF cavity.[19] All of these subjects deal with electromagnetic properties of the neutrinos, which is not a settled affair.

But even with these large enhancements, one does not reach a level which could be tested in a laboratory experiment. Even considering the vast Universe and its huge number of fossil neutrinos, the radiative decay of 20 eV neutrinos releasing monochromatic photons of 10 eV would have produced only one such photon every 10^6 km^3 in the present Universe. It is difficult to check!

SOLAR NEUTRINOS

The Solar Standard Model[20]

Solar energy comes from nuclear fusion with various chains of reactions. The main cycle consists in:

$$p + p \to e^+ + \nu_e + d$$
$$p + d \to \gamma + {}^3He$$
$${}^3He + {}^3He \to {}^4He + p + p \, .$$

Overall four protons fuse together to give:

$$4p \to {}^4He + 2e^+ + 2\nu_e$$

with the release of 27 MeV. The two neutrinos carry 570 keV. The total solar luminosity is well measured:

$$L = 3.846 \times 10^{26} \text{ W} = 2.40 \times 10^{39} \text{ MeV/s} \, .$$

Thus it is easy to calculate the total number of neutrinos liberated by the Sun: 1.8×10^{38}/s.

A secondary chain also produces neutrinos:

$$\begin{aligned}
{}^3He + {}^4He &\to \gamma + {}^7Be \\
e^- + {}^7Be &\to \nu_e + {}^7Li \\
p + {}^7Li &\to {}^4He + {}^4He \\
p + {}^7Be &\to \gamma + {}^8B \\
{}^8B &\to {}^8Be + e^- + \nu_e
\end{aligned}$$

In smaller fractions of the cases, other reactions take place.

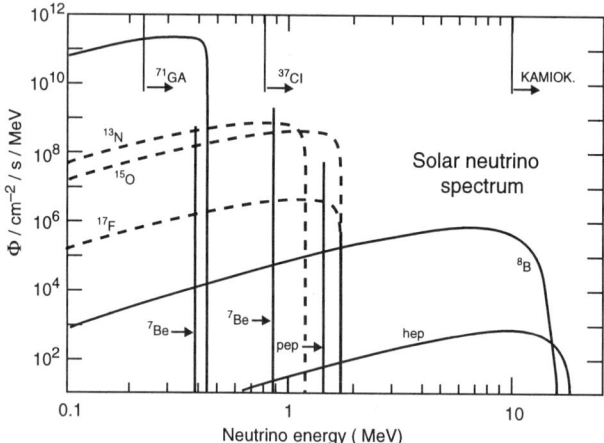

Figure 11. Energy spectrum of the ν_e emitted from the Sun.

Overall, the Sun sends to Earth some 6.4×10^{10} ν_e/cm²/s with the spectrum given in Fig. 11. The three main components are:

- the pp neutrinos representing 85% of the total flux, with a continuous spectrum up to 420 keV,
- the monoenergetic line of ^{7}Be neutrinos at 861 keV,
- the high-energy ^{8}B neutrinos with a spectrum extending up to 15 MeV.

Several experiments have now detected solar neutrinos and we will review them in historical order.

The Homestake Experiment[21]

For the past 25 years an experiment in a South Dakota gold mine has been counting neutrinos relying on the capture reaction:

$$\nu_e + {}^{37}\text{Cl} \rightarrow e^- + {}^{37}\text{Ar}.$$

The detector consists of 650 tons of C_2Cl_4 with about 125 tons of ^{37}Cl. The neutrino energy threshold for this reaction is 814 keV. As a consequence, the detector is only sensitive to ^{7}Be and ^{8}B neutrinos. It will not detect neutrinos from the main pp chain.

The capture produces ^{37}Ar atoms which are radioactive with a half-life of 34 days. Every two months, argon is extracted by means of flushing helium. The presence of ^{37}Ar is detected in a proportional counter by observing the capture:

$$e^- + {}^{37}\text{Ar} \rightarrow \nu_e + {}^{37}\text{Cl}$$

followed by the emission of X-rays or Auger electrons.

The extraction efficiency is measured by injecting a known amount of ^{37}Ar. Figure 12 shows the counting rate as a fraction of time since the beginning of the experiment.

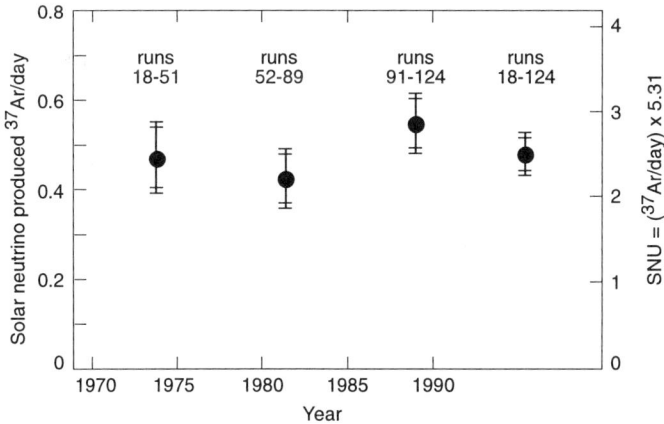

Figure 12. Rate measured in the Homestake experiment.

The average counting rate is:

$$(2.54 \pm 0.14 \pm 0.14)\,\text{SNU}$$

where 1 Solar Neutrino Unit corresponds to 10^{-36} captures/atom/s.

The rate is extremely small giving about 0.5 event per day. This is in contradiction with the SSM which predicts a rate about three times larger. This was for many years the so-called solar neutrino problem.

The Kamiokande Experiment[22]

It consists of a large tank of 2.1 kton of high purity water, for a fiducial mass of 680 tons, seen by about 1000 photomultipliers positioned all around the volume. It is situated in a mine in Japan.

Neutrinos are detected by the scattering process:

$$\nu_e + e^- \to \nu_e + e^-.$$

The emitted electrons give off Cherenkov light if they are fast enough. Practically, the threshold for detection is 7 MeV, and the experiment is limited to the ^{8}B neutrinos. But it has two main advantages:

- it is a real-time measurement;

- the electrons remember the direction of the incident neutrino, and it is possible to demonstrate the solar origin of the events as shown in Fig. 13, where the angle of arrival is plotted. $\cos\theta = 1$ corresponds to the Sun direction, and the Kamiokande experiment has proved for the first time that the Sun indeed emits neutrinos.

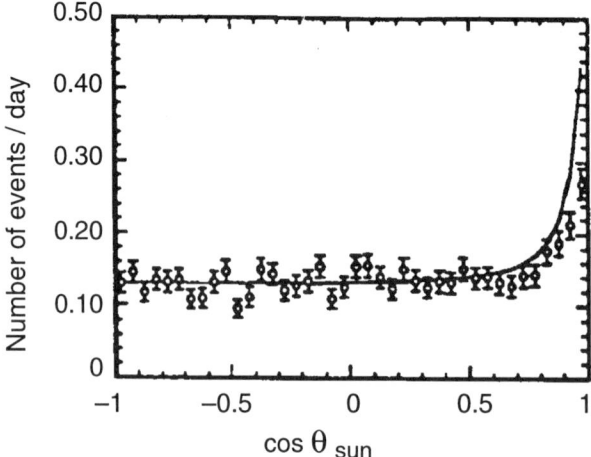

Figure 13. Solar neutrino detection in the Kamiokande experiment.

But quantitatively there is again a problem. Over more than 2000 days of data taking, Kamiokande measures a flux which is about 50% lower than the prediction.

Note that Superkamiokande has just started. It amasses more than 30 events per day and will soon overcome all statistical problems.

The Gallium Experiments[23]

Two experiments use a gallium detector: GALLEX in the Gran Sasso laboratory and SAGE in the Baksan tunnel. These are again radiochemical measurements through the reaction:

$$\nu_e + {}^{71}\text{Ga} \rightarrow {}^{71}\text{Ge} + e^-$$

with ^{71}Ge being radioactive with a half-life of 11 days.

The main interest here is the very low threshold of 233 keV which allows one to detect the neutrinos produced in the main pp chain, the ones for which the calculation is the most reliable.

The GALLEX experiment uses 30 tons of natural gallium containing about 12 tons of ^{71}Ga in the form of 101 tons of an aqueous solution of $GaCl_3$. Every three weeks an extraction takes place by means of a nitrogen flow. The presence of ^{71}Ge is detected by observing X-rays characteristic of its electron capture.

Results give:

– for SAGE $(72 \pm 11 \pm 6)$ SNU

– for GALLEX $(69.7 \pm 6.7 \pm 4)$ SNU.

The two experiments are in good agreement, but they are about 40% lower than the prediction of the SSM.

Figure 14 shows the variation of the two results over the past few years.

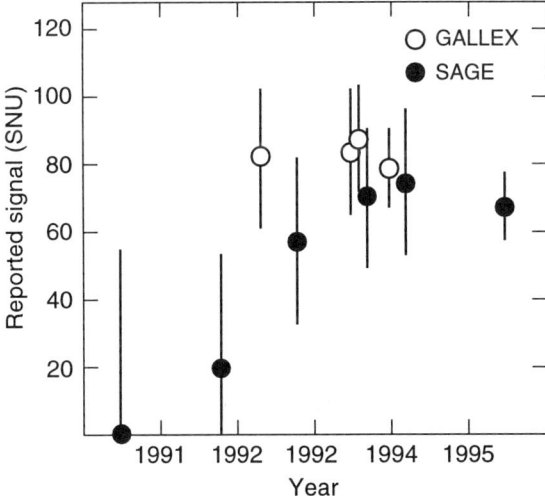

Figure 14. GALLEX and SAGE measurements.

Recently both experiments have performed a direct test of their detection method using an artificial ν_e source, a very powerful ^{51}Cr source irradiated for the occasion. This gave an increase of the number of events by an order of magnitude over the solar level, during a period of one month.

The ratios obtained between measurement and expectation for the known flux of neutrinos are:[24]

- for SAGE 1.00 ± 0.12

- for GALLEX 1.00 ± 0.10.

But GALLEX did a second test, and this one gave the ratio 0.83 ± 0.09.

Averaging the two GALLEX results gives the ratio 0.92 ± 0.07. The question arises of a correction to be applied to their solar neutrino measurement.

SUPERNOVA NEUTRINOS

On 23 February 1987 at 7:35 UT, a burst of events was recorded in the IMB and Kamiokande detectors.[25] Figure 15 shows the number of photomultipliers receiving a flash, in the second experiment. There is an abnormal accumulation over a period of 10 s, and this happened in conjunction with an optical fireworks taking place in the southern hemisphere.

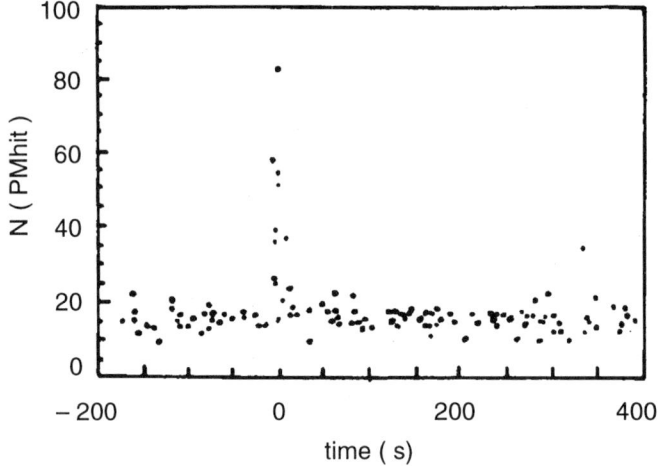

Figure 15. Burst of neutrino counts from SN1987.

In total 20 events were found, interpreted as:

$$\bar{\nu}_e + p \rightarrow e^+ + n$$

with a few possible candidates for the process $\nu_e + e^- \rightarrow \nu_e + e^-$.

A supernova had exploded 150 000 years ago in the Large Magellanic Cloud. Such a collapse of a massive star released 10^{58} neutrinos of the three flavours, leaving behind a neutron star.

These neutrinos have travelled 150 000 light-years before reaching the Earth, and this huge flight path allowed one to put limits on non-standard properties of neutrinos. In particular a non-zero mass would have spread the time of arrival, the most energetic neutrinos arriving first. This was not found and the limit obtained is:

$$m(\nu_e) < 20 \text{ eV} .$$

For the other species, their cross-sections are much smaller and a priori they were not observed.

ATMOSPHERIC NEUTRINOS

Primary cosmic rays interact in the upper layers of the atmosphere, giving hadronic showers leading to a flux of neutrinos from π and subsequent μ decays. Because of this origin, one expects about twice as many ν_μ's as ν_e's with energies in the range from 100 MeV to a few GeV. Calculations have been performed, claiming to know these fluxes to 30%. Indeed Kamiokande has seen atmospheric neutrinos as contained events in the detector, and the energy distributions for ν_e and ν_μ candidate events are shown in Fig. 16.[26] The rate of about one event per day is roughly in agreement with expectations but there seems to be a deficit of ν_μ events.

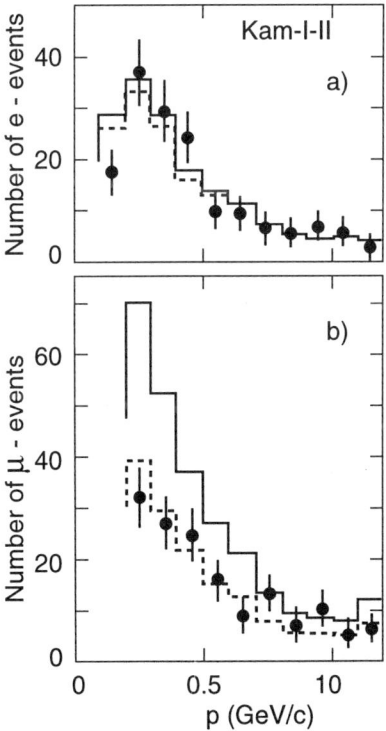

Figure 16. Spectra of atmospheric neutrinos in the Kamiokande experiment: (a) ν_e events; (b) ν_μ events.

The calculation is rather uncertain in the composition and the energy of the primary rays, also in the K/π ratio and in the value of the geomagnetic field. People prefer to consider the ratio of ν_μ/ν_e which is predicted with an uncertainty of 5% because of partial cancellation of uncertainties.

Five underground experiments have published a value for the double ratio $R = (\nu_\mu/\nu_e)_{meas}/(\nu_\mu/\nu_e)_{exp}$. If everything were according to predictions this double ratio would equal 1. The results are:[27]

- Kamiokande $R = 0.60 \pm 0.07 \pm 0.09$
- IMB $R = 0.54 \pm 0.05 \pm 0.07$
- Nusex $R = 0.99 \pm 0.40$
- Fréjus $R = 0.87 \pm 0.21$
- Soudan2 $R = 0.75 \pm 0.16$.

There is a clear deficit of ν_μ's in the most precise experiments.

Recently Kamiokande has reported the variation of the double ratio as a function of the zenith angle[28] for "multiGeV" events for which the outgoing lepton direction is correlated with the incident neutrino direction (Fig. 17).

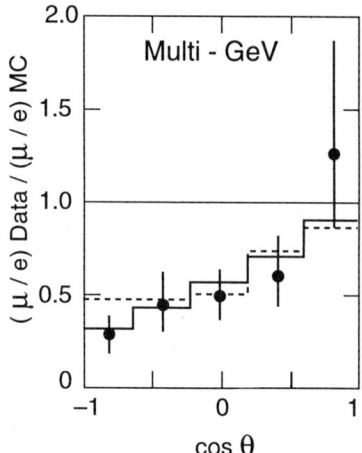

Figure 17. Zenith angle of the multiGeV atmospheric neutrino events.

For neutrinos impinging on the detector from above ($\cos\theta = 1$), there is no deficit. These neutrinos are produced a few kilometres above the detector. Upward-going neutrinos corresponding to $\cos\theta = -1$, have on the contrary traversed 13 000 km of the Earth's interior before being detected. There the deficit is manifest. Does the deficit depend on the distance travelled by the neutrinos? But there are inconsistencies in the data:

- the contained low-energy events show a deficit of ν_μ's compared to the prediction and no deficit of ν_e's with ratios of 0.97 for e events and 0.59 for μ events.

- the multiGeV events do not confirm this, having single ratios of 1.27 for e events and 0.81 for μ events.

Although the double ratios are consistent, there are still some anomalies to be understood.

OSCILLATIONS AND ASTROPHYSICS

The solar and atmospheric neutrino deficits can be readily interpreted as the occurrence of oscillations. But is it the true solution?

The Solar Neutrino Dilemma

The Sun produces a large flux of neutrinos, and only part of it seems to be detected on Earth. The deficit measured in various experiments varies with the neutrino energy. As discussed before, three techniques have been used with three very different energy thresholds for detection. The following table confronts the predictions of the SSM[29] for the main neutrino components and for each experimental technique, with the respective measurements.

	Source	SSM	Measure
Kamioka	^{8}B	1	0.5
Chlorine	^{7}Be	1.2	
	^{8}B	6.2	
	total	8.0	2.5
Gallium	pp	70.8	
	^{7}Be	35.8	
	^{8}B	13.8	
	total	131.5	70

The deficit looks larger for intermediate energies, namely for the ^{7}Be neutrinos. The different numbers are difficult to reconcile simply. The prediction for pp neutrinos is the most firm, and in fact the different groups calculating the solar flux agree on this one. The other neutrino sources are more questionable, the ^{7}Be flux varies with the central temperature of the Sun like T^8, and the ^{8}B flux like T^{18}. But the ^{8}B neutrinos have been seen, at least half of them, and they are all produced through the intermediate step of ^{7}Be. At least half of the ^{7}Be neutrinos should be seen.

Analysing these numbers, one arrives at the conclusion that:

- there is a basic flaw in the SSM

- two of the experiments are incorrect

- something happens to the neutrinos after they have been produced.

The preferred solution is obviously the last one which makes experimenters and theorists happy, specially because it requires some new physics.

Since the detectors can only see ν_e's, the puzzle is solved if part of the neutrinos change flavour. This can happen through vacuum oscillation. But the resonant matter oscillation discussed previously gives a better agreement with the data. In particular it explains the almost complete disappearance of ^{7}Be neutrinos because their energy corresponds to the resonance condition.

Figure 18 shows the favoured region in the plane $\sin^2 2\theta$, δm^2. The neutrino deficit indicates an oscillation characterized by a mass term:

$$\delta m^2 = 10^{-5} \text{ eV}^2$$

and a mixing $\sin^2 2\theta = 10^{-2}$.

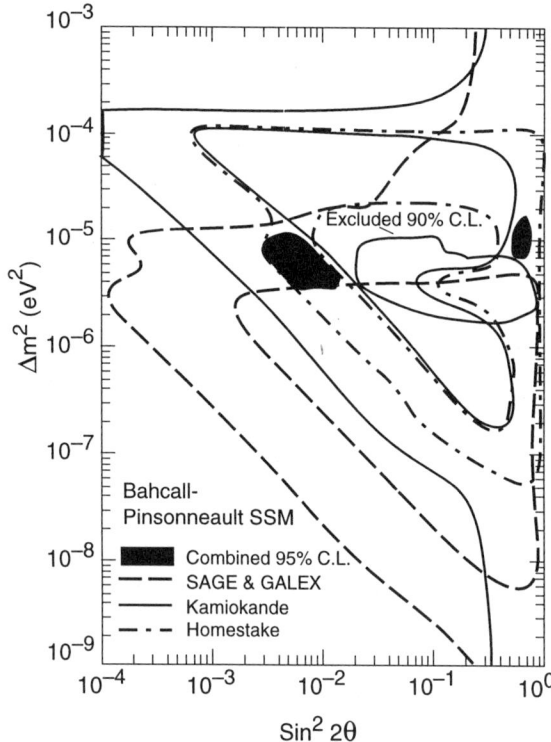

Figure 18. Parameters of the matter oscillation effect favoured by solar neutrinos.

Atmospheric Neutrinos

We have seen that the data indicate a deficit of the ν_μ flux while the ν_e flux seems normal. This could be the sign of a $\nu_\mu \nu_\tau$ oscillation.

In this framework the parameters which are retained are: $\delta m^2 = 10^{-2}$ eV^2 and a large mixing $\sin^2 2\theta > 0.6$.

OSCILLATIONS AT REACTORS AND ACCELERATORS

If neutrinos are massive they probably mix. Then the phenomenon of oscillation is to be expected. In fact it is the effect most sensitive for testing neutrino masses, and

many experiments have searched for it. As seen before, the probability of oscillation is given by:
$$P = \sin^2 2\theta \sin^2(\pi x/L)$$
with the oscillation length L (km) $= 2.5\, E$ (GeV)$/\delta m^2$(eV^2).

The principle of the search is very simple. In a beam of "green" neutrinos some of them may turn "red" over distance. One can look for the disappearance of the "green" flux with a detector sensitive only to "green" neutrinos. This is the method used in particular at reactors where a detector can be moved at different distances from the core, and fluxes and spectra compared.

One can look for the appearance of the "red" neutrinos. This requires a detector able to recognize both colours. This is the method mostly used at accelerators. It is very sensitive since a few non-ambiguous "red" events are enough to prove the phenomenon, provided that one controls the backgrounds properly. This allows one to test small $\sin^2 2\theta$ but the ratio distance/energy is up to now limited, and accelerator experiments do not reach very small δm^2.

Recent Limits

Three experiments have recently published new limits:

- Charm2 at CERN[30] looked for the appearance of ν_τ's in a ν_μ beam searching for the quasi-elastic channel:

$$\nu_\tau + N \to \tau^- + N' \quad \text{with} \quad \tau^- \to \pi^- + \nu_\tau .$$

The fine granularity of the detector allows one to look for isolated pions.

- CCFR at Fermilab[31] has also looked for the $\nu_\mu \nu_\tau$ channel by studying the ratio NC/CC as a function of the total calorimetric energy. An appearance of ν_τ would mimic an increase of NC.

- IHEP–JINR at Serpukhov[32] has looked for the disappearance of ν_e's in a beam enriched in ν_e's (5%).

All negative results are used to exclude various regions in the plane $\sin^2 2\theta$, δm^2 and Fig. 19 shows the present status in the three physical channels.

The LSND Experiment[33]

The Los Alamos beam is produced from a beam dump of 800 MeV protons. The π^+'s decay at rest giving ν_μ's of 30 MeV, then stopped muons decay giving ν_e's and $\overline{\nu}_\mu$'s in the energy range between 0 and 52 MeV. Practically no $\overline{\nu}_e$'s are produced.

The detector (Fig. 20) consists of a tank filled with 167 tons of liquid scintillator seen by 1220 photomultipliers. It has both calorimetric and imaging capabilities. It is situated 30 m from the beam stop.

LSND searches for $\overline{\nu}_e$ events through the reaction:

$$\overline{\nu}_e + p \to e^+ + n .$$

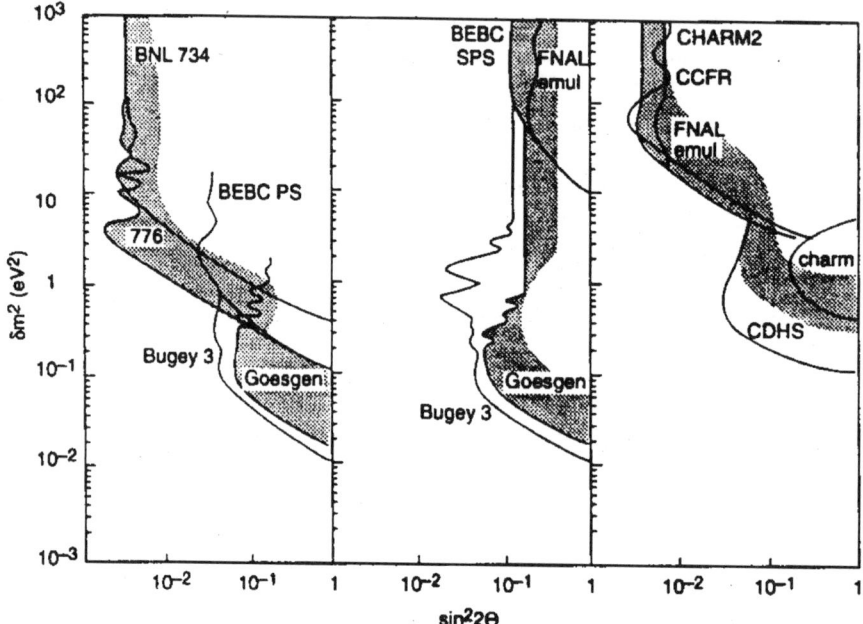

Figure 19. Limits on oscillations in the three physical channels.

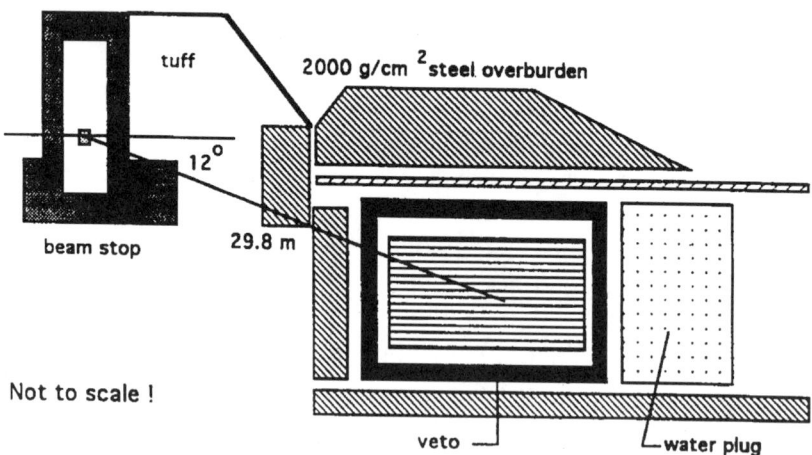

Figure 20. The LSND set-up.

The neutron is then captured and gives a delayed γ of 2.2 MeV.

The signal consists in the coincidence of a 30–50 MeV positron and a 2.2 MeV γ correlated in space (50 cm) and time (200 μs).

In 1995, nine events were reported compatible with the searched for signature for a background of 2.1. When attributed to oscillations, this excess gives a probability:

$$P = (0.34 \pm 0.20 \pm 0.07)10^{-2} \, .$$

In 1996, with twice the statistics and a better analysis, the experiment claims 22 events for 4.6 expected. The probability is now:

$$P = (0.31 \pm 0.11 \pm 0.05)10^{-2} \, .$$

From one year to the next, the result has survived and has been improved. This was not the case for the several claims of oscillation discoveries in the past.

But the allowed region favoured by LSND is almost excluded by previous experiments (Fig. 21). In particular the Karmen experiment at the ISIS spallation source has very similar beam and techniques, although lower luminosity, and does not see any effect.[34] Only a small corner remains (at 90% C.L.) around $\delta m^2 \sim 1$ eV^2.

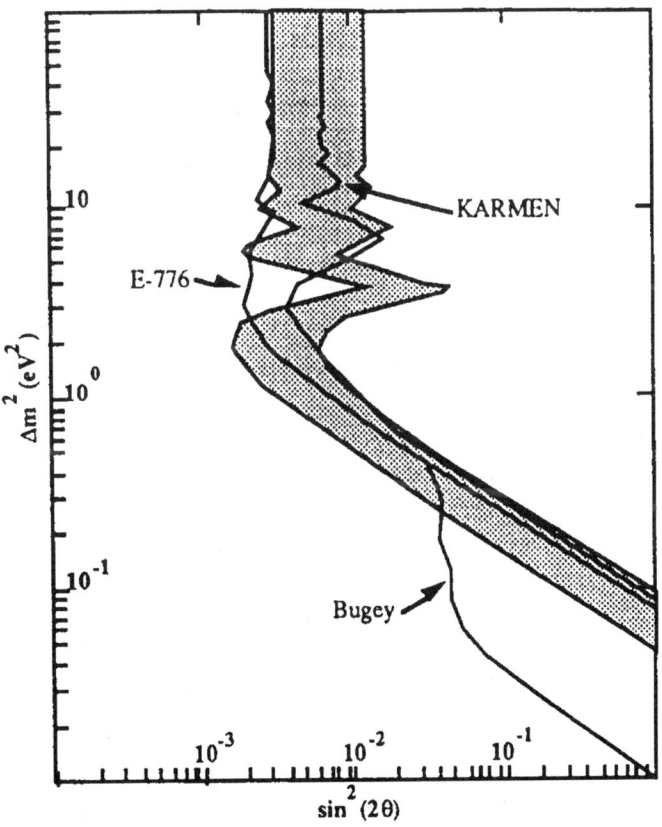

Figure 21. Region favoured by the LSND result, with other excluded regions.

HAVE OSCILLATIONS BEEN DISCOVERED?

We are left with three hints of oscillations: solar, atmospheric and LSND results. Figure 22 shows the three regions retained. They have nothing in common.

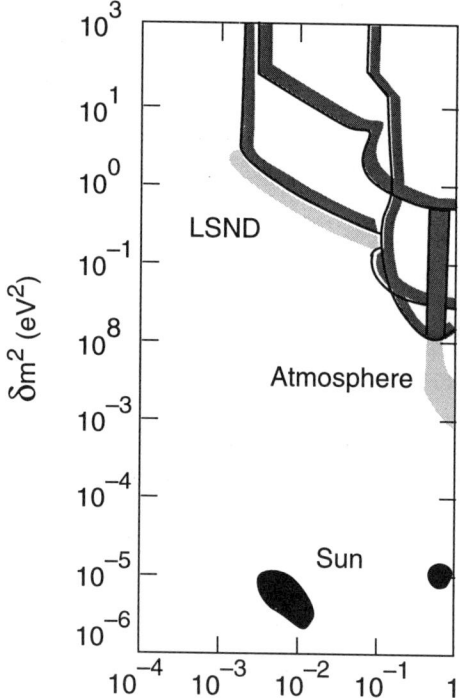

Figure 22. The three indications of oscillations.

Considering neutrino masses these results give us the following relations:

$$m^2(\nu_e) - m^2(\nu') = 10^{-5} \text{ eV}^2 \quad \text{Sun}$$
$$m^2(\nu_\mu) - m^2(\nu'') = 10^{-2} \text{ eV}^2 \quad \text{atmosphere}$$
$$m^2(\nu_e) - m^2(\nu_\mu) = 1 \text{ eV}^2 \quad \text{LSND}.$$

There are two other relations which are relevant to massive neutrinos:

- the cosmological relation for the neutrinos to be the hot component of dark matter:

$$m(\nu_e) + m(\nu_\mu) + m(\nu_\tau) = 10\text{--}20 \text{ eV}$$

- the see-saw model which predicts a strong mass hierarchy:

$$m(\nu_e) \ll m(\nu_\mu) \ll m(\nu_\tau) \,.$$

Clearly with only three different neutrinos having standard interactions, one can have only two independent δm^2. The five propositions are incompatible.

Note that the LSND and the solar results could both reflect the same $\nu_e\nu_\mu$ channel. As explained previously, LSND could probe the "indirect" oscillation generated by the mass of the third neutrino.

Can we make sense out of these conflicting requirements?

If one discards the atmospheric result, then the four remaining propositions make a consistent picture with the solution:

$$m(\nu_\tau) \simeq 1 \text{ eV} \qquad m(\nu_\mu) \simeq 3 \times 10^{-3} \text{ eV} \qquad m(\nu_e) \simeq 10^{-5} \text{ eV}.$$

The ν_τ would have a mass marginally interesting for cosmology and could not account for all of the hot dark matter component.

There is another scenario when discarding instead the LSND result. In order to give a cosmological role to the neutrinos, it is necessary to imagine a scheme with mass-degenerate particles:

$$m(\nu_\tau) \simeq m(\nu_\mu) \simeq m(\nu_\tau) \simeq 1 \text{ eV}$$

with $m^2(\nu_\nu) - m^2(\nu_e) = 10^{-5}$ eV^2 and $m^2(\nu_\tau) - m^2(\nu_\mu) = 10^{-2}$ eV^2.

The mass hierarchy is lost in this scheme.

Where to go from now on, to shed some more light on the problem? Some people suggest a great leap forward, namely long baseline experiments. These would explore the region of small δm^2. Others have advocated the improvement of the mixings between ν_μ and ν_τ, and this was the justification of the present effort at CERN with two experiments taking data.

PRESENT $\nu_\mu\nu_\tau$ SEARCHES

Motivation

The Big Bang scenario tells us that there are fossil neutrinos filling the Universe at the level of 110/cm^3/flavour. These neutrinos are extremely difficult to detect, but they are so numerous that they could have an influence on the gravitational equilibrium of the Universe.

Indeed there is a large part of the mass of the Universe which does not emit radiation and which is detected only through gravitation, either by rotation curves of celestial objects or by gravitational lens effects. This dark matter could account for 90% of the total mass.

If this dark matter were composed of neutrinos, this would imply, as previously mentioned:

$$m(\nu_e) + m(\nu_\mu) + m(\nu_\tau) = 60 \text{ eV}.$$

Present models of structure formation prefer to assign 20% or 30% of the total missing mass to the hot component. Thus:

$$\sum m(\nu) = 20 \text{ eV}.$$

If the mass hierarchy of the see-saw model is correct one then concludes:

$$m(\nu_\tau) = 20 \text{ eV}.$$

This is very small for a direct measurement, but it is well adapted for an oscillation resulting in ν_τ in a high-energy beam. For present high-energy beams, the oscillation length would be of some 100 m.

Nobody predicts the corresponding mixing. The one between quarks is 10^{-3}, and the present limit in this oscillation channel is 5×10^{-3}. It is worth while to improve it.

The Experimental Challenge

The experiment consists in detecting ν_τ's in a ν_μ beam of energy high enough to be above the threshold for ν_τ CC interactions. At 40 GeV the produced τ^- travels about 1 mm before decaying. The challenge is to detect a short track resulting in a kink close to a primary vertex which can be distributed all over a large volume. The neutrino beam is about 3 m wide at the experiments.

Two methods are used:

- CHORUS searches for a short track in stacks of emulsions.[35]

- NOMAD selects candidates by kinematical criteria.[36]

The CHORUS Experiment

The heart of the experiment is made of 800 kg of emulsions. Emulsions offer the best possible spatial resolution with an accuracy of microns. But it is a passive medium registering all charged tracks which cross it during the whole time of the experiment.

A full spectrometer (Fig. 23) follows the target. It selects candidates on the basis of kinematics and defines the region of the emulsions worth scanning with enough precision.

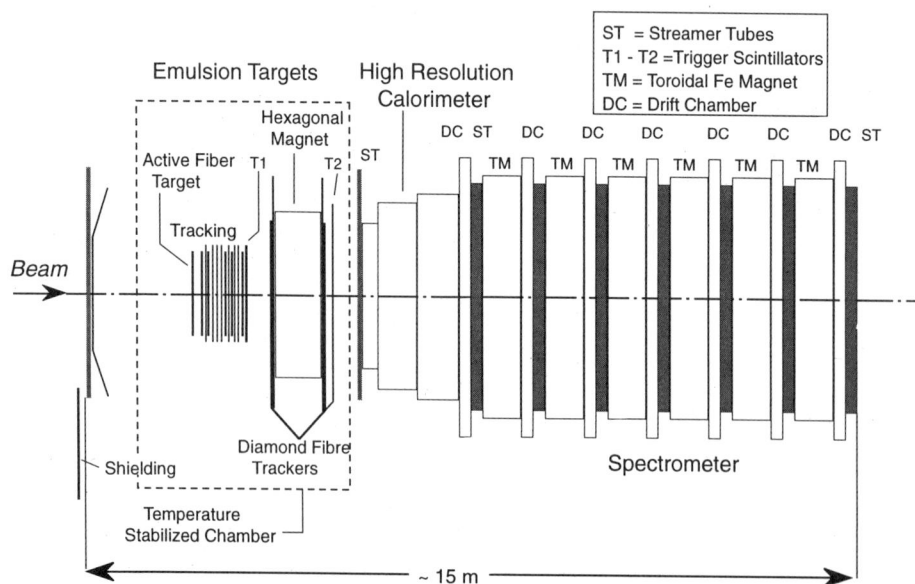

Figure 23. The CHORUS detector.

The experiment has collected 320 000 events in 1994 and 1995, and the emulsions are being scanned in Japan. Figure 24 shows a CC event with a short-lived particle of positive charge presumably a charm meson. New emulsions have been installed for data taking in 1996 and 1997. The aim of the experiment is to push the present limit to $\sin^2 2\theta < 3 \times 10^{-4}$ for $\delta m^2 > 40$ eV^2, or find oscillations.

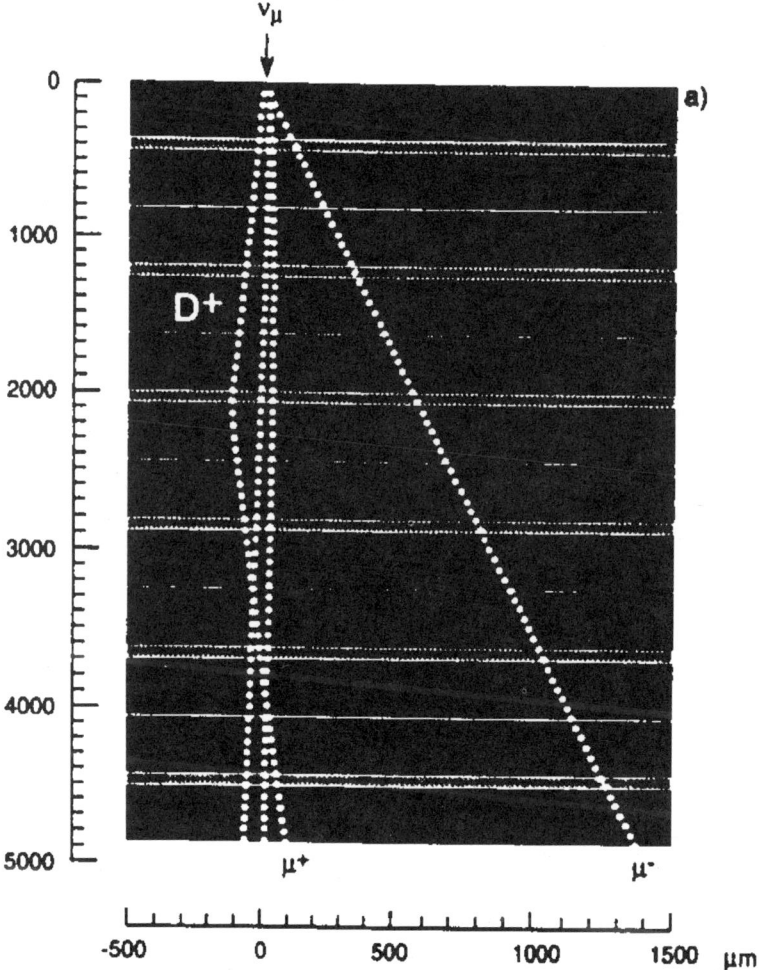

Figure 24. One event reconstructed in the emulsions.

The NOMAD Experiment

The detector is a full spectrometer built inside a large magnet which allows one to measure as precisely as possible all the tracks that are produced in an interaction and identify electrons, muons, and photons (Fig. 25). For this purpose the target must be light to avoid multiple scattering and reinteractions, but must have enough mass. It is built of self-supporting drift chambers and has the density of liquid hydrogen.

The technique is complementary to the one used by CHORUS. It does not try to identify a short track, but extracts a signal of ν_τ applying various kinematical cuts based in particular on the transverse missing momentum.

About 300 000 events were collected in 1995. One is shown in Fig. 26, demonstrating the precision of the reconstruction. The good resolution allows the reconstruction in particular of K^0's as seen in Fig. 27.

The aim is also to push the limit on mixing down to 4×10^{-4}.

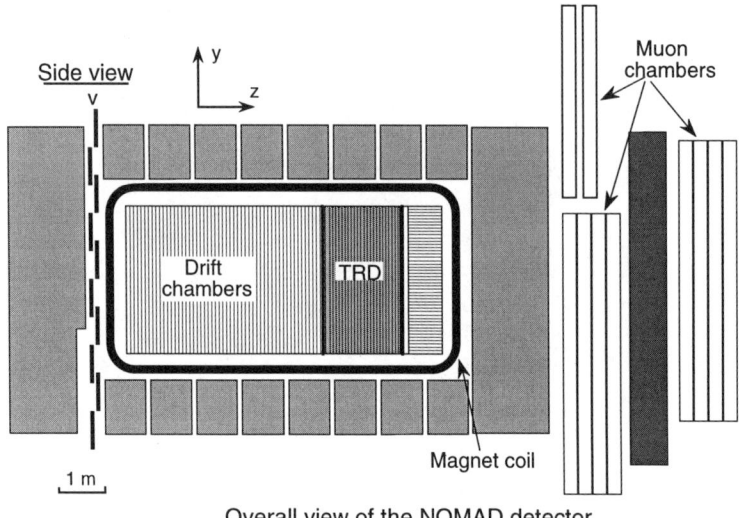

Figure 25. The NOMAD detector.

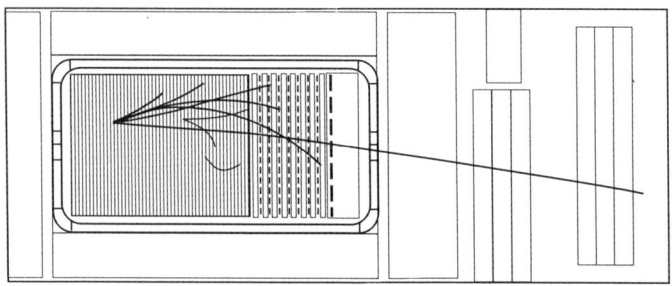

Figure 26. One event reconstructed in the NOMAD spectrometer.

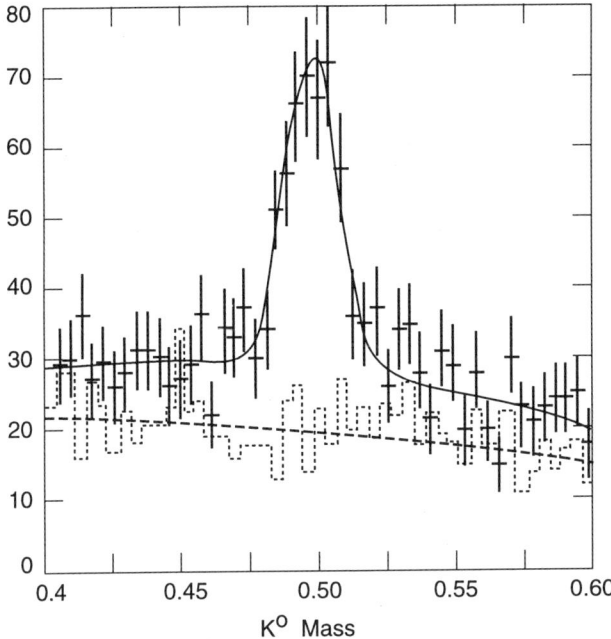

Figure 27. Reconstructed K^0 signal in the NOMAD target.

THE FUTURE OF OSCILLATION EXPERIMENTS

To widen the region explored in the $\sin^2 2\theta$, δm^2 plane, the recipe is simple:

- to push the limit in $\sin^2 2\theta$, it is essential to maximize the statistics and to lower the background;

- to push the limit in δm^2, it is enough to increase the ratio distance/energy.

The future programme is divided into these two somewhat conflicting approaches.
LSND will take data for two more years and probably double its signal.

Karmen will take data with a much improved veto shield and will be able to test the region of the LSND claim.

NOMAD and CHORUS, as already described, aim at gaining one order of magnitude in the $\nu_\mu \nu_\tau$ channel within the next two years.

CHOOZ (France) and San Onofre (California) are two new-generation reactor experiments with several kton detectors, respectively distant 1 km and 750 m from the source. They are starting to take data and should be able to push the δm^2 limit to 10^{-3} for $\sin^2 2\theta > 10\%$. This is an interesting check of the atmospheric claim.

In 1999 a beam from KEK should be aimed at the Superkamiokande detector. Again the region explored will reach down to δm^2 of 10^{-3} eV^2, but this time with a ν_μ beam.

In the more distant future, Cosmos using emulsions at Fermilab, and a possible follow-up of the NOMAD–CHORUS effort could reach the level of 10^{-5} in the mixing between ν_μ and ν_τ.

In parallel, the really long baseline experiments, Minos and Icarus, both installed 730 km away from their respective source, Fermilab and CERN, could test the region of small δm^2 with high-energy beams.

In 10 years from now, we will have reached the 10^{-5} level in mixing and the 10^{-3} level in δm^2 in laboratory experiments and for various channels.

Will the *terra incognita* still look as large as today?

EXERCISE

The present understanding of the solar neutrino deficit is that practically all of the ^{7}Be neutrinos have oscillated before escaping from the Sun. As a consequence we expect a beam of 5×10^9 $\nu'/\text{cm}^2/\text{s}$ arriving on Earth with a well-defined energy of 861 keV. The ν' is an unknown neutrino in which the ν_e has oscillated. It mixes with ν_e and its mass verifies the relation:

$$m^2(\nu') - m^2(\nu_e) = 10^{-5} \text{ eV}^2 .$$

In the scenario of mass-degenerate neutrinos, if ν' decays radiatively back to the ν_e on its way between the Sun and the Earth, the emitted γ has the energy:

$$E\gamma = E\delta m^2/2m^2 .$$

It happens that this energy falls in the range of visible photons when the neutrinos have a mass relevant for cosmology.

But the Sun sends $10^{17}\gamma/\text{cm}^2/\text{s}$ to Earth. It is necessary to shield against this high flux, and this is obtained naturally during a solar eclipse. The experiment was done on 25 October 1995 in South East Asia with a 20 cm telescope read out by a CCD camera. No excess was found and this allowed one to put a limit on the radiative decay of the ν' in the scenario of neutrino masses sketched above.

CONCLUSION

More than 60 years after its invention, the neutrino remains the most unknown of the known particles. We still have no definite answers concerning its mass and mixings, its particle/antiparticle nature, its magnetic moment, its lifetime. These theoretical questions are compounded with experimental puzzles and anomalies which, for some of them, have resisted many years. The present question marks concern the solar and atmospheric deficits and the LSND result. Will all or some of these results stand the test of time? Neutrino physics may have already given us hints of physics beyond the Standard Model. It would be good to be quite sure.

Acknowledgement

Tom Ferbel succeeded in organizing a very stimulating and very friendly physics school in a wonderful place. I wish to thank him sincerely here.

REFERENCES

1. See the Proceedings of recent neutrino conferences, in particular "Neutrino 94", *Nucl. Phys. (Proc. Suppl.)*, A. Dar, ed.;
   "Neutrino 92", *Nucl. Phys. (Proc. Suppl.)*, A. Morales ed.
2. The LEP Collaborations, *Phys. Lett.* B276:247 (1992).
3. M. Gell-Mann et al., *in* "Supergravity", P. van Nieuwenhuizen and D. Freedman, eds., North Holland, Amsterdam (1979).
4. B. Pontecorvo, *Sov. Phys. JETP* 26:984 (1968);
   S.M. Bilenki and B. Pontecorvo, *Phys. Rep.* 41:225 (1978).
5. B. Kayser, *Phys. Rev.* D24:110 (1981);
   J. Rich, *Phys. Rev.* D48:4318 (1993).
6. C.W. Kim and H. Nishiura, Preprint JHU-HET 8506 (1985);
   S.M. Bilenki et al., Preprint DFTT 25/95.
7. L. Wolfenstein, *Phys. Rev.* D17:2369 (1978);
   S.P. Mikheyev and A.Yu. Smirnov, *Nuov. Cim.* 9C:17 (1986).
8. J. Bouchez, Ecole de Gif92, p. 91, ed. A.M. Lutz.
9. V.A. Lubimov et al., *Phys. Lett.* 94B:266 (1980).
10. J. Bonn, presented at "Rencontres de Blois," June 1996; V.M. Lobashev, presented at "Neutrino 96", Helsinki, June 1996.
11. K. Assanagan et al., *Phys. Lett.* B335:2311 (1994).
12. D. Buscolic et al., CERN preprint PPE/95–03.
13. D. Lalanne, *Nucl. Phys. (Proc. Suppl.)* B35:369 (1994).
14. K. Zuber, 27th ICHEP, Glasgow, July 1994.
15. G. Bernardi et al., *Phys. Lett.* B166:479 (1986);
    J. Dorenbosch et al., *Phys. Lett.* B166:473 (1986);
    G. Bernardi et al., *Phys. Lett.* B203:332 (1988);
    P. Abreu et al., *Phys. Lett.* B274:230 (1992).
16. P. Pal and L. Wolfenstein, *Phys. Rev.* D25:766 (1982).
17. C. D'Olivo et al., *Phys. Rev. Lett.* 64:1088 (1990);
    C. Giunti et al., *Phys. Rev.* D43:164 (1991).
18. A.A. Gvozdev et al., *Phys. Lett.* B289:103 (1992).
19. M.C. Gonzales-Garcia et al., *Phys. Lett.* B373:153 (1996).
20. J.N. Bahcall, *Rev. Mod. Phys.* 50:811 (1978);
    J.N. Bahcall. "Neutrino Astrophysics," Cambridge University Press (1989).
21. R. Davis, *Prog. Part. Nucl. Phys.* 32:13 (1994).
22. K.S. Hirata et al., *Phys. Rev. Lett.* 63:6 (1989);
    Y. Suzuki, *Nucl. Phys. (Proc. Suppl.)* B38:54 (1995).
23. GALLEX Collaboration, *Phys. Lett.* B327:377 (1994);
    SAGE Collaboration, *Phys. Lett.* B328:234 (1994).
24. M. Hampel, presented at "Rencontres de Blois," June 1996.
25. H. Hirata et al., *Phys. Rev. Lett.* 58:1490 (1987);
    R.M. Bionta et al., *Phys. Rev. Lett.* 58:1494 (1987).
26. K.S. Hirata et al., *Phys. Lett.* B280:146 (1992).
27. E. Lipari, presented at "Rencontres de Blois," June 1996.
28. Y. Fukuda et al., *Phys. Lett.* B335:237 (1994).
29. J.N. Bahcall and M.H. Pinsonneault, *Rev. Mod. Phys.* 64:885 (1992);
    S. Turck-Chieze and I. Lopes, *Astrophys. J.* 408:347 (1993).
30. M. Gruwe et al., *Phys. Lett.* B309:463 (1993).
31. K.S. McFarland et al., Preprint FNAL-PUB-95/153.
32. A.A. Borisov et al., *Phys. Lett.* B369:39 (1996).
33. C. Athanassopoulos et al., *Phys. Rev. Lett.* 75:2550 (1995);
    C. Athanassopoulos et al., Preprint LA-UR-96-1326.
34. B. Zeitnitz, *Prog. Part. Nucl. Phys.* 32:351 (1994).
35. M. de Jong et al., preprint CERN-PPE/93–131 (1993).
36. P. Astier et al., CERN-SPSLC/91–21 (1991).

PROSPECTS FOR B-PHYSICS IN THE NEXT DECADE

Sheldon Stone

Department of Physics
Syracuse Univeristy
Syracuse, N.Y. 13244-1130
Email: Stone@suhep.phy.syr.edu

ABSTRACT

In these lectures I review what has been learned from studies of b-quark decays, including semileptonic decays (V_{ub} and V_{cb}), $B^o - \overline{B}^o$ mixing and rare B decays. Then a discussion on CP violation follows, which leads to a summary of plans for future experiments and what is expected to be learned from them.

1. INTRODUCTION

My assignment is to discuss "Future B Physics Experiments." But to understand what results we desire, it is necessary to understand past accomplishments and have a firm theoretical background. In this paper I will give a brief theoretical introduction to the "Standard Model," and historical introduction to the study of b quark decays. Then I will discuss in some detail the physics already found including: B lifetimes, semileptonic B decays and the CKM couplings V_{cb} and V_{ub}, $B^o - \bar{B}^o$ mixing, rare b decays, and CP violation in K_L^o decays. Following this is a pedantic discussion on CP violation in B decays, which leads into a discussion of future experiments.

1.1. Theoretical Background

The physical states of the "Standard Model" are comprised of left-handed doublets containing leptons and quarks and right handed singlets[1]

$$\begin{pmatrix} u \\ d \end{pmatrix}_L \begin{pmatrix} c \\ s \end{pmatrix}_L \begin{pmatrix} t \\ b \end{pmatrix}_L, \quad u_R,\ d_R,\ c_R,\ s_R,\ t_R,\ b_R \qquad (1)$$

$$\begin{pmatrix} e^- \\ \nu_e \end{pmatrix}_L \begin{pmatrix} \mu^- \\ \nu_\mu \end{pmatrix}_L \begin{pmatrix} \tau^- \\ \nu_\tau \end{pmatrix}_L, \quad e_R^-,\ \mu_R^-,\ \tau_R^-,\ \nu_{eR},\ \nu_{\mu R},\ \nu_{\tau R}. \qquad (2)$$

The gauge bosons, $W^\pm$, γ and Z^o couple to mixtures of the physical d, s and b

states. This mixing is described by the Cabibbo-Kobayashi-Maskawa (CKM) matrix (see below).[2]

The Lagrangian for charged current weak decays is

$$L_{cc} = -\frac{g}{\sqrt{2}} J_{cc}^\mu W_\mu^\dagger + h.c., \qquad (3)$$

where

$$J_{cc}^\mu = (\bar{\nu}_e,\ \bar{\nu}_\mu,\ \bar{\nu}_\tau)\,\gamma^\mu \begin{pmatrix} e_L \\ \mu_L \\ \tau_L \end{pmatrix} + (\bar{u}_L,\ \bar{c}_L,\ \bar{t}_L)\,\gamma^\mu V_{CKM} \begin{pmatrix} d_L \\ s_L \\ b_L \end{pmatrix} \qquad (4)$$

and

$$V_{CKM} = \begin{pmatrix} V_{ud} & V_{us} & V_{ub} \\ V_{cd} & V_{cs} & V_{cb} \\ V_{td} & V_{ts} & V_{tb} \end{pmatrix}. \qquad (5)$$

Multiplying the mass eigenstates (d, s, b) by the CKM matrix leads to the weak eigenstates (d', s', b'). There are nine complex CKM elements. These 18 numbers can be reduced to four independent quantities by applying unitarity constraints and the fact that the phases of the quark wave functions are arbitrary. These four remaining numbers are **fundamental constants** of nature that need to be determined from experiment, like any other fundamental constant such as α or G. In the Wolfenstein approximation* the matrix is written as[3]

$$V_{CKM} = \begin{pmatrix} 1 - \lambda^2/2 & \lambda & A\lambda^3(\rho - i\eta) \\ -\lambda & 1 - \lambda^2/2 & A\lambda^2 \\ A\lambda^3(1 - \rho - i\eta) & -A\lambda^2 & 1 \end{pmatrix} \qquad (6)$$

The constants λ and A are determined from charged-current weak decays. To see how this is done, first consider muon decay. The muon decays weakly into $\nu_\mu e^- \bar{\nu}_e$ as shown in Fig. 1. The decay width is given by[4]

$$\Gamma_\mu = \frac{G_F^2}{192\pi^3} m_\mu^5 \times \text{(radiative corrections)}. \qquad (7)$$

The couplings at the vertices are unity for leptons. This process serves to measure the weak interaction decay constant (Fermi constant) G_F.

A charged current decay diagram for strange quark decay is shown in Fig. 2. Here the CKM element V_{us} is present. The decay rate is given by a formula similar to equation (7), with the muon mass replaced by the s-quark mass and an additional factor of $|V_{us}|^2$. Two complications arise since we are now measuring a decay process involving hadrons, $K^- \to \pi^0 e^- \bar{\nu}$ rather than elementary constituents. One is that the s-quark mass is not well defined and the other is that we must make corrections for

---

*In higher order other terms have an imaginary part; in particular the V_{cd} term becomes $-\lambda - A^2\lambda^5(\rho + i\eta)$, which is important for CP violation in K_L^0 decay.

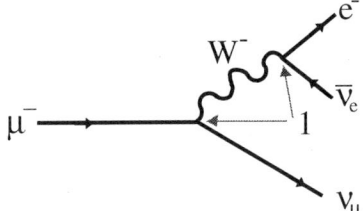

Fig. 1. Diagram for muon decay.

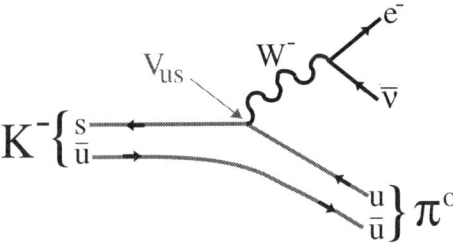

Fig. 2. Semileptonic K^- decay diagram.

the probability that the $\bar{u}$-spectator-quark indeed forms a π^o with the u-quark from the s-quark decay. These considerations will be discussed in greater detail in the semileptonic B decays section. For now[5] remember that $\lambda = V_{us} = 0.2205 \pm 0.0018$ and, $A \approx 0.8$. Constraints on ρ and η are found from other measurements. These will also be discussed later.

1.2. B Decay Mechanisms

Fig. 3 shows sample diagrams for B decays. Semileptonic decays are shown in Fig. 3(a). The name "semileptonic" is given, since there are both hadrons and leptons in the final state. The leptons arise from the virtual W^-, while the hadrons come from the coupling of the spectator anti-quark with either the c or u quark from the b quark decay. Note that the B is massive enough that all three lepton species can be produced. The simple spectator diagram for hadronic decays (Fig. 3(b)) occurs when the virtual W^- materializes as a quark-antiquark pair, rather than a lepton pair. The terminology *simple spectator* comes from viewing the decay of the b quark, while ignoring the presence of the *spectator* antiquark. If the colors of the quarks from the virtual W^- are the same as the initial b quark, then the color suppressed diagram, Fig. 3(c), can occur. While the amount of color suppression is not well understood, a good first order guess is that these modes are suppressed in amplitude by the color factor $1/3$ and thus in rate by $1/9$, with respect to the non-color suppressed spectator diagram.

The annihilation diagram shown in Fig. 3(d) occurs when the b quark and spec-

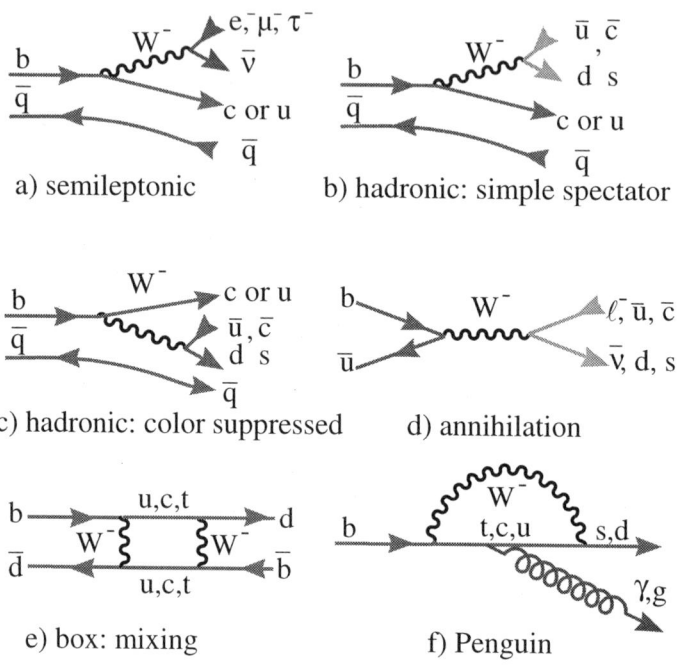

Fig. 3. Various mechanisms for B meson decay.

tator anti-quark find themselves in the same space-time region and annihilate by coupling to a virtual W^-. The probability of such a wave function overlap between the b and $\bar{u}$-quarks is proportional to a numerical factor called f_B. The decay amplitude is also proportional to the coupling V_{ub}. The mixing and penguin diagrams will be discussed later.

2. What is known

2.1. Early history

The first experimental evidence for b quarks was found at Fermilab by looking at high mass dimuon pairs in 800 GeV proton interactions on nuclear targets.[6] Their results are shown in Fig. 4 along with subsequent data from DESY using e^+e^- annihilations which shows narrow peaks at the masses of the Υ and Υ' resonances.[7]

The natural width of the peaks is narrower than the energy resolution of either experiment leading to the interpretation that these states are comprised of a bound $b\bar{b}$ quark system. The narrow decay width is similar to the situation in charmonium,

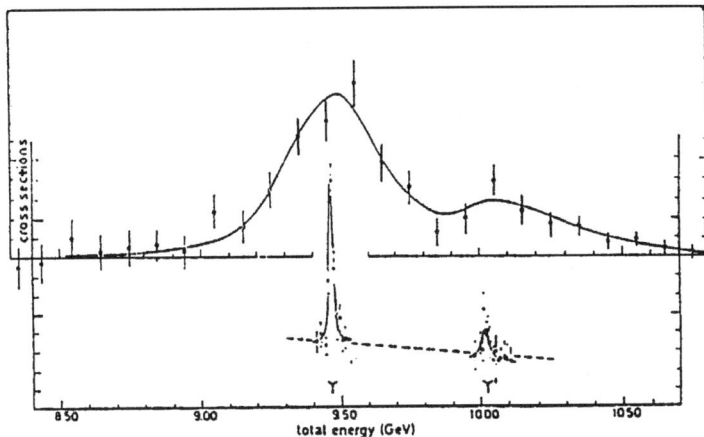

Fig. 4. The data on top is the $\mu^+\mu^-$ invariant mass from the Columbia-Fermilab-Stony Brook collaboration and the data shown below is the total e^+e^- cross-section from the DESY-Heidelberg-Hamburg-Munchen collaboration.

i.e. the decay width is proportional to the strong coupling constant α_s^3.

As the DESY machine was limited in center-of-mass energy at that time, the torch was passed to the CLEO experiment at the CESR e^+e^- storage ring. An early total cross-section scan is shown in Fig. 5(a). A new narrow state, the Υ'' (or $\Upsilon(3S)$), appears along with a state wider than the experimental resolution, the $\Upsilon(4S)$.

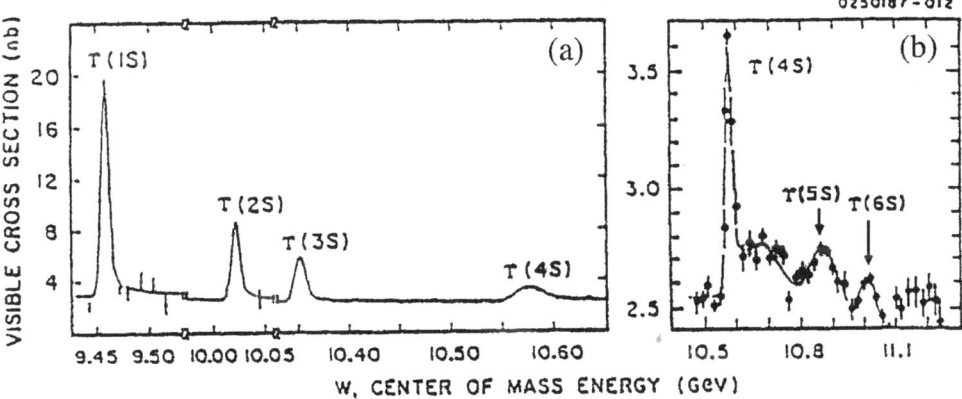

Fig. 5. Hadronic cross-section scan in the Upsilon region, (a) shows 1S-4S and (b) region above 4S.

The mechanism of b quark production in e^+e^- collisions and the subsequent production of the final states B^+B^- and $B^o\bar{B}^o$ from the $\Upsilon(4S)$ are shown in Fig. 6. Subsequent data shown in Fig. 5(b) shows that the cross-section is ≈ 1 nb and details structures in the total cross-section at higher energies.[8] Little data has been taken above the $\Upsilon(4S)$, however.

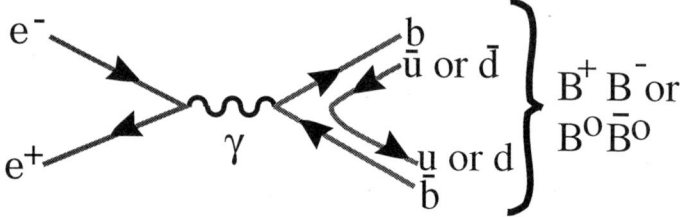

Fig. 6. B production mechanism at the $\Upsilon(4S)$.

Many properties of B meson decays have been discovered by two e^+e^- experiments operating at the $\Upsilon(4S)$ resonance, CLEO at CESR and ARGUS at DESY (the DESY machine group upgraded the energy so they could do this physics). Fully reconstructed B meson decays were first seen by CLEO and the B masses determined.[9] Now there are several thousand fully reconstructed decays in many modes allowing for branching ratio determinations. A different technique is used to reconstruct B mesons at the $\Upsilon(4S)$ than at other machines. At this resonance we have

$$e^+e^- \to \Upsilon(4S) \to B^-B^+ \quad (8)$$
$$\to B^o\overline{B}^o . \quad (9)$$

From energy conservation, the energy of each B is equal to the beam energy, E_{beam} (the center-of-mass energy is twice E_{beam}). To reconstruct exclusive B meson decays, we first require the energy of the decay products be consistent with the beam energy. Suppose the final state we are considering is $D^o\pi^-$. We require that

$$E_{D^o} + E_{\pi^-} = E_{beam}. \quad (10)$$

In practice this means that the difference between the left-hand side and the right-hand side is less than ≈ 3 times the error on the measured energy sum.

The next step is to compute the invariant mass of the candidate B^- using the well known beam energy:

$$m_B = \sqrt{E_{beam}^2 - (\vec{p}_{D^o} + \vec{p}_{\pi^-})^2}. \quad (11)$$

In practice this technique leads to large background rejections and a B mass resolution of $\sigma \approx 2.5$ MeV (at CESR) which is due mostly to the energy spread of the beam. A few sample B decay candidate mass plots are shown in Fig. 7 from the CLEO experiment.[10]

Hadronic production rates for b quarks have been measured at two $p\bar{p}$ colliders, UA1 at the SPS,[15] and CDF at the Tevatron.[16] E789 has also measured b production using an 800 GeV proton beam hitting nuclear targets.[17] CDF has reconstructed B meson decays into modes containing a ψ meson. These are shown[14] in Fig. 8.

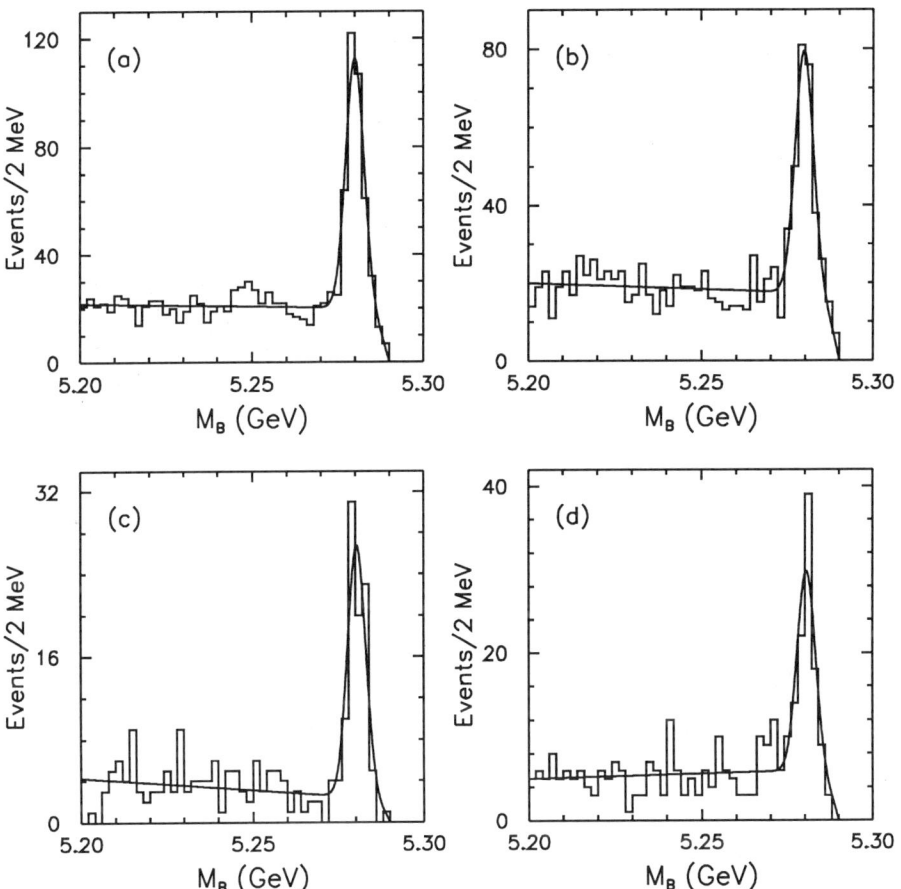

Fig. 7. Beam constrained mass distribution from CLEO for (a) $B^- \to D^o\pi^-$, (b) $B^- \to D^o\rho^-$, (c) $\overline{B}^o \to D^+\pi^-$ and (d) $\overline{B}^o \to D^+\rho^-$.

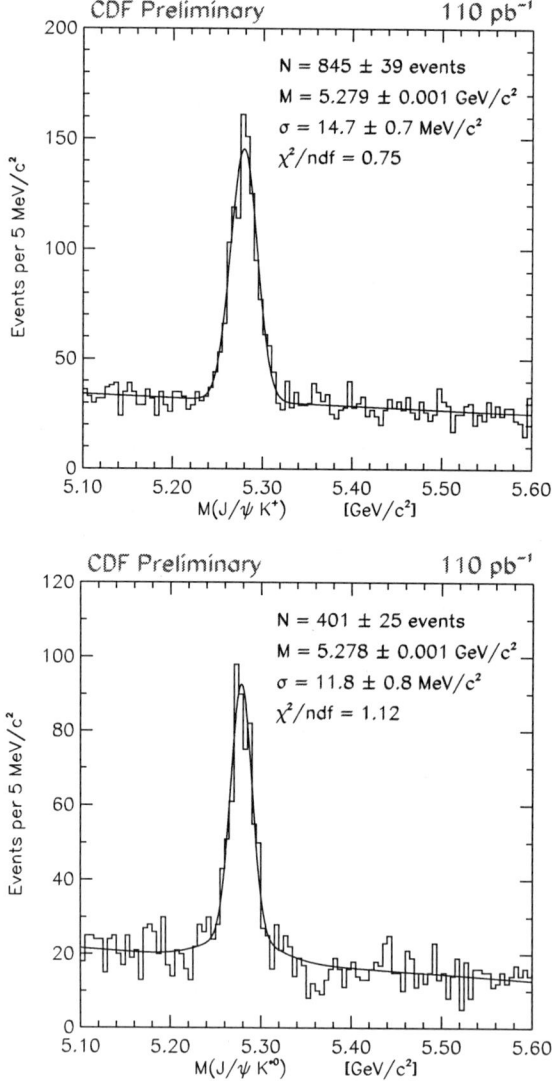

Fig. 8. Invariant mass spectra from CDF for ψK^+ and ψK^{*o} candidates.

2.2. Lifetimes

Lifetimes are a fundamental property of elementary particles. The b quark lifetime, however, was measured before the individual lifetimes of b flavored hadrons at the higher energy e^+e^- machines, PEP and PETRA.[11] More recent measurements have come from LEP, SLD and CDF.[12] The meaning of b quark lifetime is really the average of the B hadron lifetimes over the kinds of B hadrons which happen to be produced in the particular environment. The results are summarized in Table 1.[13]

Table 1. B lifetime measurements (ps)

	LEP Avg	CDF	SLD	World Avg
b quark	1.54±0.02	1.51±0.03	1.56±0.05	1.53±0.02
B^-	1.63±0.06	1.68±0.07		1.65±0.05
B^o	1.52±0.06	1.58±0.09		1.55±0.05
B^o_s	1.60±0.10	1.36±0.12		1.50±0.08
Λ_b	1.21±0.07	1.32±0.17		1.23±0.06
Ξ_b	$1.39^{+0.34}_{-0.28}$			$1.39^{+0.34}_{-0.28}$

The meson lifetimes are nearly equal implying the dominance of the spectator diagram. The Λ_b lifetime appears to be shorter, which implies the existence of other diagrams in baryon decay. This is very different from the situation in charm decay where the D^o, D^+ and Λ_c have different lifetimes.

2.3. The CKM element V_{cb}

2.3.1. Theory of semileptonic decays

The same type of semileptonic charged current decays used to find V_{us} are used to find V_{cb} and V_{ub}. The basic diagram is shown in Fig. 3(a). We can use either inclusive decays, where we look only at the lepton and ignore the hadronic system at the lower vertex, or exclusive decays where we focus on a particular single hadron. Theory currently can predict either the inclusive decay rate, or the exclusive decay rate when there is only a single hadron in the final state. The fraction of semileptonic decays into exclusive final states containing either a pseudoscalar or vector meson is given in Table 2.

Now let us briefly go through the mathematical formalism of semileptonic decays. Let us start with pseudoscalar to pseudoscalar transitions. The decay amplitude is given by[18]

$$A(\bar{B} \to m e^- \bar{\nu}) = \frac{G_F}{\sqrt{2}} V_{ij} L^\mu H_\mu, \text{ where} \qquad (12)$$

$$L^\mu = \bar{u}_e \gamma^\mu (1 - \gamma_5) v_\nu, \text{ and} \qquad (13)$$

Table 2. Fraction of $q \to x\ell\nu$ to 0^- or 1^- final states

s quark	100%	$K \to \pi\ell\nu$
c quark	>90%	$D \to (K + K^*)\ell\nu$
	?	$D \to (\pi + \rho)\ell\nu$
b quark	≈66%	$B \to (D + D^*)\ell\nu$
	?	$B \to (\pi + \rho)\ell\nu$
t quark	0%	t does not form hadrons

$$H_\mu = \langle m|J_{had}^\mu(0)|\overline{B}\rangle = f_+(q^2)(P+p)_\mu + f_-(q^2)(P-p)_\mu, \tag{14}$$

where q^2 is the four-momentum transfer squared between the B and the m, and $P(p)$ are four-vectors of the $B(m)$. H_μ is the most general form the hadronic matrix element can have. It is written in terms of the unknown f functions that are called "form-factors." It turns out that the term multiplying the $f_-(q^2)$ form-factor is the mass of lepton squared. Thus for electrons and muons (but not τ's), the decay width is given by

$$\frac{d\Gamma_{sl}}{dq^2} = \frac{G_F^2|V_{ij}|^2 K^3}{24\pi^2}|f_+(q^2)|^2, \quad \text{where} \tag{15}$$

$$K = \frac{1}{2M_B}\left[\left(M_B^2 + m^2 - q^2\right)^2 - 4m^2 M_B^2\right]^{1/2} \tag{16}$$

is the momentum of the particle m (with mass m) in the B rest frame. In principle, $d\Gamma_{sl}/dq^2$ can be measured over all q^2. Thus the shape of $f_+(q^2)$ can be determined experimentally. However, the normalization, $f_+(0)$ must be obtained from theory, for V_{ij} to be measured. In other words,

$$\Gamma_{SL} \propto |V_{ij}|^2|f_+(0)|^2 \frac{1}{\tau_B}\int K^3 g(q^2) dq^2, \tag{17}$$

where $g(q^2) = f_+(q^2)/f_+(0)$. Measurements of semileptonic B decays give the integral term, while the lifetimes are measured separately, allowing the product $|V_{ij}|^2|f_+(0)|^2$ to be experimentally determined.

For pseudoscalar to vector transistions there are three independent form-factors whose shapes and normalizations must be determined.[19]

2.3.2. $\overline{B}^o \to D^+\ell^-\bar{\nu}$

CLEO has recently measured the branching ratio and form-factor for the reaction $\overline{B}^o \to D^+\ell^-\bar{\nu}$ using two different techniques.[20] In the first method the final state is reconstructed finding only lepton and D^+ candidates, where the $D^+ \to K^-\pi^+\pi^+$ decay is used. Then, using the fact that the $B's$ produced at the $\Upsilon(4S)$ are nearly at rest the missing mass squared (MM^2) is calculated as

$$\begin{aligned} MM^2 &= E_\nu^2 - \vec{P}_\nu^2 \qquad (18)\\ &= (E_B - E_{D^+} - E_\ell)^2 - (\vec{p}_B - \vec{p}_{D^+} - \vec{p}_\ell)^2\\ &\approx (E_B - (E_{D^+} + E_\ell))^2 - (\vec{p}_{D^+} + \vec{p}_\ell)^2\\ &\approx E_{beam}^2 + m_B^2 + m_{D^+}^2 + m_\ell^2 - 2\vec{p}_{D^+} \cdot \vec{p}_\ell, \end{aligned}$$

where E refers to particle energy, m to mass and $\vec{p}$ to three-momentum. The approximation on the third line results from setting p_B to zero. This approximation causes a widening of the MM^2 distribution, giving a r.m.s. width of 0.2 GeV^2.

This analysis is done by finding the number of D^+ events with opposite sign leptons in different q^2 and MM^2 bins. The $K^-\pi^+\pi^+$ mass distributions for the interval $4 > q^2 > 2$ GeV^2 and several MM^2 bins are shown in Fig. 9.

There is also a large background from $\bar{B}^o \to D^{*+} X \ell^- \bar{\nu}$ decays where the $D^{*+} \to \pi^o D^+$. These events are reconstructed and their MM^2 distribution is directly subtracted (after correcting for efficiencies) from the candidate signal distribution. We are left with a sample that contains $D^+ \ell^- \bar{\nu}$ decays and also $D^+ X \ell^- \bar{\nu}$, where X can be a single hadron or hadrons but cannot be the result of final state with a D^{*+}. We ascertain the total number of signal events by fitting the MM^2 distribution in the different q^2 bins as shown in Fig. 10 to a $D^+ \ell^- \bar{\nu}$ signal shape and a background shape for $D^+ X \ell^- \bar{\nu}$.

The second technique reconstructs the neutrino by using missing energy and momentum measurements. Essentially all charged tracks and photons in the event are added up and since the total energy must be equal to the center of mass energy and the total three-momentum must be zero, any difference is assigned to the neutrino. Events with a second lepton or which do not conserve charge are eliminated. Furthermore, the momentum and energy measurements must be consistent. Once the neutrino four-vector is determined, the B can be reconstructed in the "usual" way as shown in Fig. 11.

The MM^2 technique gives a branching ratio of $(1.75 \pm 0.25 \pm 0.20)\%$, while the neutrino reconstruction gives $(1.89 \pm 0.22 \pm 0.35)\%$, giving a combined (preliminary) yield of $(1.78 \pm 0.20 \pm 0.24)\%$. The statistical errors in both methods are essentially uncorrelated, while the systematic error is almost completely correlated.

The q^2 distribution from the MM^2 method is shown in Fig. 12. The intercept at q^2 of zero is proportional to $|V_{cb} f_+(0)|^2$. The curve is a fit to a functional form

$$f_+(q^2) = \frac{f_+(0)}{1 - q^2/M_V^2}, \qquad (19)$$

where M_V is left unspecified but is theorized to be the mass of the vector exchange particle in the t channel, namely the B^*. The results and comparison with different models are shown in Table 3.

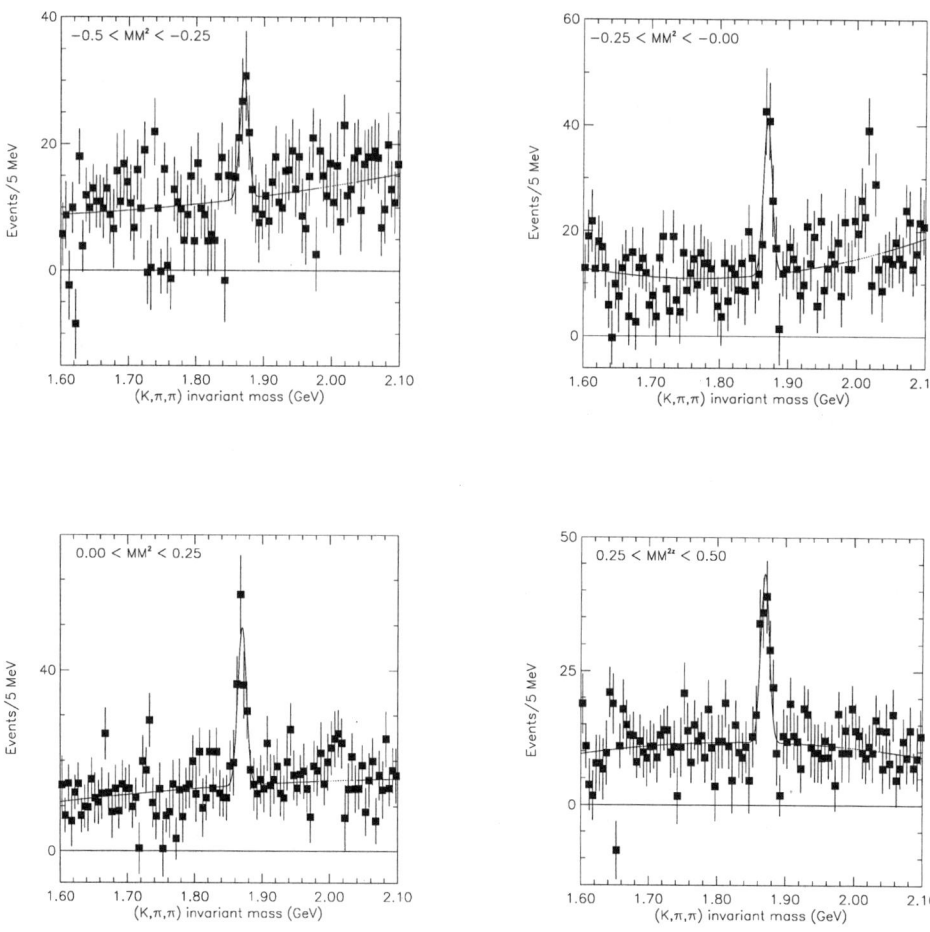

Fig. 9. Invariant $K^-\pi^+\pi^+$ mass spectra from CLEO for events with an opposite sign lepton in the interval $4 > q^2 > 2$ in four different MM^2 slices The curves are a fit to a Gaussian signal shape summed with a polynomial background.

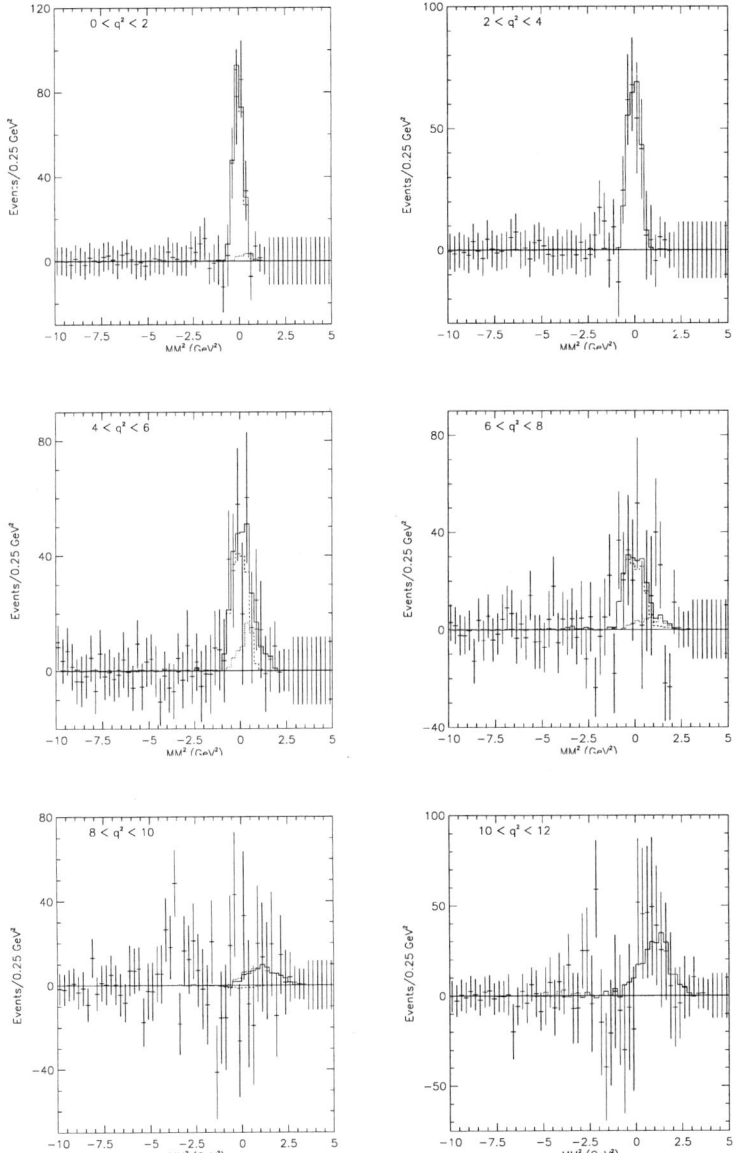

Fig. 10. Fits to the MM^2 distribution for the $D^+\ell^-\bar{\nu}$ (dashed) and $D^+X\ell^-\bar{\nu}$ (dotted) components and the sum (solid) in different q^2 intervals

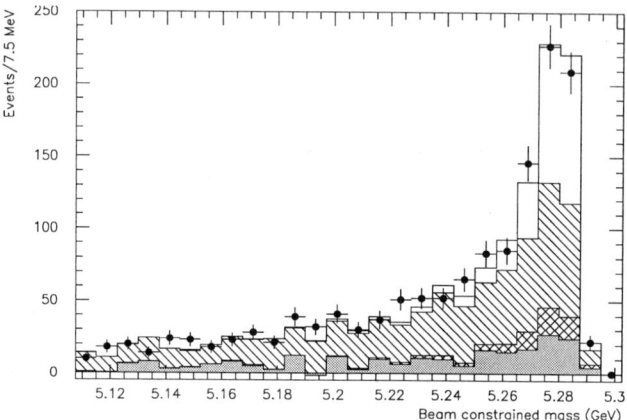

Fig. 11. Beam constrained mass spectrum for all events passing the cuts. The white area represents the signal events, the hatched area represents the combinatoric background, the crosshatched area represents the $D^{*+}\ell^-\bar{\nu}$ and the shaded area represents all the remaining backgrounds.

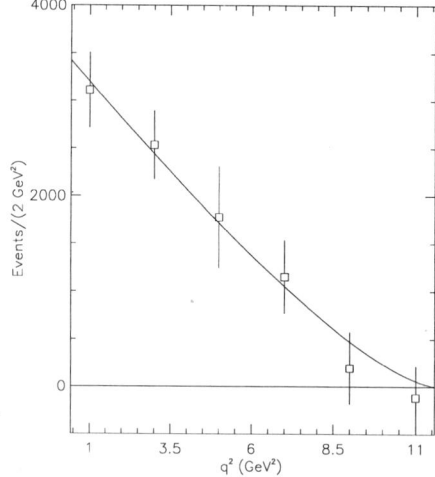

Fig. 12. The q^2 distribution for $\bar{B}^o \to D^+\ell^-\bar{\nu}$ from the MM^2 analysis.

Table 3. Results of $\bar{B}^o \to D^+\ell^-\bar{\nu}$ analysis

Model	$f_+(0)$ prediction	$|V_{cb}f_+(0)| \times 10^3$	$|V_{cb}| \times 10^3$
WSB[21]	0.70	$25.7 \pm 1.4 \pm 1.7$	$37.3 \pm 2.0 \pm 2.5$
KS[22]	0.69	$25.7 \pm 1.4 \pm 1.7$	$36.7 \pm 2.0 \pm 2.5$
Demchuk†[23]	0.68	$24.8 \pm 1.1 \pm 1.6$	$36.4 \pm 1.6 \pm 2.4$
Average			$36.9 \pm 3.7 \pm 0.5$

† A smaller statistical error is quoted for this model because M_V is specified.

For the average value for V_{cb}, the first error is the quadrature of the the systematic and statistical errors in the data, and the fact that the fraction of B^o's produced in $\Upsilon(4S)$ decay is known only as 0.49 ± 0.05.[24] The second error is due only to the model dependence.

2.3.3. Branching Ratio of $\bar{B}^o \to D^{*+}\ell^-\bar{\nu}$

We next turn to measurements of the branching ratio of the pseudoscalar to vector transition $\bar{B}^o \to D^{*+}\ell^-\bar{\nu}$, shown in Table 4.

Table 4. Measurements of $\mathcal{B}(\bar{B}^o \to D^{*+}\ell^-\bar{\nu})$

Experiment	$\mathcal{B}(\%)$
CLEO[25]	$4.1 \pm 0.5 \pm 0.7$
ARGUS[26]	$4.7 \pm 0.6 \pm 0.6$
CLEO II[24]	$4.50 \pm 0.44 \pm 0.44$
ALEPH[27]	$5.18 \pm 0.30 \pm 0.62$
DELPHI[28]	$5.47 \pm 0.16 \pm 0.67$
Average	4.90 ± 0.35

The width predictions of a collection of representative models and the resulting values of V_{cb} are given in Table 5. Here the first error on the average is the from the error on the measured branching ratio ($\pm 3.6\%$) in quadrature with the error on the lifetime ($\pm 1.6\%$) and the second error reflects the spread in the models ($\pm 5.2\%$).

2.3.4. Heavy Quark Effective Theory and $\bar{B} \to D^*\ell^-\bar{\nu}$

Our next method for finding V_{cb} uses "Heavy Quark Effective Theory" (HQET).[32] We start with a quick introduction to this theory. It is difficult to solve QCD at long distances, but its possible at short distances. Asymptotic freedom, the fact that the strong coupling constant α_s becomes weak in processes with large q^2, allows perturbative calculations. Large distances are of the order $\sim 1/\Lambda_{QCD} \sim 1$ fm, since Λ_{QCD} is about 0.2 GeV. Short distances, on the other hand, are of the order of the

Table 5. Values of V_{cb} from $\mathcal{B}(\bar{B}^o \to D^+\ell^-\bar{\nu})$

Model	Predicted $\Gamma(B \to D^*\ell\nu)$ (ps^{-1})	$\|V_{cb}\| \times 10^3$
ISGW[29]	$25.2\|V_{cb}\|^2$	35.2 ± 1.4
ISGW II[30]	$24.8\|V_{cb}\|^2$	35.5 ± 1.4
KS[22]	$25.7\|V_{cb}\|^2$	34.8 ± 1.4
WBS[21]	$21.9\|V_{cb}\|^2$	37.8 ± 1.5
Jaus1[31]	$21.7\|V_{cb}\|^2$	37.9 ± 1.5
Jaus2[31]	$21.7\|V_{cb}\|^2$	37.9 ± 1.5
Average		$36.5 \pm 1.5 \pm 1.9$

quark Compton wavelength; $\lambda_Q \sim 1/m_Q$ equals 0.04 fm for the b quark and 0.13 fm for the c quark.

For hadrons, on the order of 1 fm, the light quarks are sensitive only to the heavy quark's color electric field, not the flavor or spin direction. Thus, as $m_Q \to \infty$, hadronic systems which differ only in flavor or heavy quark spin have the same configuration of their light degrees of freedom. The following two predictions follow immediately (the actual experimental values are shown below):

$$m_{B_s} - m_{B_d} = m_{D_s} - m_{D^+} \tag{20}$$
$$(90 \pm 3) \text{ MeV} \quad (99 \pm 1) \text{ MeV , and}$$

$$m_{B^*}^2 - m_B^2 = m_{D^*}^2 - m_D^2. \tag{21}$$
$$0.49 \text{ GeV}^2 \quad 0.55 \text{ GeV}^2.$$

The agreement is quite good but not exceptional. Since the charmed quark is not that heavy, there is some heavy quark symmetry breaking. This must be accounted for in quantitative predictions, and can probably explain the discrepancies above. The basic idea is that if you replace a b quark with a c quark moving at the same **velocity**, there should only be small and calculable changes.

In lowest order HQET there is only one form-factor function ξ which is a function of the Lorentz invariant four-velocity transfer y, where

$$y = \frac{M_B^2 + M_{D^*}^2 - q^2}{2M_B M_{D^*}}. \tag{22}$$

The point y equals one corresponds to the situation where the B decays to a D^* which is at rest in the B frame. Here the "universal" form-factor function $\xi(y)$ has the value, $\xi(1) = 1$, in lowest order. This is the point in phase space where the b quark changes to a c quark with zero velocity transfer. The idea is to measure the decay rate at this point, since we know the value of the form-factor, namely unity, and then apply the hopefully small and hopefully well understood corrections. Although this analysis can be applied to $\bar{B} \to D\ell^-\nu$, the vanishing of the decay rate at y equals 1, (maximum q^2, see Fig. 12), makes this inaccurate.[20]

The corrections are of two types: quark mass, characterized as some coefficient times Λ_{QCD}/m_Q, and hard gluon, characterized as η_A. The value of the form-factor can then be expressed as[33]

$$\xi(1) = \eta_A \left(1 + 0 \cdot \Lambda_{QCD}/m_Q + c_2 \cdot (\Lambda_{QCD}/m_Q)^2 + ...\right) = \eta_A(1 + \delta). \qquad (23)$$

The zero coefficient in front of the $1/m_Q$ term reflects the fact that the first order correction in quark mass vanishes at y equals one. This is called Luke's Theorem.[34] Recent estimates are 0.96 ± 0.007 and -0.55 ± 0.025 for η_A and δ, respectively. The value predicted for $\xi(1)$ then is 0.91 ± 0.03. This is the conclusion of Neubert.[33] There has been much controversy surrounding the theoretical prediction of this number.[35]

To find the value of the decay width at y equals one, it is necessary to fit data over a finite range in y and extrapolate to y of one. HQET does not predict the shape of the form-factor; hence the shape of the $d\Gamma/dy$ distribution is not specified. Most experimental groups have done the simplest thing and used a linear fit. The CLEO results with both linear and quadratic fits are shown in Fig. 13. The results from the different groups are summarized in Table 6. Also fits of the slope parameter, ρ^2, coming from the linear fit are included.

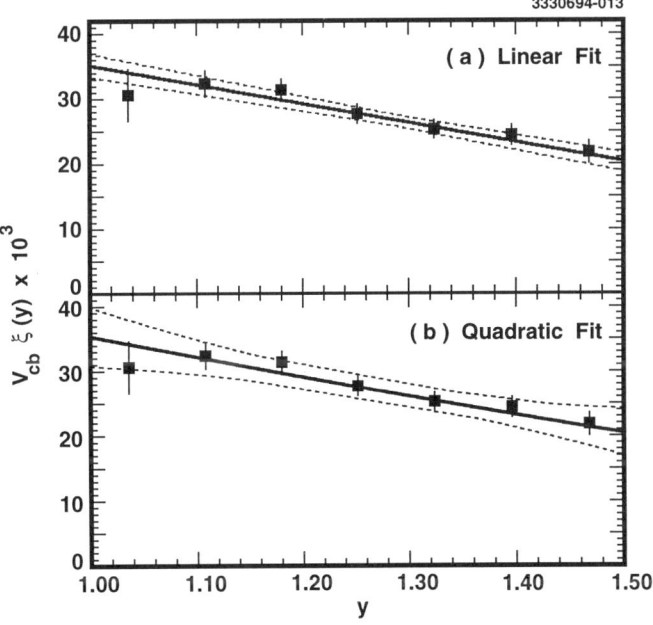

Fig. 13. Linear and quadratic fits to the CLEO data for the $D^{*+}\ell^-\bar{\nu}$ and $D^{*o}\ell^-\bar{\nu}$.

Although the shape of the function is not specified in HQET general considerations lead to the expectation that the slope is positive: there is a pole in the amplitude

Table 6. Values of $|V_{cb}|\xi(1) \times 10^3$

Experiment	$|V_{cb}|\xi(1) \times 10^3$	ρ^2
ARGUS[36]	$38.8 \pm 4.3 \pm 3.5$	$1.17 \pm 0.22 \pm 0.06$
CLEO II[37]	$35.1 \pm 1.9 \pm 2.0$	$0.84 \pm 0.12 \pm 0.08$
ALEPH[38]	$31.4 \pm 2.3 \pm 2.5$	$0.39 \pm 0.21 \pm 0.12$
DELPHI[39]	$35.0 \pm 1.9 \pm 2.3$	$0.81 \pm 0.16 \pm 0.10$
Average	34.6 ± 1.6	0.82 ± 0.09

as $y \to -1$ and $\xi(y) \to 0$ as y increases. Shapes for $\xi(y)$ are suggested by quark models. I have fit the CLEO data to different model functions as shown in Fig. 14. The results are shown in Table 7.

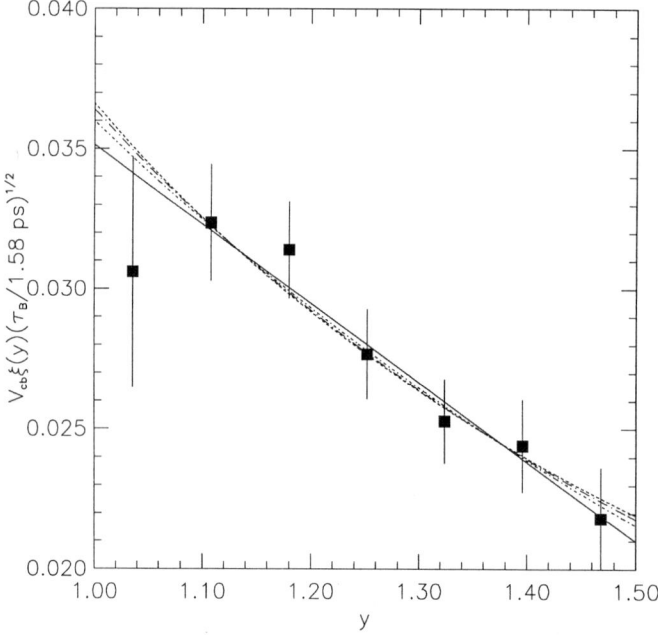

Fig. 14. Fits to the CLEO data with different shapes. The curves are linear (solid), Neubert-Reickert (NR) exponential (dashed), pole (long dash-dot) and exponential (dot-dashed).

These shapes give larger values of $|V_{cb}|\xi(1)|$ than the linear fit by $(5\pm3)\%$. I call this a model dependent error. The value then obtained for $|V_{cb}|\xi(1)|$ is $(36.3 \pm 1.6 \pm 1.0) \times 10^{-3}$, and

$$|V_{cb}| = 0.0397 \pm 0.0021 \pm 0.0017 \ . \qquad (24)$$

Table 7. Values of $|V_{cb}|\xi(1)$ for different fit shapes of CLEO II data

$\xi(y)$	name	ρ	$|V_{cb}|\xi(1) \times 10^3$
$1 - \rho^2(y-1)$	linear	0.90±0.07	0.0351±0.0018±0.0018
$\frac{2}{y+1}exp\left[-(2\rho^2-1)\frac{y-1}{y+1}\right]$	NR exp	0.90±0.12	0.0366±0.0024±0.0018
$\left(\frac{2}{y+1}\right)^{2\rho^2}$	pole	1.07±0.11	0.0364±0.0023±0.0018
$exp\left[-\rho^2(y-1)\right]$	exp	1.01±0.10	0.0360±0.0022±0.0018

2.3.5. $|V_{cb}|$ using inclusive semileptonic decays

The inclusive semileptonic branching ratio $\mathcal{B}(B \to Xe^-\bar{\nu})$ can also be used to measure V_{cb}. While this has traditionally been done by measuring the inclusive lepton momentum spectrum using only single lepton data, recently dilepton data have been used. The inclusive lepton spectrum from the latest CLEO II data[40] is shown in Fig. 15. Both electrons and muons are shown. Leptons which arise from the contin-

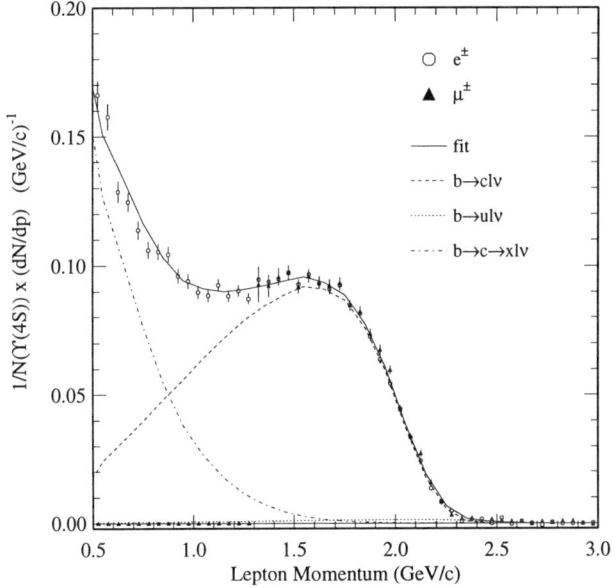

Fig. 15. Fit to the CLEO inclusive lepton spectrum with the ACM model.

uum have been statistically subtracted using the below resonance sample. The peak at low momentum is due to the decay chain $\bar{B} \to DX$, $D \to Y\ell^+\nu$. The data are fit to two shapes whose normalizations are allowed to float. The first shape is taken from models of B decay while the second comes from the measured shape of leptons from D mesons produced nearly at rest at the ψ'', which is then smeared using the measured momentum distribution of $D's$ produced in B decay. CLEO finds $\mathcal{B}_{sl}$ of 10.5±0.2% and 11.1±0.3% in the ACM[43] and ISGW* models, respectively.[40] The ACM model will be described below. The ISGW* model is a variant of the ISGW[29] model. The ISGW model includes all the exclusive single hadron modes, D, D^*, and D^{**} which contains several components. CLEO lets the normalization of the D^{**} components float in the fit, and calls this model ISGW*.

Next, I discuss how to use dilepton events to eliminate the secondary leptons at low momentum. Consider the sign of the lepton charges for the four leptons in the following decay sequence: $\Upsilon(4S) \to B^- B^+$; $B^- \to D\ell_1^- \bar{\nu}$, $B^+ \to \bar{D}\ell_3^+ \nu$; $D \to Y\ell_2^+ \nu$, $\bar{D} \to Y'\ell_4^- \bar{\nu}$. If a high momentum negative lepton (ℓ_1^-) is found, then if the second lepton is also negative it must come from the cascade decay of the B^+ (i.e. it must be ℓ_4^-). On the other hand the second lepton being positive shows that it must be either the primary lepton from the opposite B^+, (ℓ_3^+), or the cascade from the same B^-, (ℓ_2^+). However the cascades from the same B^- can be greatly reduced by insisting that the cosine of the opening angle between the two leptons be greater than zero as they tend to be aligned. The same arguments are applicable to $\Upsilon(4S) \to B^\circ \bar{B}^\circ$, except that an additional correction must be made to account for $B\bar{B}$ mixing.

The CLEO II data are shown in Fig. 16. The data fit nicely to either the ACM or ISGW* model. They find that the semileptonic branching ratio, $\mathcal{B}_{sl}$, equals $(10.36 \pm 0.17 \pm 0.40)$% with a negligible dependence on the model.[41] This result confirms that the B model shapes are appropriate down to lepton momenta of 0.6 GeV/c. ARGUS[42] did the first analysis using this technique and found $\mathcal{B}_{sl} = (9.6 \pm 0.5 \pm 0.4)$%.

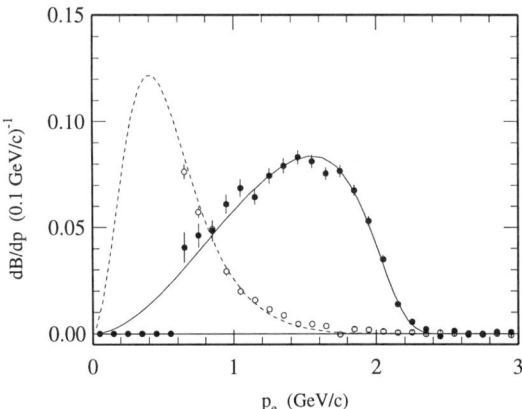

Fig. 16. The lepton momentum spectrum in dilepton events from CLEO. The solid points are for opposite sign leptons, while the open circles indicate like sign lepton pairs. The fit is to the ACM model.

The next topic is to measure V_{cb} using the inclusive lepton spectrum. Consider

$\Gamma_{sl} \equiv \Gamma(B \to Xe^-\bar{\nu})$ in the simplest parton model:

$$\Gamma_{sl} = \frac{G_F^2 m_b^5}{192\pi^3} \left(p_c |V_{cb}|^2 + p_u |V_{ub}|^2 \right) \eta_{QCD}, \tag{25}$$

where the p's are phase space factors, and the QCD correction, $\eta_{QCD} = 1 - 2\alpha_s/3\pi$. Since $|V_{ub}| \ll |V_{cb}|$, we ignore the 2nd term. To use the semileptonic width to extract $|V_{cb}|$ using this expression requires a knowledge of m_b^5, which is poorly understood. A way around this dilemma was found by Altarelli et al.[43] They make two important corrections to the simple parton model. First they treat the spectator quark in the B meson as a quasi-free particle with a Gaussian spectrum of Fermi-momentum, p:

$$f(p) = \frac{4p^2}{\sqrt{\pi} p_f^3} \exp(-p^2/p_f^2). \tag{26}$$

The average value, p_f, is a free parameter in the model. Secondly, they include the effects of gluon radiation from the quarks, which lowers the spectrum at high lepton momentum. The semileptonic width is given explicitly as:

$$\frac{d\Gamma(B \to DX\ell^-\bar{\nu})}{dx} = \frac{m_b^5 G_F^2 V_{cb}^2}{96\pi^3} \cdot [\Phi(x,\epsilon) - G(x,\epsilon)], \tag{27}$$

where $x = 2E_\ell/m_b$, E_ℓ being the lepton energy, $\epsilon = m_c/m_b$, $G(x,\epsilon)$ is a complicated gluon radiation function and

$$\Phi(x,\epsilon) = \frac{x^2(1-\epsilon^2-x)^2}{(1-x)^3} \left[(1-x)(3-2x) + (3-x)\epsilon^2 \right]. \tag{28}$$

Each value of the Fermi-momentum, p, leads to a different value of m_b and hence a different distribution for $\frac{d\Gamma}{dx}$ which must be convoluted with Eq. (27) to find the total theoretical lepton momentum spectrum. The relationship between m_b and p is just given by kinematics

$$m_b^2 = m_B^2 + m_{sp}^2 - 2m_B \sqrt{(p^2 + m_{sp}^2)}. \tag{29}$$

Here m_B is the known value of the B meson mass of 5.280 GeV and m_{sp} is the spectator quark mass. A fit to the shape of the lepton energy spectrum then is needed to determine the free parameters p_f, ϵ and m_{sp}. In turns out that one can fix m_{sp} and any latent dependence is absorbed by the other two. So a fit to the data will determine $\mathcal{B}_{sl}$, p_f and ϵ. In this way Altarelli et al. remove the explicit dependence of the m_b^5 term in the total decay rate. The ISGW and ISGW* models are also used. The resulting values are given in Table 8.

The representative value of $|V_{cb}|$ found from this analysis alone is

$$|V_{cb}| = 0.039 \pm 0.001 \pm 0.004. \tag{30}$$

Table 8. V_{cb} Values from Inclusive leptons

Model	Experiment	V_{cb}
ACM	CLEO I	0.042±0.002±0.004
ACM	ARGUS	0.039±0.001±0.003
ACM	CLEO II	0.040±0.001±0.004
ISGW	CLEO I	0.039±0.002±0.004
ISGW	ARGUS	0.039±0.001±0.005
ISGW	CLEO II	0.040±0.001±0.004
ISGW*	CLEO I	0.037±0.002±0.004
ISGW*	CLEO II	0.040±0.002±0.004

There are determinations of the inclusive B semileptonic branching ratio from LEP. These measurements average over more B species that at the $\Upsilon(4S)$. Since the lifetimes of some of these, especially the Λ_b appears to be shorter than for the ground state mesons, the semileptonic branching ratio measured at LEP should be lower than that measured on the $\Upsilon(4S)$, yet it is somewhat higher.[44] Since the measurement at LEP is far more complicated, I have chosen to leave out these results.

The results of using all four methods to find V_{cb} are shown in Fig. 17. It is remarkable that all four separate methods give such consistent results. Advocates for any particular method can choose among these results. I have chosen to average them. The errors are handled by adding the statistical and systematic errors on each method and then adding the different methods in quadrature. This should give a generous estimate of the final error. The average value of V_{cb} is 0.0381±0.0021, which gives a value for the CKM parameter

$$A = 0.784 \pm 0.043 \ . \tag{31}$$

2.4. The CKM element V_{ub}

The first evidence of a non-zero value of V_{ub} was obtained by CLEO I who saw a non-zero excess beyond the endpoint allowed for $B \to D\ell\nu$ transitions.[45] This result was quickly confirmed by ARGUS.[46] The latest evidence from CLEO II[47] is shown in Fig. 18. R_2 is the second Fox-Wolfram event shape variable,[48] which tends to zero for spherical events, such as $\Upsilon(4S)$ decays and to one for jet-like events. P_{miss} is the missing momentum in the event.

The branching ratios are small. CLEO finds that the rate in the lepton momentum interval $2.6 > p_\ell > 2.4$ GeV/c, $\mathcal{B}_u(p)$, is $(1.5 \pm 0.2 \pm 0.2) \times 10^{-4}$. To extract V_{ub} from this measurement we need to use theoretical models. It is convenient to define: $\Gamma(b \to u\ell\nu) = \gamma_u|V_{ub}|^2$, and $\Gamma(b \to c\ell\nu) = \gamma_c|V_{cb}|^2$. In addition, $f_u(p)$ is the fraction of the spectrum predicted in the end point region by different models, and $\mathcal{B}_{sl}$ is the

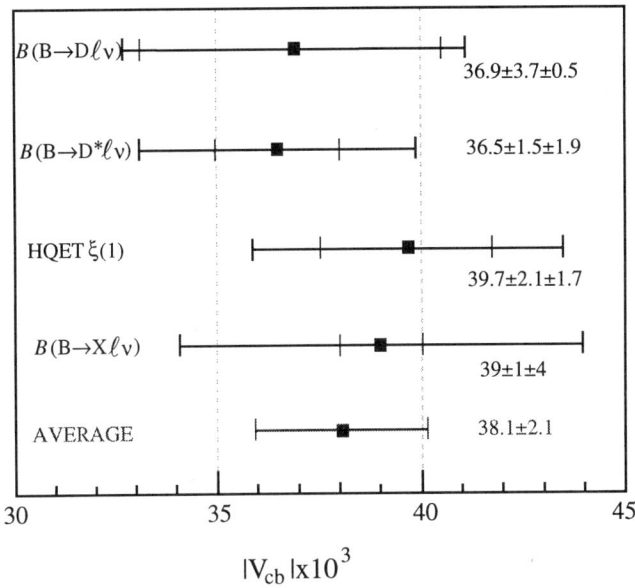

Fig. 17. Results of four different methods used to evaluate V_{cb}, and the resulting average. The horizontal lines show the values, the statistical errors out to the thin vertical lines, and the systematic errors added on linearly out to the thick vertical lines.

semileptonic branching ratio. Then:

$$\frac{|V_{ub}|^2}{|V_{cb}|^2} = \frac{\mathcal{B}_u(p)}{\mathcal{B}_{sl}} \cdot \frac{\gamma_c}{f_u(p)\gamma_u}. \tag{32}$$

These models disagree as to which final states populate the endpoint region. Most models agree roughly on values of γ_c. However, models differ greatly in the value of the product $\gamma_u \cdot f_u(p)$. There are two important reasons for these differences. First of all, different authors disagree as to the importance of the specific exclusive final states such as $\pi\ell\nu$, $\rho\ell\nu$ in the lepton endpoint region. For example, the Altarelli et al. model doesn't consider individual final states and thus can be seriously misleading if the endpoint region is dominated by only one or two final states. In fact, several inventors of exclusive models have claimed that the endpoint is dominated by only a few final states.[29,21] Secondly, even among the exclusive form-factor models there are large differences in the absolute decay rate predictions. This is illustrated in Fig. 19. The differences in the exclusive models are much larger in $b \to u$ transitions than in $b \to c$ transitions because the q^2 range is much larger.

Artuso has explicitly shown that the q^2 distributions were very different in the ACM and original ISGW model.[49] However, the new ISGW II model agrees much better with ACM (see Fig. 20).[50]

Measurement of exclusive charmless semileptonic decays can put constraints on

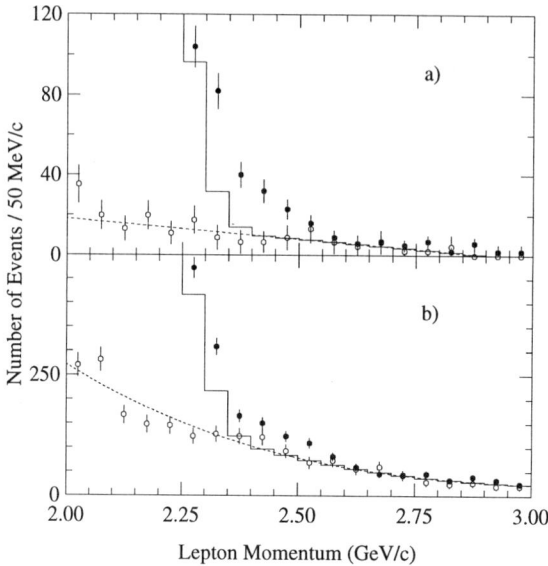

Fig. 18. Lepton yield versus momentum from CLEO II for the "strict" cut sample, $R_2 < 0.2$, $P_{miss} > 1$ GeV/c and the lepton and missing momentum direction point into opposite hemispheres, (a) and the $R_2 < 0.3$ sample (b). The filled points are from data taken on the peak of the $\Upsilon(4S)$, while the open points are continuum data scaled appropriately. The dashed curves are fits to the continuum data, while the solid histograms are predictions of the sum of $b \to c\ell\nu$ and continuum lepton production.

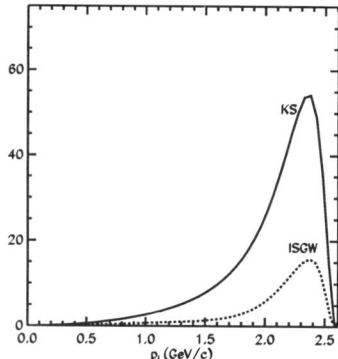

Fig. 19. Lepton momentum spectra, for $B \to \rho\ell\nu$ in the KS and the original ISGW model.

the models and therefore restrict the model dependence. In principle, the ratio of rates for $\pi\ell\nu$ and $\rho\ell\nu$ can be measured as well as the q^2 dependence of the form-factors. However, measurement of these rates is difficult. CLEO has recently succeeded in

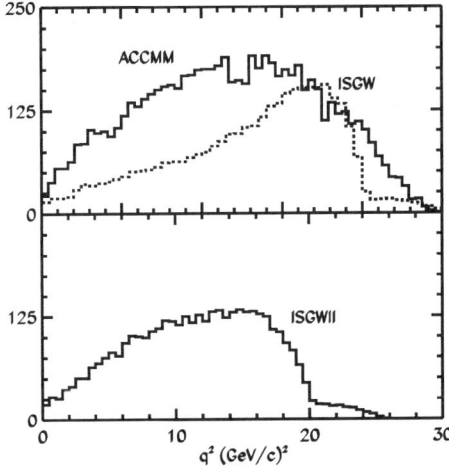

Fig. 20. q^2 distribution, for charmless semileptonic b decays in the model of Altarelli et al.(ACCMM) and the orginal ISGW model shown on top, and the new ISGW model shown on the bottom. The areas reflect the predicted widths, but the vertical scale is arbitrary. The high q^2 tails on the ISGW models arise from the $\pi\ell\nu$ final state.

measuring the branching ratios.[51]

A neutrino reconstruction technique is used. The neutrino energy and momentum is determined by evaluating the missing momentum and energy in the entire event:

$$E_{miss} = 2E_{beam} - \sum_i E_i \qquad (33)$$

$$\vec{p}_{miss} = \sum_i \vec{p}_i \ . \qquad (34)$$

Criteria are imposed to guard against events with false large missing energies. First, the net charge is required to be zero. Secondly, events with two identified leptons (implying two neutrinos) are rejected. Leptons are required to have momenta greater than 1.5 GeV/c in the case of $\pi\ell\nu$ and greater than 2.0 GeV/c in the case of $\rho\ell\nu$. In addition, the candidate neutrino mass is calculated as

$$M_\nu^2 = E_{miss}^2 - \vec{p}_{miss}^2 \ . \qquad (35)$$

Candidate events containing a neutrino are kept if $M_\nu^2/2E_{miss} < 300$ MeV. Then the semileptonic B decay candidates $(\pi^\circ, \pi^+, \rho^\circ, \omega^\circ, \rho^+)\ell\nu$ are reconstructed using the neutrino four-vector found from the missing energy measurement.[53] The beam constrained invariant mass, M_{cand} is defined as

$$M_{cand}^2 = E_{beam}^2 - \left(\vec{p}_\nu + \vec{p}_\ell + \vec{p}_{(\pi \text{ or } \rho)}\right)^2 \ , \qquad (36)$$

and with the use of the neutrino four-vector is essentially the same as any other full B reconstruction analysis done at the $\Upsilon(4S)$. The M_{cand} distributions are shown in Fig. 21.

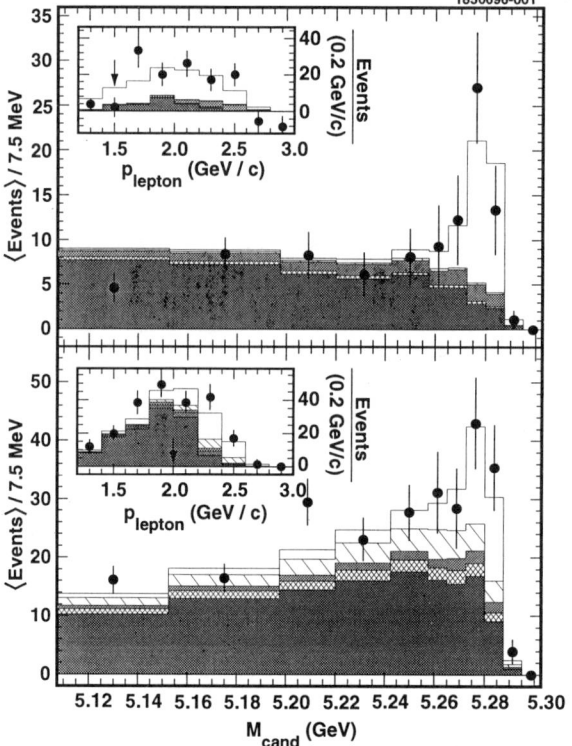

Fig. 21. The B candidate mass distributions, M_{cand}, for the sum of the scalar $\pi^+ \ell\nu$ and $\pi^0 \ell\nu$ (top) and the vector modes (ρ and ω) (bottom). The points are the data after continuum and fake background subtractions. The unshaded histogram is the signal, while the dark shaded shows the $b \to cX$ background estimate, the cross-hatched, estimated $b \to u\ell\nu$ feedown. For the π (vector) modes, the light-shaded and hatched histograms are $\pi \to \pi$ (vector→vector) and vector→ π (π →vector) crossfeed, respectively. The insets show the lepton momentum spectra for the events in the B mass peak (the arrows indicate the momentum cuts).

It is often difficult to prove that a $\pi\pi$ system indeed is dominantly resonant ρ. CLEO attempts to show ρ dominance by plotting the $\pi^+\pi^-$ and $\pi^+\pi^0$ summed mass spectrum in Fig. 22. They also show a test case of $\pi^0\pi^0\ell\nu$, which cannot be ρ, since ρ^0 cannot decay to $\pi^0\pi^0$. There is an enhancement in the $\pi^+\pi^-$ plus $\pi^+\pi^0$ sum, while the $\pi^0\pi^0$ shows a relatively flat spectrum that is explained by background. The 3π spectrum shows little evidence of resonant ω, however. More data is needed to settle this issue. CLEO proceeds by assuming they are seeing purely resonant decays in the vector channel.

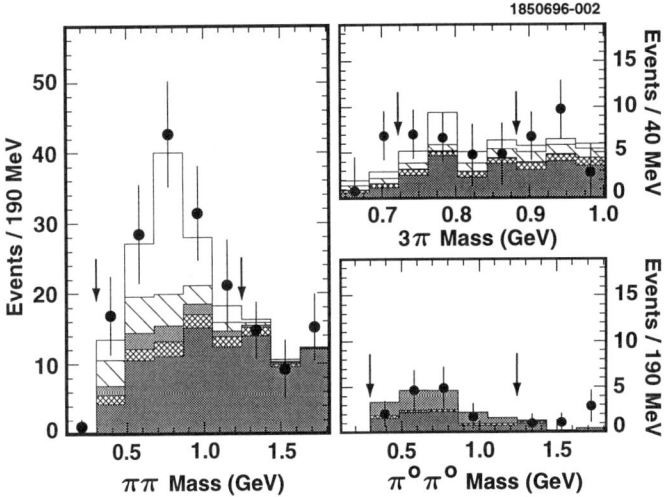

Fig. 22. Mass distributions for $\pi^+\pi^-$ plus $\pi^+\pi^o$ (left), 3π (upper right) and $\pi^o\pi^o$ (lower right), for events which are candidates $B \to x\ell\nu$ decays which satisfy all the other B candidate cuts including a cut on the B mass. The shading is the same as on the previous figure. The arrows indicate the mass range used in the analysis.

The measured branching ratio is model dependent due to different form-factor dependences on q^2 and lepton momentum. Therefore, CLEO reports different branching ratios for a selection of models. The ratio of $\rho\ell\nu/\pi\ell\nu$ is also given, see Table 9, and compared to model predictions; the errors are non-Gaussian, but the KS model has only a 0.5% likelihood of being consistent with the data.

The values of V_{ub} obtained from both the exclusive and the inclusive analyses are summarized in Fig. 23. For the inclusive analysis, results from CLEO I and ARGUS have been included in the average.[52] Since the KS model predicts the wrong pseudoscalar/vector ratio, it is excluded from the average. The ISGW model has been dropped in favor of the ISGW II model. The range of model predictions is now narrowed compared to former analyses. However, the model variations still dominate the error. A conservative estimate gives

$$\left|\frac{V_{ub}}{V_{cb}}\right| = 0.080 \pm 0.015 \ , \tag{37}$$

which provides a constraint

$$\left(\frac{1}{\lambda^2}\right)\left|\frac{V_{ub}}{V_{cb}}\right|^2 = \left(\rho^2 + \eta^2\right) = (0.36 \pm 0.07)^2 \ . \tag{38}$$

Table 9. Results from exclusive semileptonic $b \to u$ transistions

Model	$\mathcal{B}(B \to \pi\ell\nu)$ $\times 10^4$	$\mathcal{B}(B \to \rho\ell\nu)$ $\times 10^4$	$\Gamma(\rho)/\Gamma(\pi)$	$\Gamma(\rho)/\Gamma(\pi)$ predicted
ISGW II	$2.0 \pm 0.5 \pm 0.3$	$2.2 \pm 0.4^{+0.4}_{-0.6}$	$1.1^{+0.5+0.2}_{-0.3-0.3}$	1.47
WSB	$1.8 \pm 0.5 \pm 0.3$	$2.8 \pm 0.5^{+0.5}_{-0.8}$	$1.6^{+0.7+0.3}_{-0.5-0.4}$	3.51
KS	$1.9 \pm 0.5 \pm 0.3$	$1.9 \pm 0.3^{+0.4}_{-0.5}$	$1.0^{+0.5+0.2}_{-0.3-0.3}$	4.55
Melikhov†	$1.8 \pm 0.4 \pm 0.3 \pm 0.2$	$2.8 \pm 0.5^{+0.5}_{-0.8} \pm 0.4$	$1.6^{+0.7+0.3}_{-0.5-0.4} \pm 0.11$	1.53 ± 0.15

† The 3rd error arises from uncertainties in the estimated form-factors

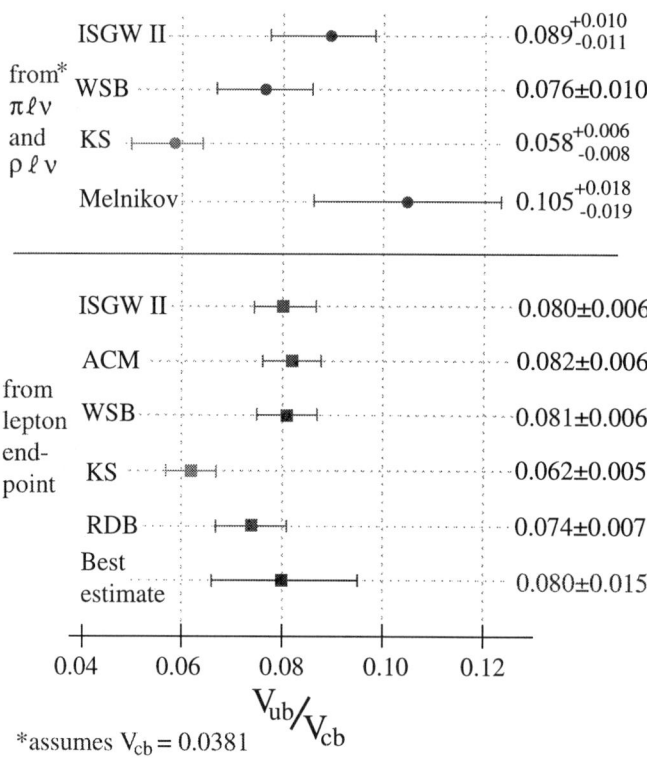

Fig. 23. Values of V_{ub}/V_{cb} obtained from the exclusive $\pi\ell\nu$ and $\rho\ell\nu$ analyses combined and taking $V_{cb} = 0.0381$, and results from the inclusive endpoint analysis. The best estimate combining all models except KS is also given.

2.5. $B_d^o - \bar{B}_d^o$ Mixing

Neutral B mesons can transform to their anti-particles before they decay. The diagrams for this process are shown in Fig. 24. Although u, c and t quark exchanges are all shown, the t quark plays a dominant role mainly due to its mass, as the amplitude of this process is proportional to the mass of the exchanged fermion. (We will discuss the phenomenon of mixing in more detail in section 3.2).

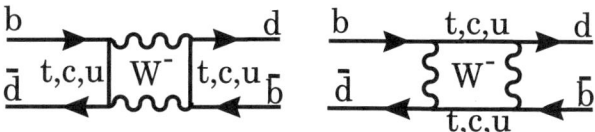

Fig. 24. The two diagrams for B_d mixing.

The probability of mixing is given by[54]

$$x \equiv \frac{\Delta m}{\Gamma} = \frac{G_F^2}{6\pi^2} B_B f_B^2 m_B \tau_B |V_{tb}^* V_{td}|^2 m_t^2 F\left(\frac{m_t^2}{M_W^2}\right) \eta_{QCD}, \quad (39)$$

where B_B is a parameter related to the probability of the d and $\bar{b}$ quarks forming a hadron and must be estimated theoretically, F is a known function which increases approximately as m_t^2, and η_{QCD} is a QCD correction, with value about 0.8. By far the largest uncertainty arises from the unknown decay constant, f_B. B_d mixing was first discovered by the ARGUS experiment.[55] (There was a previous measurement by UA1 indicating mixing for a mixture of B_d^o and B_s^o.[56] At the time it was quite a surprise, since m_t was thought to be in the 30 GeV range. Since

$$|V_{tb}^* V_{td}|^2 \propto |(1-\rho-i\eta)|^2 = (\rho-1)^2 + \eta^2, \quad (40)$$

measuring mixing gives a circle centered at (1,0) in the $\rho - \eta$ plane.

The best recent mixing measurements have been done at LEP, where the time-dependent oscillations have been measured. The OPAL data[57] is shown in Fig. 25. Averaging over all LEP experiments x=0.728±0.025.[58]

2.6. Rare B Decays

The term "rare B decays" is loosely defined. The spectator process shown in Fig. 26(a) is included since $b \to u$ doesn't occur very often ($\approx 1\%$), and the mixing process which occurs often($\approx 17\%$) is included since it involves two gauge bosons (the so called box diagram Fig. 26(b)). Other loop or box diagrams are shown in Fig. 26(d-f).

CLEO found the first unambiguous loop process, the one shown in Fig. 26(c).[59] These decays involving a loop diagram are sometimes called "penguins," an indefensible if amusing term that was injected into the literature as a result of a bet. For the Standard Model to be correct these decays must exist. In fact, penguins are expected

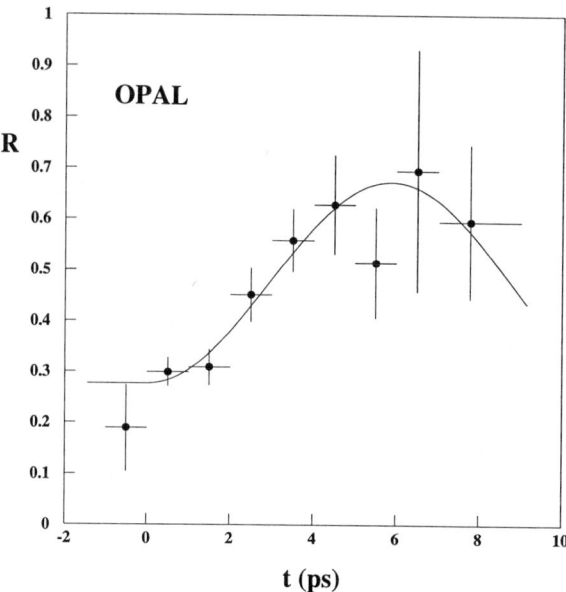

Fig. 25. The ratio, R, of like-sign to total events as a function of proper decay time, for selected $B \to D^{*+}X\ell^-\bar{\nu}$ events. The jet charge in the opposite hemisphere is used to determine the sign correlation. The curve is the result of a fit to the mixing parameter.

to play an important role in kaon decay, but there are no unique penguin final states in kaon decay. Since penguins are expected to be quite small in charm decay, it is only in B decay that penguins can clearly be discerned.

CLEO first found the exclusive final state $B \to K^*\gamma$. An updated value for the branching ratio is[60]

$$\mathcal{B}(B \to K^*\gamma) = (4.2 \pm 0.8 \pm 0.6) \times 10^{-5} \ . \tag{41}$$

This analysis uses the standard B reconstruction technique, summarized in equation (11) used at the $\Upsilon(4S)$, combined with some additional background suppression cuts. These are separated into trying to insure that one is dealing with a real K^* and trying to supress background leading to hard photons. The latter comes from initial state radiation (ISR), where one of the beams radiates a photon and then subsequently annihilates and from continuum quark-antiquark production ($Q\bar{Q}$). Suppression of ISR and $Q\bar{Q}$ is accomplished by combining event shape variables into a Fischer discriminant. A Fischer discriminant[61] is a linear combination of several variables which individually may have poor separation between signal and background, but when taken together yield acceptable background rejection, the correlations between the variables helping. The Fischer output distribution for Monte Carlo simulations of signal, ISR and $Q\bar{Q}$ backgrounds are shown in Fig. 27.

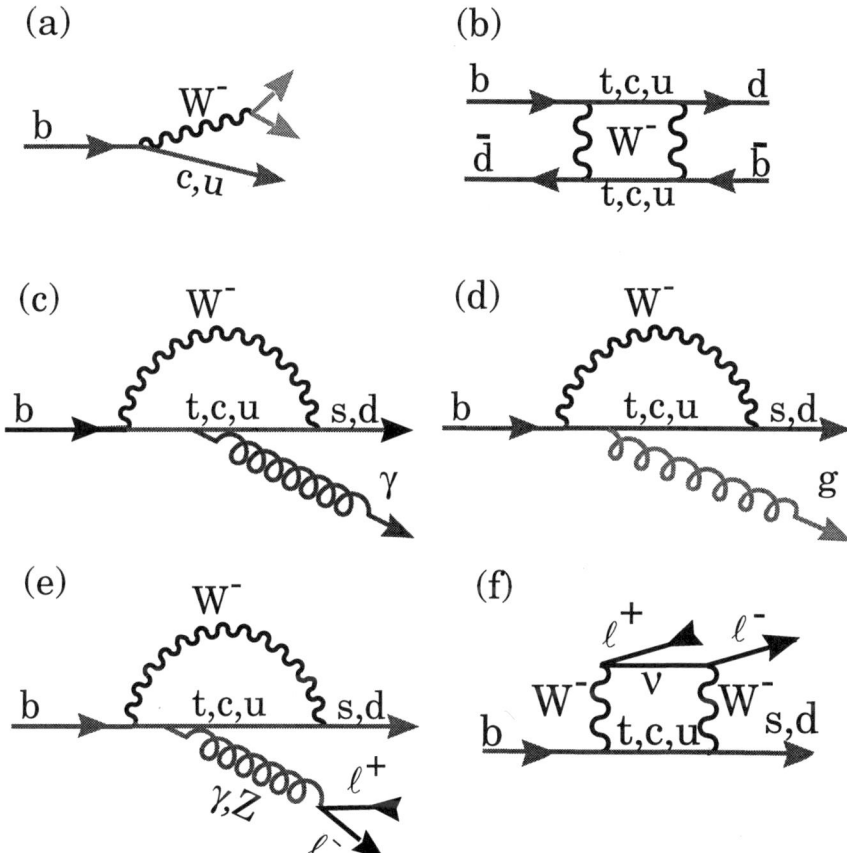

Fig. 26. A compendium of rare b decay diagrams. (a) The spectator diagram, rare when $b \to u$; (b) one of the mixing diagrams; (c) a radiative penguin diagram; (d) a gluonic penguin diagram; (e) and (f) are dilepton penguin diagrams.

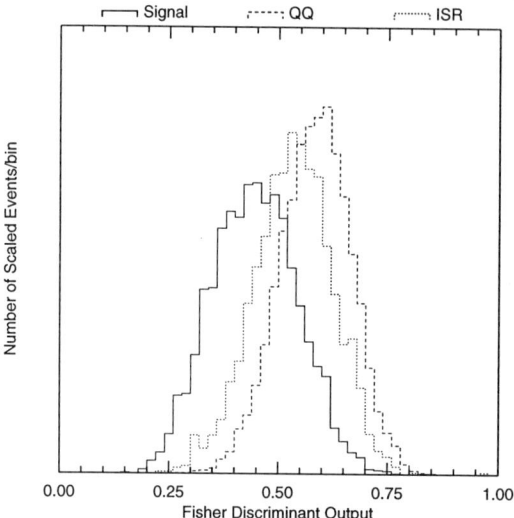

Fig. 27. The distribution of the Fischer discriminant output for Monte Carlo samples of $B^o \to K^{*o}\gamma(K^{*o} \to K^+\pi^-)$ signal, $Q\bar{Q}$ and ISR backgrounds. The histograms have equal area and the x axis has been rescaled to make the Fischer discriminant output lie between 0 and 1.

The branching ratio is extracted by making a maximum likelihood fit to four distributions, M_B, ΔE, the $K\pi$ invariant mass $m(K\pi)$, and the Fischer discriminant. To illustrate what the signal shapes look like, projection plots are made by applying restrictive selection criteria on three of the four likelihood variables and projecting the remaining events onto the axis of the fourth variable. This is shown for the $K^{*o} \to K^-\pi^+$ mode in Fig. 28.

The extraction of the inclusive rate for $b \to s\gamma$ is more difficult. There are two separate CLEO analyses.[62] The first one measures the inclusive photon spectrum from B decay near the maximum momentum end, similar to what is done to extract an inclusive $b \to X\ell\nu$ signal, but with the additional problem that the expected branching ratio is much lower. The main problem is to reduce the ISR and $Q\bar{Q}$ backgrounds. Here instead of using a Fischer discriminant, a set of event shape variables and energies formed in a series of cones parallel and antiparallel to the candidate photon direction are fed into a neural net trained on Monte Carlo. The result is shown in Fig. 29(leftside).

The second technique constructs the inclusive rate by summing up the possible exclusive final states. Since the photons are expected to be at high momentum, and therefore take away up to half the B's rest energy, the number of hadrons in the final state is quite limited. The analysis looks for the final states $B \to K$ n$\pi\gamma$ where n is allowed to be a maximum of 4, but only one can be a π^o. Only one entry per event

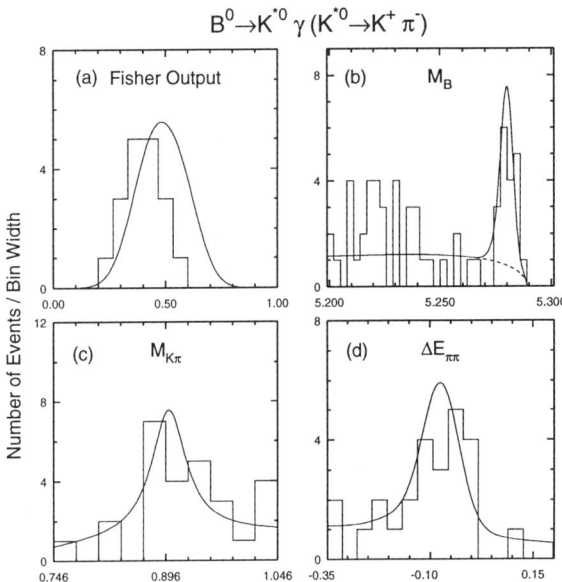

Fig. 28. Projections of $B^o \to K^{*o}\gamma(K^{*o} \to K^+\pi^-)$ data events (histograms) and maximum likelihood fits (curves) onto the four fit variables: (a) Fischer discriminant output, (b) M_B, (c) $M_{K\pi}$ and (d) $\Delta E_{\pi\pi}$, which is the difference between the candidate B energy and the beam energy assuming both charged tracks are pions.

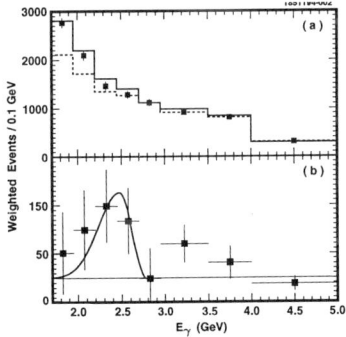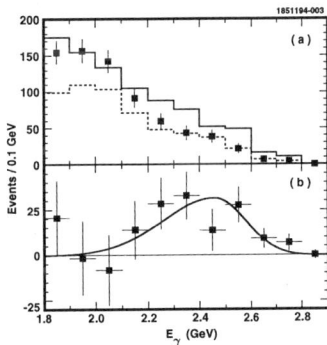

Fig. 29. Photon energy spectra from the neural net analysis shown on the left side, and from the B reconstruction analysis, shown on the right side. In (a) the on resonance date are the solid lines, the scaled off resonance data are the dashed lines, and the sum of backgrounds from off resonance data and $b \to c$ Monte Carlo are shown as the square points with error bars. In (b) the backgrounds have been subtracted to show the net signal for $b \to s\gamma$; the solid lines are fits of the signal using a spectator model prediction.

is allowed. Here background reduction is accomplished by using the full power of the exclusive B reconstruction analysis. The resulting γ energy spectrum is shown on the right side of Fig. 29.

The branching ratios found are $(1.88 \pm 0.74) \times 10^{-4}$ and $(2.75 \pm 0.67) \times 10^{-4}$ for the neural net and B reconstruction analyses, respectively. The average of the two results, taking into account the correlations between the two techniques is

$$\mathcal{B}(b \to s\gamma) = (2.3 \pm 0.5 \pm 0.4) \times 10^{-4} \ . \tag{42}$$

The theoretical prediction for the branching ratio is given by[63]

$$\frac{\Gamma(b \to s\gamma)}{\Gamma(b \to c\ell\nu)} = \left|\frac{V_{ts}^* V_{tb}}{V_{cb}}\right| \frac{\alpha}{6\pi g(m_c/m_b)} |C_7^{eff}(\mu)|^2, \tag{43}$$

where $g(m_c/m_b)$ is a known function. While C_7 is calculated perturbatively at μ equal to the W mass, the evolution to b mass scale causes $\approx 25\%$ uncertainty in the prediction, since the proper point could be $m_b/2$ or $2m_b$. In the leading log approximation the theoretical prediction is $\mathcal{B}(b \to s\gamma) = (2.8 \pm 0.8) \times 10^{-4}$,[63] while an incomplete next to leading order calculation, gives $\sim 1.9 \times 10^{-4}$.[64] A recently completed next to leading order calculation gives 3.3×10^{-4}.[65] In all cases the data are consistent with the prediction.

The second analysis also produces the mass spectrum of the $K\ n\pi$ system, shown in Fig. 30. A clear $K^*(890)$ component is observed. The best way to measure the fraction of $K^*(890)$ is to divide the exclusive result by the average inclusive result. This number can test theoretical models, but mostly we are testing the prediction of the exclusive rate which is the far more difficult calculation than the inclusive rate.

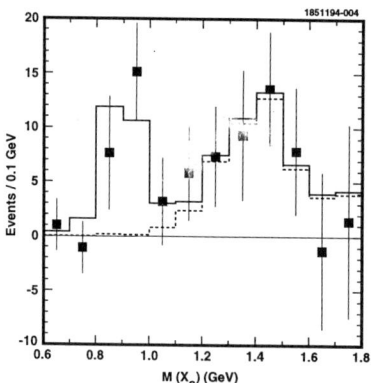

Fig. 30. The apparent $K\ n\pi$ mass distribution for the B reconstruction analysis. The points are the background subtracted data, not efficiency corrected, the solid histogram is fit to the data using several K^* resonance as input to a Monte Carlo simulation, while the dotted histogram shows all the fit components but the $K^*(890)$.

The CLEO result is[60]

$$\frac{\Gamma(B \to K^*\gamma)}{\Gamma(b \to s\gamma)} = 0.181 \pm 0.068 \ . \tag{44}$$

Model predictions vary between 4 and 40%.[68]

Rare hadronic final states have also been measured. CLEO reported a signal in the sum of $K^\pm\pi^\mp$ and $\pi^+\pi^-$ final states.[66] The particle identification could not uniquely separate high momentum kaons and pions. While the $K\pi$ mode results from a penguin diagram the $\pi\pi$ mode results mainly from a $b \to u$ spectator diagram. The reconstructed B mass plot is shown in Fig. 31, along with the results of several other searches from an updated analysis,[67] based on 2.4 fb−1 of integrated luminosity on the $\Upsilon(4S)$. Here a best guess is made as to which final state is present. The resulting rate is

$$\mathcal{B}(B^o \to K^\pm\pi^\mp + \pi^+\pi^-) = (1.8^{+0.6+0.2}_{-0.5-0.3} \pm 0.2) \times 10^{-5} \ . \tag{45}$$

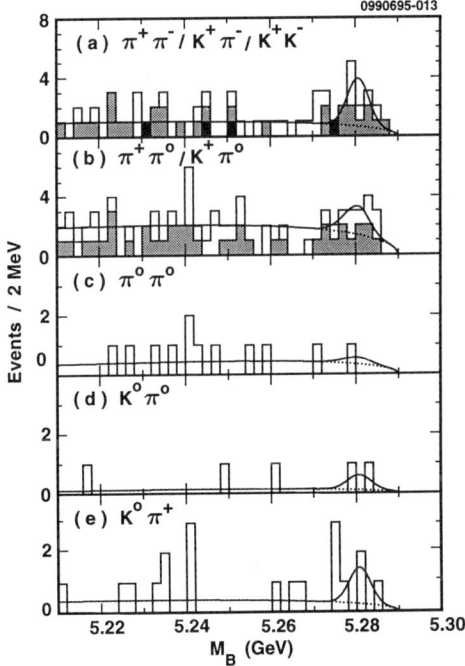

Fig. 31. M_B plots for (a) $B^o \to \pi^+\pi^-$ (unshaded), $B^o \to K^+\pi^-$, and $B^o \to K^+K^-$ (black) (b) $B^+ \to \pi^+\pi^o$ (unshaded) and $B^+ \to K^+\pi^o$ (grey), (c) $B^o \to \pi^o\pi^o$, (d) $B^o \to K^o\pi^o$, and (e) $B^+ \to K^o\pi^+$. The projection of the total likelihood fit (solid curve) and the continuum background component (dotted curve) are overlaid.

An attempt to separate the kaon and pion components using the small difference in reconstructed energy and whatever particle identification power exists leads to the dipion fraction shown in Fig. 32. The best current guess is that approximately half of the rate is due to $\pi^+\pi^-$.

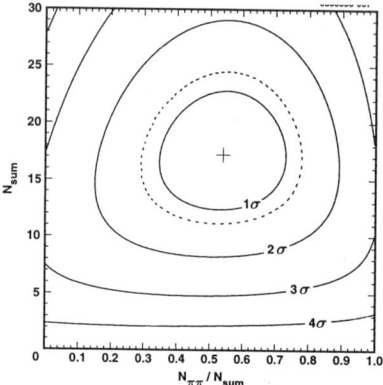

Fig. 32. The central value (+) of the likelihood fit to $N_{sum} \equiv N_{\pi\pi} + N_{K\pi}$ and the fraction $N_{\pi\pi}/N_{sum}$. The solid curves are the $n\sigma$ contours and the dotted curve is the 1.28σ contour.

CLEO also has found a signal in the sum of $\omega\pi^+$ and ωK^+ decays.[69] The B mass plot is shown in Fig. 33. The signal is 10 events observed on a background of 2 ± 0.3 events. The branching ratio is

$$\mathcal{B}(B^+ \to \omega\pi^+ + \omega K^+) = (2.8 \pm 1.0 \pm 0.5) \times 10^{-5} \ . \tag{46}$$

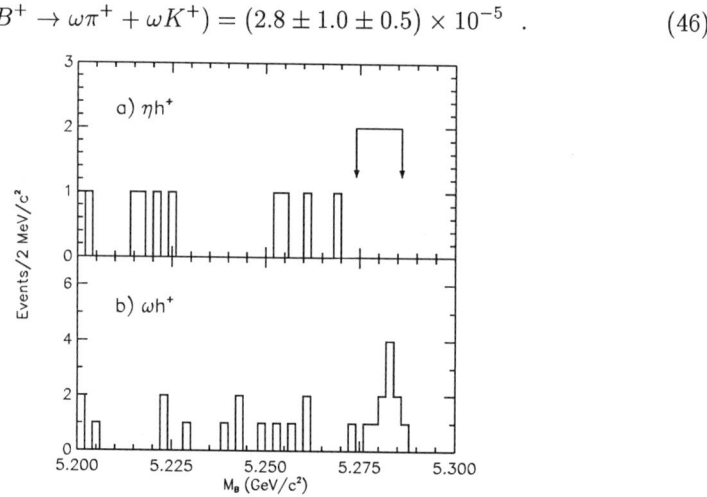

Fig. 33. The M_B projection for a) $B^+ \to \eta h^+$ and b) $B^+ \to \omega h^+$ after all other cuts, including the ΔE cut. The arrows indicate the signal region.

DELPHI also reports a signal of 11 "rare" events over a background of 1 event. The invariant mass plot is shown in Fig. 34.[70] One of these events appears to be

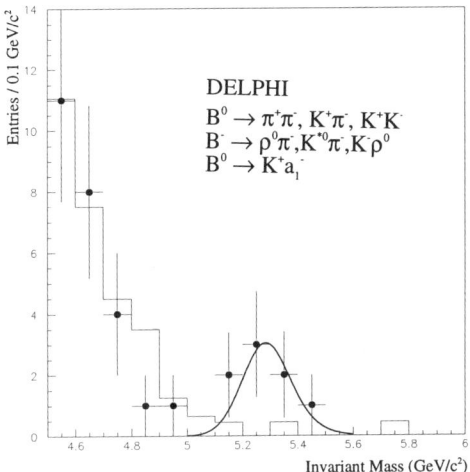

Fig. 34. Invariant mass distribution for two-body charmless hadronic B decays. The points with error bars represent the real data and the histograms the mass distributions expected in the absence of such decays as obtained from simulation. The curve represents the shape expected for the signal events normalized to the number of candidates selected in real data in the signal mass region.

uniquely identified as a $K^{*o}\pi^-$ final state and this then is an unambiguous hadronic penguin decay. The evidence is shown in Fig 35.

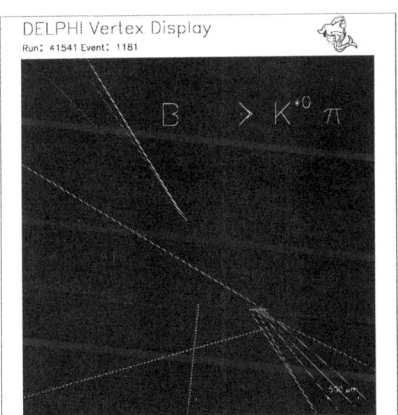
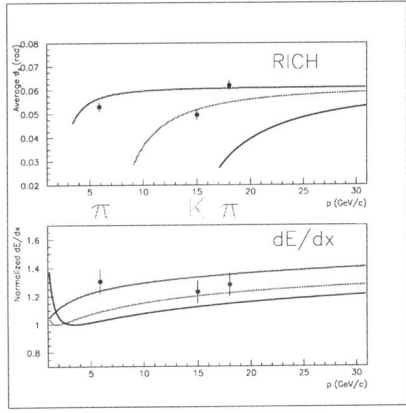

Fig. 35. The candidate $B^- \to K^{*o}\pi^-$ decay: a magnified view of the extrapolated tracks at the vertex is displayed above. The primary and secondary vertices are indicated by error ellipses corresponding to 3σ regions. The plot below summarizes the hadron identification properties. The lines represent the expected response to pions (upper), kaons (middle) and protons (lower), and the points with error bars the measured values for the reconstructed B decay products.

3. Importance of Further Study of B Decays

3.1. Tests of the Standard Model via the CKM triangle

The unitarity of the CKM matrix[†] allows us to construct six relationships. The most useful turns out to be

$$V_{ud}V_{td}^* + V_{us}V_{ts}^* + V_{ub}V_{tb}^* = 0 \ . \tag{47}$$

To a good approximation

$$V_{ud} \approx V_{tb}^* \approx 1 \text{ and } V_{ts}^* \approx -V_{cb}, \tag{48}$$

then

$$\frac{V_{ub}}{V_{cb}} + \frac{V_{td}^*}{V_{cb}} - V_{us} = 0 \ . \tag{49}$$

Since $V_{us} = \lambda$, we can define a triangle with sides

$$1 \tag{50}$$

$$\left|\frac{V_{td}}{A\lambda^3}\right| = \frac{1}{\lambda}\sqrt{(\rho-1)^2 + \eta^2} = \frac{1}{\lambda}\left|\frac{V_{td}}{V_{ts}}\right| \tag{51}$$

$$\left|\frac{V_{ub}}{A\lambda^3}\right| = \frac{1}{\lambda}\sqrt{\rho^2 + \eta^2} = \frac{1}{\lambda}\left|\frac{V_{ub}}{V_{cb}}\right|. \tag{52}$$

The CKM triangle is depicted in Fig. 36. We know two sides already: the base is

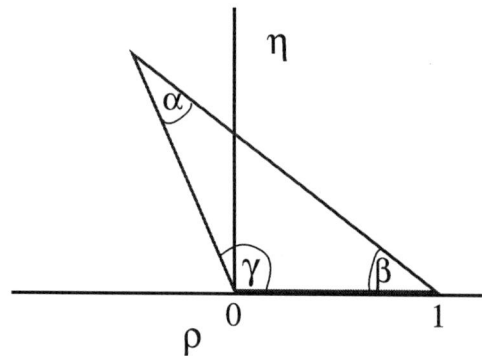

Fig. 36. The CKM triangle shown in the $\rho - \eta$ plane. The left side is determined by $|V_{ub}/V_{cb}|$ and the right side can be determined using mixing in the neutral B system. The angles can be found by making measurements of CP violation in B decays.

[†]Unitarity implies that any pair of rows or columns are orthogonal.

defined as unity and the left side is determined by the measurements of $|V_{ub}/V_{cb}|$. The right side can be determined using mixing measurements in the neutral B system. We will see, however, that there is a large error due to the uncertainty in f_B. Later we will discuss other measurements that can access this side. The figure also shows the angles as α, β, and γ. These angles can be determined by measuring CP violation in the B system. First we discuss CP violation in the K_L^o system which also provides constraints on ρ and η.

To test the Standard Model we can measure all three sides and all three angles. If we see consistency between all of these measurements we have defined the parameters of the Standard Model. If we see inconsistency, the breakdown can point us beyond the Standard Model.

3.2. CP Violation

The fact that the CKM matrix is complex allows CP violation. This is not only true for three generations of quark doublets, but for any number greater than two. Now let us explain what we mean by CP violation. C is a quantum mechanical operator that changes particle to antiparticle, while P switches left to right, i.e. $x \to -x$. Thus under a P operation, $\vec{p} \to -\vec{p}$ since t is unaffected.

Examples of CP violation have been found in the K^o system. Let us examine one such measurement. Consider the K^o to be composed of long lived and short lived components having equal weight, so the wave function is

$$|K^o\rangle = \frac{1}{\sqrt{2}}\left(|K_S\rangle + |K_L\rangle\right) \ . \tag{53}$$

In the case of neutral kaons there is a large difference in lifetimes between the short lived and long lived components. The lifetimes are 9×10^{-11} sec and 5×10^{-8} sec. Suppose we set up a detector far away from the K^o production target. Then after the K_S decay away we have only a K_L beam. We find both

$$K_L \to e^+ \nu_e \pi^- \text{ and } K_L \to e^- \bar{\nu}_e \pi^+ \tag{54}$$

are present. Now the initial state was a K^o, which contains an $\bar{s}$ quark and can only decay semileptonically into the $e^+ \nu_e \pi^-$ final state as shown in Fig. 37. Thus we have found evidence that both K^o and $\overline{K^o}$ are present. This phenomenon, $K^o \Leftrightarrow \overline{K^o}$ is called mixing and can be depicted by the diagram shown in Fig. 38, much like the diagram for $B^o \overline{B^o}$ mixing. However, here the c-quark loop has the largest amplitude, unlike the B case, where the t-quark is dominant. (This is because the CKM couplings are so much larger, i.e. V_{cs} and $V_{cd} \gg V_{ts}$ and V_{td} and this compensates for the decrease due to $(m_c/m_t)^2$.) There are also hadronic intermediate states which contribute to the real part of the mixing amplitude, such as $K^o \to \pi\pi \to \overline{K}^o$.

An example of CP violation is the measured rate asymmetry in our K_L^o detector[5]

$$\delta = 2Re(\epsilon) = \frac{\#(K_L \to e^+ \nu_e \pi^-) - \#(K_L \to e^- \bar{\nu}_e \pi^+)}{\#(K_L \to e^+ \nu_e \pi^-) + \#(K_L \to e^- \bar{\nu}_e \pi^+)} = 3.3 \times 10^{-3} \ . \tag{55}$$

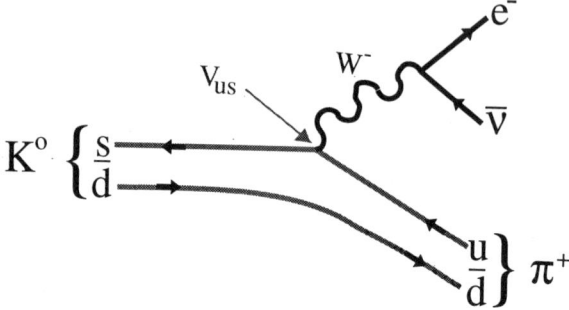

Fig. 37. Semileptonic decay of a s quark contained in a K^o meson.

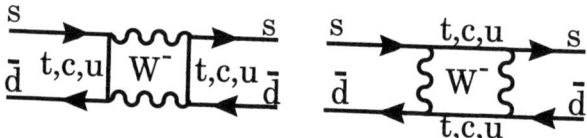

Fig. 38. $K^o - \bar{K}^o$ mixing diagrams.

Let us look at why this violates CP. In Fig. 39 the momentum and spin vectors for the two final states are shown. The CP operation transforms the $e^+\nu_e\pi^-$ to the $e^-\bar{\nu}_e\pi^+$ final state and vice-versa. Thus CP invariance would imply equal rates for the two processes, contrary to what is observed.

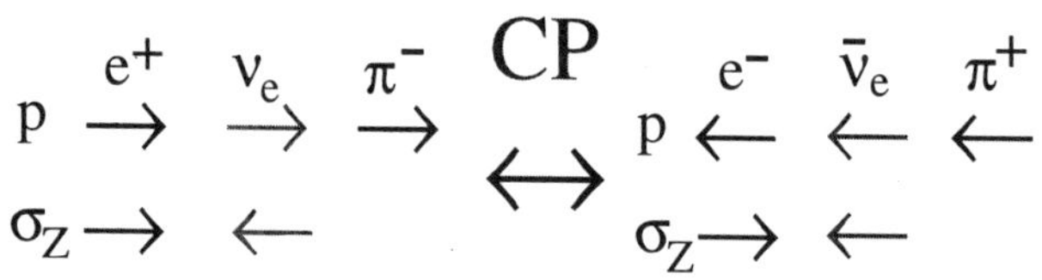

$$p \xrightarrow{} \; e^+ \xrightarrow{} \; \nu_e \xrightarrow{} \; \pi^- \xrightarrow{} \quad \overset{CP}{\longleftrightarrow} \quad p \xleftarrow{} \; e^- \xleftarrow{} \; \bar{\nu}_e \xleftarrow{} \; \pi^+$$
$$\sigma_z \rightarrow \quad \leftarrow \qquad\qquad\qquad \sigma_z \rightarrow \quad \leftarrow$$

Fig. 39. The momentum and spin orientations of the two final states in semileptonic K_L^o decay, showing that they are mapped into one another by a CP transformation.

CP violation thus far has only been seen in the neutral kaon system.[‡] If we can find CP violation in the B system we could see if the CKM model works or perhaps go beyond the model. Speculation has it that CP violation is responsible for the baryon-antibaryon asymmetry in our section of the Universe. If so, to understand the mechanism of CP violation is critical in our conjectures of why we exist.[71]

There is a constraint on ρ and η given by the K_L^o CP violation measurement (ϵ), given by[72]

$$\eta \left[(1-\rho)A^2(1.4 \pm 0.2) + 0.35\right] A^2 \frac{B_K}{0.75} = (0.30 \pm 0.06), \qquad (56)$$

where the errors arise from uncertainties on m_t and m_c. The constraints on ρ versus η from the V_{ub}/V_{cb} measurement, ϵ and B mixing are shown in Fig. 40. The width of the B mixing band is caused mainly by the uncertainty on f_B, taken here as $240 > f_B > 160$ MeV. The width of the ϵ band is caused by errors in A, m_t, m_c and B_K. The size of these error sources is shown in Fig. 41. The largest error still comes from the measurement of V_{cb}, with the theoretical estimate of B_K being a close second. The errors on m_t and m_c are less important.

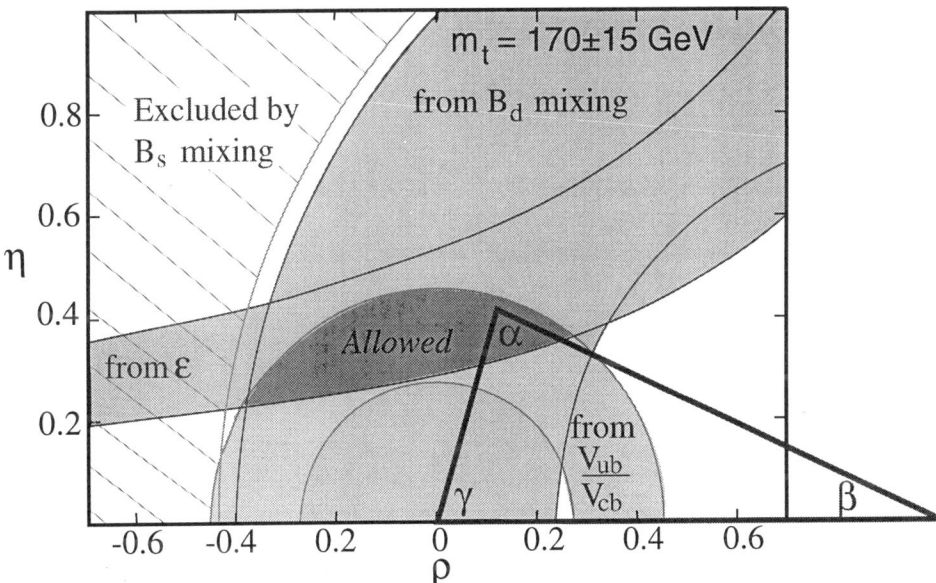

Fig. 40. The regions in $\rho - \eta$ space (shaded) consistent with measurements of CP violation in K_L^o decay (ϵ), V_{ub}/V_{cb} in semileptonic B decay, B_d^o mixing, and the excluded region from limits on B_s^o mixing. The allowed region is defined by the overlap of the 3 permitted areas, and is where the apex of the CKM triangle sits.

[‡]The other observed example of CP violation is the decay $K_L^o \to \pi\pi$.

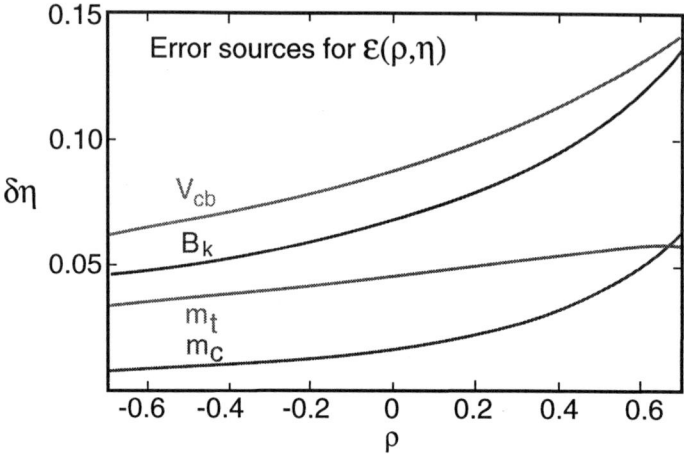

Fig. 41. Error sources in units of $\delta\eta$ on the value of η as a function of ρ provided by the CP violation constraint in K_L^o decay.

3.3. Ways of Measuring CP violation in B Decays

3.3.1. CP Violation in Charged B Decays

The theoretical basis of the study of CP violation in B decays was given in series of papers by Carter and Sanda and Bigi and Sanda.[73] We start with charged B decays. Consider the final states $f^\pm$ which can be reached by two distinct weak processes $\mathcal{A}$ and $\mathcal{B}$. Then the strong (s) and weak (w) parts are

$$\mathcal{A} = a_s e^{i\theta_s} a_w e^{i\theta_w}, \quad \mathcal{B} = b_s e^{i\delta_s} b_w e^{i\delta_w} . \tag{57}$$

Under the CP operation the strong phases remain constant but the weak phases change sign, so

$$\overline{\mathcal{A}} = a_s e^{i\theta_s} a_w e^{-i\theta_w}, \quad \overline{\mathcal{B}} = b_s e^{i\delta_s} b_w e^{-i\delta_w} . \tag{58}$$

The rate difference is

$$\Gamma - \overline{\Gamma} = |\mathcal{A} + \mathcal{B}|^2 - |\overline{\mathcal{A}} + \overline{\mathcal{B}}|^2 \tag{59}$$
$$= 2 a_s a_w b_s b_w \sin(\delta_s - \theta_s) \sin(\delta_w - \theta_w) . \tag{60}$$

A weak phase difference is guaranteed in the appropriate decay mode (different CKM phases), but the strong phase difference is not; it is very difficult to predict the magnitude of strong phase differences.

As an example consider the possibility of observing CP violation by measuring a rate difference between $B^- \to K^- \pi^o$ and $B^+ \to K^+ \pi^o$. The $K^- \pi^o$ final state can be

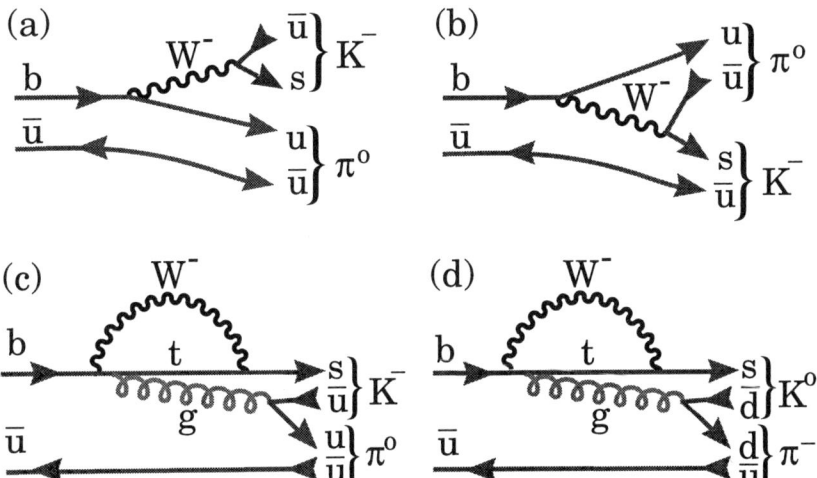

Fig. 42. Diagrams for $B^- \to K^-\pi^o$, (a) and (b) are tree level diagrams where (b) is color suppressed; (c) is a penguin diagram. (d) shows $B^- \to K^o\pi^-$, which cannot be produced via a tree diagram.

reached either by tree or penguin diagrams as shown in Fig. 42. The tree diagram has an imaginary part coming from the V_{ub} coupling, while the penguin term does not, thus insuring a weak phase difference. This type of CP violation is called "direct." Note also that the process $B^- \to K^o\pi^-$ can only be produced by the penguin diagram in Fig. 42(d). Therefore, we do not expect a rate difference between $B^- \to K^o\pi^-$ and $B^+ \to K^o\pi^+$.

3.3.2. Formalism in neutral B decays

Consider the operations of C and P:

$$C|B(\vec{p})\rangle = |\overline{B}(\vec{p})\rangle, \qquad C|\overline{B}(\vec{p})\rangle = |B(\vec{p})\rangle \qquad (61)$$
$$P|B(\vec{p})\rangle = -|B(-\vec{p})\rangle, \qquad P|\overline{B}(\vec{p})\rangle = -|B(-\vec{p})\rangle \qquad (62)$$
$$CP|B(\vec{p})\rangle = -|\overline{B}(-\vec{p})\rangle, \qquad CP|\overline{B}(\vec{p})\rangle = -|B(-\vec{p})\rangle \ . \qquad (63)$$

For neutral mesons we can construct the CP eigenstates

$$|B_1^o\rangle = \frac{1}{\sqrt{2}}\left(|B^o\rangle - |\overline{B}^o\rangle\right) , \qquad (64)$$

$$|B_2^o\rangle = \frac{1}{\sqrt{2}}\left(|B^o\rangle + |\overline{B}^o\rangle\right) , \qquad (65)$$

where

$$CP|B_1^o\rangle = |B_1^o\rangle , \qquad (66)$$

$$CP|B_2^o\rangle = -|B_2^o\rangle. \qquad (67)$$

Since B^o and $\overline{B}^o$ can mix, the mass eigenstates are a superposition of $a|B^o\rangle + b|\overline{B}^o\rangle$ which obey the Schrodinger equation

$$i\frac{d}{dt}\begin{pmatrix} a \\ b \end{pmatrix} = H\begin{pmatrix} a \\ b \end{pmatrix} = \left(M - \frac{i}{2}\Gamma\right)\begin{pmatrix} a \\ b \end{pmatrix}. \qquad (68)$$

If CP is not conserved then the eigenvectors, the mass eigenstates $|B_L\rangle$ and $|B_H\rangle$, are not the CP eigenstates but are

$$|B_L\rangle = p|B^o\rangle + q|\overline{B}^o\rangle, \quad |B_H\rangle = p|B^o\rangle - q|\overline{B}^o\rangle, \qquad (69)$$

where

$$p = \frac{1}{\sqrt{2}}\frac{1+\epsilon_B}{\sqrt{1+|\epsilon_B|^2}}, \quad q = \frac{1}{\sqrt{2}}\frac{1-\epsilon_B}{\sqrt{1+|\epsilon_B|^2}}. \qquad (70)$$

CP is violated if $\epsilon_B \neq 0$, which occurs if $|q/p| \neq 1$.

The time dependence of the mass eigenstates is

$$|B_L(t)\rangle = e^{-\Gamma_L t/2}e^{im_L t/2}|B_L(0)\rangle \qquad (71)$$
$$|B_H(t)\rangle = e^{-\Gamma_H t/2}e^{im_H t/2}|B_H(0)\rangle, \qquad (72)$$

leading to the time evolution of the flavor eigenstates as

$$|B^o(t)\rangle = e^{-(im+\frac{\Gamma}{2})t}\left(\cos\frac{\Delta mt}{2}|B^o(0)\rangle + i\frac{q}{p}\sin\frac{\Delta mt}{2}|\overline{B}^o(0)\rangle\right) \qquad (73)$$

$$|\overline{B}^o(t)\rangle = e^{-(im+\frac{\Gamma}{2})t}\left(i\frac{p}{q}\sin\frac{\Delta mt}{2}|B^o(0)\rangle + \cos\frac{\Delta mt}{2}|\overline{B}^o(0)\rangle\right), \qquad (74)$$

where $m = (m_L + m_H)/2$, $\Delta m = m_H - m_L$ and $\Gamma = \Gamma_L \approx \Gamma_H$. Note, that the probability of a B^o decay as a function of t is given by $\langle B^o(t)|B^o(t)\rangle^*$, and is a pure exponential, $e^{-\Gamma t/2}$, in the absence of CP violation.

3.3.3. Indirect CP violation in the neutral B system

As in the example described earlier for K_L decay, we can look for the rate asymmetry

$$a_{sl} = \frac{\Gamma\left(\overline{B}^o(t) \to X\ell^+\nu\right) - \Gamma\left(B^o(t) \to X\ell^-\overline{\nu}\right)}{\Gamma\left(\overline{B}^o(t) \to X\ell^+\nu\right) + \Gamma\left(B^o(t) \to X\ell^-\overline{\nu}\right)} \qquad (75)$$

$$= \frac{1 - \left|\frac{q}{p}\right|^4}{1 + \left|\frac{q}{p}\right|^4} \approx O\left(10^{-2}\right). \qquad (76)$$

These final states occur only through mixing as the direct decay occurs only as $B^o \to X\ell^+\nu$. To generate CP violation we need an interference between two diagrams. In this case the two diagrams are the mixing diagram with the t-quark and the mixing diagram with the c-quark quark. This is identical to what happens in the K_L^o case. This type of CP violation is called "indirect." The small size of the expected asymmetry is caused by the off diagonal elements of the Γ matrix in equation (68) being very small compared to the off diagonal elements of the mass matrix, i.e. $|\Gamma_{12}/M_{12}| \ll 1$. This results from the nearly equal widths of the B_L^o and B_H^o.[74]

3.3.4. CP violation for B via interference of mixing and decays

Here we choose a final state f which is accessible to both B^o and $\overline{B}^o$ decays. The second amplitude necessary for interference is provided by mixing. Fig. 43 shows the decay into f either directly or indirectly via mixing. It is necessary only that f be

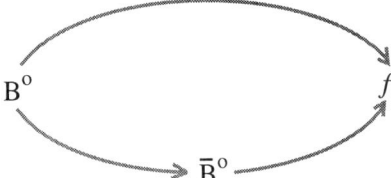

Fig. 43. Two interfering ways for a B^o to decay into a final state f.

accessible directly from either state, however if f is a CP eigenstate the situation is far simpler. For CP eigenstates

$$CP|f_{CP}\rangle = \pm|f_{CP}\rangle. \tag{77}$$

It is useful to define the amplitudes

$$A = \langle f_{CP}|\mathcal{H}|B^o\rangle, \quad \bar{A} = \langle f_{CP}|\mathcal{H}|\overline{B}^o\rangle. \tag{78}$$

If $\left|\frac{\bar{A}}{A}\right| \neq 1$, then we have "direct" CP violation in the decay amplitude, which we will discuss in detail later. Here CP can be violated by having

$$\lambda = \frac{q}{p} \cdot \frac{\bar{A}}{A} \neq 1, \tag{79}$$

which requires only that λ acquire a non-zero phase, i.e. $|\lambda|$ could be unity and CP violation can occur.

A comment on neutral B production at e^+e^- colliders is in order. At the $\Upsilon(4S)$ resonance there is coherent production of $B^o\bar{B}^o$ pairs. This puts the B's in a $C = -1$ state. In hadron colliders, or at e^+e^- machines operating at the Z^o, the B's are produced incoherently. For the rest of this article I will assume incoherent production except where explicitly noted.

The asymmetry, in this case, is defined as

$$a_{f_{CP}} = \frac{\Gamma(B^\circ(t) \to f_{CP}) - \Gamma(\overline{B}^\circ(t) \to f_{CP})}{\Gamma(B^\circ(t) \to f_{CP}) + \Gamma(\overline{B}^\circ(t) \to f_{CP})}, \qquad (80)$$

which for $|q/p| = 1$ gives

$$a_{f_{CP}} = \frac{(1 - |\lambda|^2)\cos(\Delta mt) - 2\text{Im}\lambda \sin(\Delta mt)}{1 + |\lambda|^2}. \qquad (81)$$

For the cases where there is only one decay amplitude A, $|\lambda|$ equals 1, and we have

$$a_{f_{CP}} = -\text{Im}\lambda \sin(\Delta mt). \qquad (82)$$

Only the amplitude, $-\text{Im}\lambda$ contains information about the level of CP violation, the sine term is determined only by B_d mixing. In fact, the time integrated asymmetry is given by

$$a_{f_{CP}} = -\frac{x}{1+x^2}\text{Im}\lambda = -0.48\text{Im}\lambda. \qquad (83)$$

This is quite lucky as the maximum size of the coefficient is -0.5.

Let us now find out how $\text{Im}\lambda$ relates to the CKM parameters. Recall $\lambda = \frac{q}{p} \cdot \frac{\bar{A}}{A}$. The first term is the part that comes from mixing:

$$\frac{q}{p} = \frac{(V_{tb}^* V_{td})^2}{|V_{tb} V_{td}|^2} = \frac{(1 - \rho - i\eta)^2}{(1 - \rho + i\eta)(1 - \rho - i\eta)} = e^{-2i\beta} \quad \text{and} \qquad (84)$$

$$\text{Im}\frac{q}{p} = -\frac{2(1-\rho)\eta}{(1-\rho)^2 + \eta^2} = \sin(2\beta). \qquad (85)$$

To evaluate the decay part we need to consider specific final states. For example, consider $f \equiv \pi^+\pi^-$. The simple spectator decay diagram is shown in Fig. 44. For the moment we will assume that this is the only diagram which contributes. Later I will show why this is not true. For this $b \to u\bar{u}d$ process we have

$$\frac{\bar{A}}{A} = \frac{(V_{ud}^* V_{ub})^2}{|V_{ud} V_{ub}|^2} = \frac{(\rho - i\eta)^2}{(\rho - i\eta)(\rho + i\eta)} = e^{-2i\gamma}, \qquad (86)$$

and

$$\text{Im}(\lambda) = \text{Im}(e^{-2i\beta}e^{-2i\gamma}) = \text{Im}(e^{2i\alpha}) = \sin(2\alpha). \qquad (87)$$

For our next example let's consider the final state ψK_S. The decay diagram is shown in Fig. 45. In this case we do not get a phase from the decay part because

$$\frac{\bar{A}}{A} = \frac{(V_{cb} V_{cs}^*)^2}{|V_{cb} V_{cs}|^2} \qquad (88)$$

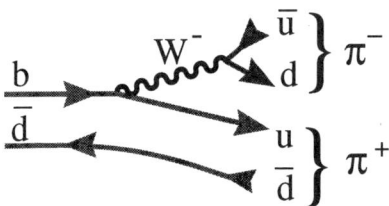

Fig. 44. Decay diagram at the tree level for $B^o \to \pi^+\pi^-$.

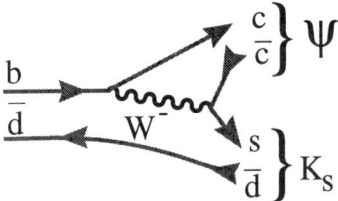

Fig. 45. Decay diagram at the tree level for $B^o \to \psi K_S$.

is real. In this case the final state is a state of negative CP, i.e. $CP|\psi K_S\rangle = -|\psi K_S\rangle$. This introduces an additional minus sign in the result for Imλ. Before finishing discussion of this final state we need to consider in more detail the presence of the K_S in the final state. Since neutral kaons can mix, we pick up another mixing phase (see Fig. 38). This term creates a phase given by

$$\left(\frac{q}{p}\right)_K = \frac{(V_{cd}^*V_{cs})^2}{|V_{cd}V_{cs}|^2}, \tag{89}$$

which is real. It necessary to include this term, however, since there are other formulations of the CKM matrix than Wolfenstein, which have the phase in a different location. It is important that the physics predictions not depend on the CKM convention.§

In summary, for the case of $f = \psi K_S$, Im$\lambda = -\sin(2\beta)$.

3.3.5. Comment on Penguin Amplitude

In principle all processes can have penguin components. One such diagram is shown in Fig. 46. The $\pi^+\pi^-$ final state is expected to have a rather large penguin

---

§Here we don't include CP violation in the neutral kaon since it is much smaller than what is expected in the B decay.

amplitude ~10% of the tree amplitude. Then $|\lambda| \neq 1$ and $a_{\pi\pi}(t)$ develops a $\cos(\Delta mt)$ term. It turns out (see Gronau[75]), that $\sin(2\alpha)$ can be extracted using isospin considerations and measurements of the branching ratios for $B^+ \to \pi^+\pi^o$ and $B^o \to \pi^o\pi^o$.

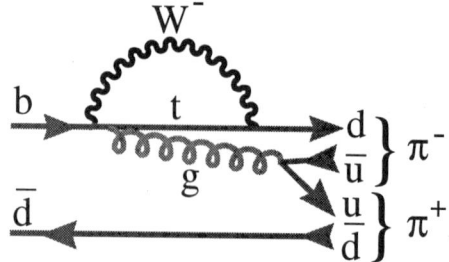

Fig. 46. Penguin diagram for $B^o \to \pi^+\pi^-$.

In the ψK_S case, the penguin amplitude is expected to be small since a $c\bar{c}$ pair must be "popped" from the vacuum. Even if the penguin decay amplitude were of significant size, the decay phase is the same as the tree level process, namely zero.

3.3.6. What actually has to be measured?

In charged B decays we only have to measure a branching ratio difference between B^+ and B^- to see CP violation. For neutral B decays we must find the flavor of the other b-quark produced in the event (this is called tagging), since we do not have any B^o beams. We then measure a rate asymmetry

$$a_{asy} = \frac{\#(f, \ell^+) - \#(f, \ell^-)}{\#(f, \ell^+) + \#(f, \ell^-)}, \tag{90}$$

where $\ell^{\pm}$ indicates the charge of the lepton from the "other" b and thus provides a flavor tag. In Fig. 47(a) the time dependence for the B^o and $\bar{B}^o$ are shown as a function of t in the B rest frame for 500 experiments of an average of 2000 events each with an input asymmetry of 0.3. In Fig. 47(b) the fitted asymmetry is shown for 500 different "experiments."

3.4. Better Measurements of the sides of the CKM triangle

One side of the triangle is determined by $|V_{ub}/V_{cb}|$. It appears that the best way to improve the values now is to measure the form-factors in the reactions $B \to \pi\ell\nu$ and $B \to \rho\ell\nu$. This will decrease the model dependence error, still the dominant errors, in the V_{ub} determination. Lattice gauge model calculations are appearing and should be quite useful.

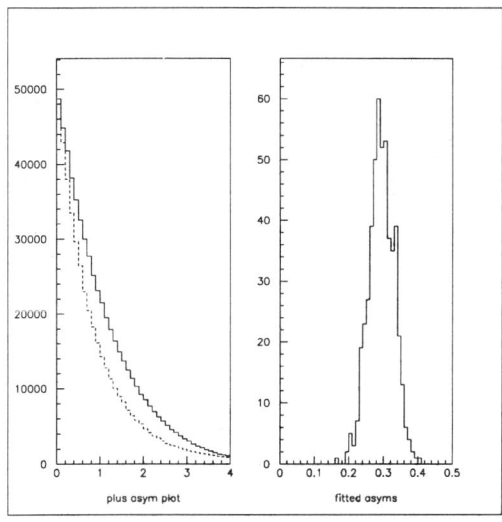

Fig. 47. (a) Time dependence of B^o and $\overline{B}^o$ decaying into a CP eigenstate, for an asymmetry of 0.3 for a total of 1 million events. The x-axis is proper time. In (b) the fitted asymmetry results are shown for 500 "experiments" of average of 2000 events each.

The other side of the triangle can determined by measuring B_s mixing, using the ratio

$$\frac{x_s}{x_d} = \left(\frac{B_s}{B}\right)\left(\frac{f_{B_s}}{f_B}\right)^2 \left(\frac{\tau_{B_s}}{\tau_B}\right)\left|\frac{V_{td}}{V_{ts}}\right|^2, \tag{91}$$

where

$$\left|\frac{V_{td}}{V_{ts}}\right|^2 = \lambda^2\left[(\rho-1)^2 + \eta^2\right]. \tag{92}$$

The large uncertainty in using the B_d mixing measurement to constrain ρ and η is largely removed as the ratio of the first three factors in equation (91) is already known to 10%.

As an alternative to measuring x_s, we can measure the ratio of the penguin decay rates

$$\frac{\mathcal{B}(B \to \rho\gamma)}{\mathcal{B}(B \to K^*\gamma)} = \xi\left|\frac{V_{td}}{V_{ts}}\right|^2, \tag{93}$$

where ξ is a model dependent correction due to different form-factors. Soni[76] has claimed that "long distance" effects, basically other diagrams spoil this simple relationship. This is unlikely for $\rho^o\gamma$ but possible for $\rho^+\gamma$.¶If this occurs, however, then

---

¶One example is the B^- decay which proceeds via $b \to uW^-$, where the $W^- \to \bar{u}d \to \rho^-$ and the u combines with the spectator $\bar{u}$ to form a photon.

it is possible to find CP violation by looking for a difference in rate between $\rho^+\gamma$ and $\rho^-\gamma$.

The CLEO II data are already background limited. The limit quoted is[60]

$$\frac{\mathcal{B}(B \to \rho\gamma)}{\mathcal{B}(B \to K^*\gamma)} < 0.19 \qquad (94)$$

at 90% confidence level.

3.5. Rare decays as Probes beyond the Standard Model

Rare decays have loops in the decay diagrams so they are sensitive to high mass gauge bosons and fermions. However, it must be kept in mind that any new effect must be consistent with already measured phenomena such as B_d^o mixing and $b \to s\gamma$.

Let us now consider searches for other rare b decay processes. The process $b \to s\ell^+\ell^-$ can result from the diagrams in Fig. 26(e or f). When searching for such decays, care must be taken to eliminate the mass region in the vicinity of the ψ or ψ' resonances, lest these more prolific processes, which are not rare decays, contaminate the sample. The result of searches are shown in Table 10.

Table 10. Searches for $b \to s\ell^+\ell^-$ decays

b decay mode	90% c.l. upper limit	Group	Ali et al. Prediction[77]
$s\mu^+\mu^-$	50×10^{-6}	UA1[78]	
$K^{*o}\mu^+\mu^-$	25×10^{-6}	CDF[80]	2.9×10^{-6}
	23×10^{-6}	UA1[78]	
	31×10^{-6}	CLEO[79]	
$K^{*o}e^+e^-$	16×10^{-6}	CLEO[79]	5.6×10^{-6}
$K^-\mu^+\mu^-$	9×10^{-6}	CLEO[79]	0.6×10^{-6}
	10×10^{-6}	CDF[80]	
$K^-e^+e^-$	12×10^{-6}	CLEO[79]	0.6×10^{-6}

B's can also decay into dilepton final states. The Standard Model diagrams are shown in Fig. 48. In (a) the decay rate is proportional to $|V_{ub}f_B|^2$. The diagram in (b) is much larger for B_s than B_d, again the factor of $|V_{ts}/V_{td}|^2$. Results of searches are given in Table 11.

4. Future Experiments

4.1. e^+e^- machines operating at the $\Upsilon(4S)$

Recall that only B meson pairs are produced at the $\Upsilon(4S)$ as shown in Fig. 6. Since each B has about 30 MeV of kinetic energy, it moves on the average only 30

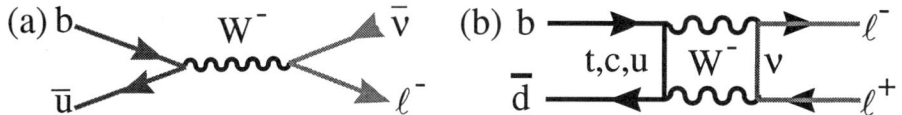

Fig. 48. Decay diagrams resulting in dilepton final states. (a) is an annihilation diagram, and (b) is a box diagram.

Table 11. Upper limits on $b \to$ dilepton decays (@90% c.l.)

	$\mathcal{B}(B^o \to \ell^+\ell^-)$		$\mathcal{B}(B_s \to \ell^+\ell^-)$	$\mathcal{B}(B^- \to \ell^-\bar{\nu})$		
	e^+e^-	$\mu^+\mu^-$	$\mu^+\mu^-$	$e^-\bar{\nu}$	$\mu^-\bar{\nu}$	$\tau^-\nu$
SM[†]	2×10^{-15}	8×10^{-11}	2×10^{-9}	10^{-15}	10^{-8}	10^{-5}
UA1[78]		8.3×10^{-6}				
CLEO[81]	5.9×10^{-6}	5.9×10^{-6}		1.5×10^{-5}	2.1×10^{-5}	2.2×10^{-3}
CDF[80]		1.6×10^{-6}	8.4×10^{-6}			
ALEPH[82]						1.8×10^{-3}

[†]SM is the Standard Model prediction.[83]

μm before it decays. Another important consequence is that the decay products mix together and do not appear in distinct jets. To measure the important time difference required in CP violation experiments via mixing, it is necessary to to give the B's a Lorentz boost which can be accomplished by using asymmetric beam energies.[84]

Let me amplify on this last statement. The asymmetry I presented

$$a_{f_{CP}} = -\text{Im}\lambda \sin(\Delta m t), \qquad (95)$$

is calculated for incoherent production of the B^o and another b quark (t is the time from production of the B^o until it decays). In e^+e^- production the B's can be produced in a coherent state. At the $\Upsilon(4S)$ $C = -1$, while at higher energies, where $B^*\bar{B}$ ($B^* \to B\gamma$) is produced, $C = +1$. For coherent production equation (95) gets modified to

$$a_{f_{CP}\ C=\pm} = -\text{Im}\lambda \sin\left(\Delta m(t \pm t')\right), \qquad (96)$$

where t refers to the decay time of f_{CP} and t' the decay time of the tagging B. In principle, $a_{f_{CP}}$ can be measured by taking a time integral. For incoherent production this works fine (see equation (83)). Here, however, the integral over the $C = -1$ case gives exactly zero, necessitating the time dependent measurement. The integral over the $C = +1$ case, does not give zero, but the measured cross-section for $B^*\bar{B}$ is about $1/7$ that of the $\Upsilon(4S)$.[85,86]

The one serious disadvantage of the $\Upsilon(4S)$ machines is that the cross-section is only 1 nb, so at a peak luminosity of 3×10^{33}, we expect only 60 million B^o's/year. For example, for a rare process with a branching ratio of 5×10^{-6} and a "typical" efficiency of 20%, we get only 60 events/year.

It is also important to note that there will not be much more B physics from LEP. The data sample has been collected and there are no current plans to get another large sample of Z^o decays to add to the brilliant b physics already done.

The CESR machine will be upgraded to produce a luminosity in excess of $2 \times 10^{33} \text{cm}^{-2}\text{s}^{-1}$, albeit with symmetric energy beams. Both the KEK laboratory in Japan and SLAC in Stanford, Cal. will construct asymmetric energy machines with planned luminosities in excess of $3 \times 10^{33} \text{cm}^{-2}\text{s}^{-1}$.

The advantages of such machines are that the b cross-section is 1/4 of the total, and the relatively clean enviornment and low interaction rates allow for superb photon detection using CsI crystal calorimeters[87] and for planned particle identification systems which should provided excellent π/K separation.[88]

4.2. Hadron machines

Let us first discuss the characteristics of hadronic b production. Hadronic b production mechanisms are shown in Fig. 49.[89] The relative contribution of the terms

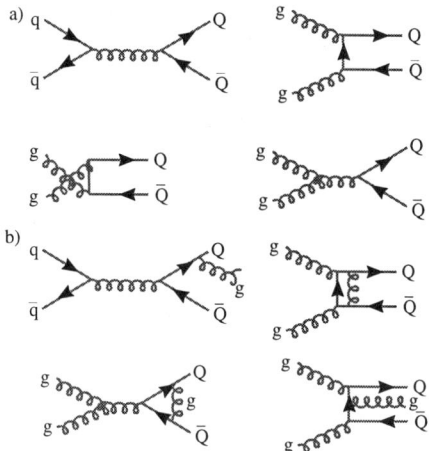

Fig. 49. Feynman diagrams for heavy quark production in hadronic collisions (a) of order α_s^2, and (b) some diagrams of order α_s^3.

proportional to α_s^2 and those proportional to α_s^3 is not well known. This is an important issue since the correlations in rapidity, η and in azimuthal angle between the b-quark and the $\bar{b}$-quark depends on the production mechanism. It is generally thought that $|\eta_b - \eta_{\bar{b}}| < 2$. In Fig. 50 I show the azimuthal opening angle distribution between a muon from a b quark decay and the $\bar{b}$ jet as measured by CDF[90] and compare with the MNR predictions.[91] The model does a good job in representing the shape which shows a strong back-to-back correlation. The normalization is about a factor of two higher in the data than the theory, which is generally true of CDF b

cross-section measurements.[92] In hadron colliders all B species are produced at the same time.

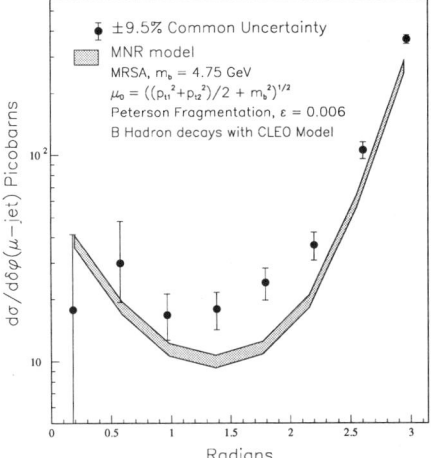

Fig. 50. The differential $\delta\phi$ cross-sections for $p_T^\mu > 9$ GeV/c, $|\eta^\mu| < 0.6$, $E_T^{\bar{b}} > 10$ GeV, $|\eta^{\bar{b}}| < 1.5$ compared with theoretical predictions. The data points have a common systematic uncertainty of $\pm 9.5\%$. The uncertainty in the theory curve arises from the error on the muonic branching ratio and the uncertainty in the fragmentation model.

The B meson transverse momentum distribution is severely limited and peaks near the B meson mass. The distribution in η, however is spread widely. In Fig. 51 I show the predicted (Pythia) distribution at the Tevatron collider. It should be realized

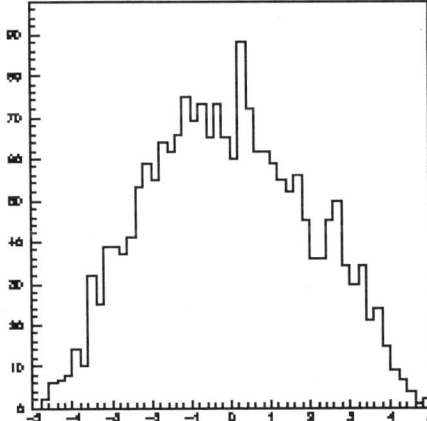

Fig. 51. The predicted distribution of B's versus η for 1.8 TeV $p\bar{p}$ collisions.

that this distribution in η reflects into a sharply peaked distribution in spatial angle $(\cos(\theta))$. The laboratory angular distributions of the B and $\overline{B}$ mesons expected at the LHC are shown in Fig. 52. Most of the events are far forward with the B and $\overline{B}$ being strongly correlated.[93]

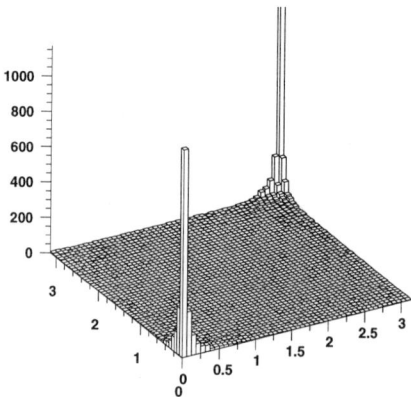

Fig. 52. Production angles of B versus production angle of the $\overline{B}$ in the laboratory (in radians) for the LHC collider calculated using PYTHIA.

Let us review some properties of current and proposed hadron b collider experiments.[94]

- The CDF and D0 detectors already exist at the Fermilab collider. The b cross-section is ~50 μb, with the ratio $\sigma(b)/\sigma(total) = 10^{-3}$. The luminosity is now close to 10^{31} and will increase with the advent of the main injector to 10^{32}. However, the restrictive trigger limits the b sample.

- The HERA-b experiment at DESY collides the HERA proton beam with fixed wire targets. The b cross-section is only ~6 nb with $\sigma(b)/\sigma(total) = 10^{-6}$. In order to produce enough b's they plan on four interactions per crossing. The goal is to measure CP violation in the ψK_S decay mode and possibly investigate other modes that are accessible by triggering on dileptons. The experiment is now under construction.

- The LHC-B experiment is being planned. At the LHC the b cross-section is ~300 μb, with the ratio $\sigma(b)/\sigma(total) = 3 \times 10^{-3}$. The experiment can run at a luminosity of 10^{32}, $\approx$240 Billion B^o/year are produced.

- Also at the LHC, the Atlas and CMS experiments will have some B capabilities.

- There is now a proposal for a dedicated B collider experiment at Fermilab called BTEV. Here $\approx$60 Billion B^o/year are produced.

4.3. Detector Considerations

For an experiment to do frontier B physics the following components appear to be necessary:

- Silicon vertex detector
- Charged particle tracking with magnetic analysis
- Cherenkov identification of charged hadrons
- Electromagnetic shower detection
- Muon detection with iron

A precision vertex detector is necessary to use the long B lifetime to reject background. Silicon is the current technology of choice; it can be realized as strips or as pixels. Charged particle tracking with magnetic analysis is important for momentum measurement as it is in most experiments. In order to pick out specific B decay modes, such as $K^+\pi^-$ from $\pi^+\pi^-$ or $\rho\gamma$ from $K^*\gamma$, it is crucial to have kaon and pion identification. Currently this is best provided using Cherenkov radiation.[88] Electromagnetic shower detection and muon identification are required to study semileptonic decays and provide flavor tags. The BELLE experiment, shown in Fig. 53 is an example of a detector that has all of these elements.

There are important constraints on all of these detection elements. Radiation damage implies various limits and certain technologies. The number of interactions per second implies a rate limit on detector elements. It appears that the maximum rate on any detector element is about 10^7/sec. The total detector readout rate is limited to about 10-100 MB/sec. (The smaller number is given by current technology and the larger number is based on expected improvement.) For an event size of 100 KB, this gives a maximum readout rate of 1000 events/sec.

Next, I will discuss the trigger. e^+e^- experiments have a distinct advantage here, since they merely trigger on everything. Experiments at hadron collider must trigger very selectively, or the data transmission rate will be swamped by background. There are several trigger strategies which have been developed. The one with the highest background rejection is $B \to \psi X$, $\psi \to \ell^+\ell^-$. Unfortunately the branching ratio for the former is only 1.1% and the latter 12%, giving a maximum triggerable B event rate of only 2.6×10^{-3}. This must be reduced by efficiency of the apparatus and kinematic cuts.

Another strategy is to trigger on semileptonic decays, where the 10% branching ratio to both muons or electrons is attractive. Furthermore, for CP violation measurements through mixing, this trigger also provides a tag. It has been traditionally easier to trigger on muons because electrons can easily be faked by photon conversions near the vertex or Dalitz decays of the π^o.

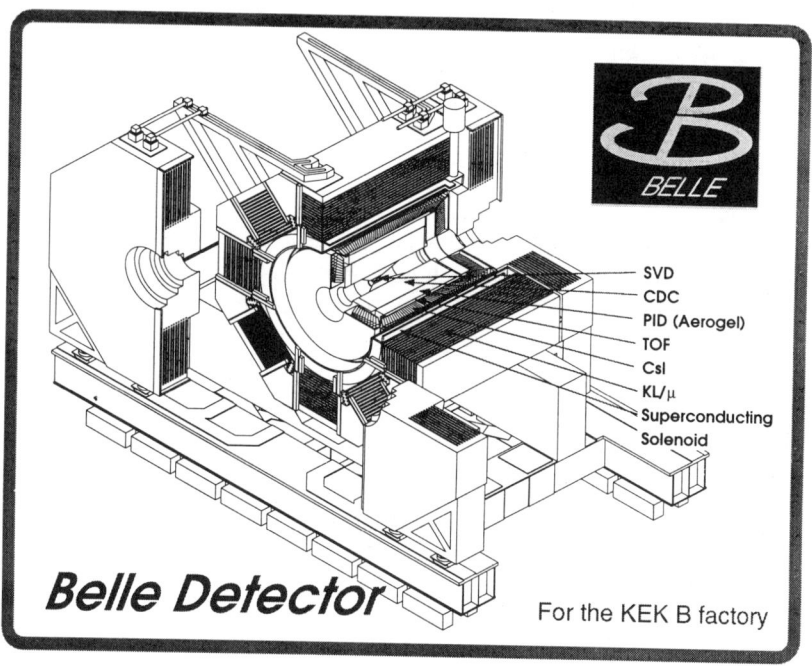

Fig. 53. Diagram of the Belle detector.

The most progressive strategy is to trigger on detached vertices. Recent simulations for BTEV have shown that it is possible to achieve a good efficiency $> 70\%$ on B decay events with a rejection on light quark background in excess of 100:1. To achieve this it is necessary to use a forward geometry with the silicon vertex detector inside the beam pipe.[95] A test of this concept was done at CERN by experiment P238.[96] A sketch of the silicon detector arrangement is shown in Fig. 54.

It is also possible to consider triggering on specific low multiplicity final states such as $B^o \to \pi^+\pi^-$ by using hadrons with $p_t >1$ GeV/c.

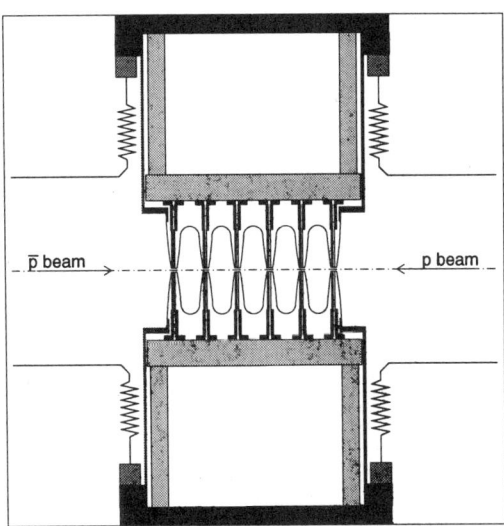

Fig. 54. Side view of the P238 silicon detector assembly and Roman pots. The 6 silicon planes are the vertical lines just above and below the beam line. The bellows (zig-zag lines) allow movements in the vertical direction of the pots, which are the thin vertical lines close to the bellows (they have 2 mm wall thickness). The edges of the 200 μm-thick aluminum RF shields closest to the beam (shown as the thin curved lines near the silicon detectors) normally ran at a distance of 1.5 mm from the circulating beams. The black horizontal pieces at top and bottom are the vacuum bulkheads bolted to the Roman pots.

The crucial issue in all of the trigger strategies is what the background rates are for a high signal efficiency. Does this give enough signal events with simultaneously rejecting background at the 100:1 level?

4.4. Hadron Geometries

There is a choice between two basic geometrical configurations that can be used for collider hadron B experiments. One is a central detector. An example is given by the planned upgraded CDF detector, shown in Fig. 55. Here the detector elements are arranged in an almost cylindrical manner about the beam pipe, so that the detector is very good near η equals zero. Notice that there are no detector elements for particle identification, though some information may be available from dE/dx measurements in the tracking chamber. An example of a forward detector is the proposed LHC-B experiment shown in Fig. 56. Here the vertex detector is inside a flared beam pipe. There are three different radiators for the RICH detectors.

In hadron colliders the most important rejection of non-B background is accomplished by seeing a detached decay vertex. In Fig. 57 I show the normalized decay length expressed in terms of L/σ where L is the decay length and σ is the error on L for the $B^o \to \pi^+\pi^-$ decay.[97] This study was done for the Fermilab Tevatron.

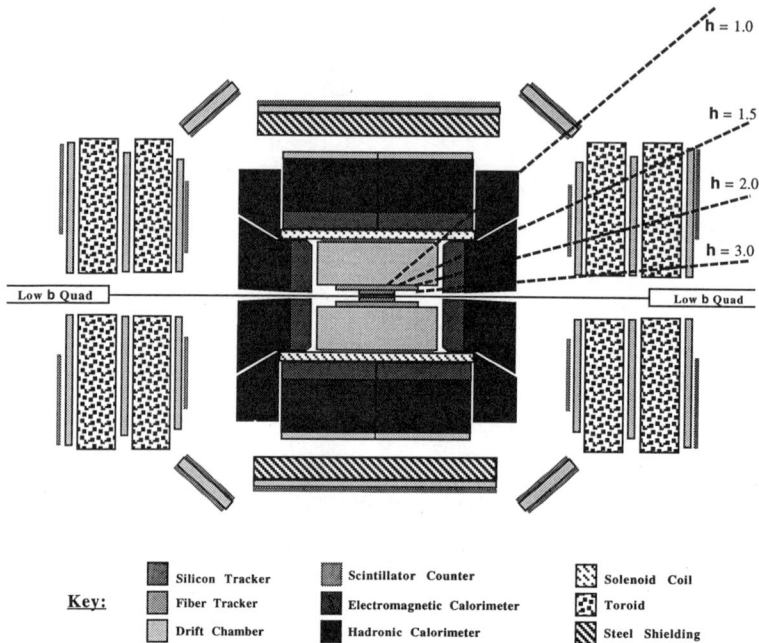

Fig. 55. A schematic diagram of the CDF upgrade. The symbol 'h' refers to rapidity. Note that the fiber tracker may change to a different technology.

The forward detector clearly has a much more favorable L/σ distribution. In Fig. 58 we show the time resolution in picoseconds for the forward and central detectors for the reaction $B_s \to \psi K s$, which has been suggested as a possible way to measure B_s mixing.[98] Remarkably the time resolution is a factor of 10 smaller for the forward detector.

A comparison of different B experiments is shown in Table 12.

5. Conclusions

B decay physics started in the 1980's and the first generation of experiments at CESR, DORIS, PEP, PETRA, LEP and CDF have made great contributions including the first fully reconstructed B's and precise measurement of the B meson masses, measurement of the B lifetimes, discovery of $B^o - \bar{B}^o$ mixing, the measurement of the CKM parameters V_{cb} and V_{ub} and the sighting of the first rare decays.

Many mysteries, however, remain to be untangled. Measuring independently all sides and angles of the CKM triangle may point us beyond the Standard Model if the data are inconsistent. This will require measuring all three CP violating angles,

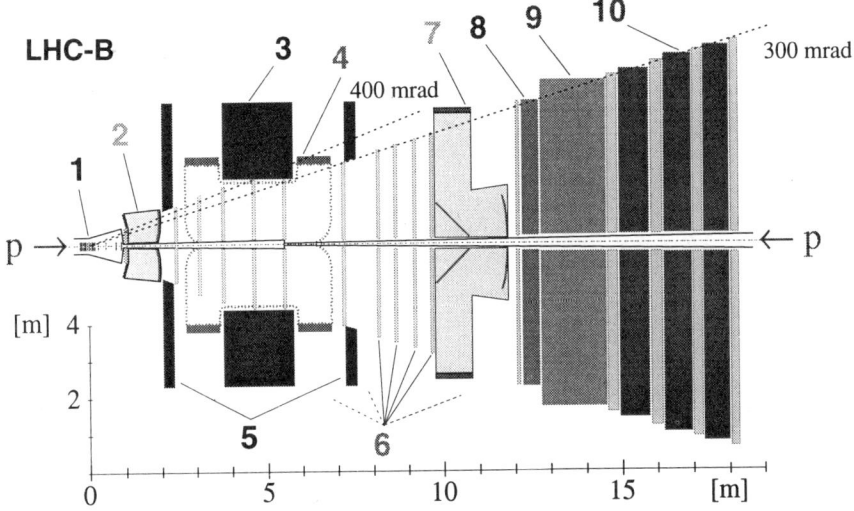

1) **Vertex detector** 2) Aerogel and Gas RICH's 3) **Magnet yoke** 4) Coils
5) **Magnetic field shielding plates** 6) Tracking chambers 7) Gas RICH
8) **Electromagnetic calorimeter** 9) Hadron calorimeter 10) Muon system

Fig. 56. A schematic diagram of the proposed LHC-B detector.

Table 12. Comparison of B decay detectors

Experiment	Particle I. D.	Vertex detection	Photon detection	$\sigma(b)$	$\dfrac{\sigma(b)}{\sigma(T)}$
Babar	Excellent	Good	Excellent	1 nb	0.25
Belle	Good	Good	Excellent	1 nb	0.25
CLEO	Excellent	Mediocre[†]	Excellent	1 nb	0.25
CDF	Poor	Good	Poor	50 μb	10^{-3}
D0	Poor	Good	Poor	50 μb	10^{-3}
HERA-B	Excellent	Excellent	Poor	6 nb	10^{-6}
LHC-B	Excellent	Excellent	Poor	300 μb	3×10^{-3}

† detector is excellent but low B velocity compromises vertex detection

measuring B_s mixing and precisely determining V_{ub}/V_{cb}. Furthermore, observation of rare B decays may also point us beyond the Standard Model.

e^+e^- threshold machines are great for future B physics. They will surely produce precision measurements of V_{ub} and V_{cb} and the important measurement of $\sin(2\beta)$ using the ψK_S decay mode. Posssibly $\sin(\gamma)$ can be measured using charged B decays and there are some who think these machines can measure $\sin(2\alpha)$, but I find that unlikely. However, these experiments are limited by the total number of B mesons.

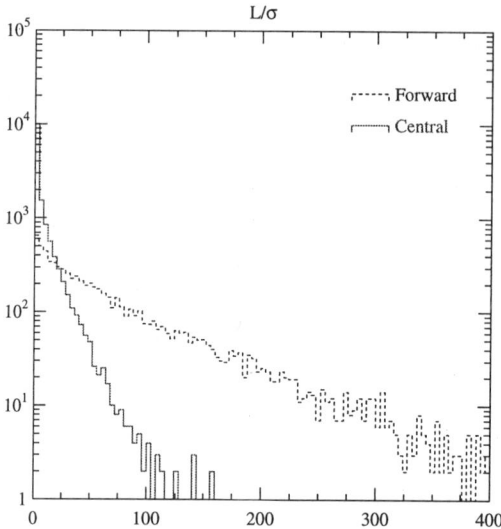

Fig. 57. Comparison of the L/σ distributions for the decay $B^o \to \pi^+\pi^-$ in central and forward detectors produced at a hadron collider with a center of mass energy of 1.8 TeV.

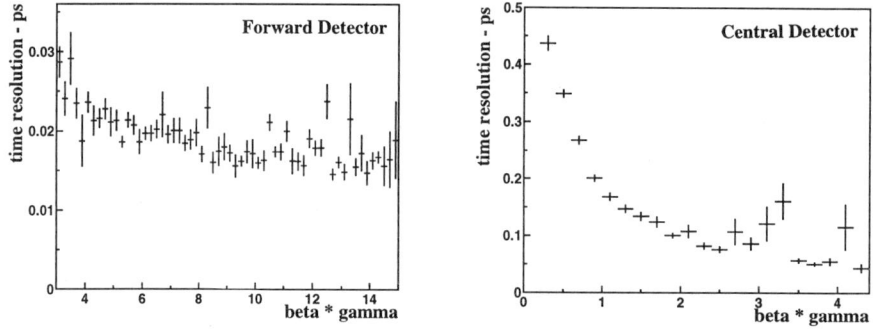

Fig. 58. The time resolution plotted as a function of $\beta\gamma$ for a forward detector $(2.0 < \eta < 4.5)$ and a central detector $(|\eta| < 1.5)$ for the decay $B_s \to \psi \overline{K}^*$ produced at a hadron collider with a center of mass energy of 1.8 TeV.

Even if these machines reach luminosities of $10^{34}\text{cm}^{-2}\text{s}^{-1}$, there are not enough B's to probe most rare phenomena. The prospects for B_s mixing, Λ_b and B_c studies are dim.

There is a fantastic potential for studying CP violation phenomena and rare B studies in hadronic machines but it's not easy. Let us consider the calculation of the error on an asymmetry measurement:

$$\sigma(a_{asy}) = \frac{1}{D\sqrt{N_{eff} \cdot \epsilon \cdot B}}, \qquad (97)$$

where

$$N_{eff} = N \frac{Signal}{Signal + Background}, \qquad (98)$$

B is the branching ratio of the final state of interest, ϵ is the overall efficiency including the tagging efficiency. D is the dilution factor caused by anything which causes a wrong-sign tag to be found, such as away side mixing, lepton misidentification etc.. A sample calculation is shown in Table 13.

Table 13. Sensitivity Calculation for Observing a CP asymmetry in ψK_S

CM energy	2 TeV
Cross-section	50 μb
Luminosity	10^{32}cm^{-2}s^{-1}
N_{B^o}/'Snowmass' year	3.75×10^{10}
$\mathcal{B}(B^o \to \psi K_S)$	5.5×10^{-4}
$\mathcal{B}(B^o \to \psi(\mu^+\mu^-)K_S(\pi^+\pi^-))$	2.2×10^{-5}
$N(B^o \to \mu^+\mu^-\pi^+\pi^-)$/year	8.2×10^5
Semi-leptonic decay of away side tag	0.10
Tagged $N(B^o \to \mu^+\mu^-\pi^+\pi^-)$/year	8.2×10^4
Triggering efficiency	0.8
Reconstruction efficiency of $\mu\mu\pi\pi$	0.25
Reconstruction efficiency μ tag	0.25
Vertex finding efficiency	0.9
Cleanup & analysis cuts	0.7
Dilution factors:	
Shape dependence D_{t-int}	0.47
mixing of muon tag	0.75
muon tag misidentification	0.9
Time resolution and cuts	0.95
Background	0.95
Total sensitivity	0.07

This calculation shows an error in the asymmetry of 7%. To see if that is in the range of interest, I show in Fig. 59 the expectations for the three CP violating angles and x_s. These plots merely reflect the "allowed" region shown in Fig. 40. It should

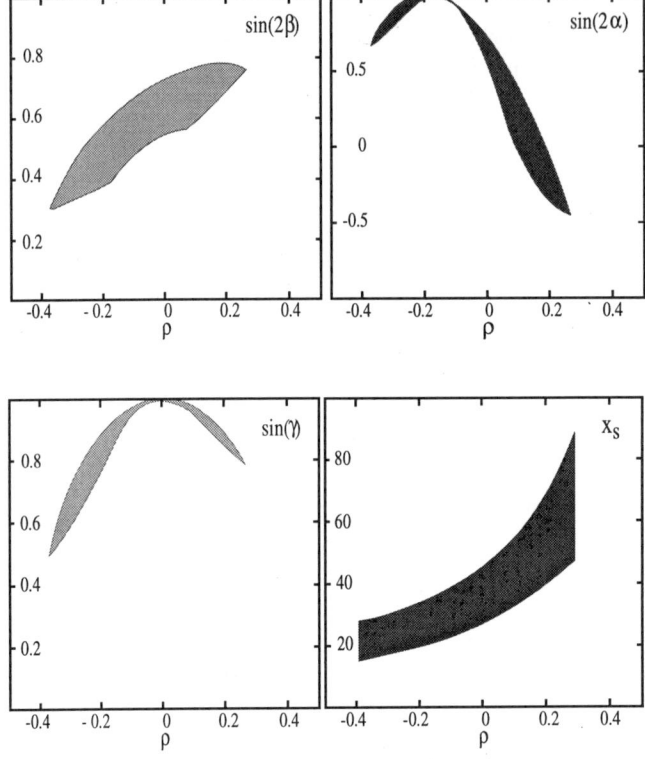

Fig. 59. The allowed values of three CP violating angles and the B_s mixing parameter x_s as a function of ρ, taken from the allowed region in Fig. 40.

be emphasized that this is not the result of a sophisticated analysis, which is difficult to do because of the non-Gaussian nature of the theoretical errors.

The decay modes which will probably be used to measure the CP violating angles are given in Table 14, with their branching ratios.

Finally, I list in Table 15 the CP violation and B_s mixing measurements of prime importance and my guess on which experiments, should they be built, are likely to perform these measurements and which could possibly perform them.

The B system challenges us with the possibility of very diverse and important measurements. Hopefully this physics will be done by the machines and experiments in the next and future decades.

Table 14. Branching ratios for decay modes used in measuring CP violation

CKM angle	Modes	$\mathcal{B}$	Product $\mathcal{B}$
β	ψK_S	0.4×10^{-3}	3.7×10^{-5}
α	$\pi^+\pi^-$	0.9×10^{-5}	0.9×10^{-5}
γ^{99}	$D^o K^-$	3.3×10^{-4}	4.0×10^{-5}
	$\overline{D}^o K^-$	4.1×10^{-6}	4.9×10^{-7}
	$D^o_{CP} K^-$	2.2×10^{-5}	2.6×10^{-6}
γ^{100}	$K^\pm \pi^o$, $\pi^\pm \pi^o$	$\approx 10^{-5}$	$\approx 10^{-5}$
	$K^o \pi^\pm$, $K^\pm \eta^{(\prime)}$	each	each

Table 15. Prospects for CP violation and B_s mixing measurements

Quantity	Modes	Possible	Likely
$\sin(2\alpha)$	$\pi^+\pi^-$	Babar, Belle	LHC-B, BTEV
$\sin(2\beta)$	ψK_S	HERA-B, CDF, CLEO	Babar, Belle, LHC-B, BTEV
$\sin(2\gamma)$	$K\pi$	Babar, Belle, CLEO	
$\sin(2\gamma)$	$D^o K^-$	Babar, Belle, CLEO	LHC-B, BTEV
x_s	ψK^*		LHC-B, BTEV

6. Acknowledgements

I have benefited greatly by physics discussion with many of my colleagues, most recently with M. Artuso, K. Berkelman, T. Skwarnicki, M. Witherell, J. Rosner and M. Gronau, M. Neubert, C. Sachrajda, A. Ali and A. Buras. I have also learned a lot from people associated with new B efforts including, J. Butler, C. Bebek, M. Procario, P. Mcbride, T. Ypsilantis and P. Schlein. I also thank Thomas Ferbel and Barbara Ferbel for arranging such an interesting school. I learned a lot and so did Julia. My special thanks to J. Rosner, K. Berkelman, M. Witherell and T. Skwarnicki for reading through the manuscript and making many useful comments.

7. REFERENCES

1. S. L. Glashow, *Nucl. Phys.* 22, 579 (1961); S. Weinberg, *Phys. Rev. Lett.* 19, 1264 (1967); A. Salam in *Elementary Particle Theory*, ed. N. Svartholm, Almqvist and Wisksell, Stockholm (1968).
2. N. Cabibbo, *Phys. Rev. Lett.* 10, 531 (1963); M. Kobayashi and K. Maskawa *Prog. Theor. Phys.* 49, 652 (1973).
3. L. Wolfenstein *Phys. Rev. Lett.* 51, 1945 (1983).
4. W. J. Marciano and A. Sirlin, *Phys. Rev.* D22, 2695 (1980).
5. R. M. Barnett et al., *Phys. Rev.* D54, 1 (1996).
6. S. W. Herb et al., (CFS Collaboration) *Phys. Rev. Lett.* 39, 252 (1977).
7. J. K. Bienlein et al., *Phys. Lett.* B78, 360 (1978). See also C. W. Darden et al., *Phys. Lett.* B78, 364 (1978).
8. D. Besson et al., *Phys. Rev. Lett.* 54, 381 (1985).
9. S. Behrends et al., *Phys. Rev. Lett.* 50, 881 (1983).
10. M. S. Alam, et al., *Phys. Rev.* D50, 43 (1994).
11. For an excellent description of the techniques used at PEP and PETRA see W. B. Atwood and J. A. Jaros, "Lifetimes," in *B* **Decays** 2nd edition revised, ed. S. Stone, World Scientific, Singapore (1994).
12. For an excellent description of the techniques used at LEP and CDF see V. A. Sharma and F. V. Weber "Recent Measurements of Lifetimes of b Hadrons," in *B* **Decays** 2nd edition revised, ed. S. Stone, World Scientific, Singapore (1994).
13. CDF results from T. Huffman, "CDF B Results and Upgrade," and LEP results from R. Hawkings, "B-Lifetimes," at "BEAUTY '96," 4th International Workshop on B-Physics at Hadron Machines June 17-21, 1996 - Roma, Italy to appear in proceedings. SLD results from K. Abe et al. *Phys. Rev. Lett.* 75, 3623 (1995).
14. CDF Collaboration, "Branching Fractions of $B^+ \to \psi(2S)K^+$ and $B^o \to \psi(2S)K^{*o}$ Decays at CDF," ICHEP96/pa01-86c, FERMILAB-Conf-96/160-E
15. C. Albajar et al., *Phys. Lett.* B186, 237 (1987); B213, 405 (1988); B256, 121 (1991).
16. K. Abe et al., *Phys. Rev. Lett.* 75, 1451 (1995).
17. D. M. Jansen et al., *Phys. Rev. Lett.* 74, 3118 (1995).
18. B. Grinstein, N. Isgur and M. B. Wise, *Phys. Rev. Lett.* 56, 258 (1986); F. J. Gilman and R. Singleton, *Phys. Rev.* D41, 142 (1990); K. Hagiwara, A. D. Martin and M. F. Wade, *Nucl. Phys.* B327, 569 (1989).
19. J. D. Richman and P. R. Burchat, *Rev .Mod. Phys.* 67, 893 (1995), and references contained therein.
20. T. Bergfeld et al., "Measurement of $\mathcal{B}(\bar{B}^o \to D^+\ell^-\bar{\nu})$ and Extraction of $|V_{cb}|$. CLEO-CONF 96-3, ICHEP-96 PA05-78 (1996).
21. M. Wirbel, B. Stech and M. Bauer *Z. Phys.* C29, 637 (1985); M. Bauer and M. Wirbel, *Z. Phys.* C42, 671(1989).

22. J. G. Korner and G. A. Schuler *Z. Phys.* C38, 511(1988); ibid, (erratum) C41 690 (1989).
23. N. B. Demchuk, I. L. Grach, I. M. Narodetski, S. Simula, " Heavy-to-heavy and heavy-to-light weak decay form factors in the light-front approach: the exclusive 0^- to 0^- case," INFN-ISS 95/18, hep-ph/9601369 (1996).
24. B. Barish et al., *Phys. Rev.* D51, 1014 (1995).
25. D. Bortoletto et al., *Phys. Rev. Lett.* 16, 1667 (1989).
26. H. Albrecht et al., *Z. Phys.* C57, 533(1993).
27. D. Buskulic et al., *Phys. Lett.* B359, 236 (1995).
28. P. Abreu et al., CERN-PPE/96-11 (1996).
29. B. Grinstein, N. Isgur and M. B. Wise, *Phys. Rev. Lett.* 56, 258 (1986).
30. N. Isgur, D. Scora, B. Grinstein, and M. B. Wise, *Phys. Rev.* D39, 799 (1989).
31. W. Jaus, *Phys. Rev* D41, 3394 (1990).
32. N. Isgur and M. B. Wise, "Heavy Quark Symmetry," in *B* **Decays** Revised 2nd Edition, World Scientific, Singapore (1994) p231; N.Isgur and M.B. Wise, *Phys. Rev.* D42 2388 (1990); N. Isgur and M. B .Wise, *Phys. Lett.* B232 113 (1989); ibidem B237, 527 (1990); M. B. Voloshin and M. A. Shifman,*Sov. J. Nucl.Phys.* 45 292 (1987); ibidem 47, 511 (1988); H. D. Politzer and M. B. Wise, *Phys. Lett.* 206B 681 (1988) 681; ibidem B208, 504 (1988); E. Eichten and B. Hill, *Phys. Lett.* B234, 511 (1990); H.Georgi, *Phys .Lett.* B240B 447 (1990); B.Grinstein, *Nucl. Phys.* B339, 253 (1990); A.F.Falk, H.Georgi, B.Grinstein and M. B .Wise, *Nucl. Phys.* B343, 1 (1990).
33. M. Neubert, *Int. J. Mod. Phys.* A11, 4173 (1996), hep-ph/9604412.
34. M. E. Luke, *Phys. Lett.* B252, 447 (1990).
35. A. Czarnecki, *Phys. Rev. Lett.* 76, 4121 (1996); M. Shifman, N.G. Uraltsev and A. Vainshtein, *Phys. Rev.* D51, 2217 (1995); M. Shifman and N. G. Uraltsev, *Int. J. Mod. Phys.* A10, 4705 (1995).
36. H. Albrecht et al., *Z. Phys.* C57, 533 (1993).
37. B. Barish et al., *Phys. Rev.* D51, 1041 (1995).
38. D. Buskulic et al., *Phys. Lett.* B359, 236 (1995).
39. P. Abreu et al., CERN-PPE/96-11 (1996), submitted to *Z. Phys. C*.
40. J. Bartelt et al., "Inclusive Measurements of *B*-meson Semileptonic Branching Fractions," submitted to Lepton Photon conf., Cornell (1993), CLEO-CONF 93-19.
41. B. Barish et al., *Phys. Rev. Lett.* 76, 1570 (1996).
42. H. Albrecht et al., (ARGUS) *Phys. Lett* B318, 397 (1993).
43. G. Altarelli, N. Cabibbo, G. Corbo and L. Maiani *Nuclear Phys.* B208, 365 (1982).
44. T. Skwarnicki, "Decays of *b* Quark," in Proc. of Lepton Photon Conf., Beijing, China August 1995, hep-ph/9512395.
45. R, Fulton et al., *Phys. Rev. Lett.* 64, 16 (1990).

46. H. Albrecht, et al., *Phys. Lett.* B234, 409 (1990).
47. J. Bartelt et al., *Phys. Rev. Lett.* 71, 4111 (1993).
48. G. Fox and S. Wolfram, *Phys. Rev. Lett.* 41, 1581 (1978).
49. M. Artuso, *Phys. Lett.* B311, 307 (1993).
50. M. Artuso, "CLEO Results," presented at **Beauty '96**, Rome, Italy, June 1996, to appear in proceedings.
51. J. Alexander et al., preprint CLNS 96/1419, CLEO 96-9 (1996).
52. R. Fulton et al. (CLEO), *Phys. Rev. Lett.* 64, 16 (1990); H. Albrecht et al. (ARGUS), *Phys. Lett.* B234, 409 (1990); a table of results and the average is given in S. Stone, "Semileptonic B Decays," in B **Decays** revised 2nd edition, ed. S. Stone, World Scientific, Singapore (1994) p349.
53. Since the resolution on is about 2.5 worse than the resolution on $|p_{miss}|$ (≈ 110 MeV), the neutrino four-vector is defined as $(E_\nu, \vec{p}_\nu) = (|\vec{p}_{miss}|, \vec{p}_{miss})$.
54. M. Gaillard and B. Lee, *Phys. Rev.* D10, 897, (1974); J. Hagelin, *Phys. Rev.* D20, 2893, (1979); A. Ali and A. Aydin, *Nucl. Phys.* B148, 165 (1979); T. Brown and S. Pakvasa, *Phys. Rev.* D31, 1661, (1985); S. Pakvasa, *Phys. Rev.* D28, 2915, (1985); I. Bigi and A. Sanda, *Phys. Rev.* D29, 1393, (1984).
55. H. Albrecht et al., *Phys. Lett.* B192, 245 (1983).
56. Using the ratio of like-sign to opposite sign dilepton events, UA1 published a mixing signal at the 3σ level that resulted from a combination of B_s and B_d, see C. Albajar et al., *Phys. Lett* B186, 245 (1987).
57. R. Akers et al., A Measurement of the B_d^0 Oscillation Frequency using Letponts and $D^{*+}\ell^-$ mesons, CERN-PPE/96-74 (1996); R. Akers et al., *Z. Phys.* C66 555 (1995).
58. C. Zeitnitz, "Oscillations and Mixing," presented at BEAUTY 96.
59. R. Ammar et al., *Phys. Rev. Lett.* 71, 674 (1993).
60. R. Ammar et al., "Radiative Penguin Decays of the B Meson," CLEO-CONF 96-6 (1996).
61. R. A. Fischer, "The Use of Multiple Measurements in Taxonomic Problems," Annals of Eugenics 7, 179 (1936); M. G. Kendall and A. Stuart, "The Advanced Theory of Statistics," Volume III, Hafner Publishing, NY 2nd edition (1968).
62. M. S. Alam et al., *Phys. Rev. Lett.* 74, 2885 (1995).
63. S. Bertolini, F. Borzumati and A. Masiero, *Phys. Rev. Lett.* 59, 180 (1987); R. Grigjanis et al., *Phys. Lett. B* 213, 35 (1988); B. Grinstein, R. Springer and M. Wise *Nucl. Phys.* B339, 269 (1990); see also N. G. Desphande, "Theory of Penguins in B Decay," in B **Decays** Revised 2nd Edition, ed. S. Stone, World Scientific, Singapore (1994).
64. M. Ciuchini et al., *Phys. Lett.* B334, 137 (1994).
65. K. Chetyrkin and M. Misiak, '$b \to s\gamma$ Decay Beyond Leading Logarithms," pa08-005, presented at ICHEP '96, Warsaw, Poland, August (1996).
66. M. Battle et al., *Phys. Rev. Lett.* 71, 3922 (1993).
67. D. Asner et al., *Phys. Rev.* D53, 1039 (1996).

68. S. Playfer and S. Stone, "Rare B Decays," *Int. Journal of Mod. Phys.* A10, No 29. 4107, (1995).
69. B. Barish et al., "Observation of the Decay $B \to \omega\pi^+$," CONF 96-23, ICHEP-96 PA05-95 (1996).
70. W. Adam et al., "Study of Rare B Decays with the DELPHI Detector at LEP," CERN PPE 96-67 (1996).
71. P. Langacker, "CP Violation and Cosmology," in **CP Violation**, ed. C. Jarlskog, World Scientific, Singapore p 552 (1989).
72. A. J. Buras, "Theoretical Review of B-physics," in **BEAUTY '95** ed. N. Harnew and P. E. Schlein, *Nucl. Instr. and Meth.* A368, 1 (1995).
73. A. Carter and A. I. Sanda, *Phys. Rev. Lett.* 45, 952 (1980); *Phys. Rev.* D23, 1567 (1981); I. I. Bigi and A. I. Sanda, *Nucl. Phys.* B193, 85 (1981); ibid B281, 41 (1987).
74. I. Bigi, V. Khoze, N. Uraltsev, in **CP Violation**, ed. C. Jarlskog, World Scientific, Singapore 175 (1989).
75. M. Gronau, *Phys. Rev. Lett.* 63, 1451 (1989); M. Gronau and D. London, *Phys. Rev. Lett.* 65, 3381 (1990).
76. D. Atwood, B. Blok and A. Soni, *Int. J. Mod. Phys.* A11, 3743 (1996); see also N. Deshpande, X. He & J. Trampetic, Preprint OITS-564-REV (1994); see also J. M. Soares, *Phys. Rev.* D53, 241 (1996). G. Eilam, A. Ioannissian R. R. Mendel and P. Singer, *Phys. Rev.* D53, 3629 (1996).
77. A. Ali, C. Greub and T. Mannel, "Rare B Decays in the Standard Model, " in Hamburg 1992, Proceedings, ECFA Workshop on a European B-meson Factory, Eds. R. Aleksan and A. Ali, p155 (1993).
78. C. Albajar et al., *Phys. Lett.* B262, 163 (1991).
79. R. Balest et al., "Search for $B \to K\ell^+\ell^-$ and $B \to K^*\ell^+\ell^-$ decays," CLEO-CONF 94-4 (1994).
80. F. Abe et al., *Phys. Rev. Lett.* 76, 2015 (1996).
81. R. Ammar et al., *Phys. Rev.* D49, 5701 (1994); M. Artuso, et al., *Phys. Rev. Lett.* 75, 785 (1995).
82. D. Buskulic et al., *Phys. Lett.* B343, 444 (1995).
83. A. Ali and T. Mannel, *Phys. Lett.* B264, 447 (1991). Erratum, ibid, B274, 526 (1992).
84. The use of asymmetric beam energies to study CP violation in e^+e^- machines operating the $\Upsilon(4S)$ was first suggested by Pierre Oddone.
85. D. S. Akerib et al., *Phys. Rev. Lett.* 67, 1692 (1991).
86. S. Stone, *Mod. Phys. Lett.* A3, 541 (1988).
87. E. Blucher et al., *Nucl. Instr. and Meth.* A249, 201 (1986); Y. Kubota et al., *Nucl. Instr. and Meth.* A320, 66 (1992).
88. M. Artuso, *Nucl. Instr. and Meth.* A371, 324 (1996); B. N. Ratcliff, *Nucl. Instr. and Meth.* A371, 309 (1996)
89. M. Gluck, J. F. Owens and E. Reya, *Phys. Rev.* D17, 2324 (1978); B. L 49 (1989); W. Beenakker et al., *Nucl. Phys.* B351, 507 (1991), *Phys. Rev.* D40, 54 (1989).

90. F. Abe et al., *Phys .Rev.* D53, 1051 (1996). See also, M. Artuso, "Experimental Facilities for b-Quark Physics, " in *B* **Decays** revised 2nd Edition, Ed. S. Stone, World Scientific, Sinagapore (1994).
91. M. Mangano, P. Nason and G. Ridolfi, *Nucl. Phys.* B373, 295 (1992).
92. F. Abe et al., *Phys. Rev. Lett.* 75, 1451 (1995). Previous UA1 measurements agreed with the theoretical predictions, see C. Albajar et al., *Phys. Lett.* B256, 121 (1991). Recent D0 measurements agree with both the CDF measurements and the high side of the theoretically allowed range. See S. Abachi et al., *Phys. Rev. Lett.* 74 3548 (1995).
93. S. Erhan, *Nucl. Instr. and Meth.* A368, 133 (1995).
94. For more information, the reader is referred to proceedings of conferences called "Beauty 93" through "Beauty 96," the proceedings of which are published by Nucl. Instru. and Methods volumes A333 (1993), A351 (1994), A368 (1995).
95. J. N. Butler, "The BTEV Experiment at Fermilab," presented at 2nd Int. Conf. on Hyperons, Charm and Beauty Hadrons, Montreal, Quebec, Canada, Aug. 1996.
96. J. Ellet et al., *Nucl. Instr. and Meth.* A317 28, (1992).
97. M. Procario, "*B* Physics Prospects Beyond the year 2000," presented at the 10th Topical Workshop on Proton-Antiproton Collider Physics, to appear in proceedings, Fermilab-BONF-95/166.
98. P. McBride and S. Stone, *Nucl. Instr. and Meth.* A368, 38 (1995).
99. A. G. Cohen et al., "B-Factory Physics from Effective Supersymmetry," UW-PT-95-22, hep-ph/9610252 (1996); J. L. Hewett and J. D. Wells, "Searching for supersymmetry in rare B decays," SLAC-PUB-7290, hep-ph/9610323 (1996); Y. Nir and H. R. Quinn, "Theory of CP Violation in *B* Decays," in *B* **Decays** ed. S. Stone, World Scientific, Singapore p520 (1994).
100. M. Gronau and D. Wyler, *Phys. Lett.* B265, 172 (1991).
101. M. Gronau, J. L. Rosner and D. London, *Phys. Rev. Lett.* 73, 21 (1994).

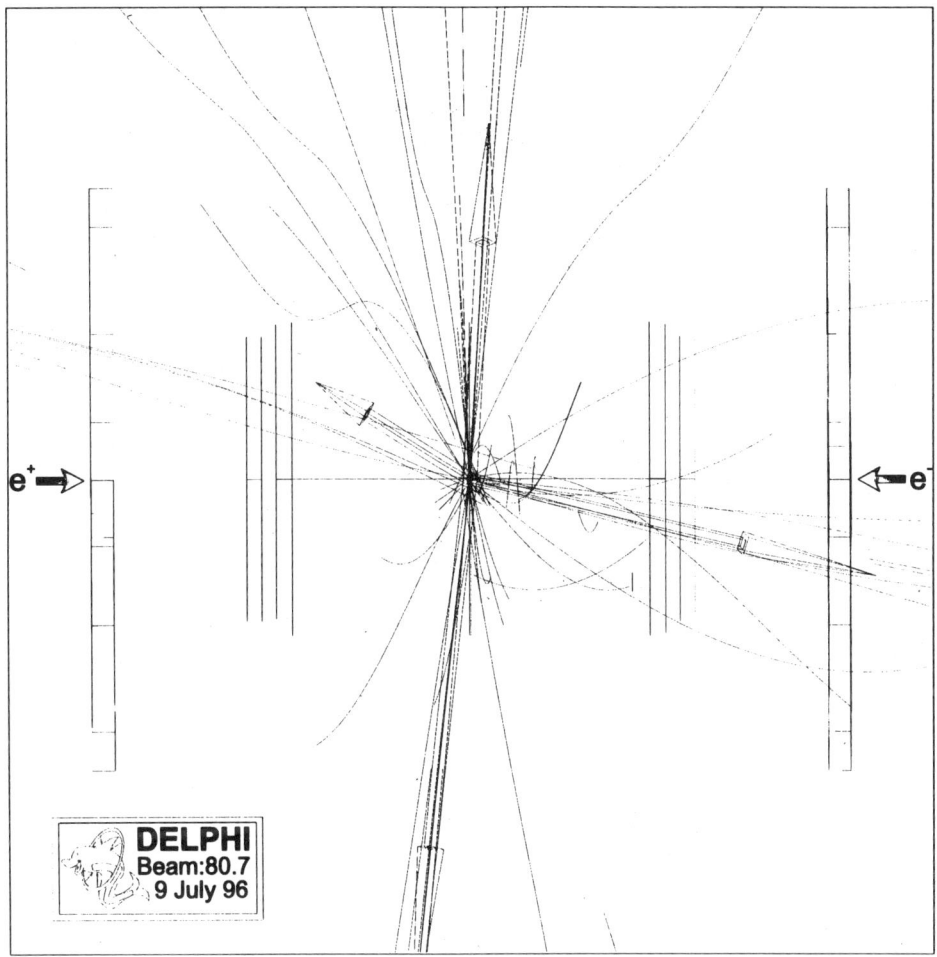

The first candidate event for the pair production of W bosons in $e^+ e^-$ collisions was recorded by the DELPHI experiment at LEP on July 9, 1996, and was presented for the first time at the NATO ASI only 48 hours later. The figure shows a projection of the event in a plane parallel to the beam pipe. Four distinct high momentum jets are observed in the event. Using the DURHAM algorithm, with a y_{cut} of 0.005, and constraining the energy and momentum in the event, leads to the best di-jet mass pairings of 88 GeV and 70 GeV. This event is therefore compatible with the pair production of W bosons and their subsequent decay to fully hadronic states. (We thank Chris Parkes for the presentation and for the figure.)

Participants at the ASI (scanning from left to right): Jiri Chudoba, Francesco Arneodo, Erhan Pesen, Owen Long, Alain Blondel, Dugan O'Neil, Gustaaf Brooijmans, Tim Bergfeld obstructed by Wendy Taylor, Tony Rooke, Bernd Wilkens, Alick McPherson, Jesse Stone, Gloria Vuagnin, Alvise Favara, Fernanda Garcia, Ronaldo Bellazzini, Peter Steinberg, Rob Griffiths, Claudia Cecchi, Francois Vannucci, Doris Rakoczy, Nick Ellis, David Kestenbaum, Guido Altarelli, Igor Volobouev, Germano Bonomi, Maria Chamizo, Paul Bright-Thomas, Barry MacEvoy, Sotirios Vlachos, Sheldon Stone, Tom Ferbel, Janus Schmidt-Sorensen, Guenther Dissertori, George Michail, Domizia Orestano, Tony Pitts, Ivan Korolko, Chris Parkes, Andreas Warburton, Maura Barone, Mark Strovink, Roberta Antolini, Uwe Bratzler, Myungyun Pang, Urs Langenegger, Didier Lacour, Sarah Truitt, Min Gao, Andrea Parri, Peter Tamburello, Rocio Vilar, Johan Blouw, Eva Wittman, Cecilia Gerber, Stephan Vandenbrink, Erich Varnes, David Reyna, Mrinmoy Bhattacharjee, Ana Henriques, Sally Dawson, Harrison Prosper, Basil Moshous, Scott Snyder, Bostjan Golob, Alain Bellerive, Jamal Tarazi, Jim Panetta, and Bob Palmer. (Missing: Greg Graham.)

INDEX

Atmospheric neutrinos 449–451

Bayesian probability 131–135, 161–172
B-physics 465–532
    CKM elements 473–492
    decays and mixing 492–518
    CP violation 502–518

Collider accelerators 183–272
Confidence intervals 145–156

Electroweak issues 5–9, 273–330, 381–430

Frequentist statistics 136–142

GUTS 53–60

Hadron asymmetries 410–414
Higgs sector 42–50, 62–69, 89–93, 200–205

LEP 24–28, 395–399
LHC 81–130, 331–343
    calorimetry 104–117
    detectors 96–121, 338–334
    muon detection 117–121
    trigger 122–130

Microgap chamber 370–376
Microstrip chambers 344–370
Minimal SUSY 21–22, 36–40, 332–335
Muon colliders 205–267
    detectors 232–246

Neutrinos 429–463
    decays 441–443
    mass 437–439
    oscillations 432–437, 451–457

Polarization asymmetries 403–404
Probability theory 157–162

Solar neutrinos 444–448
Standard Model 1–31, 273–330, 381–430, 465–468
Statistical analysis 173–180
Statistical issues 131–181
Supernova neutrinos 448–449
SUSY 21–22, 33–80, 284–286

Tau neutrinos 457–461
Top quark 273–316
Tracking chambers 331–427

W mass 317–329, 399–403

LECTURERS

G. Altarelli	CERN, Geneva, Switzerland and Rome 3 University, Italy
R. Bellazzini	University of Pisa, Pisa, Italy
A. Blondel	Ecole Polytechnique, Palaiseau, France
S. Dawson	Brookhaven National Laboratory, Upton, New York
N. Ellis	CERN, Geneva, Switzerland
R. Palmer	Brookhaven National Laboratory, Upton, New York
H. Prosper	Florida State University, Gainesville, Florida
S. Stone	Syracuse University, Syracuse, New York
M. Strovink	University of California, Berkeley, California
F. Vannucci	LPNHE, Paris, France

ADVISORY COMMITTEE

B. Barish	Caltech, Pasadena, California
L. DiLella	CERN, Geneva, Switzerland
C. Fabjan	CERN, Geneva, Switzerland
H. Georgi	Harvard University, Cambridge, Massachusetts
C. Jarlskog	Stockholm University, Stockholm, Sweden
C. Quigg	Fermilab, Batavia, Illinois
P. Soding	DESY, Zeuthen, Federal Republic of Germany
M. Tigner	Cornell University, Ithaca, New York

DIRECTOR

T. Ferbel	University of Rochester, Rochester, New York